Orthogonal Polynomials:
Theory and Practice

NATO ASI Series

Advanced Science Institutes Series

A Series presenting the results of activities sponsored by the NATO Science Committee, which aims at the dissemination of advanced scientific and technological knowledge, with a view to strengthening links between scientific communities.

The Series is published by an international board of publishers in conjunction with the NATO Scientific Affairs Division

A Life Sciences	Plenum Publishing Corporation
B Physics	London and New York
C Mathematical and Physical Sciences	Kluwer Academic Publishers
D Behavioural and Social Sciences	Dordrecht, Boston and London
E Applied Sciences	
F Computer and Systems Sciences	Springer-Verlag
G Ecological Sciences	Berlin, Heidelberg, New York, London,
H Cell Biology	Paris and Tokyo

Orthogonal Polynomials:
Theory and Practice

edited by

Paul Nevai

Department of Mathematics,
The Ohio State University,
Columbus, Ohio, U.S.A.

with the assistance of
Mourad E. H. Ismail

Department of Mathematics,
University of South Florida,
Tampa, Florida, U.S.A.

Kluwer Academic Publishers

Dordrecht / Boston / London

Published in cooperation with NATO Scientific Affairs Division

Proceedings of the NATO Advanced Study Institute on
Orthogonal Polynomials and Their Applications
Colombus, Ohio, U.S.A.
May 22 – June 3, 1989

Library of Congress Cataloging in Publication Data

NATO Advanced Study Institute on "Orthogonal Polynomials and Their
  Applications" (1989 : Ohio State University)
    Orthogonal polynomials : theory and practice : proceedings of the
  NATO Advanced Study Institute on "Orthogonal Polynomials and Their
  Applications, the Ohio State University, Columbus, Ohio, U.S.A.,
  May 22, 1989-June 3, 1989 / edited by Paul Nevai with the assistance
  of Mourad E.H. Ismail.
       p.   cm. -- (NATO ASI series. Series C, Mathematical and
  physical sciences ; vol. 294.)
     "Published in cooperation with NATO Scientific Affairs Division."
     ISBN 978-94-010-6711-9 (alk. paper)
     1. Orthogonal polynomials--Congresses.   I. Nevai, Paul G., 1948-
    II. Ismail, Mourad, 1944-   III. North Atlantic Treaty
  Organization. Scientific Affairs Division.   IV. Title.   V. Series:
  NATO ASI series. Series C, Mathematical and physical sciences ; no. 294.
  QA404.5.N38   1989
  515'.55--dc20   89-24675

ISBN-13: 978-94-010-6711-9   e-ISBN-13: 978-94-009-0501-6
DOI: 10.1007/978-94-009-0501-6

Published by Kluwer Academic Publishers,
P.O. Box 17, 3300 AA Dordrecht, The Netherlands.

Kluwer Academic Publishers incorporates the publishing programmes of
D. Reidel, Martinus Nijhoff, Dr W. Junk and MTP Press.

Sold and distributed in the U.S.A. and Canada
by Kluwer Academic Publishers,
101 Philip Drive, Norwell, MA 02061, U.S.A.

In all other countries, sold and distributed
by Kluwer Academic Publishers Group,
P.O. Box 322, 3300 AH Dordrecht, The Netherlands.

Printed on acid free paper

We proudly dedicate the proceedings
of our NATO Advanced Study Institute on
"Orthogonal Polynomials and Their Applications"
to

Dick Askey

Thank you, Dick, for all what you have done for
orthogonal polynomials and special functions!

TABLE OF CONTENTS

PREFACE

This volume contains the Proceedings of the NATO Advanced Study Institute on "Orthogonal Polynomials and Their Applications" held at The Ohio State University in Columbus, Ohio, U.S.A. between May 22, 1989 and June 3, 1989.

The Advanced Study Institute primarily concentrated on those aspects of the theory and practice of orthogonal polynomials which surfaced in the past decade when the theory of orthogonal polynomials started to experience an unparalleled growth. This progress started with Richard Askey's Regional Conference Lectures on "Orthogonal Polynomials and Special Functions" in 1975, and subsequent discoveries led to a substantial revaluation of one's perceptions as to the nature of orthogonal polynomials and their applicability. The recent popularity of orthogonal polynomials is only partially due to Louis de Branges's solution of the Bieberbach conjecture which uses an inequality of Askey and Gasper on Jacobi polynomials. The main reason lies in their wide applicability in areas such as Padé approximations, continued fractions, Tauberian theorems, numerical analysis, probability theory, mathematical statistics, scattering theory, nuclear physics, solid state physics, digital signal processing, electrical engineering, theoretical chemistry and so forth. This was emphasized and convincingly demonstrated during the presentations by both the principal speakers and the invited special lecturers.

The main subjects of our Advanced Study Institute included complex orthogonal polynomials, signal processing, the recursion method, combinatorial interpretations of orthogonal polynomials, computational problems, potential theory, Padé approximations, Julia sets, special functions, quantum groups, weighted approximations, orthogonal polynomials associated with root systems, matrix orthogonal polynomials, operator theory and group representations.

In addition to the principal lectures and invited special talks, four panel discussions were also organized on topics ranging from applications of orthogonal polynomials in coding theory (Eiichi Bannai) and in birth and death processes (Mourad E. H. Ismail) to symbolic computer algebra (George Gasper) and the history of orthogonal polynomials (Richard Askey).

I wish to thank NATO Scientific Affairs Division, the National Science Foundation, the Institute for Mathematics and Applications in Minneapolis, Minnesota, and The Ohio State University for their generous support. In particular, I am grateful to Jack Ryff (NSF), C. William Kern and Joe Ferrar (Ohio State) for their long term strong commitment to support research in orthogonal polynomials.

The Advanced Study Institute's International Advisory Board consisted of Waleed Al-Salam (Edmonton, Canada), Paul Butzer (Aachen, Federal Republic of Germany), Alphonse Magnus (Leuven-la-Neuve, Belgium), Francisco Marcellán (Madrid, Spain), and André Ronveaux (Namur, Belgium) who did a great job not only in suggesting potential participants and recommending various subjects to include in the program but also in helping us to explore the various opportunities for financial support.

My very special thanks go to Dick Askey (Madison, Wisconsin, U.S.A.), Mourad Ismail (Tampa, Florida, U.S.A.) and Dennis Stanton (Minneapolis, Minnesota, U.S.A.) for helping to organize the Advanced Study Institute, to Mrs. Marilyn Radcliff and Ms. Agnes Tuska for helping to run it, to my soon to be former student John Zhang for helping to prepare this volume, and to Judie Nevai for her moral support and for helping me to stuff thousands of envelopes which is one of the least pleasant but necessary chores when running a workshop such as ours was.

Columbus, Ohio, U.S.A.   Paul Nevai
August, 1989

Al-Salam, Waleed, Department of Mathematics, University of Edmonton, Edmonton, Alberta T6G 2G1, Canada

Bannai, Eiichi, Department of Mathematics, Ohio State University, 231 West 18th Avenue, Columbus, OH 43210-1174, U.S.A.

Bessis, Daniel, Service de Physique Théorique, Centre d'Études Nucléaires de Saclay, 91191 Gif-sur-Yvette Cédex, France

Chihara, Theodore S., Department of Mathematics, Purdue University Calumet, Hammond, IN 46323, U.S.A.

Delsarte, Philippe, Philips Laboratory, Av. Van Becelaese 2, Box 8, B-1170 Brussels, Belgium

Djrbashian, M. M., Institute of Mathematics of the Academy of Sciences of the Armenian SSR, 24 B Marshall Bagramian Street , 375019 Yerevan, Armenian SSR, U.S.S.R.

Dombrowski, Joanne M., Department of Mathematics, Wright State University, Dayton, OH 45435, U.S.A.

Gasper, George, Department of Mathematics, Northwestern University, Evanston, IL 60201, U.S.A.

Gautschi, Walter, Department of Computer Science, Purdue University, West Lafayette, IN 47907, U.S.A.

Genin, Yves, Philips Laboratory, Av. Van Becelaese 2, Box 8, B-1170 Brussels, Belgium

Haydock, Roger, Materials Science Institute, University Oregon, Eugene, OR 97403, U.S.A.

Ismail, Mourad, Department of Mathematics, University of South Florida, Tampa, FL 33620-5700, U.S.A.

Koornwinder, Tom, Centre for Mathematics and Computer Science, P.O. Box 4079, 1009 AB Amsterdam, Netherlands

Letessier, Jean, Physique Théorique et Hautes Énergies, Tour-14, 5e étage, 2, place Jussieau, 75251 Paris Cédex 05, France

Lubinsky, Doron Shaul, Department of Mathematics, Witwatersrand University, P.O. Wits 2050, South Africa

Macdonald, Ian, School of Mathematical Sciences, Queen Mary College, University of London, Mile End Road, London E1, United Kingdom

Masson, David R., Department of Mathematics, University of Toronto, Toronto, Ontario M5S 1A4, Canada

Rahman, Mizan, Department of Mathematics, Carleton University, Ottawa, Ontario K1S 5B6, Canada

Rodman, Leiba, Department of Mathematics, The College of William & Mary, Williamsburg, VA 23185, U.S.A.

Saff, Edward B., Department of Mathematics, University of South Florida, Tampa, FL 33620-5700, U.S.A.

Stahl, Herbert, Stuttgarter Platz 20, D-1000 Berlin 12, Federal Republic of Germany

Stanton, Dennis, School of Mathematics, University of Minnesota, 206 Church Street S.E., Minnesota, MN 55455, U.S.A.

Totik, Vilmos, Bolyai Institute, József Attila University, Aradi Vértanúk Tere 1, 6720 Szeged , Hungary

Valent, Galliano, Physique Théorique et Hautes Énergies, Tour-14, 5e étage, 2, place Jussieau, 75251 Paris Cédex 05, France

Van Assche, Walter, Department of Mathematics, Katholieke Universiteit Leuven, Clestijnenlaan 200 B, B–3030 Leuven (Heverlee), Belgium

CHARACTERIZATION THEOREMS FOR ORTHOGONAL POLYNOMIALS

W.A. Al-Salam
Department of Mathematics
University of Alberta
Edmonton, Canada
T6G 2G1

Abstract. We survey in this paper characterization theorems dealing with polynomial sets which are orthogonal on the real line.

1. Introduction.

In this paper we shall survey results which characterize orthogonal polynomial sets (OPS). By a polynomial set (PS) we mean a sequence of polynomials, $\{P_n(x) : n = 0, 1, 2, \ldots\}$ in which the degree of $P_n(x) = n$ for all n.

By orthogonal we mean orthogonal on the real line with respect to a positive measure $d\alpha(x)$ supported and of bounded variation on the real line with finitely or infinitely points of increase so that

$$\int_{-\infty}^{\infty} P_n(x)P_m(x)d\alpha(x) = h_n\delta_{nm},$$

where the moments of $d\alpha$, $\mu_n = \int_{-\infty}^{\infty} x^n d\alpha(x)$, exist for all n.

It is an old result that such OPS must necessarily satisfy a three term recurrence relation which we may write in one of the following forms

$$xP_n(x) = a_{n+1}P_{n+1}(x) + c_nP_n(x) + a_nP_{n-1}(x) \quad \text{(orthonormal form } h_n = 1)$$

(1.1)

$$P_{n+1}(x) = (x - c_n)P_n(x) - \lambda_nP_{n-1}(x) \quad \text{(Monic form)}$$

$$P_{n+1}(x) = (A_nx + B_n)P_n(x) - C_nP_{n-1}(x) \quad \text{(unrestricted form)}$$

$$P_0(x) = 1, \ P_{-1}(x) = 0. \ \lambda_n > 0, \quad A_nA_{n-1}C_n > 0.$$

The converse of this is known as Favard's Theorem [45]. Of course Favard's Theorem is a characterization theorem. However we shall deal here only with those results that characterize special classes of orthogonal polynomials, i.e., we shall deal only with OPS's that have certain additional properties. Polynomial sets which are obtainable from one another by a linear change of the independent variable and/or a rescaling of the dependent variable will be considered equivalent.

1

P. Nevai (ed.), Orthogonal Polynomials, 1–24.
© *1990 by Kluwer Academic Publishers.*

To illustrate our purpose let us consider an old and easy characterization of the Hermite polynomials, which may be defined by means of the generating function

$$(1.2) \qquad e^{2xt-t^2} = \sum_{n=0}^{\infty} H_n(x)\frac{t^n}{n!} \qquad \text{so that} \qquad H_n'(x) = 2nH_{n-1}(x).$$

These two relations remind us of Appell polynomial sets which, in general, are defined by

$$(1.3) \qquad A(t)e^{xt} = \sum_{n=0}^{\infty} p_n(x)t^n, \qquad A(t) = \sum_{n=0}^{\infty} a_n t^n, \qquad a_0 = 1,$$

Examples of Appell polynomials are: The powers, $\{x^n/n!\}$, the Hermite PS $\{H_n(x)/2^n n!\}$, and the Bernoulli PS $\{B_n(x)/n!\}$.
It follows from (1.3) that a PS, $\{p_n(x)\}$, is Appell iff for $n = 0,1,2,\ldots$

$$(1.4) \qquad (a) \;\; p_n(x) = \sum_{k=0}^{n} a_{n-k}\frac{x^k}{k!}, \qquad \text{or} \qquad (b) \;\; p_n'(x) = p_{n-1}(x).$$

holds. Thus we state the problem:

Problem: Find all Appell PS which are also orthogonal.

As far as I know the solution to this problem was first given by Angelesco [25], and later by other authors. (See Hahn [67], Meixner [94], Shohat [104], Webster [114], Toscano [108], Carlitz [36]).

We sketch the solution. Let $\{p_n(x)\}$ be orthogonal and satisfies (1.4-a) so that

$$p_{n+1}(x) = \frac{1}{n+1}(x + c_n)p_n(x) - \lambda_n p_{n-1}(x).$$

Put (1.4-a) in this recurrence and then equate coefficients of x^k. We get, for $k = 0,1,2,\ldots,n$,

$$(1.5) \qquad \frac{a_{n+1-k}}{k!} = \frac{1}{n+1}\left(\frac{a_{n+1-k}}{(k-1)!} + c_n\frac{a_{n-k}}{k!}\right) - \lambda_n \frac{a_{n-k-1}}{k!}$$

Putting $k = n$ in (1.5) results in $c_n = a_1$, a constant. Similarly if we put $k = n-1$ we get
$$\lambda_n = \frac{a_1^2 - 2a_2}{n+1} = \frac{c}{n+1}.$$
The two arbitrary constants a_1 and a_2 must be chosen so that $c > 0$.
Put $R_n(x) = (2/c)^{n/2}p_n(x)$. We get

$$R_0(x) = 1, \; R_1(x) = 2\left(\frac{x+a_1}{\sqrt{2c}}\right), \qquad R_{n+1}(x) = 2\left(\frac{x+a_1}{\sqrt{2c}}\right)R_n(x) - 2nR_{n-1}(x),$$

Thus $R_n(x) = H_n(\frac{x+a_1}{\sqrt{2c}})$, $\qquad n = 0,1,2,\ldots$

Two remarks are in order here. First we did not use the fact that $\lambda_n > 0$ we only used that $\lambda_1 > 0$. Secondly we did not need to use the full force of the recurrences (1.5); we only used the necessity that they must hold for $k = n$, $k = n-1$. At this stage one can verify directly that the resulting PS is orthogonal or that they satisfy a three term recurrence relation like (1.1) or the rest of the equations in (1.5).

Extensions and similar results will be described below. We shall assume that the reader is familiar with hypergeometric and q-hypergeometric notations. (See Rahman's paper in this volume.)

2. The Sheffer Polynomials

Let us consider polynomial sets generated by

$$(2.1) \qquad A(t)e^{xu(t)} = \sum_{n=0}^{\infty} S_n(x)t^n$$

where $A(0) = 1, u(0) = 0,\ u'(0) = 1$.

Meixner [94] looked for all OPS's with a generating function of the form (2.1) and later Sheffer [103] made a complete study of polynomial sets which satisfy (2.1) and asked the same question as Meixner did. We state the problem formally.

Problem: Find all polynomial sets which satisfy (2.1) and which are also orthogonal.

To solve this problem they defined a differential operator of infinite order which commutes with the differentiation operator D and which acts on the polynomials $\{p_n(x)\}$ like D on the powers. Take the formal inverse of $u(t)$ defined by $u(J(t)) = J(u(t)) = t$ so that we have from (2.1) $J(D)S_n(x) = S_{n-1}(x)$.
Both Meixner and Sheffer proved that $\{S_n(x)\}$ is an OPS and satisfies (2.1) if and only if $\{S_n(x)\}$ satisfies the recurrence relation:

$$S_{n+1}(x) = [x - (an + b)]S_n(x) - n(cn + d)S_{n-1}(x),$$

in which case

$$J'(u) = 1 - ct + \kappa t^2 = (1 - \alpha t)(1 - \beta t) = \frac{1}{u'(t)}, \qquad \frac{A'(t)}{A(t)} = \frac{\lambda_2 t}{(1 - \alpha t)(1 - \beta t)}.$$

By considering all possible cases we get that $\{S_n(x)\}$ must be one of the following cases which we list below rescaled so that the polynomials are put in a standard form:

    (a). $\alpha = \beta = 0$. This gives the Hermite PS. (1.2)
    (b). $\alpha = \beta \neq 0$. This case leads to the Laguerre PS with the generating function

$$(2.2) \qquad (1 - t)^{-\alpha-1} \exp\{\frac{xt}{1-t}\} = \sum_{n=0}^{\infty} L_n^{(\alpha)}(x)t^n.$$

    (c). $\alpha \neq 0$, $\beta = 0$, yields the Charlier PS with

$$(2.3) \qquad e^t \left(1 - \frac{t}{a}\right)^x = \sum_{n=0}^{\infty} c_n(x; a)\frac{t^n}{n!}.$$

    (d). $\alpha \neq \beta$ (both real). We get the Meixner PS for which the generating function is

4

$$(2.4) \qquad \left(1 - \frac{t}{c}\right)^x (1-t)^{-x-\beta} = \sum_{n=0}^{\infty} m_n(x;\beta,c)\frac{t^n}{n!}.$$

(e). $\alpha \neq \beta$ (complex-conjugate of each other), results in the Meixner-Pollaczek PS generated by

$$(2.5) \qquad \left(1 + te^{i\phi}\right)^{-\lambda+ix} \left(1 + te^{-i\phi}\right)^{-\lambda-ix} = \sum_{n=0}^{\infty} P_n^{(\lambda)}(x;\phi)t^n,$$

so that

$$A(t) = (1 + 2t\cos\phi + t^2)^{-\lambda}, \qquad u(t) = 2\tan^{-1}\left(\frac{t\sin\phi}{1 + t\cos\phi}\right).$$

The conditions of this problem can be weakened (without getting new OPS's). It was shown in [8] that the left hand side of (2.1) can be replaced by $\exp\{Q(x,t)\}$ where $Q(x,t)$ is a polynomial in x with coefficients that are functions of t. The resulting OPS's are the same as those obtained by Meixner [94] and by Sheffer [103].

3. The Brenke PS

Before we consider the next case let us recall the Geronimus problem. Put $\omega_k(x) = (x - x_1)(x - x_2)\cdots(x - x_k)$ and $\{x_k\}$, $\{a_k\}$ are arbitrary, $b_k \neq 0$, $a_0 = b_0 = 1$. Geronimus [64] raised the question:

Problem: Find all OPS $\{P_n(x)\}$ where

$$(3.1) \qquad P_n(x) = \sum_{k=0}^{n} a_{n-k} b_k \omega_k(x) \qquad n = 0,1,2,...$$

He did not solve this problem. However he gave some necessary and sufficient conditions on the coefficients, $\{a_n\}$, $\{b_n\}$, and the sequence $\{x_k\}$. Then the polynomials in (3.1) are orthogonal, i.e., they satisfy

$$\frac{P_{n+1}(x)}{b_{n+1}} = [x - c_n]\frac{P_n(x)}{b_n} - \lambda_n \frac{P_{n-1}(x)}{b_{n-1}}, \qquad n \geq 1,$$

if and only if

$$(3.2) \qquad a_{k+1}(B_{n-k} - B_{n+1}) = a_1 a_k(B_n - B_{n+1}) + \frac{\lambda_n}{B_{n-1}}a_{k-1} + a_k(x_{n+1} - x_{n-k+1}),$$

for $k = 0,1,\ldots,n$ where $B_n = b_{n-1}/b_n$, $B_0 = 0$.
This problem has remained unsolved in its full generality. However some special choices of the sequence $\{x_k\}$ lead to complete solutions. For example in case $x_k = 0$ the resulting PS is known as Brenke PS [41] and may alternatively be defined by

$$(3.3) \qquad A(t)B(xt) = \sum_{n=0}^{\infty} P_n(x)t^n.$$

T.S. Chihara [41;42] solved the following

Problem: Find all Brenke type polynomials which are also orthogonal.

To get his results he put $\Delta_n = B_{n-1} - B_n$ then $k = 2$ and $x_k = 0$ in (3.2) to get

(3.5) $$(a_3 - 2a_1a_2 + a_1^3)\Delta_{n+1} + (a_3 - a_1a_2)\Delta_n + a_3\Delta_{n-1} = 0.$$

Solving this difference equation he found all OPS in this class. They are
  (a). The Laguerre PS: $L_n^{(\alpha)}(x)/(1+\alpha)_n$ for which

$$a_n = \frac{1}{n!} \quad \text{and} \quad b_n = \frac{(-1)^n}{n!(1+\alpha)_n}.$$

  (b). The generalized Hermite PS: $\{H_n^{\mu}(x)\}$ where

$$H_{2n}^{\mu}(x) = \frac{L_n^{(\mu-\frac{1}{2})}(x^2)}{(\mu+\frac{1}{2})_n} \quad \text{and} \quad H_{2n+1}^{\mu}(x) = \frac{xL_n^{(\mu+\frac{1}{2})}(x^2)}{(\mu+\frac{1}{2})_n}.$$

  (c). The Wall PS: (Wall [112], Hahn [67]) $W_n(x,b,q)/(q;q)_n(b;q)_n$, where

(3.6) $$W_n(x,b,q) = (-1)^n(b;q)_n q^{\frac{1}{2}n(n+1)} {}_2\Phi_1 \left[\begin{matrix} q^{-n}, 0; \\ b \end{matrix} q, x \right].$$

  (d). The generalized Stieltes-Wigert (q-Laguerre) PS: These are essentially the Wall PS with $q \to 1/q$. More specifically (see (7.3))

$$L_n^{(\alpha)}(x;q) = W_n(\frac{x}{q}, q^{-\alpha-1}, \frac{1}{q}).$$

  (e). Some symmetric PS's related to the Wall and Stieltjes-Wigert PS's.

(3.7) $\qquad Y_{2n}(x,b,q) = W_n(x^2;b,q), \qquad Y_{2n+1}(x,b,q) = xW_n(x^2;bq,q).$

(3.8) $\qquad T_{2n}(x,p,q) = S_n(x^2;p,q), \qquad T_{2n+1}(x,p,q) = q^{-n}xS_n(qx^2;pq,q).$

  (f). A non-symmetric PS related to the Wall PS

$$Z_{2n}(x) = W_n(x^2;b,q) + (1-q^2)xW_{n-1}(x^2;qb,q),$$
$$Z_{2n+1}(x) = = xW_n(x^2;bq,q) + (1-q)W_n(x^2;b,q).$$

  (g). The Al-Salam and Carlitz PS : $\{U_n^{(a)}(x;q)\}$, and $\{V_n^{(a)}(x;q)\} = \{U_n^{(a)}(x;q^{-1})\}$

(3.9) $\quad U_{n+1}^{(a)}(x;q) = [x - (1+a)q^n]U_n^{(a)}(x;q) + aq^{n-1}(1-q^n)U_{n-1}^{(a)}(x;q) \qquad (a < 0).$

It is generated by

(3.10) $$\frac{(t;q)_\infty(at;q)_\infty}{(xt;q)_\infty} = \sum_{n=0}^{\infty} U_n^{(a)}(x;q)\frac{t^n}{(q;q)_n}.$$

We remark that the polynomials in (g) can be regarded as q-Appell PS, i.e, those that has a generating function of the form

(3.11)
$$\frac{A(t)}{(xt;q)_\infty} = \sum_{n=0}^{\infty} P_n(x)t^n.$$

They can be equivalently defined by

(3.12) $\quad D_q P_n(x) = (1 - q^n)P_{n-1}(x), \qquad$ where $\qquad D_q f(x) = \dfrac{f(x) - f(qx)}{x}.$

Thus q-Appell PS's belong to the Brenke class. Now Chihara's results show that the polynomials (3.9) are the only OPS satisfying (3.11) or (3.12). This also implies that the PS (3.9) is an example of a PS $\{P_n(x)\}$ which is orthogonal and at the same time $\{D_q P_n(x)\}$ is also orthogonal.

4. Another Gerominus PS

Al-Salam and Verma considered in [20] the case $x_k = q^{-k+1}$. Then (3.2) leads to

$$a_1(a_1 a_{k-2} - a_{k-1})\Delta_n - a_2 a_{k-2}(\Delta_n + \Delta_{n-1}) + a_k(\Delta_n + \Delta_{n-1} + ... + \Delta_{n-k+1}) = C_k q^{-n},$$

$n \geq 3$, where $C_k = a_1 q(1 - q)a_{k-2} - q(1 - q^{k-1})a_{k-1}$. In particular if $k = 3$ we have

(4.1) $\quad (a_1^3 - 2a_1 a_2 + a_3)\Delta_n + (a_3 - a_1 a_2)\Delta_{n-1} + a_3\Delta_{n-2} = C_3 q^{-n}, \qquad (n \geq 3).$

The general solution of (4.1) is of the form

$$\Delta_n = (k_1 + k_2 n)\lambda_1^n + k_3\lambda_2^n + (k_4 + k_5 n + k_6 n^2)q^{-n}, \qquad (n \geq 1)$$

where λ_1, and λ_2 satisfy

(4.2) $$(a_1^3 - 2a_1 a_2 + a_3)\lambda^2 + (a_3 - a_1 a_2)\lambda + a_3 = 0.$$

Some of the constants $k_i = 0$ depending on the character of the roots of (4.2). Considering all possible cases the following theorem was proved.

Theorem: Let $x_k = q^{-k+1}$, and let $\{P_n(x)\}$ be of the form (3.1). Then it is an OPS if and only if it is one of the following:

(a). Any of the OPS's of Brenke type ($q = 1$).
(b). The OPS (Al-Salam and Verma [21])

(4.3) $$p_n(x) = [x - (-1)^n \alpha]p_{n-1}(x) - \lambda_n p_{n-2}(x)$$

where $\lambda_{2n} = Eq^n(1 - \gamma q^{n-1})$, $\lambda_{2n+1} = E\gamma q^n(1 - q^n)$, $E > 0, \gamma > 0, 0 < q < 1$. They appeared [21] as the only solution of the Geronimus problem when $x_{2k+1} = x_1$ and $x_{2k} = x_2$.

(c). A special Askey-Wilson OPS defined by [29]

$$P_n(x) = {}_3\Phi_2 \begin{bmatrix} q^{-n}, 0, x; \\ q\alpha, q\beta \end{bmatrix} q, q \end{bmatrix}.$$

(d). A q-Meixner PS (Hahn [67])

(4.4)
$$M_n(x,\beta,d) = {}_2\Phi_1\left[\begin{matrix} q^{-n}, x, q^{n+1} \\ \beta q \qquad d \end{matrix}\right].$$

(e). The PS obtained in [20]

$$R_n(x;\beta) = \frac{1-q+q^{1+n}}{a^n(q;q)_n}\, {}_3\Phi_2\left[\begin{matrix} q^{-n}, -(1-q)q^{-n}, x, \\ q\beta, -(1-q)q^{-n-1} \end{matrix} q,(1+\beta)q^{n+1}\right]$$

$$= \frac{1}{a^n(q;q)_n}\left\{(1-q)\,{}_2\Phi_1\left[\begin{matrix} q^{-n}, x, \\ \beta q \end{matrix} (1+\beta)q^{n+2}\right] + q^{n+1}\,{}_2\Phi_1\left[\begin{matrix} q^{-n}, x, \\ \beta q \end{matrix} (1+\beta)q^{n+1}\right]\right\}.$$

The monic set satisfies

$$\widehat{R}_{-1}(x) = 0, \quad \widehat{R}_0(x) = 1, \qquad \widehat{R}_n(x) = (x+c_n)\widehat{R}_{n-1}(x) - \lambda_n\widehat{R}_{n-2}(x), \quad n \geq 1$$

where

$$\lambda_n = \frac{q^{5-4n}}{(1+\beta)^2}(1-q^{n-1})(1-\beta q^{n-1})[1-(1+\beta)q^{n-1}]$$

$$c_n = \frac{1}{1+\beta}\{q^{-2n}(1+q^3) - (1+\beta)(1+q^2)q^{-n}\}$$

In this case $a_n = \dfrac{1-q+q^n}{a^n(q;q)_n}$, $b_n = \dfrac{q^{n(n+1)}}{b^n(\beta q;q)_n(q;q)_n}$, $\dfrac{a}{b} = \beta+1$.

For proper choices of q and β these polynomials are orthogonal with respect to the moments

$$\mu_n = \mathcal{L}(x^n) = \frac{1}{q}\sum_{k=0}^{n}\frac{(q^{-n};q)_k(\beta q;q)_k}{(q;q)_k\,(\beta+1)^k}\{1-(1-q)q^k\}.$$

Formal orthogonality relation is given by

$$\mathcal{L}\{R_n(x)R_m(x)\} = \frac{((1+\beta)q;q)_n q^n}{a^{2n}(\beta q;q)_n(q;q)_n}\delta_{nm}.$$

These polynomials belong to a determined moment problem. However we have been unable to find the measure $\alpha(x)$ associated with these OPS's.

5. The Classical OPS's

Normally by the classical orthogonal polynomials one refers to the PS's of Hermite, Laguerre and Jacobi. These have several properties common to all of them, namely,

(a). They all satisfy a second order linear differential equation of the Sturm-Liouville type

(5.1)
$$\sigma(x)y''(x) + \tau(x)y'(x) + \lambda_n y(x) = 0.$$

where $\sigma(x)$ is a polynomial of degree ≤ 2 and $\tau(x)$ is a linear polynomial, both independent of n, and λ_n is independent of x.

(b). They all have derivatives which form OPS's.

(c). They all possess a Rodriques' type formula

$$(5.2) \qquad P_n(x) = \frac{1}{K_n w(x)} D^n \{w(x)\sigma^n(x)\} \qquad n = 0, 1, 2, \ldots$$

where $w(x)$ is a function which is non-negative on a certain interval and where $\sigma(x)$ is a polynomial in x and independent of n.

(d). They are all orthogonal with respect to a weight function that satisfies a Pearson differential equation

$$(5.3) \qquad \frac{w'(x)}{w(x)} = \frac{N(x)}{\sigma(x)}, \qquad (\sigma(x)w(x))' = \tau(x)w(x), \qquad N(x) = \tau(x) - \sigma'(x).$$

(e). They all satisfy the differential-difference relation

$$(5.4) \qquad \pi(x)p_n'(x) = (\alpha_n x + \beta_n)p_n(x) + \gamma_n p_{n-1}(x).$$

(f). They all satisfy a non-linear equation of the form [93]

$$(5.5) \qquad \sigma(x)\frac{d}{dx}\{p_n(x)p_{n-1}(x)\} = (\alpha_n x + \beta_n)p_n(x)p_{n-1}(x) + \gamma_n p_n^2(x) + \delta_n p_{n-1}^2(x)$$

where $\alpha_n, \beta_n, \gamma_n$ and δ_n are independent of x.

Every one of the above properties has a converse, i.e., any PS which satisfies any one of the above properties must necessarily be one of the classical OPS's.

Bochner [31] (see also Routh [102]) determined all polynomial solutions in case (a). He found that, essentially, the only OPS's which fall within this type are the classical orthogonal polynomials. However others which are orthogonal in the complex plane also arise. We summarize the results

(i) $\sigma(x) = 0 \Rightarrow$ non-orth. system (ii) $\sigma(x) = 1 \Rightarrow$ The Hermite PS

(iii) $\sigma(x) = x \Rightarrow$ The Laguerre PS (iv) $\sigma(x) = (x-1)(x+1) \Rightarrow$ The Jacobi PS

(v) $\sigma(x) = (x^2 + a^2)$

Case (v) yields a PS which is orthogonal in the complex plane. The special case $a = 0$ leads to the Bessel PS.

Other authors who looked at this problem are J. Aczél [2;3], Beale [30], Cryer [47], Á. Császár [48], L. Feldman [59;60], E.H. Hildebrandt [72], Lesky [87;88], Peeples [99], and Tricomi [109].

The converse of (b) was given by W. Hahn [67] who proved that the only orthogonal polynomials whose derivative forms an OPS are the classical PS.

His method of proof goes like this. Assume that $\{P_n(x)\}$ and $\{P_n'(x)\}$ satisfy respectively the three term recurrence relations

$$P_{n+1}(x) = (x - c_n)P_n(x) - \lambda_n P_{n-1}(x), \qquad \frac{P_{n+1}'(x)}{n+1} = (x - \gamma_n)\frac{P_n'(x)}{n} - \mu_n \frac{P_{n-1}'(x)}{n-1}$$

After some manipulations he was able to show that $\{P_n(x)\}$ satisfies (5.1) and hence is in the Bochner class. See also Krall [77;78], Webster [114], Beale [30], and Geronimus [63]. Geronimus [65] states that Sonin [105] was the first to solve this problem. Luzin [90] raised a more general problem of finding orthogonal functions whose derivatives are also orthogonal.

The converse of property (c) was first shown by Hildebrandt [72] who proved that (c) implies (a) (see also Brenke[35]). It was also given by Tricomi [108] and Ebert [52]

Hildebrandt also proved that (d) with $\deg N = 1$, $\deg \sigma \leq 2$ implies (5.1), (5.2) and (5.4).

In [13] it was shown that OPS's satisfying (5.4) are necessarily one of the classical OPS's.

A more general problem was raised by Askey who asked what OPS $\{p_n(x)\}$ whose derivatives satisfy a relation of the form

$$(5.6) \qquad \pi(x)p'_n(x) = \sum_{k=n-t}^{n+s} \alpha_{nk}p_k(x).$$

This problem was solved by Maroni [91;92] who called such OPS semi-classical. See also [101]. He stated the problem in this way

Problem: Find all OPS's whose derivatives are quasi-orthogonal.

A similar problem was initiated and solved by Bonan, Lubinsky, and Nevai [33] who gave necessary and sufficient conditions on the measures of two OPS's $\{P_n(x)\}$ and $\{q_n(x)\}$ so that they are related by a formula of the form

$$R(x)P^{(j)}_{n+j}(x) = \sum_{k=n-t}^{n+s} c_{nk}q_k(x) \qquad \text{for some integers } j, \ t, \ s.$$

Another problem was given and solved by Bonan and Nevai [32] is

Problem: Find all OPS whose derivatives are linear combinations of at most two polynomials of the same system.

They concluded that such OPS's must have a weight function

$$w(x) = D \exp\{-\frac{c}{4}(x-a)^4 - \frac{k}{2}(x-b)^2\} \qquad D, a, b, c, k. \text{ are real constants.}$$

On the other hand property (f) was used by McCarthy [93] to characterize the classical OPS's.

6. Discrete Classical OPS

The Bochner problem was generalized by considering a finite difference operator Δ_h

$$\Delta_h f(x) = \frac{f(x+h) - f(x)}{h}$$

so that $\lim_{h\to 0} \Delta_h f(x) = f'(x)$. This point is not really important at this time. One can always take such a limit. We shall take without loss of generality $h = 1$ and write Δ for Δ_1.

Instead of (5.1) we consider a difference equation of the second order of the Sturm Liouville type,

(6.1) $$\sigma(x)\Delta^2 y(x) + \tau(x)\Delta y(x) + \lambda_n y(x+1) = 0.$$

Let ∇ be backward difference operator $\nabla f(x) = f(x) - f(x-1)$. Then it follows that $[\Delta\nabla + \nabla]f(x) = \Delta f(x)$ and $E\Delta\nabla = \Delta^2$. Thus (6.1) has the equivalent form

(6.2) $$\sigma_1(x)\Delta\nabla y(x) + \tau_1(x)\Delta y(x) + \lambda_n y(x) = 0$$

where $\sigma_1(x+1) = \sigma(x) - \tau(x)$, $\tau_1(x+1) = \tau(x)$. In both equations (6.1) and (6.2) σ and τ are polynomials independent of n. The polynomial solutions of (6.1) were explicitly stated for the first time by Lancaster [82] and later by Lesky [85;86]. We summarize these solutions as

(a) The Charlier PS (2.3)

(6.3) $$c_n(x;a) = {}_2F_0(-n, -x; -; -\frac{1}{a}) \qquad (a > 0).$$

(b) The Meixner PS (2.4)

(6.4) $$m_n(x;\beta,c) = (\beta)_{n2}F_1(-n, -x, \beta; 1 - \frac{1}{c}) \qquad (\beta > 0, \ 0 < c < 1).$$

(c) The Krawtchouk polynomials ($\beta = -N$ in the Meixner polynomials)

(6.5) $$K_n(x) = {}_2F_1(-n, -x; -N; \frac{1}{p}) \qquad (0 < p < 1, \ N, \text{ a positive integer})$$

(d) The Hahn polynomials

(6.6) $$h_n(x;\alpha,\beta) = {}_3F_2(-n, n + \alpha + \beta + 1, -x; -N, \alpha + 1; 1) \qquad n = 0, 1, 2, \dots, N.$$

All these polynomial sets share with the classical OPS's properties (a) through (e) listed in §5. Although we credited Lancaster as the first to give an explicit solution to (6.1), it appears that Hildebrandt [72] made an analogous study of these discrete OPS. He did not state his theorems as characterization theorems, however he had all the necessary ingredients for them. He started with the finite difference analog of (5.3)

(6.7) $$\frac{\Delta u(x)}{u(x)} = \frac{a_0 + a_1 x}{b_0 + b_1 x + b_2 x^2} \equiv \frac{N(x)}{D(x)} \qquad \text{so that } \sigma(x) = N(x+1) + D(x+1),$$

and then defined

(6.8) $$P_n(x) = \frac{1}{u(x)}\Delta^n\{D_n(x - n)u(x)\}$$

where $D_n(x) = D(x)D(x+1)\dots D(x+n-1)$. He then showed that $\{P_n(x)\}$ satisfy difference equations of the form (6.1) (see also [98]) and that they have a property analogous to (5.4).

A difference analog of property (f) of §5 should hold but I have not seen this done.

7. The Hahn Problem

Motivated by the properties mentioned above, Hahn [67] considered the operator

$$Lf(x) = \frac{f(qx + \omega) - f(x)}{(q-1)x + \omega}.$$

Note that when $q = 1$, $L = \Delta_\omega$, the finite difference operator. On the other hand when $q \neq 1$ $\omega = 0$ then L is the q-derivative.

$$Lf(x) = \frac{f(qx) - f(x)}{(q-1)x}.$$

Hahn then showed that, without loss of generality, ω may be take to be equal to zero. Thus he posed and solved the following problems.

Problems: Find all OPS such that one of the following holds

1. $\{LP_n(x)\}$ is also OPS
2. $\{P_n(x)\}$ satisfy the functional equation

$$\sigma(x)L^2 P_n(x) + \tau(x)LP_n(x) + \lambda_n P_n(x) = 0.$$

3. $P_n(x)$ has the representaion

$$P_n(x) = \frac{1}{w(x)}L^n\{f_0(x)f_1(x)\cdots f_{n-1}(x)w(x)\},$$

where $f_k(x) = f_{k+1}(qx)$
4. If $P_n(x) = \sum a_{nk}x^k$ then $a_{nk}/a_{n,k-1}$ is a rational function of q^n and q^k.
5. The moments associated with $\{P_n(x)\}$ satisfy

$$M_n = \frac{a + bq^n}{c + dq^n} M_{n-1}, \qquad \text{where } (ad - bc \neq 0).$$

Here Hahn meant by moments those that are either the power moments (moments against x^k) or a generalized moments (moments against $(x;q)_k$).
He showed that all the above conditions lead to the same OPS. They are
I. The q-Hahn PS

(7.1)
$$Q_n(x; \alpha, \beta, \gamma; q) = {}_3\Phi_2 \left[\begin{matrix} q^{-n}, \alpha\beta q^{n+1}, x; \\ \alpha q, \gamma q \end{matrix} q, q \right].$$

In case $\gamma q = q^{-N}$ the orthogonality is discrete. If $\gamma q = -\delta$ then the resulting PS are called by Andrews and Askey the Big q-Jacobi PS. Andrews and Askey [24] worked out their orthogonality.
II. The (little) q-Jacobi PS

(7.2)
$$p_n(x; \alpha, \beta; q) = {}_2\Phi_1 \left[\begin{matrix} q^{-n}, \alpha\beta q^{n+1}; \\ \alpha q \end{matrix} qx \right].$$

III. The q-Meixner PS (4.4).

IV. The q-Laguerre PS (or its equivalent Wall PS):

$$(7.3) \qquad L_n^{(\alpha)}(x;q) = \sum_{k=0}^{n} \frac{(q^{-n};q)_k q^{\frac{1}{2}k(k+1)}}{(q;q)_k(q^{\alpha+1};q)_k}(-xq^{\alpha+n})^k.$$

V. The (discrete) q-Hermite PS $\{h_n(x,q)\}$. These are special cases of the polynomials (3.7) or (3.9), viz.,

$$(7.4) \qquad h_n(x,q) = Y_n(x,q,q^2) \equiv q^n U_n^{(-1)}(qx;q).$$

They satisfy the three term recurrence [12]

$$h_0(x,q) = 1, \; h_{-1}(x,q) = 0, \quad h_{n+1}(x,q) = xh_n(x,q) - q^{n-1}(1-q^n)h_{n-1}(x,q), \qquad (n \geq 1).$$

VI. The q-Charlier OPS (these are the polynomials (3.9); see also [12])

$$(7.5) \qquad U_n^{(a)}(x,q) = (-a)^n q^{\frac{1}{2}n(n-1)}(\frac{x}{a}q^{1-n};q)_n \; {}_1\Phi_1\left[\begin{matrix} q^{-n}; & q \\ \frac{x}{a}q^{1-n} & a \end{matrix}\right], \qquad (a < 0).$$

Both (7.4) and (7.5) were inaccurately stated by Hahn in [67].

8. Convolutions of OPS

Al-Salam and Chihara [14] noted that

$$(8.1) \qquad 2^{n/2} H_n(x+y) = \sum_{k=0}^{n} \binom{n}{k} H_k(2^{\frac{1}{2}}x)H_{n-k}(2^{\frac{1}{2}}y),$$

$$(8.2) \qquad L_n^{(\alpha+\beta+1)}(x+y) = \sum_{k=0}^{n} L_k^{(\alpha)}(x)L_{n-k}^{(\beta)}(y)$$

$$(8.3) \qquad (a+b)^n c_n(x+y,a+b) = \sum_{k=0}^{n} \binom{n}{k} a^k b^{n-k} c_k(x,a)c_{n-k}(y,b)$$

$$(8.4) \qquad m_n(x+y;\alpha+\beta,c) = \sum_{k=0}^{n} \binom{n}{k} m_k(x;\alpha,c)m_{n-k}(y;\beta,c)$$

$$(8.5) \qquad P_n^{\lambda+\mu}(x+y,\phi) = \sum_{k=0}^{n} P_k^{\lambda}(x;\phi)P_{n-k}^{\mu}(y;\phi).$$

All these formulas are consequences of their generating functions which are of the Meixner-Sheffer type

$$A(t)e^{xB(t)} = \sum_{n=0}^{\infty} P_n(x)t^n, \qquad A(0) = 1, \; B(0) = 0, B'(0) = 1$$

Formulas (8.1)-(8.5) show that the convolution of a pair of certain OPS's is also an OPS. Thus we state the problem

Problem: Find all OPS's $\{p_n(x)\}$, and $\{q_n(x)\}$ such that their convolution

$$(8.6) \qquad Q_n(x,y) = \sum_{k=0}^{n} p_k(x) q_{n-k}(y)$$

is also an OPS in x.

The solution of this problem was given in [14] by the theorem

Theorem: If $\{p_n(x)\}$, $\{q_n(x)\}$, OPS's in x and if $Q_n(x,y)$ as defined by (8.6) is an OPS in x for infinitely many values of y then the three polynomial sets are either the Meixner-Sheffer OPS (8.1)-(8.5) or the (at that time) new polynomial set $P_n(x) = P_n(x; q, ; a, b, g)$ given by

$$(8.7) \qquad P_{n+1}(x) = [x - aq^n] P_n(x) - (g - bq^{n-1})(1 - q^n) P_{n-1}(x),$$
$$P_{-1}(x) = 0, \quad P_0(x) = 1.$$

The convolution property reads
$$(8.8)$$
$$P_n(x; q; y, k, g) = \sum_{k=0}^{n} \frac{(q;q)_n}{(q;q)_k (q;q)_{n-k}} (-1)^k q^{k(k-1)/2} P_{n-k}(x; q; a, b, g) P_k(y; q^{-1}; a, b, k).$$

Its generating function is

$$(8.9) \qquad \Phi(x,t) = \sum_{n=0}^{\infty} P_n(x) \frac{t^n}{(q;q)_n} = \prod_{k=0}^{\infty} \frac{(1 - \alpha t q^k)(1 - \beta t q^k)}{(1 - \gamma t q^k)(1 - \delta t q^k)},$$

where $1 - at + bt^2 = (1 - \alpha t)(1 - \beta t)$, $1 - xt + gt^2 = (1 - \gamma t)(1 - \delta t)$.
The case $a = b = 0$, $g \neq 0$ leads to the continuous q-Hermite PS.

$$(8.10) \qquad \frac{1}{(e^{i\theta}t; q)_\infty (e^{-i\theta}t; q)_\infty} = \sum_{n=0}^{\infty} H_n(x|q) \frac{t^n}{(q;q)_n}$$

The case $g = 0$, $ab \neq 0$ lead to the PS (3.9).
The general case was shown later by Askey and Ismail [28] to be a special case of the Askey-Wilson PS. They gave another proof of its orthogonality.
We note that the generating function (8.9) is of the form

$$(8.11) \qquad A(t) \prod_{k=0}^{\infty} [1 - xK(tq^k)]^{-1} = \sum_{n=0}^{\infty} P_n(x) t^n,$$

where

$$A(t) = \prod_{k=0}^{\infty} \frac{1 - qtq^k + bt^2 q^{2k}}{1 + gt^2 q^{2k}}, \quad K(t) = \frac{t}{1 + gt^2}.$$

Andrews and Askey [24] conjectured that if $\{p_n(x)\}$ is an OPS and has a generating function of the form (8.11) then it is one of the PS found in [14].
In [15] Al-Salam and Chihara then considered

$$(8.12) \qquad A(t) \prod_{k=0}^{\infty} \frac{1 - \delta x H(tq^k)}{1 - \theta x K(tq^k)} = \sum_{n=0}^{\infty} Q_n(x)t^n.$$

where $H(0) = K(0) = 0$, $H'(0) = K'(0) = 1$, $\theta^2 + \delta^2 \neq 0$.
The case $\theta\delta = 0$ leads to all OPS's found by Al-Salam and Chihara in [14]. The case $\theta\delta \neq 0$ leads to The q-Pollaczek PS (Askey-Ismail [28])

$$(1 - q^{n+1})Q_{n+1}(x) = [(1 - aq^n)x + bq^n]Q_n(x) - (1 - cq^{n-1})Q_{n-1}(x).$$

In this case

$$A(t) = \prod_{k=0}^{\infty} \frac{1 - btq^k + ct^2 q^{2k}}{1 + t^2 q^{2k}}, \qquad H(t) = \frac{t}{(1 - bt + ct^2)}, \qquad K(t) = \frac{t}{(1 + t^2)}.$$

We note that if we put $y = x$ in (8.1)-(8.5) the left hand sides are also orthogonal polynomials as polynomials in x. This motivates the following problem

Unsolved Problem: Find all OPS's $\{P_n(x)\}$ and $\{Q_n(x)\}$ so that the convolution

$$R_n(x) = \sum_{k=0}^{n} P_k(x)Q_{n-k}(x)$$

also forms an OPS.

9. The Feldheim-Lanczewisky Problem

Recall that the Legendre PS can be defined by

$$(1 - 2xt + t^2)^{-1/2} = \sum_{n=0}^{\infty} P_n(x)t^n = \left| \frac{1}{\sqrt{1 - te^{i\theta}}} \right|^2 \qquad (x = \cos\theta)$$

Fejér [55] introduced the "generalized Legendre " by means of

$$(9.1) \qquad \left| F(te^{i\theta}) \right|^2 = \sum_{n=0}^{\infty} F_n(x)t^n, \qquad \text{so that} \qquad F_n(x) = \sum_{k=0}^{n} g_k g_{n-k} T_{n-2k}(x)$$

where $F(u) = \sum_{k=0}^{\infty} g_k u^k$ and $T_n(x)$ is the Tchebycheff polynomial of the first kind. This led to

Problem (Feldheim [56], Lanzewizky [83]): Find all OPS with a generating function (9.1)
They independently solved this problem by calculating the coefficients in the three term recurrence relation. However they did not identify the resulting polynomials or calculate their

orthogonality relation. Askey and Ismail [27] identified them as the Rogers' q-ultraspherical polynomials. They found the weight function, $W_\beta(x)$, with respect to which these polynomials are orthogonal. They put the recurrence relation in the form

(9.2) $\qquad (1 - q^{n+1})C_{n+1}(x; \beta|q) = 2x(1 - \beta q^n)C_n(x; \beta|q) - (1 - \beta^2 q^{n-1})C_{n-1}(x; \beta|q)$

The case $\beta = 0$ gives the Continuous q-Hermite polynomials.
The function $F(z)$ that generates the polynomials (9.2) is

$$F(z) = \frac{(\beta z; q)_\infty}{(z; q)_\infty} = \prod_{k=0}^{\infty} \frac{1 - \beta z q^k}{1 - z q^k}.$$

It seems that both Feldheim and Lanzewwisky missed an important limiting case, viz., the sieved ultraspherical polynomials of the second kind [11]. This can be obtained from (9.2) by putting $\omega_k = e^{2\pi i/k}$, $\beta = s^{\lambda k+1}\omega_k$ $q = s\omega_k$, then dividing by $1 - s\omega_k^{n+1}$ then letting $s \to 1$. We get

(9.3) $\qquad 2x B_n^\lambda(x; k) = B_{n+1}^\lambda(x; k) + B_{n-1}^\lambda(x; k) \qquad (n + 1 \neq mk)$

$\qquad\qquad 2(m + \lambda)x B_{mk-1}^\lambda(x; k) = m B_{mk}^\lambda(x; k) + (m + 2\lambda) B_{mk-2}^\lambda(x; k)$

$\qquad\qquad B_0^\lambda(x; k) = 1, \quad B_1^\lambda(x; k) = 2x$

They are orthogonal on $(-1, 1)$ with respect to the weight function

$$W(x) = (1 - x^2)^{\lambda+\frac{1}{2}}|U_{k-1}(x)|^{2\lambda}$$

The function $F(z)$ and the coefficients g_n are given by

$$F(z) = (1 - z)^{-1}(1 - z^k)^{-\lambda} \qquad g_{mk+j} = \frac{(\lambda + 1)_m}{m!}, \qquad (j = 0, 1, 2, ..., k - 1).$$

10. Sieved Ultraspherical PS.

Another representation of the Legendre polynomials is due to Heine

(10.1) $\qquad P_n(x) = \frac{4}{\pi} \frac{2^{2n}(n!)^2}{(2n + 1)!} \sum_{k=0}^{\infty} f_{kn}^{(\frac{1}{2})}(1 - x^2)^{\frac{1}{2}} U_{n+2k}(x) \qquad |x| < 1$

Szegö [106] generalized this to

(10.2) $\qquad (1 - x^2)^{\lambda-\frac{1}{2}}C_n^{(\lambda)}(x) = \frac{2^{2-2\lambda}\Gamma(n + 2\lambda)}{\Gamma(\lambda)\Gamma(n + \lambda + 1)} \sum_{k=0}^{\infty} f_{kn}^{(\lambda)}(1 - x^2)^{\frac{1}{2}} U_{n+2k}(x)$

where

$$f_{kn}^{(\lambda)} = \frac{(1 - \lambda)_k (n + k)!}{k! n! (n + \lambda + 1)_k}$$

Askey and Ismail [27] generalized Szegö's formula further to

$$W_\beta(x)C_n(x;\beta|q) = \sum_{k=0}^{\infty} a(n,k)W_\gamma(x)C_{n+2k}(x;\gamma|q)$$

Allaway, in his thesis [23], considered a characterization problem motivated by (10.1) ans (10.2). An equivalent form of the problem was stated in [10;11]

PROBLEM: Find all symmetric OPS, $\{A_n(x)\}$, which can be represented by

$$W(x)A_n(x) = \sum_{k=0}^{\infty} a_k b_{n+k} U_{n+2k}(x)$$

The solutions were either the Roger's q-ultrasherical PS (9.2) and some limiting cases of them which was called "sieved ultraspherical of the first kind."

$$2xc_n^\lambda(x;k) = c_{n+1}^\lambda(x;k) + c_{n-1}^\lambda(x;k) \quad (n \neq mk)$$

(10.3) $\qquad 2x(m+k)c_{mk}^\lambda(x;k) = (m+2\lambda)c_{mk+1}^\lambda(x;k) + mc_{mk-1}^\lambda(x;k),$

where $c_0^\lambda(x;k) = 1$, $c_1^\lambda(x;k) = x$. They are orthogonal on (-1,1) with respect to the weight function

$$w(x) = (1-x^2)^{\lambda-\frac{1}{2}}|U_{k-1}(x)|^{2\lambda}$$

where $U_n(x)$ are the Tchebycheff polynomials of the second kind.

11. Miscellaneous Results

In [16] Al-Salam and Ismail proved the following theorem

Theorem: A necessary and sufficient condition that a symmetric PS $\{P_n(x)\}$ be orthogonal and $\{P_n(q^n x)\}$ be also orthogonal is that

$$P_0(x) = 1, \ P_1(x) = cx, \qquad P_n(x) = xP_{n-1}(x) - \lambda_n P_{n-2}(x) \qquad (c \neq 0)$$

where λ_2 is arbitrary and $\lambda_n = q^{2n-4}$ for $n \geq 3$. The measure for these orthogonal polynomials were given in [16]. The corresponding characterization for non-symmetric OPS has not been found yet.

A generalization of the Brenke PS's are the Boas-Buck PS's. They have generating functions of the form

$$A(t)B(xu(t)) = \sum_{n=0}^{\infty} P_n(x)t^n.$$

It is not known what are the OPS's which are also Boas-Buck. The case $u(t) = t$ leads to the Brenke PS (see §3). Another special case was done by Ismail [73] in which he took $B(t) = \{(t;q)_\infty\}^{-1}$. Ismail proved that the only OPS's with generating function of the form

$$\frac{A(t)}{(xu(t);q)_\infty} = \sum_{n=0}^{\infty} P_n(x)t^n,$$

are

(a) The Meixner-Sheffer OPS's (q=1)   (b) The OPS (3.9) if $q(1-q) \neq 0$

(c) If $q = 0$ the polynomial set given by

$$P_0(x) = 1, \quad P_1(x) = U_1(x) + 2a, \qquad P_n(x) = U_n(x) + 2aU_{n-1} - cU_{n-2}(x) \qquad n \geq 2$$

The PS in (c) was first encountered by Geronimus who calculated the weight function only in the case when the measure is absolutely continuous. The general case was obtained by by Allaway [23].

In [21] Al-Salam and Verma considered another Boas-Buck PS, namely,

$$\frac{A(t)}{(1 - xu(t))^{\nu}} = \sum_{n=0}^{\infty} P_n(x)t^n.$$

They obtained all OPS in this class. The case $\nu = 1$ reduces to the case (c) of the previous problem. The case $\nu \neq 1$ leads to several OPS's (see [22])

Toscano [108] proved that $\{P_n(x)\}$ is OPS and $\{x^n P_n(1/x)\}$ is Appell if and only if $P_n(x)$ is the Laguerre PS.

In [1] and [4] characterizations of certain orthogonal hypergeometric polynomials were obtained. For example in [1] it was proved that the only orthogonal polynomials of the form

$$_{p+1}F_p \left[\begin{array}{c} -n, \alpha_1, \alpha_2, \ldots, \alpha_p; \\ \beta_1, \beta_2, \ldots, \beta_p \end{array} x \right]$$

where all the $\alpha's$ are independent of x and n is the Laguerre PS.

Similarly it was proved in [4] that the only orthogonal polynomials of the form

$$_{p+3}F_q \left[\begin{array}{c} -n, n + \gamma, x, \alpha_1, \ldots, \alpha_p; \\ \beta_1, \beta_2, \ldots, \beta_q \end{array} 1 \right]$$

is the Pasternack (or Hahn) PS. Other results of this type can be found in [4].

12.   The Askey and Wilson PS.

We conclude by stating a characterization problem that Askey [26] posed. First let us introduce the Askey-Wilson OPS [29]

$$(12.1) \qquad P_n(x, a, b, c, d|q) = a^{-n}(ab; q)_n(ac; q)_n(ad; q)_n {}_4\Phi_3 \left[\begin{array}{c} q^{-n}, abcdq^{n-1}, az, a/z; \\ ab, ac, ad \end{array} q \right]$$

where $z = e^{i\theta}$ and $x = \cos\theta = \frac{1}{2}(z + z^{-1})$.

A special case of it is the q-Racah polynomials (note the change in notation)

$$(12.2) \qquad R_n(\lambda(x)) = {}_4\Phi_3 \left[\begin{array}{c} q^{-n}, abq^{n+1}, q^{-x}, cdq^{x+1}; q \\ aq, cq, q^{-N} \end{array} \right] \qquad n = 0, 1, 2, \ldots, N.$$

They are orthogonal on $x = 0, 1, 2, \ldots, N$.

The polynomials (12.1) include many of the OPS's that have been mentioned earlier including Rogers' continuous q-ultraspherical polynomials (9.2). Askey and Wilson made comprehensive study of them in their memoir [29]. They are orthogonal on (-1,1) with respect to the weight function

$$w(x, a, b, c, d) = \frac{(z^2; q)_\infty (z^{-2}; q)_\infty}{2i(z - z^{-1})} [h(z, a)h(z, b)h(z, c)h(z, d)]^{-1}$$

where $h(z, a) = \prod_{k=0}^{\infty} (1 - (z + z^{-1})aq^k + a^2 q^{2k})$. Among other results, Askey and Wilson introduced the operator

(12.3)
$$\Delta_q f(z) \equiv \frac{f(q^{\frac{1}{2}} z) - f(q^{-\frac{1}{2}} z)}{z - z^{-1}},$$

so that in particular

(12.4)
$$\Delta_q (az; q)_k (\frac{a}{z}; q)_k = aq^{-\frac{1}{2}}(1 - q^k)(aq^{\frac{1}{2}} z; q)_{k-1}(\frac{aq^{\frac{1}{2}}}{z}; q)_{k-1}.$$

Formula (12.4) can be used to show

(12.5)
$$\Delta_q P_n(x, a, b, c, d|q) = k_n(z - z^{-1})P_{n-1}(x, aq^{\frac{1}{2}}, bq^{\frac{1}{2}}, cq^{\frac{1}{2}}, dq^{\frac{1}{2}}|q).$$

This is a formula that also shows that $\{P_n(x, a, b, c, d)\}$ is an OPS and at the same time $\{\Delta_q P_n(x, a, b, c, d)\}$ is also an OPS.
Other properties that were given in [29] are

(12.6)
$$\Delta_q^n [w(x; aq^{\frac{1}{2}n}, bq^{\frac{1}{2}n}, cq^{\frac{1}{2}n}, dq^{\frac{1}{2}n}) = B_n w(x; a, b, c, d)P_n(x, a, b, c, d|q)$$

which is an analog of Rodriques' formula, and

(12.7)
$$\Delta_q [w(x, aq^{\frac{1}{2}}, bq^{\frac{1}{2}}, cq^{\frac{1}{2}}, dq^{\frac{1}{2}})\Delta_q P_n(x)] + \lambda_n P_n(x) = 0$$

where $\lambda_n = -4(1 - q^{-n})(1 - q^{n-1}abcd)(1 - q)^2$, which is an analog of the second order differential equation (5.1) or the finite difference equation (6.1).
Properties (12.5), (12.6), and (12.7) are similar to those satisfied by the classical OPS of §5, the discrete OPS of §6, and the q-polynomials sets of §7. This prompted Andrews and Askey [24] to suggest that the polynomials (12.1) should be called the classical OPS. This seems reasonable because they contain all those polynomials of §5, §6 and §7 and share with them their most important properties.
Leonard [84] gave a theorem which shows that a PS satisfying certain conditions and (12.7) must be (12.2). Askey in [26] raises the problem of obtaining results by means of which (12.1) are characterized by means of any one of the properties (12.5) or (12.6). This has not been resolved and may prove to be an interesting question to try to settle.

References

1. N.A. Abdul-Halim and W.A. Al-Salam, A characterization of the Laguerre polynomials, Rend. del Seminario Mat. Univ. Padova, **34**(1964), 176-179.

2. J. Aczél, Eine Bemerkung über Charakterisierung der "klassische" orthogonalpolynome. Acta Math. Acad. Sci. Hung., 4(1953), 315-321.

3. J. Aczél, Sur l'equation différentielle des polynomes orthogonaux classiques, Annales Univ. Scient. Budapest, **2**(1959), 27-29.

4. N.A. Al-Salam, Orthogonal polynomials of hypergeometric type, Duke Math. J., **33**(1966), 109-122.

5. N.A. Al-Salam and W.A. Al-Salam, Some characterizations of the ultraspherical polynomials,Canad. Math. Bulletin, **11**(1968), 457-464.

6. W.A. Al-Salam, Characterization of certain classes of orthogonal polynomials related to elliptic functions, Annali di Matematica pura ed applicata (IV), **LXVII**(1965), 75-94.

7. W.A. Al-Salam, On a characterization of orthogonality, Math. Mag., **31**(1957), 41-44.

8. W.A. Al-Salam, On a characterization of Meixner's polynomials, The Quart. J. of Mathematics (Oxf)(2), **17**(1966), 7-10.

9. W.A. Al-Salam, On a characterization of a certain set of orthogonal polynomials, Boll. Unione Mat. Ital.(3), **19**(1964), 448-450.

10. W.A. Al-Salam ,W. Allaway and R. Askey, A characterization of the continuous q-ultraspherical polynomials, Canad. Math. Bull., **27(3)**(1984), 329-336.

11. W.A. Al-Salam, W.R Allaway, and R. Askey, Sieved ultrashperical orthogonal polynomials, Trans. Amer. Math. Soc., **284**(1984), 39-55.

12. W.A. Al-Salam and L. Carlitz, Some orthogonal q-polynomials, Math. Nachr., **30** (1965), 47-61.

13. W.A. Al-Salam and T.S. Chihara, Another characterization of the classical orthogonal polynomials, SIAM J. Math. Anal., **3**(1972), 65-70.

14. W.A. Al-Salam and T.S. Chihara, Convolution of orthogonal polynomials, SIAM J. Math. Anal., **7**(1976), 16-28.

15. W.A. Al-Salam and T.S. Chihara, q-Pollaczek polynomials and a conjecture of Andrews and Askey, SIAM J. Math. Anal., **18**(1987), 228-242.

16. W.A. Al-Salam and M. Ismail, Orthogonal polynomials associated with the Rogers-Ramanujan continued fractions, Pacific J. of Math., **104**(1983), 269-283.

17. W.A. Al-Salam and A. Verma, Some orthogonality preserving operators, Proc. Amer. Math. Soc., **23**(1969), 136-139.

18. W.A. Al-Salam and A. Verma, Orthogonality preserving operators I, Rendiconti Acad. Naz. dei Lincei(8), **LVIII**(1975), 833-838.

19. W.A. Al-Salam and A. Verma, Orthogonality preserving operators II, Rendiconti Acad. Naz, dei Lincei(8), **LIX**(1976), 26-31.

20. W.A. Al-Salam and A. Verma, On the Geronimus polynomial sets, Proc. Orthogonal Polynomials and Their Applications, Segovia 1986. Lecture Notes in Mathematics #1329 Springer-Verlag (pp. 193-202).

21. W.A. Al-Salam and A. Verma, On an orthogonal polynomial set, Indagationes Mathematicae, **44**(1982), 335-340.

22. W.A. Al-Salam and A. Verma, Some sets of orthogonal polynomials, Rev. Téc. Ing., Univ. Zulia, **9**(1986), 83-88.

23. W.R. Allaway, The identification of a class of orthogonal polynomial sets, Ph.D. thesis, University of Alberta, Edmonton, Canada., 1972.

24. G.E. Andrews and R. Askey, Classical orthogonal polynomials, Polynômes Orthogonaux et Applications- Proc Bar-le-Duc 1984, Lecture Notes in Math. # 1171, Springer-Verlag

25. A. Angelesco, Sur les polynomes orthogonaux en rapport avec d'autre polynomes, Buletinul Societâtii Stiite din Cluj, 1(1921), 44-59.

26. R. Askey, Divided difference operators and classical orthogonal polynomials, Rocky Mountain J. Math, (1989) To appear.

27. R. Askey and M. Ismail, A generalization of the ultraspherical polynomials, Studies in Pure Mathematics, edited by P. Erdös, Birkhäuser, Basel, 1983, 55-78.

28. R. Askey and M. Ismail, Recurrence relations, continued fractions and orthogonal polynomials, Memoirs of the Amer. Math. Soc. #300, 1984.

29. R. Askey and J. Wilson, Some basic hypergeometric orthogonal polynomials that generalize Jacobi polynomials, Memoirs AMS #319, 1985

30. F.S. Beale, On a certain class of orthogonal polynomials, Annals of Math. Statistics, 12(1941), 97-103.

31. S. Bochner, Über Sturm-Liouvillesche Polynomsysteme, Math. Zeit., 29(1929), 730-736.

32. S. Bonan and P. Nevai, Orthogonal polynomials and their derivatives, I, J. Approximation Theory, 40(1984), 134-147.

33. S. Bonan, D. Lubinsky, and P. Nevai, Orthogonal polynomials and their derivatives, II, SIAM J. Math. Anal., 18(1987), 1163-1176.

34. A. Boukhemis et P. Maroni, Une caractérisation des polynômes strictement 1/p orthogonaux de type Sheffer. Etude du cas $p = 2$. J. of Approximation Theory, 54(1988), 67-91.

35. W.C. Brenke, On polynomial solutions of a class of linear differential equations of the second order, Bull. Amer. Math. Soc., 36(1930), 77-84.

36. L. Carlitz, Characterization of certain sequences of orthogonal polynomials, Portugaliae Math., 20(1961), 43-46.

37. L. Carlitz, Characterization of the Krawtchouk polynomials, Revista Mat. Hisp-Amer. (4), 21(1961), 79-84.

38. L. Carlitz, Characterization of certain sequences of orthogonal polynomials, Portugaliae Math., 20(1961), 43-46.

39. L. Carlitz, Note on Legendre polynomials, Bull. Calcutta Math. Society, 46(1954), 93-95.

40. L. Carlitz, Characterization of the Laguerre polynomials, Monatshefte für Mathematik, 66(1962), 389-392.

41. T.S. Chihara, Orthogonal polynomials with Brenke type generating function, Duke Math. J., 35(1968), 505-518.

42. T.S. Chihara, Orthogonality relations for a class of Brenke polynomials, Duke Math. J., 38(1971), 599-603.

43. T.S. Chihara, A characterization of a class of distribution functions for the Stieltjes-Wigert polynomials, Canadian Math. Bull., 13(1970), 529-532.

44. T.S. Chihara, On generalized Stieltjes-Wigert and related orthogonal polynomials, Journal of Computational and Applied Mathematics, 5(1979), 291-297.

45. T.S. Chihara, An Introduction to Orthogonal Polynomials, Gordon and Breach Pub., 1976.

46. T.S. Chihara and M. Ismail, Orthogonal polynomials suggested by queueing model, Advances in Mathematics, 3(1982), 441-462.

47. C.W. Cryer, Rodriques' formulas and the classical orthogonal polynomials, Boll. Unione Mat. Ital. (3), 25(1970), 1-11.

48. Á Császár, Sur les polynômes orthogonaux classiques. Annales Univ. Sci. Budapest sec. Math., 1(1958), 33-39.

49. A. Danese, On a characterization of ultraspherical polynomials, Boll. U.M.I. (3), 21(1966), 1-3.

50. D. Dickinson, On quasi-orthogonal polynomials, Proc. Amer. Math. Soc., 12(1961), 185-194.

51. G.K. Eagleson, A characterization theorem for positive definite sequences on the Krawtchouk polynomials, Australian J. Statistics, 2(1969), 29-38.

52. G. Ebert, Über Polynomsysteme mit Rodriquessher Darstellung. Dissertation, Cologne, 1964.

53. K. Endl, On the involutory property Laguerre polynomials, Colloquia Math. Soc. János Bolyai, 19(1976) Budapest.

54. A. Erdélyi et al., Higher Transcendental Functions, Vol. 2, McGraw-Hill Pub., 1953.

55. L. Fejér, Absch. für die Legendreschen und verwandte Polynome, Math. Zeit., 24 (1925), 285-298.

56. E. Feldheim, Sur les polynomes généralisés de Legendre, Bull. de l'Academie des Science de l'URSS, 5(1941), 241-248.

57. E. Feldheim, Sur une propriété des polynomes orthogonaux, J. London Math. Soc., 13 (1938), 44-53.

58. E. Feldheim, Une propriété caractéristique des polynomes de Laguerre, Comment. Math. Helv., 13(1940), 6-10.

59. L. Feldmann, On a characterization of the classical orthogonal polynomials, Acta Sc. Math., 17(1956), 129-133.

60. L. Feldmann, Über durch Sturm-Liouvillesche Differentialgleichungen charakterisierte orthogonale Polynomsysteme. Pub. Math. (Debrecen), 3(1954), 297-304.

61. B.M. Gagaev, Sur quelques classes de fonctions orthogonales (in Russian), Bulletin (Izv.) de l'Academie des Sciences de l'URSS, 10(1946), 197-206.

62. B. Gabutti, Some characteristic property of Meixner polynomials, J. Math. Analysis and Applications, 95(1983), 265-277.

63. Ja. L. Geronimus, On polynomials orthogonal with respect to numerical sequences and on Hahn's theorem, Izv. Akad. Nauk, 4(1940), 215-228.

64. Ja. L. Geronimus, The orthogonality of some systems of polynomials, Duke Math. J., 14(1947), 503-510.

65. Ja. L. Geronimus, Orthogonal polynomials, Appendix, Amer. Math. Soc. translations (2), 108, 37-130.

66. B. Gnedenko, Sur l'unicité du système de fonctions orthogonales invariant par rapport à la dérivation, Compt. Rend.s (Dokl.) de l'Acad. des Sci. de l'URSS, 14(1937), 159-161.

67. W. Hahn, Über Orthogonalpolynome, die q-Differenzengleichungen genugen, Math Nach., 2(1949), 4-34.

68. W. Hahn, Über die Jacobischen Polynome und zwei verwandte Polynomklassen, Math. Zeit., **39**(1935), 634-638.

69. W. Hahn, Über höhere Ableitungen von Orthogonalpolynomen, Math. Zeit., **43**(1937), 101.

70. W. Hahn, Über Polynome, die gleichzweitig zwei verschiedenen Orthogonalsystemen angehoren, Math. Nach., **2**(1949), 263-278.

71. W. Hahn, Über Orthogonalitätserhaltende Operatoren, Math. Ves., **12**(1975), 337-339.

72. E.H. Hildebrandt, Systems of polynomials connected with the Charlier expamsion and the Pearson differential equation, Ann. Math. Statistics, **2**(1931), 379-439.

73. M. Ismail, Orthogonal polynomials in a certain class of polynomials, Bull.Inst. Polit din Iasi, **20**(1974), 45-50

74. S. Karlin and G. Szegö, On certain determinants, Journal d'Analyse Math. **8**(1960), 1-157.

75. A.M. Krall, Chebyshev sets of polynomials which satisfy an ordinary differential equation, SIAM Review, **22**(1980), 436-441.

76. H.L. Krall, On orthogonal polynomials satisfying certain fourth order differential equations, Pennsylvania State College Studies #6, State College,Pa., 1940.

77. H.L. Krall, On derivatives of orthogonal polynomials, Amer. Math. Soc. Bull., **42**(1936), 423-428.

78. H.L. Krall, On higher derivatives of orthogonal polynomials, Amer. Math. Soc. Bull., **42**(1936), 867-870.

79. H.L. Krall and I.M. Sheffer, A characterization of orthogonal polynomials, J. Math. Anal. Appl., **8**(1964), 232-244.

80. H.L. Krall and I.M. Sheffer, Differential equations of infinite order for orthogonal polynomials, Annali di Mat. (4), **74**(1966), 136-172

81. H.L Krall and I.M. Sheffer, On pairs of related orthogonal polynomial sets, Math. Zeit., **86**(1965), 425-450

82. O.E. Lancaster, Orthogonal polynomials defined by difference equations, American Journal of Mathematics, **63**(1941), 185-207.

83. I.L. Lanzewizky, Über Die Orthogonalität der Fejér-Szegöschen polynome, Comptes Rendus (Doklady) de l'Académie des Sciences de l'URSS, **31**(1941), 199-200.

84. D.A. Leonard, Orthogonal polynomials, duality and association schemes, SIAM J. Math. Anal., **13**(1982), 656-663.

85. P. Lesky, Über Polynomsysteme, die Sturm-Liouvilleschen Differenzengleichungen genügen, Math. Zeit., **78**(1962), 439-445.

86. P. Lesky, Orthogonale Polynomsysteme als Lösungen Sturm-Liouvilleschen Differenzengleichungen, Monatshefte Für Math., **66**(1962), 203-214.

87. P. Lesky, Polinomi ortogonali classici caratterizzati mediante equazioni differenziali del tipo Sturm-Liouville, Simposio di didatties della matematica, 1964.

88. P. Lesky, Die Charakteisierung der klassischen orthogonalen Polynome durch Sturm-Liouvillesche Differentialgleichungen, Archiv for Rational Mechanics and Analysis, **10** (1962), 341-351.

89. D.C. Lewis, Orthogonal functions whose derivatives are also orthogonal, Rend. Circ. Mat. Palermo (2), **2**(1953), 159-168.

90. N.N. Luzin, Integral and Trigonometric series in "Collected Works of N.N. Luzin" vol. 1, 48-212, Acad. of Sciences of the USRR, 1953.

91. P. Maroni, Une caracterisation des polynômes orthogonaux semi-classique , C.R. Acad. Sci. Paris, 301,ser.1 (1985), 269-272.

92. P. Maroni, Prologomenes a l'etude des polynômes orthogonaux semi- classique, Ann. Mat. pura ed Appl. (4), 149(1987), 165-184.

93. P. J. McCarthy, Characterization of the classical orthogonal polynomials, Portugaliae Mathematica, 20(1961), 47-52.

94. J. Meixner, Orthogonale Polynomsysteme mit einer besonderen Gestalt der erzeugenden Funktionen, J. London Math. Soc., 9(1934), 6-13.

95. J. Meixner, Symmetric systems of orthogonal polynomials, Arch. Rat. Mech. Analysis, 44(1972), 69-75.

96. J.C. Merlo, On orthogonal polynomials and second order linear difference operators, Annales Polonici Math., 19(1967), 69-79.

97. M. Mikolás, Common characterization of the Jacobi, Laguerre and Hermite-like polynomials (in Hungarian), Matematikai Lapok, 7(1956), 238-248.

98. A. Nikiforov and V. Uvarov, Special Functions of Mathematical Physics, Birkhäuser Verlag, 1988 (transl. by R.P. Boas)

99. G. Peebles, Some characterizations of the theory of orthogonal polynomials, Duke Math J., 6(1940), 89-100.

100. M. Perlstadt, A property of orthogonal polynomial families with polynomial duals, SIAM J. Math Anal., 15(1984), 1043-1054.

101. A. Ronveaux, Polynôme orthogonaux dont les polynômes derives sont quasi orthogonaux, Comptes Rendus Acad. Sci. Paris, Ser. A, 289(1979), 433-436.

102. E.J. Routh, On some properties of certain solutions of a differential equation of the second order, Proc London Math Soc., 16(1885), 245-261.

103. I.M. Sheffer, Some properties of polynomials of type zero, Duke Math. J., 5(1939), 590-622

104. J. Shohat, The relation of the classical orthogonal polynomials to the polynomials of Appell, Amer. J. Math., 58(1936), 453-464.

105. N. Ja. Sonin, Über die angenäherte Berechnung der bestimmten Integrale und über die dabei vorkommenden ganzen Functionen. Warsaw Univ. Izv., 18(1887), 1-76. Jbuch. Fortschritte Math. 19, p. 282.

106. G. Szegö, Orthogonal Polynomials, 4th ed., Amer. Math Soc Colloqu. Pub., vol. 23 , Providence, R.I. 1975.

107. M.S. Šun, On a characteristic property of classical orthogonal polynomials, Trudy Harkov Aviction Inst. 15(1954), 25-26.

108. L. Toscano, Polinomi ortogonali o reciproci di ortogonali nella classe di Appell, Le Matematica 11(1956), 168-174

109. F. Tricomi, Equazioni differenziali, Torino, 1948.

110. A.K. Varma, A new characterization of Hermite polynomials, Acta Math. Hung., 49(1987), 169-172

111. A. Verma and J. Prasad, Characterization of some orthogonal polynomials, Pub. de l'Institut Math. (Beograd), 9(1969), 177-180.

112. H.S. Wall, A continued fraction related to some partition formulas of Euler, Amer. Math. Monthly, 48(1941), 102-108.

113. M. Weber and A. Erdélyi, On the finite difference analog of Rodrigues' formula, Amer. Math. Monthly, 59(1952), 163-168.

24

114. M. Webster, Orthogonal polynomials with orthogonal derivatives, Bull. Amer. Math. Soc., **44**(1938), 880-888.
115. K.P. Williams, A uniqueness theorem for the Legendre and Hermite polynomials, Trans Amer. Math. Soc., **26**(1924), 441-445.

ORTHOGONAL POLYNOMIALS IN CODING THEORY
AND ALGEBRAIC COMBINATORICS

Eiichi Bannai *
Department of Mathematics
The Ohio State University
Columbus, OH 43210, U.S.A.

ABSTRACT. This paper surveys the role of orthogonal polynomials in Algebraic Combinatorics, an area which includes association schemes, coding theory, design theory, various
theories of group representation, and so on. The main topics discussed in this paper include the following: The connection between orthogonal polynomials and P-polynomial (or
Q-polynomial) association schemes. The classification problem for P- and Q-polynomial
association schemes and its connection with Askey-Wilson orthogonal polynomials. Delsarte
theory of codes and designs in association schemes. The nonexistence of perfect e-codes
and tight t-designs through the study of the zeros of orthogonal polynomials. The possible
importance of multi-variable versions of Askey-Wilson polynomials in the future study of
general commutative association schemes.

Introduction

This paper is a survey of the role of orthogonal polynomials in Algebraic Combinatorics, which includes association schemes, coding theory, design theory, and so on, and
possibly, if we interpret it broadly, various theories of group representation and even much
of finite group theory. As the role of orthogonal polynomials in (algebraic) combinatorics is
extremely broad and diverse, I will not try to cover all the connections; rather I will focus
on the topics with which I am familiar and in which I am strongly interested. Therefore,
this survey is heavily biased by my mathematical interest. I apologize to the reader about
this limitation of this survey.

This paper is aimed at nonspecialists in combinatorics, so some material well known
to combinatorialists is included here.

---

* This material is based upon research supported by the National Science Foundation
under grant number DMS-8703075.

P. Nevai (ed.), Orthogonal Polynomials, 25–53.

Table of Contents

Chapter 1. Orthogonal polynomials in association schemes

1.1. ASSOCIATION SCHEMES

It seems that the concept of association scheme is fundamental in algebraic combinatorics. Besides being important in their own mathematical right, association schemes also provide an excellent framework for coding and design theories. This last mentioned fact was first clearly noticed and effectively used in Delsarte's epochal thesis (Delsarte (1973)). I think this work of Delsarte marked the establishment of Algebraic Combinatorics.

We begin this survey by explaining the definition and basic properties of association schemes, aiming at the reader who is not familiar with the concept of association scheme. The reader is referred to Bannai-Ito (1984) for the details about association schemes, as well as for historical remarks on the study of association schemes.

Definition 1. Let X be a finite set and let $R_i (0 \leq i \leq d)$ be a set of relations on X satisfying the following conditions (1), (2), (3), (4):

    (1) $R_0 = \{(x, x) \mid x \in X\}$.

    (2) $R_0, R_1, \ldots, R_d$ gives a partition of $X \times X$.

    (3)   For each $i \in \{0, 1, \ldots, d\}$, $\,^t R_i := \{(y, x) \mid (x, y) \in R_i\} = R_j$ for some $j \in \{0, 1, \ldots, d\}$.

    (4)   For each fixed $i, j, k \in \{0, 1, \ldots, d\}$, the cardinality of the set $\{z \in X \mid (x, z) \in R_i, \ (z, y) \in R_j\}$ is a constant $(= p_{ij}^k)$ for all $(x, y) \in R_k$.

Then $\chi = (X, \{R_i\}_{0 \leq i \leq d})$ is called an association scheme of class d.

Remark. An association scheme $\chi = (X, \{R_i\}_{0 \leq i \leq d})$ is called commutative if $p_{ij}^k = p_{ji}^k$ for all i, j, k. An association scheme is called symmetric if ${}^t R_i = R_i$ for all i. It is easy to see that a symmetric association scheme is a commutative association scheme. But the reverse implication is not true.

Let A_i be the adjacency matrix of the relation R_i. That is,

$$A_i = (a_{xy})_{x \in X, y \in X} \quad \text{with} \quad a_{xy} = \begin{cases} 1 & \text{if } (x, y) \in R_i \\ 0 & \text{otherwise.} \end{cases}$$

Then we have $A_i A_j = \sum_{k=0}^d p_{ij}^k A_k$. The algebra $\mathcal{A} = \langle A_0, A_1, \ldots, A_d \rangle$ spanned by the $A_i (0 \leq i \leq d)$, which is a subalgebra of the full matrix algebra of size $|X| \times |X|$ over the complex number field, is called the Bose-Mesner algebra of the association scheme χ.

In what follows we assume that χ is a commutative association scheme, unless the contrary is stated. The algebra $\mathcal{A}$ is commutative and semi-simple, and so there exists a uniquely determined set $E_0, E_1, \ldots, E_d$ of primitive idempotents of $\mathcal{A}$. In other words, let $V = CX$ by the vector space with the basis the elements of X, then $A_0, A_1, \ldots A_d$ acts on V. Since the A_i's are all normal matrices commuting each other, they are diagonalized simultaneously by a unitary matrix. In fact we can see that there are exactly $d+1$ maximal common eigenspaces $V_i (0 \leq i \leq d)$ of V for $A_0, A_1, \ldots, A_d$. Each E_i is the projection from V to V_i. Without loss of generality, we may assume that $E_0 = (1/|X|)J$, where J is the matrix whose entries are all 1. Both $A_0, A_1, \ldots, A_d$ and $E_0, E_1, \ldots, E_d$ are bases of $\mathcal{A}$ (as a vector space). Let P be the base-change matrix defined by $(A_0, A_1, \ldots, A_d) = (E_0, E_1, \ldots, E_d)P$. The matrix P is called the first eigenmatrix or the *character table* of the commutative association scheme χ. The matrix Q, called the second eigen-matrix, is defined by $PQ = QP = |X|I$.

There are many known examples of association schemes. Many of them are obtained from permutation groups, but there are also many association schemes which do not directly come from groups.

Definition 2. Let $\chi = (X, \{R_i\}_{0 \leq i \leq d})$ be a symmetric association scheme. χ is called a P-polynomial association scheme if there exist polynomials $v_i(x)$ (with real coefficients) of degree exactly $i (0 \leq i \leq d)$ such that $A_i = v_i(A_1)$ $(0 \leq i \leq d)$. (To be precise, the P-polynomial property is defined for the given ordering of the relations $R_0, R_1, \ldots, R_d$. That is, χ may have another P-polynomial structure with respect to a different ordering of the relations.)

Let B_i be the intersection matrix, defined by $B_i = (p_{ij}^k)_{0 \leq j \leq d, \ 0 \leq k \leq d}$. Then it is easy to see that χ is a P-polynomial association scheme if and only if B_1 is a tri-diagonal matrix,

$$B_1 = \begin{bmatrix} 0 & 1 & & & & & 0 \\ k & a_1 & c_2 & & & & \\ & b_1 & a_2 & \ddots & & & \\ & & b_2 & \ddots & & c_{d-1} & \\ & & & \ddots & & a_{d-1} & c_d \\ & 0 & & & & b_{d-1} & a_d \end{bmatrix},$$

with $b_i \neq 0, c_i \neq 0$ for all i.

Note that then the $v_i(x)$ satisfy the following three-term recurrence relation (hence they are orthogonal polynomials):

$$x v_i(x) = b_{i-1} v_{i-1}(x) + a_i v_i(x) + c_{i+1} v_{i+1}(x) \quad (0 \leq i \leq d-1).$$

(Note that $a_{-1} = a_0 = 0$, $b_0 = k$, $c_1 = 1$, $v_0(x) = 1, v_1(x) = x$.)

Remark. If $\chi = (X, \{R_i\}_{0 \leq i \leq d})$ is a symmetric P-polynomial association scheme, the graph $\Gamma = (X, R_1)$ is distance-regular. Conversely, it is easily seen that from a distance-regular graph, a P-polynomial association scheme is canonically constructed (cf. Bannai-Ito (1984), Section 3.1)). So the two concepts of P-polynomial association scheme and distance-regular graph are essentially the same concept. The study of distance-regular graphs is a very important and fascinating topic in graph theory and combinatorics, because they are graphs with extra ordinarily rich regularity yet they are very rich in examples. The reader is referred to Bannai-Ito (1984) and Brouwer-Cohen-Neumaier (1989) for details of this subject. Here, let us just mention some connection with orthogonal polynomials. Let k be the valency of a distance-regular graph, let $\theta_0 = k$, $\theta_1, \ldots, \theta_d$ be the eigenvalues of the adjacency matrix $A = A_1$, and let m_i be the multiplicity of θ_i in A. (The θ_i are all real numbers because A is a symmetric matrix.) Define $k_i = p_{ii}^0 = |\{y|(x,y) \in R_i\}|$. Then the m_i are expressed as follows:

$$m_i = |X| / \sum_{j=0}^{d} v_j^2(\theta_i)/k_j \quad \text{(for all } i\text{)}$$

and

$$m_i = \frac{|X| k b_1 \cdots b_{d-1} c_2 \cdots c_{d-1}}{(k - \theta_i) F_d'(\theta_i) F_{d-1}(\theta_i)} \quad \text{(for } i \geq 1\text{)}$$

where $F_i(x) = c_2 c_3 \cdots c_i (v_0(x) + v_1(x) + \cdots + v_i(x))$.

(The reader will notice that these numbers are in fact the Christoffel numbers of the orthogonal polynomials.) The name P-polynomial, due to Delsarte, comes form the fact that the entries of the matrix $P = (p_{ij}) = (p_j(i))$ are expressed by $p_i(j) = v_i(\theta_j)$ with $\theta_j = p_1(j)$.

Let us mention that the polynomials $v_i(x)$ satisfy the following orthogonality relations:

$$\sum_{\nu=0}^{d} m_\nu v_i(\theta_\nu) v_j(\theta_\nu) = |X| \cdot k_i \cdot \delta_{ij}.$$

Remark. In the study of distance-regular graphs, it is very useful to study the spectrum, i.e., the eigenvalues θ_i and their multiplicities m_i, of the adjacency matrix A of the graph. What we are hoping and trying to do is to narrow down the possibilities for the tri-diagonal matrix B_1 corresponding to a distance-regular graph. From the viewpoint of orthogonal polynomials, the matrix B_1 is completely arbitrary. But a B_1 correspnding to a distance-regular graph has strong restrictions. For example, we can prove combinatorially, relatively easily, that the c_i's are non-decreasing and the b_i's are non-increasing. But the following combinatorial condition, due to Ivanov(1983), is very deep and highly nontrivial: in a distance-regular graph, if $(c_{s-1}, a_{s-1}, b_{s-1}) \neq (c_s, a_s, b_s) = (c_{s+1}, a_{s+1}, b_{s+1}) = \ldots = (c_{s+t-1}, a_{s+t-1}, b_{s+t-1})$, then we have $t \leq s$ if $s > 1$. I believe that further combinatorial restrictions remain to be found. Probably in orthogonal polynomials all real numbers are similar, but this is not the case in combinatorics. For example if $m_i \neq m_j$ then the two numbers θ_i and θ_j are not algebraically conjugate over the rational numbers. Combining facts like these together, we can study distance-regular graphs fairly deeply. For example, we now have a complete classifications of distance-regular graphs with $k = 3$ and $k = 4$. (For this topic the reader is referred to Bannai-Ito (1986, 1987).)

Now, let us return to an arbitrary symmetric association scheme $\chi = (X, \{R_i\}_{0 \leq i \leq d})$. For two elements of $\mathcal{A}$, define the Hadamard product $\circ$ to be the entry-wise product. Since $A_i \circ A_j = \delta_{ij} A_i$ the algebra $\mathcal{A}$ is closed under Hadamard multiplication. Therefore $E_i \circ E_j$ is also in $\mathcal{A}$, so there exist numbers q_{ij}^k (called Krein parameters) such that $E_i \circ E_j = (1/|X|) \sum_{k=0}^{d} q_{ij}^k E_k$. (The q_{ij}^k enjoy similar properties to these of the p_{ij}^k, though the q_{ij}^k may not be integers and may not even be rational numbers.)

Now, another important property (the Q-polynomial property) which is again closely related to orthogonal polynomials is defined as follows.

Definition 3. Let $\chi = (X, \{R_i\}_{0 \leq i \leq d})$ be a symmetric association scheme. χ is called a Q-polynomial association scheme (with respect to the ordering $E_0, E_1, \ldots, E_d$ of primitive idempotents) if there exist polynomials $v_i^*(x)$ of degree exactly $i (0 \leq i \leq d)$ such that $|X|E_i = v_i^*(|X|E_1)$ $(0 \leq i \leq d)$, where the multiplications are the Hadamard product.

It is easy to see that χ is a Q-polynomial association scheme if and only if the dual intersection matrix, defined by $B_1^* = (q_{1j}^k)_{0 \leq j \leq d, \, 0 \leq k \leq d}$, is a tridiagonal matrix:

$$
B_1^* = \begin{bmatrix}
0 & 1 & & & & & 0 \\
m_1 & a_1^* & c_2^* & & & & \\
& b_1^* & a_2^* & \ddots & & & \\
& & b_2^* & \ddots & c_{d-1}^* & & \\
& & & \ddots & a_{d-1}^* & c_d^* & \\
0 & & & & b_{d-1}^* & a_d^* &
\end{bmatrix},
$$

with $b_i^* \neq 0, c_i^* \neq 0$ for all i.

Note that the $v_i^*(x)$ satisfy the following three-term recurrence relation (hence they are orthogonal polynomials):

$$xv_i^*(x) = b_{i-1}^* v_{i-1}^*(x) + a_i^* v_i^*(x) + c_{i+1}^* v_{i+1}^*(x) \quad (0 \le i \le d-1).$$

(Note that $a_{-1}^* = a_0^* = 0$, $b_0^* = m_1$, $c_1^* = 1$, $v_0^*(x) = 1$, $v_1^*(x) = x$.) Also, let us remark that the name Q-polynomial, due to Delsarte, comes from the fact that the entries of the matrix $Q = (q_{ij}) = (q_j(i))$ are expressed by polynomials as $q_i(j) = v_i^*(\theta_j^*)$ with $\theta_j^* = q_1(j)$.

1.2. HAMMING SCHEMES $H(d,q)$ AND JOHNSON SCHEMES $J(v,d)$

In the previous section, we explained the basic concepts of association schemes in an abstract way. Here we pause for a while to discuss some examples.

Definition 4. (Hamming scheme) Let S be a finite set with $|S| = q \ge 2$. Let $X = S \times S \times \ldots \times S$ (direct product of d $S's$). For $x = (x_1, x_2, \ldots, x_d)$, $y = (y_1, y_2, \ldots, y_d) \in X$, let $(x, y) \in R_i$ if and only if $\#\{j \mid x_j \ne y_j\} = i$. Then $\chi = (X, \{R_i\}_{0 \le i \le d})$ is a symmetric association scheme of class d; it is called the Hamming (association) scheme $H(d,q)$.

Remark. The Hamming scheme $H(d,q)$ can also be defined from the action of the group $G \cong S_q \int S_d$ (wreath product, $|G| = (q!)^d \cdot d!$) on the set $X = G/H$ of cosets of $H \cong S_{q-1} \int S_d$. Then $|X| = |G : H| = q^d$.

Remark. $H(d,q)$ is a P-polynomial association scheme. The matrix B_1 is given by

$$B_1 = \begin{bmatrix} 0 & 1 & & & & & 0 \\ d(q-1) & q-2 & \ddots & & & & \\ & (d-1)(q-1) & \ddots & i & & & \\ & & \ddots & (q-2)i & \ddots & & \\ & & & (d-i)(q-1) & \ddots & d & \\ & 0 & & & \ddots & (q-2)d \end{bmatrix}.$$

We obtain

$$v_i(x) = K_i(d, u; q) = \sum_{j=0}^{i} (-1)^j (q-1)^{i-j} \binom{u}{j} \binom{d-u}{i-j}$$

with $x = dq - (q-1)u$. (here $K_i(d, u; q)$ is the Krawtchouk polynomial, and $\theta_i = dq - (q-1)i$ $(0 \le i \le d)$. $H(d,q)$ is also a Q-polynomial association scheme, because $P = Q$ if we take a suitable ordering of the primitive idempotents. Therefore, we have $B_1 = B_1^*$, $v_i(x) = v_i^*(x)$, and $k_i = m_i = \binom{d}{i}(q-1)^i$.

Definition 5. (Johnson scheme) Let V be a finite set with $|V| = v$. Let $d \le v/2$, and let X be the set of all the d-element subsets of V. (Thus, $|X| = \binom{v}{d}$.) For $x, y \in X$, define $(x,y) \in R_i$ if and only if $|x \cap y| = d - i$ $(0 \le i \le d)$. Then $\chi = (X, \{R_i\}_{0 \le i \le d})$ is a symmetric association scheme of class d; it is called the Johnson scheme $J(v,d)$.

Remark. The Johnson scheme $J(v,d)$ can also be defined from the action of the group $G = S_v$ acting on the cosets of $H = S_d \times S_{v-d}$, with $X = G/H$.

Remark. $J(v,d)$ is a P-polynomial association scheme, with B_1 given by

$$
B_1 = \begin{bmatrix}
0 & 1 & & & & & & 0 \\
d(v-d) & v-2 & \ddots & & & & & \\
 & (d-1)(v-d-1) & \ddots & & i^2 & & & \\
 & & \ddots & & i(v-2i) & & & \\
 & & & & (d-i)(v-d-i) & \ddots & & \\
 & & & & & \ddots & d^2 & \\
 & & 0 & & & & \ddots & d(v-2d)
\end{bmatrix}.
$$

We can also obtain

$$
v_i(x) = E_i(x) = \sum_{t=0}^{i} (-1)^t \binom{u}{t} \binom{d-u}{i-t} \binom{v-d-u}{i-t}
$$

with $x = u^2 - (v+1)u + d(v-d)$. ($E_i(x)$ is called a dual Hahn polynomial. It is a polynomial of degree i in x and degree $2i$ in u). Also we have $\theta_i = i^2 - (v+1)i + d(v-d)$ $(0 \le i \le d-1)$.

Furthermore, it can be proved that $J(v,d)$ becomes a Q-polynomial association scheme for a natural ordering of the primitive idempotents, and that

$$
\frac{p_i(j)}{k_i} = \frac{q_j(i)}{m_j} = Q_j(i) = {}_3F_2\left(\begin{matrix} -i, -j, v-1+j \\ -d, -v+d \end{matrix} ; 1 \right)
$$

where $k_i = \binom{d}{i}\binom{v-d}{i}$ and $m_j = \binom{v}{j} - \binom{v}{j-1}$. So we have

$$
v_j^*(x) = m_j Q_j(y)
$$

where $Q_j(y)$ is a Hahn polynomial of degree j in y and $x = (v-1)\left(1 - \frac{vy}{d(v-d)}\right)$. We have $\theta_i^* = (v-1)\left(1 - \frac{vi}{d(v-d)}\right)$ $(0 \le i \le d)$.

1.3. ASKEY-WILSON POLYNOMIALS AND THE CLASSIFICATION PROBLEM FOR P- AND Q-POLYNOMIAL ASSOCIATION SCHEMES

In the previous section, we remarked that $H(d,q)$ and $J(v,d)$ have both P-polynomial and Q-polynomial structures. Are there many association schemes which have both structures? We call such a symmetric association scheme a P- and Q-polynomial association scheme. We will discuss the classification problem of these schemes. This problem is not just one of the most important and interesting problems in Algebraic Combinatorics but also very interesting for its connection with orthogonal polynomials.

Let $\chi = (X, \{R_i\}_{0 \leq i \leq d})$ be a symmetric association scheme. Then the relation

$$p_i(\theta_j)/k_i = q_j(\theta_i^*)/m_j$$

holds for all i, j. Therefore, if χ is a P- and Q-polynomial association scheme, in addition to the condition that $\{v_i(x)\}$ and $\{v_i^*(x)\}$ are orthogonal polynomials (i.e., satisfy the 3-term recurrence relations) they must satisfy the relation

$$v_i(\theta_j)/k_i = v_j^*(\theta_i^*)/m_j$$

where the numbers $\theta_j, \theta_i^*, k_i$, and m_j are all determined from the $\{v_i(x)\}$ and $\{v_i^*(x)\}$. It seems that this is a very strong restriction on the possible parameters of P- and Q-polynomial association schemes.

Besides $H(d, q)$ and $J(v, d)$, about ten years ago several other families of P- and Q-polynomial association schemes were known. While I was getting interested in these schemes I heard from R. Askey about the work of D. Stanton, in which, extending earlier work by Dunkl, he calculated the parameters of dual polar spaces (i.e., association schemes coming from the action of classical groups on maximal isotropic subspaces) and the association schemes coming from classical forms (i.e., bilinear, alternate, hermite, and quadratic forms), and proved that they are P- and Q-polynomial association schemes. This encouraged me to believe that P- and Q-polynomial association schemes are very interesting classes of association schemes and that the classification problem is attackable, as I had the feeling that most P- and Q-polynomial association schemes were known. So I felt a strong need to attack P- and Q-polynomial association schemes systematically. Doug Leonard took up this problem, and made the following breakthrough.

Theorem 1. (D. Leonard) *Let χ be a P- and Q-polynomial symmetric association scheme. Then its parameters, as well as the polynomials $\{v_i(x)\}$ and $\{v_i^*(x)\}$, are expressed by using 5 free parameters.*

If we consider the P-polynomial or Q-polynomial property at the level of orthogonal polynomials, the number of free parameters is 2d-1. Thus, the result of Leonard that the combined property reduces the number of free parameters to only 5 is a very substantial accomplishment. When I mentioned this result of Leonard in a conference, R. Askey immediately noticed that the Askey-Wilson polynomials (i.e., q-Racah orthogonal polynomials obtained as balanced ${}_4\phi_3$'s) also satisfy the same properties and also have 5 free parameters. Inspired by this remark of Askey, Leonard obtained the following theorem.

Theorem 2. (D. Leonard) *Let $\{v_i(x)\}$ $(0 \leq i \leq d)$ and $\{v_i^*(x)\}$ $(0 \leq i \leq d)$ be orthogonal polynomials satisfying the P- and Q-property (i.e., the relation $v_i(\theta_j)/k_i = v_j^*(\theta_i^*)/m_j$ for all i,j where the numbers $\theta_j, \theta_i^*, k_i$ and m_j are all canonically determined by $\{v_i(x)\}$ and $\{v_i^*(x)\}$). Then the polynomials $\{u_i(x) = v_i(x)/k_i\}$ and $\{u_i^*(x) = v_i^*(x)/m_i\}$ can be expressed as Askey-Wilson orthogonal polynomials or certain of their limiting cases.*

Remark. More precise descriptions of possible forms of the polynomials $\{u_i(x)\}$ and $\{u_i^*(x)\}$ are given in Bannai-Ito (1984, Section 3.5).

At the level of orthogonal polynomials, the result of Leonard gives a complete solution. Now, the important problem from the viewpoint of combinatorics is to complete the classification of P- and Q-polynomial association schemes (of large diameters d) starting from Leonard's theorem. This is not yet completely solved at the time of this writing, but there are remarkable advances, notably by Paul Terwilliger. As this work is more combinatorial or geometric in nature, we will not discuss the details in this paper. The reader is referred to Bannai-Ito (1986) and Terwilliger (to appear). There are some very recent developments. Two new families of P- and Q- polynomial association schemes were discovered by Ivanov-Muzichuk-Ustimenko (to appear) and by Hemmeter (to appear). They have the same parameters (and hence the same polynomials $\{v_i(x)\}$ and $\{v_i^*(x)\}$) as some already known ones. There are also some recent developments in the characterization of known P- and Q-polynomial association schemes by their parameters. Ivanov-Shpectorov (to appear) proved that the association schemes of hermitian forms are characterized by their parameters with an additional condition on the local structure, and I heard that this additional condition was shown to be unnecessary by Neumaier and Terwilliger. Also I heard from Terwilliger that he has completed the classification of the almost dual bipartite case, i.e., $a_1^* = a_2^* = \ldots = a_{d-1}^* = 0$. Although it is not clear whether the complete classification of P- and Q-polynomial association schemes of large d is imminent or not, I believe that there is a good chance that it will be accomplished in the near future.

The work of Leonard mentioned above is extremely significant, as it really shows the importance of the role of orthogonal polynomials in Algebraic Combinatorics. Also I feel that it is a remarkable coincidence that Askey-Wilson polynomials had just been discovered in the field of orthogonal polynomials, when Leonard grasped them through the study of P- and Q- polynomial association schemes.

Andrews-Askey (1985) gives the definition of "classical orthogonal polynomials" as the polynomials obtained either as special cases or limiting cases of Askey-Wilson polynomials. I think this is a very correct definition. P- and Q-polynomial association schemes are really classical association schemes, closely connected with classical groups and classical forms. Thus, the P- and Q-property really corresponds to what is "classical" in a broad area of mathematics.

1.4. NON-SYMMETRIC P- AND Q-POLYNOMIAL ASSOCIATION SCHEMES

In the previous sections, we considered only symmetric P- and Q-polynomial association schemes. These concepts are also defined for non-symmetric (commutative) association schemes. Namely, a non-symmetric association scheme $\chi = (X, \{R_i\}_{0 \leq i \leq d})$ is P-polynomial if there exist (real coefficient) polynomials $v_i(x)$ of degree exactly i such that

$A_i = v_i(A_1)$ $(0 \leq i \leq d)$, and it is Q-polynomial if there exist polynomials $v_i^*(x)$ of degree exactly i such that $|X|E_i = v_i^*(|X|E_1)(0 \leq i \leq d)$ (with multiplication by the Hadamard product). Note that these definitions are exactly the same definitions as in the symmetric case. However, from the non-symmetry of χ, the matrices B_1 and B_1^* cannot be tri-diagonal, hence $\{v_i(x)\}$ and $\{v_i^*(x)\}$ cannot be orthogonal polynomials on an interval in the real line. Nevertheless, we still have orthogonality relations (coming from the orthogonality of the character table of the association scheme), and we can regard $\{v_i(x)\}$ and $\{v_i^*(x)\}$ as orthogonal polynomials on certain curves (or finitely many points) in the complex plane. In this sense, it seems that the study of P-polynomial non-symmetric association schemes (or equivalently, distance-regular digraphs) and/or Q-polynomial non-symmetric association schemes are interesting from both combinatorial and orthogonal-polynomial viewpoints. In particular, the existence of such combinatorial objects will give us interesting orthogonal polynomials on a curve in the complex plane.

In P-polynomial non-symmetric association schemes, or in distance-regular digraphs, we have either $d = g - 1$ (short digraph) or $d = g$ (long digraph), by the result of Lam-Damerell, where g is the girth of the digraph. It is also known that a short digraph is obtained from a long one, and vice versa. There had been very few known examples of distance-regular digraphs and none for $g \geq 4$ until recently, except for the trivial ones coming from the ordinary directed n-gons. A new family of distance-regular digraphs with $g = d + 1 = 4$ was recently discovered by Liebler-Mena (1988), and remarkably they are both P- and Q- polynomial non-symmetric schemes. Still, we don't yet know any examples of non-trivial distance-regular digraphs with $g \geq 5$. Leonard (to appear) tried to determine the polynomials $\{v_i(x)\}$ and $\{v_i^*(x)\}$ of P- and Q- non-symmetric schemes. He obtained many relations among the parameters and eigenvalues, and he proved such association schemes are self-dual, i.e., $P = Q$. Munemasa pushed the work of Leonard forward, proving that if a non-symmetric scheme is P- and Q-polynomial, then it must be from a short digraph, and that there exist no nontrivial ones with $g = d + 1 = 5$. It seems that the work of Munemasa (and the arguments in his paper) strongly suggest that the number of free parameters of P- and Q-polynomial non-symmetric association scheme of class d is very small, probably at most one, even at the level of orthogonal polynomials, though this claim is not yet completely verified at the time of this writing.

Chapter 2. Orthogonal polynomials in coding theory and design theory

2.1. DELSARTE THEORY OF CODES AND DESIGNS IN ASSOCIATION SCHEMES

The usual coding theory is the study of codes in the Hamming association scheme. Let $\chi = (X, \{R_i\}_{0 \leq i \leq d})$ be the Hamming scheme $H(d, q)$ defined in Section 1.2. A code in $H(d, q)$ is just a subset C of X. C is called an e-error correcting code (or e-code for short) if the sets $B_e(x)$ are disjoint for $x \in C$, where $B_e(x) = \{y \in X \mid (x, y) \in R_i$ with $i \leq e\}$, or equivalently the minimum distance of C is at least $2e + 1$, that is for $x, y \in C$, $(x, y) \notin R_i$ for $i = 1, 2, \ldots, 2e$. If the set $\{B_e(x) \mid x \in C\}$ covers the set X, i.e., if they give a partition of X, then C is called a *perfect e-code*.

In coding theory, we sometimes assume that the set S is a finite field $GF(q)$ and so the set X has the structure of a vector space of dimension d, and that C is a vector subspace of X. Such a code C is called a *linear code*. However, here we consider S which are not necessarily fields and codes C which are not necessarily linear.

It seems that one of the main purposes of coding theory is to construct codes C with the following two opposing properties: to make the e of the e-error correcting property as large as possible and to make the size $|C|$ as large as possible (cf. standard text books in coding theory such as MacWilliams-Sloane (1977), McEliece (1977), etc.). In this sense, perfect codes are the best codes conceivable, though they are rare, as will be explained in the next section.

The concept of e-error correcting code is very naturally generalized for (symmetric) P-polynomial association schemes as follows. Let $\chi = (X, \{R_i\}_{0 \leq i \leq d})$ be a symmetric P-polynomial association scheme (with respect to the natural ordering $R_0, R_1, \ldots, R_d$). Then a subset Y of X is called an e-code in χ if the sets $B_e(x)$ for $x \in Y$ are disjoint. Moreover, if the $B_e(x)$ for $x \in Y$ give a partition of X, then Y is called a perfect e-code in χ. (Here we remark that more general codes are defined for arbitrary commutative association schemes, but the natural definition of e-code for integers e is for P-polynomial symmetric association schemes.)

We recall the general theory of e-codes in P-polynomial association schemes, due to Delsarte (1973). Let $\chi = (X, \{R_i\}_{0 \leq i \leq d})$ be a symmetric P-polynomial association scheme. Let Y be a subset of X. We define $a_i = \frac{1}{|Y|}|R_i \cap (Y \times Y)|$. The vector $\underline{a} = (a_0, a_1, \ldots, a_d)$ is called the distance-distribution (or inner distribution) of the code Y. Note that we have $a_0 + a_1 + \ldots + a_d = |Y|$ and $a_0 = 1$. Now, we define the dual distribution $\underline{b} = (b_0, b_1, \ldots, b_d)$ of Y by the relation:

$$(b_0, b_1, \ldots, b_d) = \frac{1}{|Y|}(a_0, a_1, \ldots, a_d) \cdot Q$$

where Q is the second eigen-matrix defined by $PQ = QP = |X|I$.

Remark. As we have remarked in Section 1.2, the entries of the matrix $Q = (q_j(i))$ for the Hamming scheme $H(d, q)$ are given by Krawtchouk polynomial:

$$q_i(j) = K_i(d, j; q) \quad (= p_i(j)).$$

If X is a vector space over $GF(q)$ and if C is a linear code, the dual distribution $\underline{b}$ is the distance distribution of the dual code $C^\perp = \{y \in X \mid (x, y) = 0 \text{ for all } x \in C\}$ where $(\ ,\ )$ is the usual inner product in the vector space X over $GF(q)$. In this case, the linear transformation Q is called the MacWilliams transformation (or finite Fourier transformation).

Two important numbers of a code Y in a P-polynomial association scheme χ are defined as follows.

The minimum distance of Y: $f = \text{Max}\{i \mid a_1 = a_2 = \ldots = a_{i-1} = 0\}$.

The external distance of Y: $r = \#\{i \mid i \geq 1 \text{ and } b_i \neq 0\}$.

The main results, due to Delsarte (1973), are summarized as follows.

Theorem 3. *Let* $\chi = (X, \{R_i\}_{0 \leq i \leq d})$ *be a symmetric* P *-polynomial association scheme (with respect to the natural ordering* $R_0, R_1, \ldots, R_d)$.
(i) If Y *is of minimum distance* f , *then we have*

$$\frac{|X|}{|Y|} \geq 1 + k_1 + \ldots + k_{[\frac{f-1}{2}]}.$$

(ii) If Y *is of external distance* r , *then we have*

$$1 + k_1 + \ldots + k_r \geq \frac{|X|}{|Y|}.$$

(iii) If Y *is of minimum distance* f *or of external distance* r , *then we have*

$$r \geq \left[\frac{f-1}{2}\right].$$

(iv) If the hypothesis and equality hold in one of the above statements (i), (ii), (iii), then f *must be an odd number and the hypotheses and conclusions in the other two of (i), (ii), (iii) hold with equality. (Such a* Y *is called a perfect* e *-code with* $e = r = (f-1)/2$.)
(v) If there exists a perfect e *-code in* χ *which is nontrivial (i.e.* $|Y| \neq 1$) *then the* e *zeros (all of which are simple) of the polynomial* $v_0(x) + v_1(x) + \ldots + v_e(x)$ *of degree* e *lie in the set* $\{\theta_1, \theta_2, \ldots, \theta_d\}$. *(See Section 1.1 for the definitions of* $v_i(x)$ *and* θ_i .*)*

Some parts of the above proof are trivial, but some parts are nontrivial. Linear programming is useful in the proofs, but it is also possible to prove these claims without using linear programming (see Delsarte (1973) for proofs).

Remark. In $H(d, q)$, we have $k_i = \binom{d}{i}(q-1)^i$, $\theta_i = p_1(i) = dq - (q-1)i$ $(0 \leq i \leq d)$, and

$$v_0(x) + v_1(x) + \ldots + v_e(x) = \sum_{i=0}^{e} K_i(d, u; q) = K_e(d-1, u-1; q)$$

with $x = dq - (q-1)u$ (cf. Section 1.2). Therefore,

$$1 + k_1 + \ldots + k_r = \sum_{i=0}^{r} \binom{d}{i}(q-1)^i,$$

and the condition (v) in Theorem 3 becomes: all the e zeros of the polynomial

$$K_e(d-1, X-1; q) = \sum_{i=0}^{e}(-1)^i(q-1)^{e-i}\binom{d-X}{e-i}\binom{X-1}{i}$$

are in the set $\{1, 2, \ldots, d\}$.

As the usual coding theory is the study of subsets of X in the Hamming scheme $H(d,q)$, ordinary design theory is the study of subsets Y of X in the Johnson scheme $J(v,d)$. (The Johnson scheme was defined in Section 1.2.) Usually, a t-(v,d,λ) design (or t-design for short) is defined as a pair consisting of a set V and a collection $\mathcal{B}$ of d-element subsets of V with the property that for each t-element subset of V there are exactly λ elements of $\mathcal{B}$ which contain the t-element subset. Delsarte (1973) noticed that this combinatorial concept of t-design is equivalent to the following property of a subset Y of X in the Johnson scheme $J(v,d)$.

Let $\underline{a} = (a_0, a_1, \ldots, a_d)$ be the distance distribution of Y, and let $\underline{b} = (b_0, b_1, \ldots, b_d)$ be the dual distribution of Y. (Note that these concepts are defined for any commutative association scheme.) Now, for a subset Y of X in the Johnson scheme (or in any Q-polynomial association scheme) the following two numbers are defined.

$$\text{The degree of } Y: \quad s = \#\{i \mid i \geq 1, \ a_i \neq 0\}$$

and

$$\text{The strength of } Y: \quad t = \max\{i \mid b_1 = b_2 = \ldots = b_i = 0\}.$$

A subset Y with strength at least t is called a t-design. In $J(v,d)$, t-designs in this algebraic sense are equivalent to the above mentioned combinatorial (or classical) t-designs by taking $\mathcal{B} = Y$.

By this algebraic definition t-designs are defined in any Q-polynomial association schemes. (Note that in defining them, the natural ordering $E_0, E_1, \ldots, E_d$ of primitive idempotents was important.) For a general commutative association scheme, one can define the concept of T-design for a subset $T \subset \{E_1, E_2, \ldots, E_d\}$ as follows: Y is a T-design if and only if $b_i = 0$ for all i with $E_i \in T$. A t-design for an integer t is defined only for Q-polynomial schemes because it requires the natural ordering of the primitive idempotents; it is a T-design with $T = \{E_1, E_2, \ldots, E_t\}$. We note that a t-design Y in a Q-polynomial association scheme is better when t is larger and the cardinality of Y is smaller. We summarize the main results (due to Delsarte) on t-designs in Q-polynomial association schemes as follows.

Theorem 4. Let $\chi = (X, \{R_i\}_{0 \leq i \leq d})$ be a symmetric Q-polynomial scheme (with respect to the ordering $E_0, E_1, \ldots, E_d$).
(i) If Y is a t-design in χ, then we have

$$|Y| \geq 1 + m_1 + \ldots + m_{[\frac{t}{2}]}.$$

(ii) If $Y \subset X$ is of degree s, then we have

$$|Y| \leq 1 + m_1 + \ldots + m_s.$$

38

(iii) If Y is a t-design and of degree s, then we have

$$s \geq \left[\frac{t}{2}\right].$$

(iv) If both the hypothesis and equality hold in one of the above (i), (ii), (iii), then t must be even, and Y must satisfy the hypothesis and equality in the other two of (i), (ii), (iii). (Such a Y is called a tight $2s$-design with $t = 2s$.)

(v) If there is a tight $2s$-design in χ which is nontrivial (i.e. $Y \neq X$), then the s zeros (all of which are simple) of the polynomial $v_0^*(x) + v_1^*(x) + \ldots + v_s^*(x)$ of degree s lie in the set $\{\theta_1^*, \theta_2^*, \ldots, \theta_d^*\}$. (See Section 1.1 for the definitions of $v_i^*(x)$ and θ_i^*.)

Remark. In $J(v,d)$, we have

$$m_i = \binom{v}{i} - \binom{v}{i-1},$$

and so $m_0 + m_1 + \ldots + m_s = \binom{v}{s}$. Also we have

$$\theta_i^* = q_1(i) = (v-1)\left(1 - \frac{vi}{d(v-d)}\right)$$

and

$$v_0^*(x) + v_1^*(x) + \ldots + v_s^*(x) = \sum_{i=0}^{s}(-1)^{s-i}\frac{\binom{v-s}{i}\binom{d-i}{s-i}\binom{d-1-i}{s-i}}{\binom{s}{i}}\binom{u}{i}$$

with $x = (v-1)\left(1 - \frac{vu}{d(v-d)}\right)$. So the condition (v) in Theorem 4 becomes: all the s zeros of the polynomial

$$\sum_{i=0}^{s}(-1)^{s-i}\frac{\binom{v-s}{i}\binom{d-i}{s-i}\binom{d-1-i}{s-i}}{\binom{s}{i}}\binom{X}{i}$$

are in $\{1, 2, \ldots, d\}$.

2.2. NONEXISTENCE OF PEREFCT e-CODES IN $H(d,q)$

In this section we discuss the (non)existence of perfect e-codes in $H(d,q)$. As we have discussed in Theorem 3 (v), if there exists a nontrivial perfect e-code C in the Hamming scheme $H(d,q)$, then the e zeros of the polynomial

$$\sum_{i=0}^{e}(-1)^i(q-1)^{e-i}\binom{d-X}{e-i}\binom{X-1}{i}$$

must all be integers. (This theorem is called a Lloyd theorem, as this was first proved by Lloyd (1957) for $H(d,2)$. This necessary condition for the existence of perfect e-codes is called the Lloyd condition. There is another necessary condition for the existence of a perfect e-code, called the sphere packing condition: $|B_e(x)|$ divides $|X| = q^d$. The necessity of this is obvious from the definition of perfect e-code.

We remark that the following perfect codes in $H(d, q)$ are known.

(1) Trivial codes: $|C| = 1$ and $e \geq d$.
(2) Almost trivial codes (binary repetition codes): $q = 2$, $d = 2e + 1$, $|C| = 2$, e.g. $C = \{(0, 0, \ldots, 0), (1, 1, \ldots, 1)\}$.
(3) Hamming perfect 1-codes: $e = 1$, $d = (q^m - 1)/(q - 1)$, $|C| = q^{d-m}$, and other non-linear codes with these parameters.
(4) The binary Golay code: $e = 3, q = 2, d = 23, |C| = 2^{12}$.
(5) The ternary Golay code: $e = 2$, $q = 3$, $d = 11$, $|C| = 3^6$.

The study of perfect e-codes has a long history. The first breakthrough was obtained by Tietavainen (1973), after some preliminary work by van Lint. Tietavainen (1973) proved that if q is a *prime power*, and if there exists a nontrivial perfect e-code with $e \geq 2$, then C must be one of the known ones listed above as (1), (2), (4) or (5). Tietavainen used the Lloyd condition and the sphere packing condition very effectively. In his proof the fact that q is a prime power was very crucial, because the key point of his proof was to carefully study the p-share of e integer zeros $x_i (1 \leq i \leq e)$ of the Lloyd polynomial (which in this case is a Krawtchouk polynomial) to get a contradiction. Note that

$$\prod_{i=1}^{e} x_i = |B_e(x)| \cdot e!/q^e$$

and $|B_e(x)|$ must divide q^d because of the sphere packing condition.

Since the fact of q being prime power was very crucial in Tietavainen's proof, it was thought difficult to generalize the nonexistence proof to arbitrary integers q. But this difficulty was overcome by Bannai (1977a). The key idea there was to notice that the zeros of the Lloyd (i.e., Krawtchouk) polynomial are approximated asymptotically very exactly by using the zeros of the ordinary Hermite polynomials. To be more precise, for a fixed e, if

$$\beta = \sqrt{(d - e)(q - 1)}/q \to \infty,$$

then the zeros relabelled as $x_{(i)}$ (so that

$$x_{(-[\frac{e}{2}])} < \cdots < x_{(-1)} < (x_{(0)} <)x_{(1)} < \cdots < x_{([\frac{e}{2}])})$$

have limits

$$x_{(i)} \to \alpha + \beta + \lambda_{(i)},$$

where α is a constant, β is given above, and

$$\lambda_{(i)} = \frac{q-2}{q} \left(\frac{e-1}{6} - \frac{\xi_{(i)}^2}{6} \right)$$

where the $\xi_{(i)}$ are the corresponding zeros (arranged in increasing order from $-\left[\frac{e}{2}\right]$ to $\left[\frac{e}{2}\right]$) of the ordinary Hermite polynomial $H_e(x)$ defined by

$$H_e(x) = \sum_{r=0}^{\left[\frac{e}{2}\right]} \binom{e}{r} (2r-1)!! x^{e-2r}.$$

This asymptotic result implies that if $q \neq 2$ and if $\beta \to \infty$ then the zeros of the Lloyd (Krawtchouk) polynomial are not exactly symmetric with respect to $x = \alpha$, but *almost symmetric*. Moreover, we can evaluate the values of $\lambda_{(i)}$ precisely by using information on the zeros of Hermite polynomial. In fact we can show that

$$x_{(1)} + x_{(-1)} - x_{(2)} - x_{(-2)} \to \{\frac{q-2}{q} \cdot \frac{2}{6}(\xi_2^2 - \xi_1^2)\},$$

and the value in the right hand side is not an integer (or more precisely, it is strictly between 0 and 1 except for some special small values of parameters). Therefore, we can conclude that if $\beta \to \infty$, at least one zero $x_{(i)}$ is not an integer, which is a contradiciton. Thus, β must be bounded by some number β_0. Now it is easy to show that for bounded β there are only finitely many possible values of d and q for which all the zeros $x_{(i)}$ could be integers. (Note that for $q = 2$, the locations of $x_{(i)}$ are actually symmetric, and so we cannot use this technique. But the case $q = 2$ was already treated by Tietavainen.)

This idea of Bannai (1977a) was crucially and successfully used by Best (1982) to obtain the complete nonexistence of unknown perfect e-codes in $H(d,q)$ for all $e \geq 3$. Best evaluated the possible range of the interval in which each zero $x_{(i)}$ lies very precisely, instead of the asymptotic result explained above. This made it possible for Best to show the nonexistence of perfect e-codes in $H(d,q)$ except for $e = 6$ and $e = 8$. The remaining two cases were settled by Hong (1984, 1986) by making the idea of Bannai/Best even more precise. Combining all these efforts, the following result was finally obtained.

Theorem 5. *For $e \geq 3$. there exists no unknown perfect e-codes in $H(d,q)$.*

Remark. The perfect e-code problem for $e = 2$ is still open, though it is very unlikely that any new example exists. It is known that q must have at least 4 prime factors, due to Tietavainen and Laakso. There are many nonisomorphic perfect 1-codes (of the parameters of the Hamming perfect 1-code), and their complete classification seems to be impossible, i.e., too chaotic.

2.3. PERFECT CODES AND TIGHT DESIGNS IN P- AND Q- POLYNOMIAL ASSOCIATION SCHEMES

The techniques described in the previous section can also be used to study tight $2s$-designs in $J(v,d)$. As mentioned in Theorem 4 (v), if there exists a non-trivial tight $2s$-design, then the following polynomial has all its zeros in $\{1, 2 \dots, d\}$:

$$\sum_{i=0}^{s} (-1)^{s-i} \frac{\binom{v-s}{i}\binom{d-i}{s-i}\binom{d-1-i}{s-i}}{\binom{s}{i}} \binom{X}{i}.$$

The zeros of this polynomial are asymptotically approximated by using the zeros of the Hermite polynomial $H_s(x)$ in a way very similar to how we approximated the zeros of the Krawtchouk polynomial in the previous section. The results for tight $2s$-designs in $J(v, d)$ are summarized as follows.

Theorem 6. *(i) (Enomoto-Ito-Noda (1979), supplemented by a result on a diophantine equation) The Witt design* $4 - (23, 7, 1)$ *and its complementary design* $4 - (23, 16, 52)$ *are the only nontrivial tight* 4 *-designs.*

(ii) (Peterson (1977)) There exist no nontrivial tight 6-designs.

(iii) (Bannai (1977b)) For each fixed $s \geq 4$, there are only finitely many nontrivial tight $2s$-designs.

Remark. It is expected that there exist no nontrivial tight $2s$-designs for $s \geq 4$, and that a more precise calculation than the asymptotic one in Bannai (1977b) for the location of the zeros will certainly finish the problem, but this is not yet completed.

As we discussed in Theorem 3 and Theorem 4 in Section 2.1, perfect e-codes are defined for any symmetric P-polynomial scheme and tight t-designs are defined for any Q-polynomial scheme. Here we will mention non-existence results for perfect codes and tight t-designs in known P- and Q- polynomial association schemes.

(1) *Tight $2s$-designs in* $H(d, q)$. In $H(d, q)$ the polynomial $v_i(x)$ is equal to the polynomial $v_i^*(x)$. Therefore, if there exists a nontrivial tight $2s$-design in $H(d, q)$, then the Lloyd condition is exactly the same as the Lloyd condition for perfect e-codes. In the nonexistence proof for perfect e-codes in $H(d, q)$ discussed in Section 2.2, we proved nonexistence just by using the Lloyd condition (without using the sphere packing condition) for $q > 2$. So we have the following theorem, cf. Hong (1986).

Theorem 7. *Let* $q > 2$. *Then there exist no nontrivial tight $2s$-designs in* $H(d, q)$ *for* $s \geq 3$. *(The classification is still open for* $q = 2$ *and also for* $s \leq 2$.)

(2) *Perfect e-codes in* $J(v, d)$. Each polynomial $v_i(x)$ is a dual Hahn polynomial; and the Lloyd polynomial $v_0(x) + v_1(x) + \ldots + v_e(x)$ can also be expressed as a dual Hahn polynomial. This time, the Lloyd condition is very difficult to use, becuase the zeros of the Lloyd polynomial are always symmetrically located so the method of evaluating the deviation from symmetry (as used in Section 2.2) cannot be used. There are many partial results about the nonexistence of perfect e-codes in $J(v, d)$, proved by using combinatorial properties (cf. Roos (1982)) and Lloyd condition. However, it seems that the main part of this problem is still essentially open.

(3) *Perfect e-codes and tight $2s$-designs in other known P- and Q-polynomial association schemes.*

The problems of the existence of perfect e-codes and tight $2s$-designs have been studied for many of the known (i.e., classical) P- and Q- polynomial association schemes. In particular, for those classical association schemes whose character tables (which are equivalent to spherical functions) are described by Askey-Wilson polynomials using the basic hypergeometric series with base $q > 1$, where q corresponds to the q of $GF(q)$, very systematic and complete results were obtained by Chihara (1987). In this case, the Lloyd polynomials

$$v_0(x) + v_1(x) + \ldots + v_e(x) \quad \text{and} \quad v_0^*(x) + v_1^*(x) + \ldots + v_s^*(x)$$

are also described by using Askey-Wilson polynomials (with parameters modified from those of $v_i(x)$ and $v_i^*(x)$). In these cases θ_i and θ_i^* are typically expressed as $(1 - q^{-i+1})(1 - abq^{i+2})$. It is interesting that in order to show the nonexistence of perfect e-codes and tight $2s$-designs in these cases, a relatively easy calculation is sufficient, and usually it is not necessary to investigate the locations of the zeros very precisely. The results are summarized as follows.

Theorem 8. *(Chihara (1987)). (i) There are no perfect e-codes for $e \geq 1$ in the following classical association schemes: q-analogues of Johnson i.e., spaces of type A_{v-1}; association schemes of the dual polar spaces $^2A_{2d}$, $^2A_{2d-1}, D_d$, $^2D_{d+1}$; and association schemes of bilinear forms, hermitian forms, alternating bilinear forms, and quadratic forms. There are no perfect e-codes for $e \geq 2$ in the schemes of the dual polar spaces of type B_d, C_d. (The existence problem for perfect 1-codes in the dual polar spaces of type B_d and C_d is an interesting open problem.)*

(ii) *There are no tight $2s$-designs in the schemes of the spaces A_{v-1}, $^2A_{2d}$, $^2A_{2d-1}$, B_d, C_d, D_d, $^2D_{d+1}$, and the spaces of bilinear, schemes of the hermitian, alternating bilinear, and quadratic forms with the exception of tight $2s$-designs with $s = (d+1)/2$ in $^2D_{d+1}$.*

We conclude this section by mentioning generalizations of the problems of perfect e-codes and tight t-designs in two directions. One direction is to study a slightly weaker combinatorial objects by using orthogonal polynomials $v_i(x)$ and $v_i^*(x)$. For example, there were inequalities in Theorem 3 and Theorem 4 in Section 2.1. In Q-polynomial schemes we have $s \geq [t/2]$ and equality implied tightness. The question is what happens if $t = 2s - 1$, say. If there exists such a combinatorial object, then we can see (in general, though some minor modification may be necessary) that there exists a (real) number α so that the polynomial

$$v_0^*(x) + v_1^*(x) + \ldots + v_{s-1}^*(x) + \alpha v_s^*(x)$$

must have s zeros in the set $\theta_1^*, \theta_2^*, \ldots, \theta_d^*$. It seems that the question is not yet answered for $H(d,q)$, unless q is a prime power. So there is a close connection here between number theoretical problems of diophantine equations and orthogonal polynomials.

Related questions are the determinations of the Galois group of orthogonal polynomials. A classical result by Schur solves this question for Hermite and Laguerre polynomials, but the question is open for many interesting families of orthogonal polynomials, in particular Chebycheff polynomials. In this context, the Lloyd-type condition discussed here is equivalent to the question of when the Galois group of the polynomial is the identity group.

A similar question is whether there exists at least one integer zero for a given polynomial. This question was studied for Krawtchouk polynomials in connection with the existence of the inverse of the Radon transform and its connection with probability theory (cf. Diaconis (1988) and the references therein) and the existence of multiple perfect codes (cf. Clayton (to appear)). Chihara-Stanton (to appear) gives some examples of integer zeros of Krawtchouk polynomials. Habsieger-Stanton (to appear) gives a very detailed study for Krawtchouk polynomials of degree 4. It seems that the case where there is an integral zero is very rare. Even the question of when the binary Krawtchouk polynomial $K_e(d-1, x-1; 2)$ has all its zeros integral is still open. A solution would resolve the question of the existence of tight $2s$-designs in $H(d, 2)$ (cf. Hong (1986)). Chihara-Stanton (to appear) shows that q-Krawtchouk polynomials cannot have any θ_i as zero, in general. There are many interesting open questions about number theoretical aspects of orthogonal polynomials coming from coding theory and design theory.

2.4. SPHERICAL t-DESIGNS AND GENERALIZATIONS

An analogous theory to that of t-designs in Q-polynomial association schemes holds for t-designs in the unit sphere S^d, and more generally in compact symmetric spaces of rank 1. Here, Gegenbauer polynomials and Jacobi polynomials play an important role.

Let $S^d = \{(x_1, x_2, \ldots, x_{d+1}) \in R^{d+1} \mid x_1^2 + x_2^2 + \cdots + x_{d+1}^2 = 1\}$ be the unit sphere. We have the decomposition

$$L^2(S^d) = \bigoplus_{i \geq 0} \text{Harm}(i),$$

where Harm(i) is the space of homogeneous harmonic polynomials of degree i, and

$$\dim \text{Harm}(i) = \binom{d+i}{i} - \binom{d+i-2}{i-2} (= m_i).$$

Let $\{f_{i1}(x), \ldots, f_{im_i}(x)\}$ be an orthonormal basis of Harm(i). Then we have the following well known Addition Theorem:

$$\sum_{\ell=1}^{m_i} f_{i\ell}(x) f_{i\ell}(y) = c \cdot Q_i((x, y)),$$

where c is a certain constant, $Q_i(x) = \frac{d+2i-1}{d-1} \cdot C_i^{(d-1)/2}(x)$, $C_i^{(d-1)/2}(x)$ being the usual Gegenbaer polynomial, and (x, y) is the usual Euclidean inner product in R^{d+1}. This addition theorem plays a very important role on the study of finite sets in S^d.

The concept of a t-design in S^d (or spherical t-design) was defined by Delsarte-Goethals-Seidel (1977).

Definition 6. A finite nonempty set X in S^d is called a t-design in S^d if

$$\frac{1}{|X|} \sum_{x \in X} f(x) = \frac{1}{|S^d|} \int_{S^d} f(x) dw(x)$$

for all polynomials $f(x) = f(x_1, x_2, \ldots, x_{d+1})$ of degree $\leq t$.

Remark. There are several different but equivalent definitions of t-designs in S^d. One equivalent definition is that X is a t-design in S^d if

$$\sum_{x \in X} f(x) = 0$$

for $f \in \text{Harm}(1) \oplus \text{Harm}(2) \oplus \cdots \oplus \text{Harm}(t)$.

For any subset X of S^d, we define

$$A(X) = \{(x,y) \mid x,y \in X,\ x \neq y\}.$$

We say that X is an s-distance set (or of degree s) if $|A(X)| = s$.

The following theorem, which is analogous to Theorem 4 in Section 2.1, was obtained by Delsarte-Goethals-Seidel (1977).

Theorem 9. *(i) If X is a t-design in S^d, then*

$$|X| \geq \sum_{i=0}^{\left[\frac{t}{2}\right]} m_i = \binom{d + \left[\frac{t}{2}\right]}{\left[\frac{t}{2}\right]} + \binom{d + \left[\frac{t}{2}\right] - 1}{\left[\frac{t}{2}\right] - 1}.$$

(ii) If X is an s-distance set in S^d, then

$$|X| \leq \sum_{i=0}^{s} m_i = \binom{d + s}{s} + \binom{d + s - 1}{s - 1}.$$

(iii) If X is a t-design as well as an s-distance set in S^d, then $t \leq 2s$.

(iv) If X is a subset of S^d which satisfies both the specific hypothesis and equality in one of the above three statements (i), (ii) and (iii), then X also satisfies the specific hypothesis and equality in each of the other two of (i), (ii) and (iii). (If this happens, then t must be even and $t = 2s$. Such an X is called a tight $2s$-design in S^d.)

(v) If X is tight $2s$-design in S^d, then the set $A(X)$ must coincide with the set of s zeros of the polynomial $Q_0(x) + Q_1(x) + \cdots + Q_s(x)$.

Remark. It can be shown that the polynomial $Q_0(x) + Q_1(x) + \ldots + Q_s(x)$ is a Jacobi polynomial.

A subset X in S^d is called antipodal if $X = -X$. The following is a result for antipodal X in S^d.

Theorem 10. *(i) If X is a t-design in S^d, then*

$$|X| \geq 2 \binom{d + \left[\frac{t}{2}\right]}{\left[\frac{t}{2}\right]} = m_{\left[\frac{t}{2}\right]} + m_{\left[\frac{t}{2}\right] - 2} + m_{\left[\frac{t}{2}\right] - 4} + \cdots.$$

(ii) If X is antipodal and if it is an s-distance set in S^d, then

$$|X| \leq 2 \binom{d + s - 1}{s - 1} = m_s + m_{s-2} + m_{s-4} + \cdots.$$

(iii) If X is antipodal and if X is a t-design as well as an s-distance set in S^d, then $t \leq 2s - 1$.

(iv) If X is a subset of S^d which satisfies both the specific condition and equality in one of the above three statements (i), (ii) and (iii), then X also satisfies the specific condition and equality in each of the other two of (i), (ii) and (iii). (If this happens, then t must be odd and $t = 2s - 1$. We call such an X a tight $(2s - 1)$-design in S^d.

(v) If X is a tight $(2s - 1)$-design in S^d, then $A(X)$ coincides with the set of s zeros of the polynomial $(x + 1)C_{s-1}(x)$, where $C_{s-1}(x) = Q_{s-1}(x) + Q_{s-3}(x) + Q_{s-5}(x) + \cdots$. Further more, except for trivial cases, all the elements in $A(X)$ must be rational numbers.

Remark. The polynomial $C_{s-1}(x)$ is a Gegenbauer polynomial.

An antipodal X can be regarded as a design in the real projective space. A tight $(2s - 1)$-design in S^d is actually a tight $(s - 1)$-design in the real projective space.

By using the Lloyd condition in (v) in Theorem 9 and Theorem 10, we can study the (non)existence of tight t-designs in S^d.

Theorem 11. *Tight t-designs in $S^d(d \geq 2)$ exist only for $t \leq 5, t = 7$, and $t = 11$.*

Theorem 11 was proved for even t by Bannai-Damerell (1979), where we first proved that each element in $A(X)$ must be a reciprocal of an integer, then studied the deviation from symmetry of the zeros of the reciprocal polynomials of the Jacobi polynomial. The proof for odd t was obtained by Bannai-Damerell (1980), by using more number theoretical arguments, i.e., studying the decomposition of the reciprocal polynomial of $(x + 1)C_{s-1}(x)$ over the 2-adic integers. It is interesting that the polynomial $(x + 1)C_5(x)$ has the zeros $-1, \pm\frac{1}{2}, \pm\frac{1}{4}, 0$ and this corresponds to the $A(X)$ of the set X of minimum vectors (with $t = 11$ and $|X| = 196560$) of the Leech lattice in R^{24}.

The concept of spherical t-design generalizes naturally to other compact symmetric spaces of rank one (i.e., projective spaces over real, complex, quaternion and Cayley numbers.) Because of the limitations of space, I will not discuss this in detail, but there is a similar theory to that in the previous examples in S^d, the polynomials appearing there are ordinary Jacobi polynomials, and the addition theorem (on the space) plays a very important role. Bannai-Hoggar (1989) proves the nonexistence of tight t-designs for $t \geq 6$ in projective spaces, and Hoggar (1989) classifies tight 4- and 5-designs in projective spaces. There are many survey articles on this and related topics. The reader is referred to Bannai (1988), Hoggar (to appear), Neumaier (1981) as well as the original papers Delsarte-Goethals-Seidel (1975, 1977). There are many interesting related topics, e.g. the addition formula for real hyperbolic spaces (cf. Bannai-Blokhuis-Delsarte-Seidel (1984)). Also, the problem of constructing spherical t-designs is very interesting from the viewpoint of numerical analysis as well as analysis. (Cf. Bannai (1988) and Seymour-Zaslavsky (1984).) Again it should be emphasized that there is a close connection between the theory of spherical harmonics or its generalizations and the algebraic (i.e., character theoretical) studies of association schemes, and in both cases, orthogonal polynomials are very important tools with which to study combinatorics.

Chapter 3. Speculation on the role of orthogonal polynomials in the future study of Algebraic Combinatorics

3.1. COMMUTATIVE ASSOCIATION SCHEMES AND THEIR CHARACTER TABLES

The most important problem in the future of Algebraic Combinatorics is, I believe, the classification problem of primitive commutative association schemes.

An association scheme $\chi = (X, \{R_i\}_{0 \le i \le d})$ is called primitive if the graph (X, R_i) is connected for every $i \ge 1$. It is known that (commutative) primitive association schemes are the building blocks of general (commutative) association schemes, in a similar sense to how the simple groups are the building blocks of general finite groups, though the composition of association schemes is far more complicated than that of finite groups (cf. Bannai-Ito (1984), Section 2.9).

For each finite group G, we can canonically construct the group association scheme $\chi(G) = (X, \{R_i\}_{0 \le i \le d})$ by setting $(x, y) \in R_i$ if and only if $yx^{-1} \in C_i$, where $C_0 = \{1\}$, $C_1, \ldots, C_d$ are all the conjugacy classes of the group G. The group association scheme $\chi(G)$ is always commutative and has very nice properties among general commutative association schemes. For example, all the Krein parameters q_{ij}^k are rational numbers. The association scheme $\chi(G)$ is primitive if and only if the group G is simple. In this sense, the classification of commutative primitive association schemes contains the classification of finite simple groups as a special case, though it seems that the former is far more difficult than the latter, and the complete solution to the former classification problem is quite unlikely to be obtained in the near future. Nevertheless, I believe it is extremely important and worthwhile in Algebraic Combinatorics to try to approach the classification problem of commutative primitive association schemes, just as it was very important to try to classify finite simple groups even in the earlier stage of finite group theory when the possibility of classifying finite simple groups was not in sight.

It seems that the character theory of finite groups was very effective in the study of finite groups, in particular at the earlier stage. I think it is realistic and worthwhile to study association schemes from an algebraic (i.e., character theoretic) viewpoint, particularly at the present stage. (The theory of association schemes is not yet fully developed, and should be compared with the earlier study of finite groups.) With this understood, we started collecting examples of commutative association schemes, as many as possible, putting our emphasis on knowing their character tables. The details of these examples are described in the survey paper Bannai (to appear), which was written mainly for finite group theorists. We will not repeat all the details of the examples and observations in that paper; we just point out that there are many important examples coming from permutation groups, though there are also many association schemes which are not directly obtained from permutation groups. Also, many of these important examples are counterparts of compact symmetric spaces or prehomogeneous vector spaces.

One of the most interesting observations is that in many examples of commutative association schemes, there exist smaller association schemes whose character tables control the character tables of the bigger association schemes, rather like how the Weyl groups control the Chevalley groups, or q-analogues are obtained from the case $q = 1$. In a series of papers (cf. Bannai (to appear), Bannai-Hao-Song (to appear)), we observed that character tables of the association schemes obtained from the action of classical groups on the set of non-isotropic points (or lines in the symplectic case) are controlled by the character tables of the group association scheme $\chi(PSL(2,q))$ or various association schemes obtained from the action of the group $PSL(2,q)$ on certain cosets.

The most interesting and deepest result in this direction is the observation that the character table of the association scheme $\chi = GL(2n,q)/Sp(2n,q)$ is controlled by the character table of the group association scheme $\chi(GL(n,q))$; see Bannai-Kawanaka-Song (to appear). Roughly speaking, the former character table is obtained from the latter character table by replacing q by q^2 (and with appropriate adjustments). In Bannai-Kawanaka-Song (to appear) the character table of $\chi = GL(2n,q)/Sp(2n,q)$ was calculated, where the two-variable polynomials due to Macdonald (cf. Macdonald (1979 and forthcoming new edition)) were used.

Macdonald's polynomials generalize Schur's symmetric functions, Hall polynomials, Jack polynomials, etc. It seems that current work by Macdonald and others on multivariable polynomials is very much relevant to what I am going to describe in the next section.

3.2. A MULTI-VARIABLE VERSION OF ASKEY-WILSON POLYNOMIALS?

In the previous section, we mentioned that there is a close connection between the character table of the association scheme $\chi = GL(2n,q)/Sp(2n,q)$ and the character table of the group association scheme $\chi(GL(n,q))$. I believe this is not an accident, because an analogous fact is observed in the following examples of compact symmetric spaces. (The compact symmetric spaces were classified by E. Cartan in 1920's.)

The following three spaces are compact symmetric spaces of rank $\ell - 1$.

    (a) $SU(\ell)$. (The group $SU(\ell)$ itself is regarded as a symmetric space in the same way as the group association scheme is obtained from a group.)

    (b) $SU(\ell)/SO(\ell)$.

    (c) $SU(2\ell)/Sp(2\ell)$.

These three compact symmetric spaces are all related to the root system of type $A_{\ell-1}$. The spherical functions (which are equivalent to the character table in commutative association schemes) have been calculated for all the compact symmetric spaces (cf. Koornwinder (1975), Vretare (1984), etc.) It is found that the spherical functions are described by using generalized Jacobi polynomials of several variables where the number of variables is equal to the rank of the symmetric space. These generalized Jacobi polynomials are described by using root systems, and in the above three examples (a), (b) and (c) they all come from the same root system of type $A_{\ell-1}$. The differences among the spherical functions of these three cases come from what weight functions are given to the same root system for each case (cf. Vretare (1984)). This situation is very similar to how the character table of $\chi = GL(2n,q)/Sp(2n,q)$ is, roughly speaking, obtained from the character table of the group association scheme of $\chi(GL(n,q))$ by the replacement $q \rightarrow q^2$.

Many examples and theories suggest that many of the important known commutative association schemes are analogous in many senses to compact symmetric spaces of arbitrary rank, though there are certainly other examples not related to compact symmetric spaces.

In other words, they are finite versions of compact symmetric spaces, and their character tables may be expected to be obtained in a systematic way, as were the spherical functions of compact symmetric spaces. In fact, what we expect and hope is that, at least for nice classes of general commutative association schemes (though certainly not for all general commutative association schemes), the spherical functions (i.e., the character tables) should be obtained by using orthogonal polynomials of several variables where the number of variables is some number corresponding to the rank of the root system.

Although we know many examples of character tables of commutative association schemes, we still do not know exactly how these examples of character tables are described by multi-variable discrete orthogonal polynomials. As the P- and Q-polynomial association schemes can be regarded as the case of rank one (hence their character tables are expressed by ordinary one-variable Askey-Wilson polynomials), it would not be unreasonable to expect that the character tables of nice classes of association schemes are described by some multi-variable version of the Askey-Wilson orthogonal polynomials.

I learned from several people that many experts in orthogonal polynomials are interested in finding multi-variable versions of Askey-Wilson polynomials. (It seems there are several formulations of multi-variable Askey-Wilson polynomials.) Therefore, it would be extremely nice if we could formulate the right definition of P- and Q-polynomial association schemes *of general rank* and obtain a characterization of the right multi-variable version of Askey-Wilson orthogonal polynomials in terms of association schemes, much as the theorem of Leonard (Theorem 2 in Section 1.3) was obtained for the one-variable case.

References

G.E. Andrews and R. Askey (1985): Classical orthogonal polynomials, Orthogonal polynomials and applications (Bar-le-Duc, 1984), 36-69, Springer Lecture Note Series 1171, 1985.

R. Askey and J. Wilson (1979): A set of orthogonal polynomials that generalize the Racah coefficients of $6 - j$ symbols, SIAM J. Math. Anal. 10(1979), 1008–1016.

R. Askey and J. Wilson: Some basic hypergeometric orthogonal polynomials that generalize Jacobi polynomials, Mem. Amer. Math. Soc. 54(1985), No. 319.

E. Bannai (1977a): On perfect codes in the Hamming scheme $H(n,q)$ with q arbitrary, J. of Combinatorial Theory (A), 23(1977), 52–67.

E. Bannai (1977b): On tight designs, Quart. J. Math. (Oxford), 28(1977), 433–448.

E. Bannai (1988): On extremal finite sets in the sphere and other metric spaces; in Algebraic, extremal and metric combinatorics, London Math. Lecture Note Series, No. 131, 1988, 13–38.

E. Bannai (to appear): Character tables of commutative association schemes, to be published in Proceedings of the conference "Finite Buildings and Related Geometries" held in Pingree Park, Colorado, July, 1988, Oxford Univ. Press.

E. Bannai, A. Blokhuis, P. Delsarte and J. J. Seidel: An addition formula for hyperbolic space, J. of Combinatorial Theory (A), 36(1984), 332–341.

E. Bannai and R. M. Damerell (1979): Tight spherical designs I, J. Math. Soc. Japan, 31(1979), 199–207.

E. Bannai and R. M. Damerell (1980): Tight spherical designs II, J. of London Math. Soc., 21(1980), 13–30.

E. Bannai, S. Hao and S.Y. Song (to appear): Character tables of association schemes of finite orthogonal groups acting on the nonisotropic points, to appear in J. of Combinatorial Theory (A).

E. Bannai and S.G. Hoggar (1989): Tight designs and squarefree integers, Europ. J. of Combinatorics, 10(1989), 113–135.

E. Bannai and T. Ito (1984): Algebraic Combinatorics I, Association Schemes, Benjamin/Cummings, Menlo Park. California, 1984.

E. Bannai and T. Ito (1986): Current research on algebraic combinatorics, Graphs and Combinatorics, 2(1986), 287–308.

E. Bannai and T. Ito (1987): The study of distance-regular graphs from the algebraic (i.e., character theoretical) viewpoint, Proc. Symp. in Pure Math. (AMS) 47(1987), 343–349.

E. Bannai, N. Kawanaka and S.Y. Song (to appear): The character table of the Hecke algebra $H(GL_{2n}(F_q), Sp_{2n}(F_q))$, to appear in J. of Algebra.

E. Bannai and S.Y. Song (1989): The character tables of Paige's simple Moufang loops and their relationship to the character tables of $PSL(2,q)$, Proc. London Math. Soc. 58(1989), 209–236.

M. R. Best (1982): A contribution to the non-existence of perfect codes Ph. D. thesis, Amsterdam, 1982.

N. L. Biggs (1973): Perfect codes in graphs, J. of Combinatorial Theory (B), 15(1973), 289–296.

N. L. Biggs (1974): Algebraic Graph Theory, Cambridge Univ. Press, 1974.

N. L. Biggs, A. G. Boshier and J. Shaw-Taylor (1986): Cubic distance-regular graphs, J. London Math. Soc. 33(1986), 385–394.

A. E. Brouwer, A. Cohen and A. Neumaier (1989): Distance-regular graphs, Springer, 1989.

L. Chihara (1987): On the zeros of the Askey-Wilson polynomials with applications to coding theory, SIAM J. Math. Anal. 18(1987), 183–207.

L. Chihara and D. Stanton (1986): Association schemes and quadratic transformations for orthogonal polynomials, Graphs and Combinatorics, 2(1986), 101–112.

L. Chihara and D. Stanton (to appear): Zeros of generalized Krawtchouk polynomials, to appear in J. Approx. Theory.

R. Clayton (to appear): Perfect multiple coverings in metric spaces, to appear in IMA Proceedings.

J. H. Conway and N.J.A. Sloane (1988): Sphere Packings, Lattices and Groups, Springer, 1988.

R. M. Damerell (1981): Distance-transitive and distance-regular digraphs, J. of Combinatorial Theory (B), 31(1981), 46–53.

P. Delsarte (1973): An algebraic approach to the association schemes of coding theory, Philips Res. Rep. Suppl. No 10, 1973.

P. Delsarte (1974): The association schemes of coding thoery; in Combinatorics (M. Hall, Jr. and J. H. van Lint, eds.), Mathematical Center Tracts 55, Amsterdam, 1974, 139–157.

P. Delsarte (1978): Hahn polynomials, discrete harmonics, and t-designs, SIAM J. Appl. Math. 34(1978), 157–166.

P. Delsarte, J. M. Goethals and J. J. Seidel (1975): Bounds for systems of lines, and Jacobi polynomials, Philips Res. Repts 30(1975), 95*–105* (Bouwkamp volume).

P. Delsarte, J. M. Goethals and J. J. Seidel (1977): Spherical codes and designs, Geom. Dedicata, 6(1977), 363–388.

P. Diaconis (1988): Group Representations in Probability and Statistics, Institute of Mathematical Statistics, Lecture Note-Monograph Series, No. 11.

C. F. Dunkl (1976): A Krawtchouk polynomial addition theorem and wreath products of symmetric groups, Indiana Univ. Math. J. 25(1975), 335–358.

C. F. Dunkl (1977): An addition theorem for q-Hahn polynomials, Monats. Math. 85(1977), 5–37.

C. F. Dunkl (1978): An addition theorem for Hahn polynomials: The spherical functions. SIAM J. Math. Anal. 9(1978), 627–637.

C. F. Dunkl (1979): Discrete quadrature and bounds on t-designs, Mich. Math. J. 26135(1979), 81–102.

C. F. Dunkl (1980): Orthogonal polynomials in two variables of q-Hahn and q-Jacobi type, SIAM J. Alg. Disc. Math. 1(1980), 137–151.

C. F. Dunkl (1988): Reflection groups and orthogonal polynomials on the sphere, Math. Z. 197(1988), 33–60.

H. Enomoto, N. Ito and R. Noda (1979): Tight 4-desings, Osaka J. Math. 16(1979), 39–43.

J. M. Goethals and H. C. A. van Tilborg (1975), Uniformly packed codes. Philips Res. Rep. 30(1975), 9–36.

L. Habsieger and D. Stanton (to appear): More zeros of Krawtchouk polynomials, to appear.

J. Hemmeter (to appear): A new family of distance-regular graphs, to appear.

S. G. Hoggar (1982): t-designs in projectives spaces, Europ. J. Combinatorics 3(1982), 233-254.

S. G. Hoggar (1989): Tight 4- and 5-designs in projectives spaces, Graphs and Combinatorics, 5(1989), 87-94.

S. G. Hoggar (to appear): t-designs in Delsarte spaces, to appear in IMA Proceedings.

Y. Hong (1984): On the non existence of unknown perfect 6- and 8-codes in Hamming schemes $H(n,q)$ with q arbitrary, Osaka J. Math., 21(1984), 687–700.

Y. Hong(1987): On the nonexistence of nontrivial perfect e-codes and tight $2e$-designs in Hamming schemes $H(n,q)$ with $e \geq 3$ and $q \geq 3$, Graphs and Combinatorics, 2(1986), 145–164.

A. A. Ivanov (1983): Bounding the diameter of distance-regular graphs, Soviet Math. Dokl. 38(1983), 149-152.

A. A. Ivanov, M. E. Muzichuk and V. A. Ustimenko (to appear), On a new family of (P and Q)-polynomial schemes, to appear in Europ. J. Combinatorics.

A. A. Ivanov and S. V. Shpectorov (to appear): A characterization of the association schemes of Hermitian forms, to appear.

A. A. Klyachko (1984): Models for the complex representations of the group $GL(n,q)$, Math. USSR Sbornik, 48(1987), 365-379.

T. H. Koornwinder (1973): The addition formula for Jacobi polynomials and spherical harmonics, SIAM J. Appl. Math. 25(1973), 236–246.

T. H. Koornwinder (1975): Two variable analogues of the classical orthogonal polynomials; in R. Askey (ed.) Theory and Application of Special Functions, Academic Press, N.Y., 1975, pp. 435–495.

D. Leonard (1984): Parameters of association schemes that are both P- and Q-polynomial, J. of Combinatorial Theory (A) 36(1984), 355-363.

D. Leonard (1982): Orthogonal polynomials, duality, and association schemes, SIAM J. Math. Anal. 13(1982), 656-663.

D. Leonard (to appear): Non-symmetric, metric, cometric association schemes, to appear in J. of Combinatorial Theory (B).

D. Leonard (to appear): Non-symmetric, metric, cometric association schemes are self-dual, to appear.

R. A. Liebler and R. A. Mena (1988): Certain distance-regular digraphs and related rings of characteristic 4, J. of Combinatorial Theory (A), 47(1988), 111–123.

S. P. Lloyd (1957): Binary block ciding, Bell System Tech. J., 36(1957), 517-535.

52

I. G. Macdonald (1987): Commuting differential operators and zonal spherical functions; in "Algebraic Groups, Utrecht 1986", Lecture Notes in Mathematics No. 1271, Springer, 1987.

I. G. Macdonald (Chapter VI of the second edition of 1979): Symmetric Functions and Hall polynomials, Oxford Univ. Press, Oxford, 1979. (2nd edition, to appear).

J. F. MacWilliams and N. J. A. Sloane (1977): The Theory of Error Correcting Codes, North-Holland, Amsterdam, 1977.

R. J. McEliece (1977): The theory of infomation and coding, in Encyclopedia of Math. and its Application, Addison-Wesley, 1977.

A. Munemasa (to appear): On nonsymmetric P- and Q-polynomial association schemes, to appear.

A. Neumaier (1981): Combinatorial configurations in terms of distances, T.H.E. Memorandum 81 1981, No. 09, p. 98, Eindhoven Univ. Tech.

C. L. Peterson (1977): On tight 6-designs, Osaka J. Math., 14(1977), 417-435.

D. K. Ray-Chaudhuri and R. M. Wilson (1975): On t-designs, Osaka J. Math. 12(1975), 737-744.

H. F. H. Reuvers (1977): Some nonexistence theorems for perfect codes over arbitrary alphabets, Ph. D. thesis, Tech. Univ. Eindhoven, 1977.

C. Roos (1982): Some remarks on perfect subsets in distance-regular graphs, Delfte Progress Report, 7(1982), 90–94.

I. Schur (1930): Gleichungen ohne Affekt; Sitz. der Preuss Acad. Wiss. (1930), 443-449. Also in Gesammelte Abhandlungen, vol. 3, 191–197.

P. Seymour and T. Zaslavsky (1984): Averaging sets: a generalization of mean values and spherical designs, Advs. in Math. 52(1984), 213–240.

N. J. A. Sloane: An introduction to association schemes and coding theory; in Theory and Application of Special Functions (R. Askey, ed.), Academic Press, N. Y. 1975, 225–260.

D. Stanton (1980): Some q-Krawtchouk polynomials in Chevalley groups, Amer. J. Math. 102(1980), 625–662.

D. Stanton (1981): Three addition theorems for some q-Krawtchouk polynomials, Geom. Dedicata 10(1981), 403–425.

D. Stanton (1984): Orthogonal polynomials and Chevalley groups; in Special Functions: Group Theoretical Aspects and Applications (R. A. Askey et al. eds) Reidel, Boston, 1984, pp. 87–125.

D. Stanton (1986): t-designs in classical association schemes, Graphs and Combinatorics, 2(1986), 283–286.

H. Tarnanen, M. J. Aaltonen and J. M. Goethals (1985): On the nonbinary Johnson Scheme, Europ. J. Combinatorics (1985), 279–285.

P. Terwilliger (1987): A characterization of P- and Q-polynomial association schemes, J. of Combinatorial Theory (A), 45(1987), 8–26.

P. Terwilliger (to appear): The incidence algebra of a uniform poset, to appear.

A. Tietavainen (1973): On the nonexistence of perfect codes over finite fields, SIAM J. Appl. Math. 24(1973), 88–96.

A. Tietavainen (1977): Nonexistence of nontrivial perfect codes in case $q = p_1^s p_2^t, e \geq 3$, Discrete Math. 17(1977), 199–205.

L. Vretare (1976): Elementary spherical functions on symmetric spaces. Math. Scand. 39(1976), 343-358.

L. Vretare (1984): Formula for elementary spherical functions and generalized Jacobi polynomials. SIAM J. Math. Anal. 15(1984), 805–833.

ORTHOGONAL POLYNOMIALS, PADÉ APPROXIMATIONS AND JULIA SETS [*]

D. BESSIS
Service de Physique Théorique de Saclay
Laboratoire de l'Institut de Recherche Fondamentale
du Commissariat à l'Energie Atomique
91191 Gif-sur-Yvette Cedex, France

ABSTRACT. The iterates of any polynomial are shown to form a subset of a family of orthogonal polynomials. This allows the transformation of a non linear problem, the subsitution of a polynomial into itself, into a linear one, the three terms recursive relation fulfilled by any orthogonal polynomial system. To those iterates of a polynomial is associated in a natural manner an Hilbert space operator, the Jacobi matrix $\mathbf{J}$ generated by the three term recursive relation. The spectrum of this operator is shown to be the Julia set associated to those iterations. Furthermore the coefficients of the Jacobi matrix are quasi-periodic functions of their indices, a consequence of an exact renormalization group equation satisfied by the Jacobi matrix. This allows the construction of exactly solvable chaotic quantum systems.

The paper is self-contained and introduces Julia sets and invariant measures on Julia sets. Then the general theory of orthogonal polynomials on a complex contour is given. The $\mathbf{J}$ matrix is introduced as well as its approximations, continued fractions and Padé approximations.

Constructive schemes for orthogonal polynomials associated to Julia sets are built using functional equations connected to an exact group renormalization property. The quasi-periodic properties of the coefficients of the $\mathbf{J}$ matrix are discussed for the logistic map. Finally an outlook is given for the applications to image processing.

I. INTRODUCTION

An intriguing question about orthogonal polynomials on the real line is the following. Given a family of monic orthogonal polynomials

$$P_n(x) = x^n + ... \tag{I.1}$$

and a fixed monic polynomial $T(x)$ of degree d

$$T(x) = x^d + t_1 x^{d-1} + ... + t_d, \tag{I.2}$$

---

[*] The reader is kindly requested to accept the author's somewhat informal style. It would have meant a delay of unacceptable proportions if the entire paper were completely rewritten by the editors. Luckily, neither the occasional misprints nor the informal style detract from enjoying reading of what follows.

P. Nevai (ed.), Orthogonal Polynomials, 55–97.

let us substitute $T(x)$ for x in $(I.1)$ and define a family of polynomials

$$P_n(T(x)) = P_{nd}^T(x) \qquad (I.3)$$

of degree nd. What are the properties of the set $P_{nd}^T(x)$, $n = 0, 1, \ldots$? We shall see that they form a subset of new families of orthogonal polynomials, but no more in general on the real line. A fundamental problem, will be to construct the missing ones, and define precisely those new families. Another problem, which may seem, a priori, to have little connection with the previous one, is to study the family of iterates of $T(x)$

$$\begin{cases} T^{(0)}(x) = x \\ T^{(1)}(x) = T(x) \\ T^{(2)}(x) = T[T(x)] \\ \quad\cdots\cdots\cdots\cdots\cdots \\ T^{(n)}(x) = T\left[T^{(n-1)}(x)\right] \\ \quad\cdots\cdots\cdots\cdots\cdots \end{cases} \qquad (I.4)$$

These form a set of polynomials of degrees d^n. Again, we shall see that the $T^{(n)}(x)$ are a subset of orthogonal polynomials on a set of the complex plane, which is nothing but the Julia set[1],[2] associated to the polynomial $T(x)$. Again we shall construct the missing orthogonal polynomials.

The substitution of a polynomial into itself is a highly non linear operation. It is well known that orthogonal polynomials satisfies a three term recursive relation which is linear[3]. Therefore, at the expense of introducing the missing polynomials, we can replace the non linear iterative process, by a linear one, which gives a new insight on the process of iteration of polynomials[4],[5],[6],[7],[8]. With the three terms recursive relation, satisfied by any orthogonal polynomial family, comes a tridiagonal infinite matrix: the Jacobi matrix, J associated to it[9]. The spectrum of J, invariant under $T(x)$, will appear to be the Julia set itself. Furthermore the coefficients of the Jacobi matrix are quasi-periodic functions of their indices[6],[10],[11]. These properties permit the construction of exactly solvable models for quasi-periodic Schrödinger operators in Quantum Mechanics[10],[12].

The measures with respect to which the polynomials are orthogonal (invariant measures) will generate moment problems[9]. The generating function of the moments is the mean value of the resolvant of the J matrix. A fundamental functional equation is satisfied by this generating function as well as by its truncated continued fraction expansion, the $[p-1/p]$ Padé approximation to it, and plays a central role in the exposition[13],[14].

The paper will be organized in the following way:

- Section II, an introduction to Julia sets.

- Section III, an introduction to the moment problem on the complex line, with the associated, orthogonal polynomials, Jacobi matrix, continued fractions and Padé approximations.

- Section IV, the theory of orthogonal polynomials on Julia sets, the properties of quasi-periodicity of the coefficient of the three terms relation, the renormalization group invariance.

- Section V presents the conclusion and gives an outlook.

II. POLYNOMIAL JULIA SETS

II.1. A Simple Question

Suppose one wants to solve the following second degree equation

$$z^2 - z - \lambda = 0 . \qquad (II.1)$$

Using fixed point iteration method, one introduces the transformation

$$T(z) = z^2 - \lambda \qquad (II.2)$$

and starting from z_0 in the complex plane, builds the iterates $z_1, z_2, ..., z_n, ...$

$$\begin{cases} z_1 = T(z_0) \\ z_2 = T(z_1) \\ \text{........} \\ z_n = T(z_{n-1}) \\ \text{........} \end{cases} \qquad (II.3)$$

An important question is as to where z_0 should *not* be in order to insure that z_n tends toward a root of (II.1) as $n \longrightarrow \infty$. Clearly, if we choose z_0 large in modulus, z_1 will be even larger, and so on, therefore z_n will tend to infinity. Infinity is an attractive point of $T(z)$. This is clearly true for any polynomial $T_d(z)$ of degree d.

There is a region in the complex plane surrounding the *point at ∞* such that if we start from any z_0 in this region, we converge toward infinity. This is called the basin of attraction of the point at infinity for the polynomial (with complex coefficients) $T_d(z)$. An intriguing question is finding the shape of the border of this domain. Consider for instance the simple case

$$T(z) = z^2. \qquad (II.4)$$

Let us define the successive iterates of an arbitrary polynomial $T(z)$ by

$$\begin{cases} T^{(0)}(z) = z \\ T^{(1)}(z) = T(z) \\ \text{................} \\ T^{(n)}(z) = T\left[T^{(n-1)}(z)\right] . \end{cases} \qquad (II.5)$$

$T^{(n)}(z)$ is a polynomial of degree d^n. In the case of (II.4) we simply have

$$T^{(n)}(z) = z^{2^n} \qquad (II.6)$$

and the border of the basin of attraction of the point at infinity is a simple algebraic curve: the unit circle. One would expect that for

$$T(z) = z^2 - 1 \qquad (\lambda = 1) \qquad (II.7)$$

one also gets a simple algebraic curve. It is a great surprise that the border is a complicated fractal[15]. In fact for all values of λ complex or real except $\lambda = 0$ and $\lambda = 2$ the border of the basin of attraction of the point at infinity is conjectured to be a fractal.

II.2. Definition of Polynomial Julia Sets

Let $T(z)$ be a polynomial transformation of degree d, namely

$$z' = T_d(z) \equiv t_0 z^d + t_1 z^{d-1} + \dots + t_d \qquad t_0 \neq 0 \qquad (II.8)$$

very often we shall consider monic transformations with $t_0 = 1$. We shall always assume $d \geq 2$.

The Julia set J is defined as being the boundary of the basin of attraction of the point at infinity. Let us point out some properties of J[16][17].

- J is *totally* invariant under T, which means that if $z \in J$, then $T(z) \in J$ (forward invariance), and also $T_i^{-1}(z) \in J$, where T_i^{-1}, $i = 1, 2, \dots d$ are the d inverse functions of T (backward invariance).

- J is the closure of the set of the repulsive fixed points of any order of T. The fixed points of order n of T are the d^n solution of

$$T^{(n)}(z) = z, \qquad n > 0, \qquad (II.9)$$

where $T^{(n)}$ is the nth iterate of T. A fixed point, will be repulsive if the derivative $[T^{(n)}(z)]'$ computed at this point is in modulus greater than 1. We shall also define the filled-in Julia set K (sometimes called a Fatou domain, not to be confused with Fatou's dust defined on the next page) as the complement of the basin of attraction of the point at infinity. K is clearly a compact set in C. We shall also sometime need a *canonical* form for $T_d(z)$. By making a linear conjugacy on T, it can be brought to the canonical form

$$\hat{T}_d(z) = z^d + 0.z^{d-1} + \hat{t}_2 z^{d-2} + \dots + \hat{t}_d \qquad (II.10)$$

where

$$\hat{T}_d = \mathcal{L}^{-1} o \, T \, o \, \mathcal{L} \qquad (II.11)$$

and $\mathcal{L}$ is the linear transformation

$$\mathcal{L}(z) = z t_0^{-1/d-1} - t_1 t_0^{-1} d^{-1}. \qquad (II.12)$$

For the case where $T(z)$ is a second degree polynomial

$$\hat{T}_2(z) \equiv z^2 + c . \qquad (II.13)$$

An interesting question is to know if K_c, the filled-in Julia set for (II.13), is connected or not. In this case the answer is known: if $c \in \mathcal{M}$ (the Mandelbrot set) K_c is connected, otherwise K_c is a completely disconnected set (Cantor set) and is called a Fatou's dust. The Mandelbrot set as well as filled-in Julia sets are easy to generate on a micro computer. [A few years ago it took a long time to made $\mathcal{M}$, now there are excellent algorithms]. K_c is always inside the circle Γ of center origin and radius $\frac{1+\sqrt{1+4|c|}}{2}$. To draw a filled-in Julia set on a microcomputer, take $z_0 \in \Gamma$, iterate say fifty times, and reject it, if the fiftieth iterate

Fig. 0. *Julia sets corresponding to the transformation $z' = z^2 - \lambda$ for various values of λ.*

is outside Γ. Nice pictures of Julia sets or $\mathcal{M}$−sets as well as more details on the computer generation are available in Ref. [17],[18],[19]. Figure 0 shows some filled in Julia sets.

II.3. Invariant Measures

On the Julia set for T_d lives a natural invariant measure [2][20], which can be intuitively described as follows. Take an arbitrary point z_0 in the complex plane, compute its d preimages, that is the d solutions of

$$T_d(z) = z_0 . \qquad (II.14)$$

call them $z_1^{(i)}$, $i = 1, ...d$. Compute the preimages of each $z_1^{(i)}$, call them $z_2^{(i)}$, $i = 1, 2, ...d^2$ and so on. At order k, there are d^k preimages of z_0, called $z_k^{(i)}$, $i = 1, 2, ..d^k$. For any $k \geq 0$ define a discrete measure μ_k having point masses at each $z_k^{(i)}$ with equal weights d^{-k}, in such a way that the total mass be one. This sequence of probability measures can be proved[2][20] to have a (weak) limit independent of z_0. (There may exist at most two exceptional z_0, for which this statement is not true). This limiting process is the basis of a computer algorithm for constructing plots of Julia sets: take a repulsive fixed point of order one on J, compute its preimages, plot them, then choose one of them at random (with equal probability) and consider this point as the new starting point and repeat the procedure.

Before we describe some properties of this measure which is called the "invariant balanced measure" on the Julia set J, we must introduce polynomial transformations on a measure in the complex plane.

Let μ be a bounded complex measure on the complex plane with bounded support. Let $T_i^{-1}(z)$, $i = 1, 2, ...d$ be a complete assignment of branches of the inverse of the polynomial function $T(z)$. In addition let $\gamma_i(x)$, $i = 1, 2, ..., d$, be measurable complex functions on the complex plane, such that

$$\sum_{i=1}^{d} \gamma_i(x) = 1. \tag{II.15}$$

To the measure μ, we associate the transformed measure μ^T by

$$\int f(x)d\mu^T = \sum_{i=1}^{d} \int \gamma_i(x)f\left[T_i^{-1}(x)\right] d\mu \tag{II.16}$$

for any measurable f. For any Borel set E, $T_i^{-1}(E) \cap T_j^{-1}(E)$, $i \neq j$ is contained in the finite set C^T of critical values of T. The points x such that

$$T'(x) = 0 \tag{II.17}$$

are called the critical points x_c, while $T(x_c)$ is called a critical value. We shall also assume that

$$\mu(x_c) = 0 \tag{II.18}$$

(no mass on the critical points). Choosing

$$f(x) = \chi_{T_i^{-1}(E)}(x) \tag{II.19}$$

where $\chi_B(x)$ is the characteristic function of the set B. We get

$$\mu^T\left[T_i^{-1}(E)\right] = \int_E \gamma_i(x)d\mu = \int \gamma_i(x)\chi_E(x)d\mu. \tag{II.20}$$

Summing upon the index i, we have

$$\mu^T(T^{-1}(E)) = \mu(E). \tag{II.21}$$

In the particular case where $\mu^T \equiv \mu$, μ will be an invariant measure. Under the transformation T

$$\mu(T^{-1}(E)) = \mu(E). \tag{II.22}$$

In the special case where the place dependent functions $\gamma_i(x)$ are all equal, the invariant measure is called "balanced". In that case

$$\gamma_i(x) = \frac{1}{d} \tag{II.23}$$

and from (II.20) furthermore

$$\frac{1}{d}\mu(E) = \mu(T_i^{-1}(E)), \quad i = 1, 2, ..., d. \tag{II.24}$$

In that case there is a unique solution to (II.24) and the measure μ can be normalized by

$$\mu(J) = 1. \qquad (II.25)$$

Then μ is a probability measure (positive and normalized). It is the limit of the process introduced at the begining of the paragraph. The support of μ is J.

II.4. Electrostatic Description. Moments. Green Function

When z is large, $T(z)$ behaves like z^d, for simplicity let us assume T to be monic. For there to exist a change of coordinates

$$u = B(z) \qquad (II.26)$$

such that the action of T appears as the action of z^d we must have $B(T(z)) = [B(z)]^d$, where $B(z)$ is the so-called Böttcher function. The Böttcher[1] function $B(z)$ makes the diagram below commute

$$
\begin{array}{ccc}
 & T & \\
z & \longrightarrow & z' \\
B \downarrow & & \downarrow \\
u & \longrightarrow & u^d
\end{array}
\qquad (II.27)
$$

Clearly $B(z)$ must behave like z, for z large, furthermore it can be shown that (II.27) allows an extension of $B(z)$ into an analytic function inside the basin of attraction of the point at infinity.

$B(z)$ for $|z|$ large enough is given by

$$B(z) = \lim_{n \to \infty} \left[T^{(n)}(z) \right]^{d^{-n}}. \qquad (II.28)$$

In the case where the filled in Julia set K is connected $B(z)$ maps conformally the complement of K, $A(\infty)$ (the basin of attraction of the point at infinity) onto the exterior of the unit disk. In this case $B(z)$ is holomorphic in $A(\infty)$. If the Julia sets is disconnected, $B(z)$ is still analytic in $A(\infty)$ but algebraic branch points occur at the critical values of T and its backward iterates and $B(z)$ becomes multivalued. However $|B(z)|$ is still unique and $\ln|B(z)|$ is an harmonic function, vanishing on the Julia set. Hence $\ln|B(z)|$ is the Green function[21] for the basin of attraction of the point at infinity.

From classical results in harmonic analysis, one checks that $\ln|B(z)|$ is the potential generated by the equilibrium measure on the Julia set J, and that the equilibrium measure is the invariant balanced measure previously introduced. Introducing the generalized Green function (its real part is the usual Green function)

$$G(z) = \ln\, B(z) \qquad (II.29)$$

we see from (II.27) that $G(z)$ satisfies the functional equation

$$\frac{1}{d} G(T(z)) = G(z). \qquad (II.30)$$

If in (II.16) we choose for μ the balanced measure and for $f(x)$

$$f(x) = \ln(z - x) \qquad (II.31)$$

we get

$$\int \ln(z-x)\mathrm{d}\mu = \frac{1}{d}\int \ln \prod_{i=1}^{d} [z - T_i^{-1}(x)]\,\mathrm{d}\mu \qquad (II.32)$$

$$= \frac{1}{d}\int \ln[T(z) - x]\mathrm{d}\mu \qquad (II.33)$$

(here T is choosen to be monic). (II.33) shows that

$$G(z) = \int \ln(z-x)\mathrm{d}\mu(x) \qquad (II.34)$$

taking into account the uniqueness of the solution of (II.30). Here $\mu(x)$ is the invariant balanced measure. If we take the real part of (II.34) we get

$$\operatorname{Re} G(z) = \int \ln|z-x|\mathrm{d}\mu(x). \qquad (II.35)$$

$\operatorname{Re} G(z)$ is the usual electrostatic potential generated by the charge distribution μ on the Julia set J. Taking the derivative of (II.34) to get rid of the logarithmic singularity in at infinity of $G(z)$, we introduce

$$g(z) = \frac{\mathrm{d}}{\mathrm{d}z}G(z) = \int \frac{\mathrm{d}\mu}{z-x} = \frac{\mu_0}{z} + \frac{\mu_1}{z^2} + \dots + \frac{\mu_k}{z^{k+1}} + \dots \qquad (II.36)$$

where

$$\mu_k = \int x^k \mathrm{d}\mu \qquad (II.37)$$

are the moments of the balanced measure. We shall need to introduce moments of a complex measure on the complex plane, as well as their moment generating function. This will be the object of the next section. To end this paragraph, let us notice that we have differentiating (II.30)

$$g(z) = \frac{1}{d}T'(z)g(T(z)) \qquad (II.38)$$

which allows to compute recursively all the moment of the balanced measure.

III. ORTHOGONAL POLYNOMIALS ON A COMPLEX CONTOUR[4]

We shall extend the theory of orthogonal polynomials on the real line to complex contours. Most of the classical properties will still be true, in particular the important fact that any sequence of orthogonal polynomials satisfies a linear three term recursive relation. However properties linked to the positivity of the measure, for instance the localization of the zeros, will be lost: for orthogonal polynomials associated to a Julia set, the problem of the localization of the zeros has received only a partial answer.

We shall be interested in the class of holomorphic functions $H(\infty)$ around the point at infinity. More precisely $g \in H(\infty)$ if: i) g is holomorphic in $A(\infty)$, a domain surrounding the point at infinity. For convenience we shall call J the boundary of $A(\infty)$. ii) g has the expansion around ∞

$$g(z) = \frac{\mu_0}{z} + \frac{\mu_1}{z^2} + \frac{\mu_2}{z^3} + \ldots + \frac{\mu_k}{z^{k+1}} + \ldots \qquad \mu_0 = 1 \qquad (III.1)$$

the series $\sum_{k=0}^{\infty} \mu_k w^k$ having a positive radius of convergence R. For $g \in H(\infty)$ we can write

$$g(z) = \frac{1}{2i\pi} \oint_{\Gamma_\infty^+} \frac{g(v)}{z - v} dv \qquad (III.2)$$

where Γ_∞^+ is a contour circling J (counter clockwise), and belonging to $A(\infty)$ and z is outside Γ_∞ (see Figure 1).

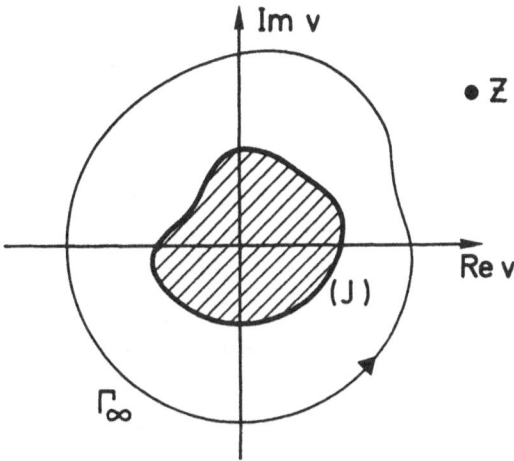

Fig. 1. *A closed contour belonging to $A(\infty)$.*

This is easy to prove by deforming Γ_∞ into $\bar{\Gamma}_\infty^{(1)}$ and $\bar{\Gamma}_\infty^{(2)}$ as shown on Figure 2, and letting $\bar{\Gamma}_\infty^{(1)}$ grow. The contribution coming from $\bar{\Gamma}_\infty^{(1)}$ is zero due to the fact that (III.1) implies that for $|z|$ sufficiently large, there exists $M > 0$ such that

$$|g(z)| < \frac{M}{|z|}. \qquad (III.3)$$

Expanding (III.2) in power of z (remember $|z| > |v|$)

$$g(z) = \sum_{k=0}^{\infty} \frac{\mu_k}{z^{k+1}} \qquad (III.4)$$

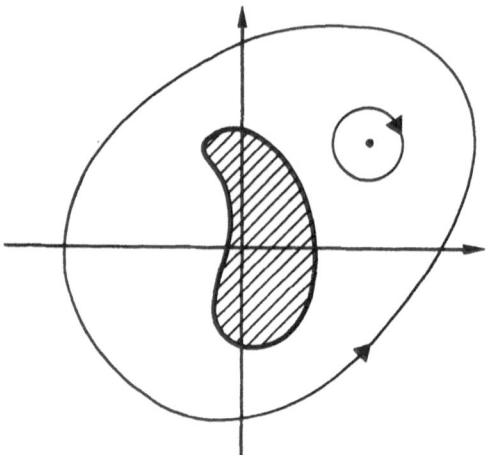

Fig. 2. *Deforming the closed contour of Fig. 1.*

where

$$\mu_k = \frac{1}{2i\pi} \oint_{\Gamma_\infty^+} v^k g(v) dv \tag{III.5}$$

is called the kth moment of $g(v)$. In the special case where J reduces to a piece of the real axis we can squeeze Γ_∞^+ on the real line and get (see Figure 3)

$$\mu_k = \int_a^b x^k \left\{ -\frac{1}{2i\pi}[g(x+i\epsilon) - g(x-i\epsilon)] \right\} dx$$

$$+ \frac{1}{2\pi i} \int_{-\epsilon}^{+\epsilon} b^k g(b+it) dt - \frac{1}{2\pi i} \int_{-\epsilon}^{+\epsilon} a^k g(a+it) dt \tag{III.6}$$

If $g(v)$ is a real function, that is,

$$g(v) = g^*(v^*) \tag{III.7}$$

then

$$\mu_k = \int_a^b x^k \left[-\frac{1}{\pi} \text{Im } g(x+i\epsilon) \right] dx. \tag{III.8}$$

If

$$d\mu(x) = \lim_{\epsilon \to 0^+} -\frac{1}{\pi} \text{Im } g(x+i\epsilon) dx \tag{III.9}$$

exists and is a positive measure, we are back to the ordinary theory of moments, we shall call $g(z)$ the moment generating function. Notice that we can always choose Γ_∞ to be as near as we want to J. Note that when $g(z)$ is a Stieltjes function, Stieltjes inversion formula gives this measure always, also when it is singular.

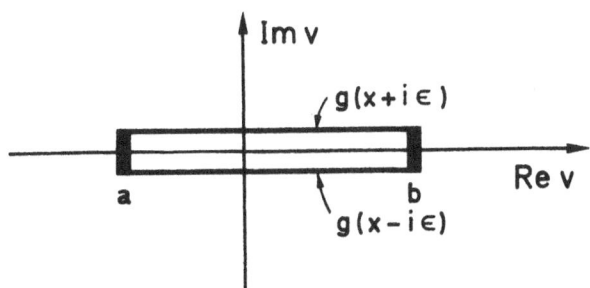

Fig. 3. *Squeezing Γ_∞^+ on the real line.*

III.1. Definition of Orthogonal Polynomial on a Complex Contour

We shall introduce the following (indefinite) scalar product

$$(f_1, f_2) = \frac{1}{2i\pi} \oint_{\Gamma_\infty^+} f_1(v) f_2(v) g(v) dv \qquad (III.10)$$

where f_1 and $f_2 \in H(\infty)$. A set of orthogonal polynomial with respect to $g(v)$ is defined by

$$(P_n, P_m) = \frac{1}{2i\pi} \oint_{\Gamma_\infty^+} P_n(v) P_m(v) g(v) dv = h_n \delta_{n,m} \qquad (III.11)$$

where

$$P_n(v) = v^n + ... \qquad (III.12)$$

is a monic polynomial of degree n and h_n is the norm of the polynomial. We shall also introduce the Hankel-Hadamard determinants

$$\Delta_n = \begin{vmatrix} \mu_0 & \mu_1 & \cdots & \mu_{n-1} \\ \mu_1 & \mu_2 & \cdots & \mu_n \\ \cdots & \cdots & \cdots & \cdots \\ \mu_{n-1} & \mu_n & \cdots & \mu_{2n-2} \end{vmatrix}, \quad \Delta_1 = \mu_0 = 1. \qquad (III.13)$$

We shall always suppose that none of these determinants vanish. The $P_n(v)$ being monic are linearly independent. Therefore for $n \leq m$ (III.11) can be replaced by

$$(P_n, P_m) = \frac{1}{2i\pi} \oint_{\Gamma_\infty^+} v^n P_m(v) g(v) dv = h_n \delta_{n,m} \qquad (III.14)$$

It is not difficult to prove that if $\Delta_n \neq 0$ for any n, the polynomials P_n are unique. Furthermore,

$$\Delta_n P_n(v) = \begin{vmatrix} \mu_0 & \mu_1 & \cdots & \mu_{n-1} & \mu_n \\ \mu_1 & \mu_2 & \cdots & \mu_n & \mu_{n+1} \\ \cdots & \cdots & \cdots & \cdots & \cdots \\ \mu_{n-1} & \mu_n & \cdots & \mu_{2n-2} & \mu_{2n-1} \\ 1 & v & v^2 & \cdots & v^n \end{vmatrix}. \qquad (III.15)$$

(III.15) shows that $P_n(v)$ is a monic polynomial of degree n. Furthermore

$$\Delta_n \oint_{\Gamma_\infty^+} v^k P_n(v) g(v) dv = \Delta_n \oint_{\Gamma_\infty^+} v^k \begin{vmatrix} \mu_0 & \mu_1 & \cdots & \mu_n \\ \mu_1 & \cdots & \cdots & \mu_{n+1} \\ \cdots & \cdots & \cdots & \cdots \\ \mu_{n-1} & \mu_n & \cdots & \mu_{2n-1} \\ 1 & v & \cdots & v^n \end{vmatrix} g(v) dv$$

$$= \Delta_n \begin{vmatrix} \mu_0 & \mu_1 & \cdots & \mu_n \\ \mu_1 & \mu_2 & \cdots & \mu_{n+1} \\ \cdots & \cdots & \cdots & \cdots \\ \mu_{n-1} & \mu_n & \cdots & \mu_{2n-1} \\ \mu_k & \mu_{k+1} & \cdots & \mu_{k+n} \end{vmatrix}. \qquad (III.16)$$

The last determinant in (III.16) is always zero for $k = 0, 1, ..., n-1$ which proves that $P_n(v)$ is the orthogonal polynomial we are looking for. For $k = n$ we get

$$\begin{cases} \Delta_n h_n = \Delta_{n+1} & n \geq 1 \\ h_0 = \mu_0 = 1. \end{cases} \qquad (III.17)$$

We note that, while for the classical moment problem, all norms h_n are clearly strictly positive which, in turn, implies that the Δ_n should also all be strictly positive, in this more general context the h_n can be complex, as well as the Δ_n. When $g(z)$ is real analytic, that is

$$g^*(z^*) = g(z) \qquad (III.18)$$

the μ_k are real, as well as the Δ_n and h_n. However the sign of the h_n can be arbitrary (indefinite metric).

III.2 Recursive Relations

Let us expand $P_{n+1}(v)$ in the following manner

$$P_{n+1}(v) = (v - A_n) P_n(v) - R_n P_{n-1}(v) + \sum_{k=0}^{n-2} \epsilon_k P_k(v). \qquad (III.19)$$

Taking the scalar product of both sides of (III.19) with $P_i(v)$, $i = 0, 1, ..., n-2$. We get

$$0 = (P_i, vP_n) + \epsilon_i = (vP_i, P_n) + \epsilon_i \qquad (III.20)$$

but vP_i is a polynomial of degree less or equal to $n-1$, and therefore orthogonal to P_n. We conclude that all ϵ_i are identically zero. Therefore our orthogonal polynomials satisfy the standard three term recursive relation

$$P_{n+1}(v) = (v - A_n) P_n(v) - R_n P_{n-1}(v). \qquad (III.21)$$

If we take the scalar product of (III.21) with $P_{n-1}(v)$, we get

$$\begin{aligned} 0 &= (P_{n-1}, vP_n) - R_n h_{n-1} = (vP_{n-1}, P_n) - R_n h_{n-1} \\ &= (v^n, P_n) - R_n h_{n-1} = (P_n, P_n) - R_n h_{n-1} = h_n - R_n h_{n-1}. \end{aligned} \qquad (III.22)$$

Therefore

$$h_n = R_n h_{n-1}. \qquad (III.23)$$

This shows that for the usual case, when h_n is an ordinary norm, the R_n have to be positive real numbers. Using (III.17) we get

$$R_n = \frac{h_n}{h_{n-1}} = \frac{\Delta_{n+1}}{\Delta_n} \cdot \frac{\Delta_{n-1}}{\Delta_n} = \frac{\Delta_{n-1} \cdot \Delta_{n+1}}{(\Delta_n)^2}. \qquad (III.24)$$

In the same way one gets

$$A_n = (vP_n, P_n). \qquad (III.25)$$

III.3. Zeros of Orthogonal Polynomials

In the special case where J is a piece of the real axis, and $g(z)$ is a real analytic function and the limiting value

$$\lim_{\epsilon \to 0+} -\frac{1}{\pi} \text{Im } g(x + i\epsilon) \qquad (III.26)$$

defines a positive measure, we are back to the classical orthogonal polynomial on the real line, all norms h_n are real and positive. For that case, we have the following simple properties of the zeros of $P_n(x)$. We shall make use of the fact that

$$\int_a^b Q_m(x) P_n(x) \mathrm{d}\mu = 0 \qquad (III.27)$$

for any polynomial $Q_m(x)$ of degree smaller than n
  (i) All zeros are real and inside the convex hull of J.
Let $[a, b]$ be the convex hull of J, let $\alpha + i\beta$ be a zero of $P_n(x)$. Write

$$P_n(x) = (x - \alpha - i\beta) S_{n-1}(x). \qquad (III.28)$$

choose $Q_m(x)$ to be $\bar{S}_{n-1}(x)$. (Complex conjugate polynomial). We have

$$\int_a^b (x - \alpha - i\beta) |S_{n-1}(x)|^2 \, \mathrm{d}\mu = 0. \qquad (III.29)$$

Therefore

$$\begin{cases} \beta \displaystyle\int_a^b |S_{n-1}(x)|^2 \, \mathrm{d}\mu & = 0 \\[2mm] \displaystyle\int_a^b (x - \alpha) |S_{n-1}(x)|^2 \, \mathrm{d}\mu & = 0. \end{cases} \qquad (III.30)$$

(III.30) shows that $\beta = 0$ and $a < \alpha < b$. [We shall always suppose that $\mathrm{d}\mu$ is not a finite sum of mass point measures].
  (ii) All zeros are simple.
Suppose there is a double zero. Write

$$P_n(x) = (x - \alpha)^2 S_{n-2}(x), \qquad (III.31)$$

and choose $Q_m(x)$ to be $\bar{S}_{n-2}(x)$.

(iii) In a subset of $[a, b]$ where the measure μ is constant, there is at most one zero.

Let (c, d) be the subset, and suppose there would be two zeros in (c, d), α and β. We write

$$P_n(x) = (x - \alpha)(x - \beta)S_{n-2}(x) \qquad (III.32)$$

and we choose $Q_m(x)$ to be $\bar{S}_{n-2}(x)$. We have

$$0 = \int_a^c (x - \alpha)(x - \beta)|S_{n-2}(x)|^2 \, d\mu + \int_d^b (x - \alpha)(x - \beta)|S_{n-2}(x)|^2 \, d\mu. \qquad (III.33)$$

However both integral in (III.33) are positive, which shows that the hypothesis made is absurd.

(iv) The zeros of $P_n(x)$ and $P_{n+1}(x)$ interlace.

This is easily proved, using the three terms recursive relation, and the fact that all zeros are real and simple, as well as the fact that R_n is positive. There are no known extensions to complex contour of those very simple properties of the classical orthogonal polynomials. We shall discuss later what can be said for the case of orthogonal polynomials on Julia sets.

III.4. J Matrix

To the three terms recursive relation

$$P_{n+1}(v) = (v - A_n) P_n(v) - R_n P_{n-1}(v) \qquad (III.34)$$

we associate a tridiagonal matrix $\mathbf{J}$, the semi-infinite Jacobi matrix

$$\mathbf{J} = \begin{bmatrix} A_0 & 1 & 0 & 0 & 0 & \dots \\ R_1 & A_1 & 1 & 0 & 0 & \dots \\ 0 & R_2 & A_2 & 1 & 0 & \dots \\ \dots & \dots & \dots & \dots & \dots & \dots \end{bmatrix}. \qquad (III.35)$$

The non zero elements of $\mathbf{J}$ are

$$J_{ii} = A_i, \quad J_{i,i+1} = 1, \quad J_{i,i-1} = R_i, \quad i = 0, 1, 2, \dots \qquad (III.36)$$

If we introduce the vector

$$\psi^T(v) = [P_0(v), P_1(v), P_2(v), \dots, P_n(v), \dots] \qquad (III.37)$$

we have

$$\mathbf{J} \, \psi(v) = v \, \psi(v). \qquad (III.38)$$

Due to the particular structure of the matrix $\mathbf{J}$, all products involved contain only sums with a finite number of terms, therefore no specification on the growth of the components of the vector is needed at this stage (Eigenvalue problem).

Let us introduce a basis $\{e_n\}$ with

$$e_n = (0, 0, \dots, 1, 0, \dots) \qquad (III.39)$$

having a one in the nth component. One has

$$h_n e_n = P_n(\mathbf{J})e_0 \qquad (III.40)$$

where $P_n(\mathbf{J})$ is obtained by substitution of the matrix $\mathbf{J}$ in place of the variable x, and

$$e_n = P_n\left(\mathbf{J}^T\right)e_0 \qquad (III.41)$$

where h_n is the norm of P_n and $\mathbf{J}^T$ is the transposed $\mathbf{J}$ matrix. (III.40) and (III.41) are proved by induction, and using the identity

$$P_{n+1}(\mathbf{J}) = [\mathbf{J} - A_n \mathbf{I}] P_n(\mathbf{J}) - R_n P_{n-1}(\mathbf{J}). \qquad (III.42)$$

All matrices $P_n(\mathbf{J})$ are well defined because the matrix elements of $\mathbf{J}^k$ are computable from those of $\mathbf{J}$ through finite sums only.

For convenience, we shall introduce the *indefinite* scalar product of two vectors a, b

$$\begin{cases} a^T &= [a_0, a_1, ..., a_n, ...] \\ b^T &= [b_0, b_1, ..., b_n, ...] \end{cases} \qquad (III.43)$$

as

$$(a, b) = a_0 b_0 + a_1 b_1 + ... + a_n b_n + ... \qquad (III.44)$$

We suppose $a_n = b_n = 0$ for $n \geq N$. Note the absence of complex conjugation! Furthermore, we will have never to worry about convergence of such sums, because our sums will always involve a finite number of terms. From (III.38) we get

$$\mathbf{J}^m \psi(v) = v^m \psi(v) \qquad (III.45)$$

or

$$(e_0, \mathbf{J}^m \psi(v)) = v^m \qquad (III.46)$$

(recall that $P_0 = 1$). Let us multiply both sides of (III.46) by $g(v)$ and integrate on Γ_∞^+; we get (using III.5)

$$\mu_m = \frac{1}{2i\pi} \oint_{\Gamma_\infty^+} (e_0, \mathbf{J}^m \psi(v)) g(v) dv. \qquad (III.47)$$

Notice that $\mathbf{J}^m$ is a $(2m + 1)$ banded matrix and therefore $(e_0, \mathbf{J}^m \psi(v))$ is a polynomial in v of degree m.

$$(e_0, \mathbf{J}^m \psi(v)) = \sum_{k=0}^{k=m} P_k(v) \ (e_0, \mathbf{J}^m e_k). \qquad (III.48)$$

From the orthogonality property of the P_k, we see that after integration, only the $(e_0, \mathbf{J}^m e_0)$ term survives in the sum, and since $\mu_0 = +1$, we get

$$\mu_m = (e_0, \mathbf{J}^m e_0). \qquad (III.49)$$

Therefore the moment generating function can be written

$$g(z) = \sum_{k=0}^{\infty} \frac{\mu_k}{z^{k+1}} = \sum_{k=0}^{\infty} \frac{(e_0, \mathbf{J}^k e_0)}{z^{k+1}} \qquad (III.50)$$

or in compact form

$$g(z) = \left(e_0, [z - \mathbf{J}]^{-1} e_0\right).\qquad (III.51)$$

(III.51) is only a condensed way to write (III.49). If one wants to give a full content to (III.51), it would be necessary to introduce a metric space. We shall not pursue this question here, for the classical moment problem, the resolvant (III.51) exists in the ℓ^2 Hilbert space[9].

III.5. J-Matrix Approximations

Let us contemplate the truncated approximations of the $\mathbf{J}$ – matrix (III.35) at order N, $\mathbf{J}_N$

$$\mathbf{J}_N = \begin{bmatrix} A_0 & 1 & 0 & 0 & 0 & \cdots & 0 \\ R_1 & A_1 & 1 & 0 & 0 & \cdots & \cdots \\ 0 & R_2 & A_2 & 1 & 0 & \cdots & \cdots \\ \cdots & 0 & R_k & A_k & 1 & 0 & \cdots \\ \cdots & \cdots & \cdots & \cdots & \cdots & \cdots & \cdots \\ 0 & \cdots & 0 & R_{N-2} & A_{N-2} & 1 & 0 \\ 0 & \cdots & 0 & 0 & R_{N-1} & A_{N-1} & 1 \\ 0 & \cdots & 0 & 0 & 0 & R_N & A_N \end{bmatrix}. \qquad (III.52)$$

This is an $(N+1) \times (N+1)$ tridiagonal matrix. Introducing the resolvant $g^{(N)}(z)$ associated with $\mathbf{J}_N$, we have

$$g^{(N)}(z) = \left(e_0, [z - \mathbf{J}^N]^{-1} e_0\right) = \frac{\mu_0^{(N)}}{z} + \frac{\mu_1^{(N)}}{z^2} + \cdots + \frac{\mu_k^{(N)}}{z^{k+1}} + \cdots \qquad (III.53)$$

where

$$\mu_k^{(N)} = \left(e_0, \mathbf{J}_N^k e_0\right) \qquad (III.54)$$

However $\mathbf{J}$ being a tridiagonal matrix, the calculation of

$$\mu_k = \left(e_0, \mathbf{J}^k e_0\right) \qquad (III.55)$$

involves only the matrix elements of $\mathbf{J}_N$, provided that $k \leq 2N + 1$.

Therefore the $(2N + 2)$ first moments of the expansion of $g(z)$ and $g^{(N)}(z)$ are identical

$$\mu_k^{(N)} \equiv \mu_k \qquad k = 0, 1, ..., 2N + 1. \qquad (III.56)$$

The difference

$$g(z) - g^{(N)}(z) = \frac{\left[\mu_{2N+2} - \mu_{2N+2}^{(N)}\right]}{z^{2N+3}} + \cdots + \frac{\left[\mu_{2N+1+\ell} - \mu_{2N+1+\ell}^{(N)}\right]}{z^{2N+2+\ell}} + \cdots \qquad (III.57)$$

is an element of $H(\infty)$, but with the first $2N + 2$, coefficients missing, we shall write a short hand of (III.57) by using the notation

$$g(z) - g^{(N)}(z) = O\left(z^{-(2N+3)}\right). \qquad (III.58)$$

It means that for z sufficiently large $g(z)$ and $g^{(N)}(z)$ differ only by terms of order $z^{-(2N+3)}$.

III.6. The Continued Fraction Approximation

J_N being a finite matrix, its resolvant $[z - J_N]^{-1}$ exist everywhere except when z belongs to the spectrum of J_N. Therefore for $|z|$ sufficiently large, we can safely contemplate the identity

$$[z - J_N]^{-1} [z - J_N] = I \qquad (III.59)$$

where I is the identity matrix in the $N + 1$ dimensional space. Introducing the matrix elements

$$g_k^{(N)} = \left(e_0, g^{(N)} e_k \right)$$

in the relation deduced from (III.59)

$$\left(e_0, (z - J_N)^{-1} [z - J_N] e_n \right) = \delta_{0,n} \qquad (III.60)$$

[$\delta_{i,j}$ is the usual Kronecker symbol, equal to 1 if $i = j$, 0 otherwise], we get

$$\left. \begin{aligned}
(z - A_0) g_0^{(N)} - g_1^{(N)} R_1 &= 1 \\
(z - A_1) g_1^{(N)} - g_2^{(N)} R_2 &= g_0^{(N)} \\
.... \qquad\qquad\qquad &\quad ... \\
(z - A_\ell) g_\ell^{(N)} - g_{\ell+1}^{(N)} R_{\ell+1} &= g_{\ell-1}^{(N)} \\
... \qquad\qquad\qquad &\quad ... \\
(z - A_{N-1}) g_{N-1}^{(N)} - g_N^{(N)} R_N &= g_{N-2}^{(N)} \\
(z - A_N) g_N^{(N)} &= g_{N-1}^{(N)}
\end{aligned} \right\} \qquad (III.61)$$

Introducing the ratios

$$\rho_\ell = \frac{g_{\ell-1}^{(N)}}{g_\ell^{(N)}} \qquad \ell = 1, 2, ..., N \qquad (III.62)$$

we rewrite (III.61) in the following way

$$\left. \begin{aligned}
g_0^{(N)} &= \frac{1}{z - A_0 - R_1 \rho_1^{-1}} \\
\rho_1 &= z - A_1 - R_2 \rho_2^{-1} \\
... \quad ... \\
\rho_\ell &= z - A_\ell - R_{\ell+1} \rho_{\ell+1}^{-1} \\
... \quad ... \\
\rho_{N-1} &= Z - A_{N-1} - R_N \rho_N^{-1} \\
\rho_N &= z - A_N
\end{aligned} \right\} \qquad (III.63)$$

This defines the continued fraction expansion of $g_0^{(N)}$

$$g_0^{(N)} \equiv g^{(N)}(z) = \cfrac{1}{z - A_0 - \cfrac{R_1}{z - A_1 - \cfrac{R_2}{z - ....}}}$$

$$.................................$$

$$.... - \cfrac{R_{N-2}}{z - A_{N-2} - \cfrac{R_{N-1}}{z - A_{N-1} - \cfrac{R_N}{z - A_N}}} \qquad (III.64)$$

Clearly $g^{(N)}(z)$ is a rational fraction whose numerator is a polynomial of degree N and its denominator a polynomial of degree $N + 1$. We shall write

$$g^{(N)}(z) = \frac{N_N(z)}{D_{N+1}(z)}. \qquad (III.65)$$

Formally, we can associate to the semi-infinite $\mathbf{J}$ matrix, an infinite continued fraction called a J-fraction, by letting $N \longrightarrow \infty$ in (III.64). The question of the convergence of such process will not be discussed here[9].

III.7. Padé Approximations

For $g \in H(\infty)$, the expansion

$$g(z) = \frac{\mu_0}{z} + \frac{\mu_1}{z^2} + ... + \frac{\mu_k}{z^{k+1}} + ... \qquad (III.65)$$

around the point at ∞ will converge outside a circle whose radius is fixed by the singularity of $g(z)$ of largest modulus.

However, if we know in advance that $g(z)$ is holomorphic in a domain larger than this circle, we are able, using for instance the Weierstrass construction of the star of holomorphy, to continue analytically inside this domain everywhere until the natural boundary of $g(z)$ if it exists.

The construction of the star of holomorphy is very painful, furthermore it requires the knowledge of all the coefficients μ_k, $k = 0, 1, ...$. If we restrict the class of $g(z)$, for instance, to the simplest case, when $g(z)$ is a rational fraction, there exist a simple method for analytically continuing $g(z)$ to its complete domain of analyticity, knowing only a finite number of μ's. To achieve this goal we have to introduce, the continued fraction expansion of $g(z)$.

We therefore expand $g(z)$ in continued fraction, by computing successively A_0, R_1, A_1, $R_2, ..., A_k, R_k$, from $\mu_0, \mu_1, \mu_2, ...$ We stop when we find $R_{N+1} = 0$. The continued fraction will converge everywhere in the complex plane (except of course on the poles) and will define the analytic continuation inside the circle corresponding to the pole of largest modulus, this analytic continuation can be made knowing only a finite number of moments.

The relations between the μ's and the A's and R's is $\mu_k = (e_0, \mathbf{J}^k e_0)$ which allows to compute the μ_k, knowing $\mathbf{J}$, it also allows the unique computation of $\mathbf{J}$ knowing the μ_k. [See for instance Ref.[9] for a practical algorithm]. We can think to apply this process to any function belonging to $H(\infty)$, we shall therefore define a succession of approximation which are rational fractions with numerator of degree N and denominator of degree $N + 1$, we shall use for them the standard and self-explanatory notation

$$[0/1]_{g(z)} \;\; ; \;\; [1/2]_{g(z)} \;\; ; \;\; [N/N + 1]_{g(z)}, ... \qquad (III.66)$$

$[N/N + 1]_{g(z)}$ is a rational fraction with $N + 1$ poles, whose expansion round ∞, coincide with that of $g(z)$ up to order $2N + 2$, that is

$$g(z) - [N/N + 1]_{g(z)} = O\left(z^{-(2N+3)}\right). \qquad (III.67)$$

The poles of those fractions can cluster for N large on points, or curves, and build essential singularities, or logarithmic singularities or natural boundaries. Therefore one expects that those approximations could be very powerfull to analytically continue a function inside its circle of convergence.

Let us now quote one of the most remarkable properties of such approximations.

Theorem 1. The denominator (monic) of the Padé approximation $[N-1/N]_{g(z)}$ is the orthogonal polynomial $Q_N(z)$ associated to the complex measure $g(z)$ itself.

To prove this fundamental result, we write

$$g(v) - [N - 1/N]_{g(v)} = \frac{R_N(v)}{v^{2N+1}} \qquad (III.68)$$

where $R_N(v)$, is for $|v|$ sufficiently large an holomorphic function of v, and furthermore, because

$$g(v) - [N - 1/N]_{g(v)} = O\left(\frac{1}{v^{2N+1}}\right), \qquad (III.69)$$

we have

$$|R_N(v)| < C \quad \text{for} \quad |v| \quad \text{sufficiently large.} \qquad (III.70)$$

Let us write

$$[N - 1/N]_{g(v)} = \frac{N_{N-1}(v)}{D_N(v)}. \qquad (III.71)$$

Then

$$g(v)D_N(v) - N_{N-1}(v) = R_N(v)D_N(v)v^{-(2N+1)}. \qquad (III.72)$$

Multiplying (III.72) by v^k and integrating on Γ_∞, gives

$$\frac{1}{2\pi}\oint_{\Gamma_\infty^+} v^k g(v)D_N(v)dv - \frac{1}{2i\pi}\int_{\Gamma_\infty^+} v^k N_{N-1}(v)dv$$
$$= \frac{1}{2i\pi}\oint_{\Gamma_\infty^+} R_N(v)D_N(v)v^{k-2N-1}dv, \qquad k = 0,1,2,...,N-1. \qquad (III.73)$$

The second integral on the left hand side is identically zero, the one on the right hand side is also zero if $k \le N - 1$ because, integrating on a large circle of radius R we can bound the right hand side by

$$C\ R^{k-2N} \sup_{|v| = R} D_N(v) < C'\ R^{k-2N+N} = C'\ R^{k-N}. \qquad (III.74)$$

If $k \le N - 1$, this last bound tends to zero when $R \longrightarrow \infty$.

A direct consequence of this result is that, the zeros of $Q_{N+1}(z)$ are the poles of $g^N(z)$ which is the resolvant of $\mathbf{J}_N$, those poles are the eigenvalues of $\mathbf{J}_N$, therefore we get the following fundamental result.

Theorem 2. The zeros of the orthogonal polynomial $Q_{N+1}(z)$ are the eigenvalues of the truncated $\mathbf{J}$ – matrix $\mathbf{J}_N$.

This produces an easy way to compute the zeros of an orthogonal polynomial on a complex contour by diagonalizing a tridiagonal matrix.

III.8. An Error Formula

When $g(z)$ was a rational fraction, we saw that the continued fraction of $g(z)$ provided the analytic continuation of $g(z)$ everywhere. To be able to extend this result to other classes of analytic functions belonging to $H(\infty)$, we need to estimate

$$\epsilon_N(z) = g(z) - g^{(N)}(z) = g(z) - [N/N + 1]_{g(z)} \qquad (III.75)$$

where $|\epsilon_N(z)|$ can be thought as the "error".

To describe a useful expression for $\epsilon_N(z)$, we proceed in the following manner. $P_n(z)$ being the monic orthogonal polynomial associated to $g(z)$. We write

$$P_n(z)g(z) = \frac{1}{2i\pi} \oint_{\Gamma_\infty^+} \frac{g(v)P_n(z)}{z - v} dv \qquad (III.76)$$

$$= \frac{1}{2i\pi} \oint_{\Gamma_\infty^+} \frac{P_n(z) - P_n(v)}{z - v} g(v)dv + \frac{1}{2i\pi} \oint_{\Gamma_\infty^+} \frac{P_n(v)}{z - v} g(v)dv. \qquad (III.77)$$

Let us introduce the notations

$$\rho_n(z) = \frac{1}{2i\pi} \oint_{\Gamma_\infty^+} \frac{P_n(z) - P_n(v)}{z - v} g(v)dv. \qquad (III.78)$$

Clearly $\rho_n(z)$ is a polynomial in z of degree $n - 1$. Now, we have the following identity

$$\frac{1}{2i\pi} \oint_{\Gamma_\infty^+} \frac{z^n P_n(v)}{z - v} g(v)dv \equiv \frac{1}{2i\pi} \oint_{\Gamma_\infty^+} \frac{v^n P_n(v)}{z - v} g(v)dv \qquad (III.79)$$

for z outside Γ_∞^+. To prove (III.79) write

$$z^n = [(z - v) + v]^n \qquad (III.80)$$

and expand this last expression in the left hand side of (III.79), then after integration, only the contribution coming from v^n survives, all the other contributions giving zero by orthogonality. Therefore we can write

$$g(z) = \frac{\rho_n(z)}{P_n(z)} + \frac{1}{z^n} \frac{1}{2i\pi} \frac{1}{P_n(z)} \oint_{\Gamma_\infty^+} \frac{v^n P_n(v)}{z - v} g(v)dv. \qquad (III.81)$$

We notice that $\frac{\rho_n(z)}{P_n(z)}$ is a rational fraction of degree $n - 1$, over n. For z sufficiently large, we can expand the remainder and we see that

$$g(z) = \frac{\rho_n(z)}{P_n(z)} + \frac{h_n}{z^{2n+1}} + \sum_{k=0}^{\infty} \frac{\rho_k}{z^{2n+1+k}}. \qquad (III.82)$$

(III.82) shows that $\frac{\rho_n(z)}{P_n(z)}$ is the $[n - 1/n]_{g(z)}$ Padé approximation to $g(z)$. Because

$$g(z) - \frac{\rho_n(z)}{P_n(z)} = O\left(\frac{1}{z^{2n+1}}\right) \qquad (III.83)$$

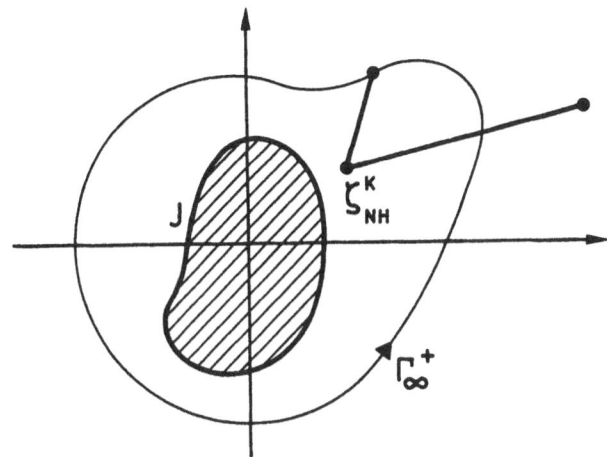

Fig. 4. *The location of the zeros S_{N+1}^k of $P_{N+1}(z)$.*

and $\frac{\rho_n(z)}{P_n(z)}$ is rational fraction of degree $n-1$ over n. (It is trivial to show that the $[n-1/n]_{g(z)}$ Padé approximant is unique). Therefore $\rho_n(z)$ is the numerator of the Padé approximant and is given by (III.78). Furthermore, (III.78) gives us a bonus: the $\rho_n(z)$ satisfy the same three term recursive relation as the $P_n(z)$. This is proved using (III.21) and the orthogonality property of $P_n(z)$. The initial conditions for the ρ_n are

$$\begin{cases} \rho_0(z) &= 0 \\ \rho_1(z) &= 1, \end{cases} \qquad (III.84)$$

while for the P_n they are

$$\begin{cases} P_{-1} &= 0 \\ P_0 &= 1. \end{cases} \qquad (III.85)$$

Coming back to (III.81), we see that

$$\epsilon_N(z) = \frac{1}{z^{N+1}} \frac{1}{P_{N+1}(z)} \frac{1}{2i\pi} \oint_{\Gamma_\infty^+} \frac{v^{N+1} P_{N+1}(v)}{z-v} g(v) dv. \qquad (III.86)$$

This is the fundamental error formula for the $[N/N+1]_g(z)$. Padé approximation to $g(z)$.
Let us suppose that all the zeros of $P_{N+1}(z)$ are inside Γ_∞^+, let us call those zeros S_{N+1}^K, $K = 0, ..., N$. Then

$$|\epsilon_N(z)| \le \frac{1}{|z|^{N+1}} \frac{1}{2\pi} \oint_{\Gamma_\infty^+} \frac{|v|^{N+1}}{|z-v|} \prod_{k=0}^{N} \left| \frac{v - S_{N+1}^k}{z - S_{N+1}^k} \right| |g(v)||dv|. \qquad (III.87)$$

Let δ be the diameter of Γ_∞^+. Then

$$\left| v - S_{N+1}^k \right| \leq \delta \tag{III.88}$$

$$\left| z - S_{N+1}^k \right| \geq \Delta \tag{III.89}$$

where Δ is the distance of z to Γ_∞^+. $g(v)$ being analytic on Γ_∞^+ is bounded by

$$|g(v)| < M \tag{III.90}$$

and

$$|v| < k \tag{III.91}$$

where k is the modulus of the most distant point of Γ_∞^+ to the origin. Then (L being the length of Γ_∞^+)

$$|\epsilon_N(z)| \leq \frac{L}{2\pi} M \frac{k^{N+1}}{\Delta} \left(\frac{\delta}{\Delta} \right)^{N+1} \frac{1}{|z|^{N+1}} \leq \frac{LM}{2\pi\Delta} \left(\frac{k\delta}{\Delta|z|} \right)^{N+1}. \tag{III.92}$$

By choosing z sufficiently large, we see that, if all the zeros of $P_n(z)$ stay inside a compact bounded by Γ_∞^+ the error goes to zero geometrically fast. With N and uniformly in z for $|z| > B$. Therefore the set of holomorphic function $g^{(N)}(z)$ converges uniformly toward $g(z)$ for $|z|$ sufficiently large.

III.9. Convergence Considerations

We shall consider here, a simple class of functions belonging to $H(\infty)$, for which we shall prove the uniform convergence of the continued fraction of $g(z)$ everywhere outside of the convex hull of the singularities of $g(z)$. This class is the class of Stieltjes functions

$$g(z) = \int_a^b \frac{d\mu(x)}{z - x} \tag{III.93}$$

where $d\mu(x)$ is a positive measure.

In this case, the set of orthogonal polynomials associated to $g(z)$ being the set of polynomials orthogonal with respect to $d\mu$, all the zeros belong to (a,b). Therefore from (III.8), the continued fraction converges to $g(z)$ for $|z|$ sufficiently large.

To show our statement we need, the result that $g^{(N)}(z)$ is the resolvant of a real *symmetric* matrix $\hat{\mathbf{J}}_N$:

$$g^{(N)}(z) = \left(e_0, \left[z - \hat{\mathbf{J}}_N \right]^{-1} e_0 \right) \tag{III.94}$$

We already know from (III.53) that

$$g^{(N)}(z) = \left(e_0, (z - \mathbf{J}_N)^{-1} e_0 \right). \tag{III.95}$$

However $\mathbf{J}_N$ is not symmetric.

We introduce the diagonal matrix $L^{(N)}$ with coefficient

$$L_{ij}^{(N)} = \delta_{ij} R_1^{-1/2} R_2^{-1/2} ... R_j^{-1/2} \tag{III.96}$$

(recall that all R_i's are positive in this case, because the norms h_n are positive). One easily checks that

$$\hat{\mathbf{J}}_N = L^{(N)} \mathbf{J}_N L^{(N)-1} \qquad (III.97)$$

is real symmetric. Furthermore, because

$$L^{(N)} e_0 = e_0, \qquad (III.98)$$

one gets (III.94). It then follows that, using the spectral theorem,

$$g^{(N)}(z) = \sum_{k=0}^{k=N} \frac{\gamma_{N+1}^k}{z - x_{N+1}^{(k)}} \qquad (III.99)$$

where all γ_{N+1}^k are positive and the x_{N+1}^k are the eigenvalues of $\hat{\mathbf{J}}_N$ or $\mathbf{J}_N$, which are the zeros of the orthogonal polynomial $P_{N+1}(x)$. For any z outside the convex hull of the singularities of $g(z)$, that is outside the interval $[a, b]$, we have

$$|g^N(z)| \leq \frac{1}{\delta} \qquad (III.100)$$

where δ is the distance of z to $[a, b]$. This is so because

$$\sum_{k=0}^{N} \gamma_{N+1}^k = \lim_{z \to 0} z\, g^{(N)}(z) = \mu_0 = 1. \qquad (III.101)$$

The $g^{(N)}(z)$ being holomorphic function of z outside $[a, b]$, uniformly bounded, and converging to $g(z)$ for $|z|$ sufficiently large, will converge (normal families) uniformly to $g(z)$ everywhere outside $[a, b]$. Therefore the analytic continuation we were looking for $g(z)$, is correctly provided in this case by the continued fraction expansion of $g(z)$.

IV. ORTHOGONAL POLYNOMIALS ON JULIA SETS

IV.1. An Introductory Example

Let us consider the transformation

$$z_1 = T(z) \equiv z^2 - 2, \qquad z \in \mathbf{C}. \qquad (IV.1)$$

Iterating the transformation, we get

$$z_2 = T(z_1) = T^{(2)}(z) \equiv z^4 - 4z^2 + 2. \qquad (IV.2)$$

After n iterations, we get a monic polynomial of degree 2^n

$$z_n = T(z_{n-1}) = z_{n-1}^2 - 2 = T^{(N)}(z) = z^{2^n} + \dots \qquad (IV.3)$$

What can be said about $T^{(n)}(z)$? To clarify this question, let us make the change of variable

$$\begin{cases} z & = 2\cos\theta_0 \\ z_k & = 2\cos\theta_k, \qquad k = 1, 2, \dots \end{cases} \qquad (IV.4)$$

(IV.3) becomes

$$\theta_n = 2\theta_{n-1} \qquad\qquad (IV.5)$$

or

$$\theta_n = 2^n\theta_0. \qquad\qquad (IV.6)$$

Going back to the z variable we get

$$z_n = T^{(n)}(z) = 2\cos 2^n\theta_0 = 2\, T_{2^n}(\cos\theta_0) = 2\, T_{2^n}\left(\frac{z}{2}\right) \qquad\qquad (IV.7)$$

where $T_k(z)$ is the classical Chebyshev polynomial of degree k. If we introduce the *monic* Chebyshev polynomials $\bar{T}_k(z)$ defined on $[-2, +2]$, we see that

$$T^{(n)}(z) \equiv \bar{T}_{2^n}(z). \qquad\qquad (IV.8)$$

For this simple case, we notice that the iterates of $z^2 - 2$, form a subsequence of orthogonal polynomials (in this case the classical Tchebycheff polynomials). Will this property extend to any iterate of any given polynomial? And if this is the case, how can one construct the missing ones? We shall see that the answer to the first question is yes. There is not an unique answer to the second question, however we shall provide an explicit construction scheme for a class of useful interpolations.

IV.2. Functional Equations

In order to study the properties of the iterates of a given polynomial $T_d(z)$ of degree d (to which is associated a Julia set, see section I.2), we remark that, because $T_d(z)$ behaves for z large like to z^d, it is natural to introduce a change of coordinates

$$u = B(z) \qquad\qquad (IV.9)$$

such that, in the u–complex plane, the transformation $T(z)$ looks exactly like u^d. We have described this process in (II.4) by introducing the Böttcher function $B(z)$. $B(z)$ is the unique function, analytic at infinity, which solves the functional equation

$$B\left(T_d(z)\right) = [B(z)]^d \qquad\qquad (IV.10)$$

and behaves like z, when z goes to infinity. This function is in fact holomorphic in $A(\infty)$, the basin of attraction of the point at infinity of $T_d(z)$.

Associated to the Böttcher function, we have introduced, in the electrostatic analogy, the Green function

$$G(z) = \ell n\, B(z) \qquad\qquad (IV.11)$$

which satisfies the functional equation

$$\frac{1}{d}G\left(T_d(z)\right) = G(z). \qquad\qquad (IV.12)$$

(IV.11) is not very convenient to deal with, because it has a logarithmic branch point at infinity.

However, its derivative (see II.36) the moment generating function, is holomorphic in $A(\infty)$ and satisfies

$$g(z) = \frac{1}{d}T_d'(z)g\left(T_d(z)\right) \qquad (IV.13)$$

(see (II.38)) with the following convergent asymptotic expansion

$$g(z) = \frac{\mu_0}{z} + \frac{\mu_1}{z^2} + ... + \frac{\mu_k}{z^{k+1}} + ... \qquad \mu_0 = 1. \qquad (IV.14)$$

We will slightly generalize equation (IV.13) which defines the moment generating functions of the balanced measure (see section II.3). Recall equation (II.3)

$$\begin{cases} \int f(x)d\mu^T & = \displaystyle\sum_{i=1}^{d} \int \gamma_i(x)f\left[T_i^{-1}(x)\right] d\mu \\ \displaystyle\sum_{i=1}^{d} \gamma_i(x) & = 1 \end{cases} \qquad (IV.15)$$

which defines μ^T to be the transformed measure of μ, associated with the polynomial $T(x)$ of degree d and the place dependent weights $\gamma_i(x)$,, and equation (II.22)

$$\mu^T\left(T^{-1}(E)\right) = \mu(E). \qquad (IV.16)$$

which shows that the support of μ^T is the union of the preimages of the support of μ.

A convenient way to deal with (IV.15) is to choose for $f(x)$ the Cauchy kernel $\frac{1}{z-x}$ and introduce

$$\begin{cases} g(z) & = \displaystyle\int \frac{d\mu}{z-x} \\ g^T(z) & = \displaystyle\int \frac{d\mu^T}{z-x}. \end{cases} \qquad (IV.17)$$

Then (IV.15) becomes

$$g^T(z) = \int \sum_{i=1}^{d} \frac{\gamma_i(x)}{z - T_i^{-1}(x)} d\mu = \int \frac{W(z,x)}{T(z) - x} d\mu \qquad (IV.18)$$

where $W(z,x)$ is a polynomial in z of degree $(d-1)$ with highest coefficient to z^{d-1}. To get this result, we have used the identity

$$T(z) - x \equiv t_0 \prod_{i=1}^{d} \left(z - T_i^{-1}(x)\right). \qquad (IV.19)$$

We shall suppose $W(z,x)$ to be independent of x, this defines a special family of function weighted polynomial transformations on a measure. An even more general case, would be the case of $W(z,x)$ being a polynomial in x, for which straightforward generalizations could be done. We shall restrict ourselves here, to this simplest case[14]

$$W(z,x) \equiv W(z). \qquad (IV.20)$$

Then, defined τ as the transformation

$$g^T(z) = W(z)g(T(z)) = \tau g(z). \qquad (IV.21)$$

This corresponds to the following choice for the place dependent weights γ_i

$$\gamma_i(x) = \frac{W\left(T_i^{-1}(x)\right)}{T'\left(T_i^{-1}(x)\right)}. \qquad (IV.22)$$

We may ask, what should $W(z, x)$ be for the equal weight case

$$\gamma_i(x) = \frac{1}{d}. \qquad (IV.23)$$

In this case we have

$$\frac{W(z, x)}{T(z) - x} = \frac{1}{d}\sum_{i=1}^{d}\frac{1}{z - T_i^{-1}(x)} \qquad (IV.24)$$

and

$$W(z, x) \equiv W(z) = \frac{1}{d}T'(z). \qquad (IV.25)$$

(IV.25) can be obtained starting from the identity

$$\ell n[T(z) - x] = \sum_{i=1}^{d}\ell n\left(z - T_i^{-1}(x)\right) + \ell n\ t_0. \qquad (IV.26)$$

Differentiating with respect to z, we get

$$\frac{T'(z)}{T(z) - x} = \sum_{i=1}^{d}\frac{1}{z - T_i^{-1}(x)}. \qquad (IV.27)$$

Therefore, the important case (see II.38), of the balanced invariant measure, or electrostatic measure, for which

$$g^T(z) \equiv g(z) = \frac{1}{d}T'(z)g(T(z)) \qquad (IV.28)$$

is included in our generalization. (This case has already been analyzed in the first section). We shall *therefore* be interested in solving functional equations of the form

$$g(z) = W(z)g(T(z)) \qquad (IV.28')$$

where W and T are given polynomials of degree $d - 1$ and d respectively, with the same leading coefficient and $g(z)$ is a member of $H(\infty)$.

Similar functional equation appear in many areas of mathematics such as the theory of iteration of polynomials[17], the analysis of the geometrical properties of invariants sets[8], the orthogonality properties of iterated polynomials[4][7], and in theoretical physics in the study of the almost periodic discrete Schrödinger equation[10][12], in the renormalization group approach of critical behaviour in some statistics mechanic models[23][24][25] and in the vibration spectrum of fractal structures[26].

IV.3. Constructive Iteration Scheme

To construct the solution of (IV.28') which is in $H(\infty)$. We look for fixed points of the transformation

$$(\tau g)(z) = W(z)g(T(z)). \tag{IV.29}$$

We shall use $T(z)$ for $T_d(z)$ in the sequel. We have the following result.

Theorem 3. (IV.29) has a unique fixed point $g_\infty(z)$ in $H(\infty)$ satisfying

$$(\tau g_\infty)(z) = W(z)g_\infty(T(z)). \tag{IV.30}$$

Furthermore, the iteration scheme,

$$g_{n+1}(z) = \tau g_n(z) = W(z)g_n(T(z)) \tag{IV.31}$$

starting from any $g_0 \in H(\infty)$, satisfies

$$\mu_k^{(n)} = \mu_k^{(\infty)} \quad \text{for} \quad k < d^n \tag{IV.32}$$

where

$$\begin{cases} g_n(z) &= \displaystyle\sum_{k=0}^{\infty} \mu_k^{(n)} z^{-(k+1)} \\ g_\infty(z) &= \displaystyle\sum_{k=0}^{\infty} \mu_k^{(\infty)} z^{-(k+1)}. \end{cases} \tag{IV.33}$$

In other words,

$$g_\infty(z) - g_n(z) = O\left(z^{-(1+d^n)}\right) \tag{IV.34}$$

when $z \to \infty$.

We shall call $T^{(k)}(z)$ the kth iterate of $T(z)$.

Proof: Iterating (IV.31), we get

$$\begin{aligned} g_n(z) &= W(z)W(T(z))W\left(T^{(2)}(z)\right)\cdots W\left(T^{(n-1)}(z)\right) g_0\left(T^{(n)}(z)\right) \\ &= R_n(z)T^{(n)}(z)g_0\left(T^{(n)}(z)\right) \end{aligned} \tag{IV.35}$$

where we have defined $R_n(z)$ to be the rational fraction of degree $d^n - 1/d^n$ (notice that it is *not* in general a Padé approximations to $g_\infty(z)$)

$$R_n(z) \equiv \frac{W(z)W(T(z))...W\left(T^{(n-1)}(z)\right)}{T^{(n)}(z)}. \tag{IV.36}$$

Using the fact that, in leading order,

$$\begin{cases} T^{(k)}(z) &\sim t_0^{\frac{d^k-1}{d-1}} z^{d^k}, & z \to \infty \\ W\left(T^{(k)}(z)\right) &\sim t_0^{d^k-1} z^{d^{k+1}-d^k}, & z \to \infty. \end{cases} \tag{IV.37}$$

We check that $R_n(z) \in H(\infty)$ that is

$$R_n(z) \sim \frac{1}{z}. \tag{IV.38}$$

On the other hand,

$$T^{(n)}(z)g_0\left(T^{(n)}(z)\right) = 1 + O\left(\frac{1}{T^{(n)}(z)}\right) = 1 + O\left(z^{-d^n}\right). \tag{IV.39}$$

Therefore

$$g_n(z) = R_n(z)\left[1 + O\left(z^{-d^n}\right)\right] = R_n(z) + O\left(z^{-[1+d^n]}\right). \tag{IV.40}$$

We remark that the rational fractions $R_n(z)$ does *not* depend on $g_0(z)$, it is explicitly constructed out of $W(z)$ and $T(z)$ only by (IV.36). (It is the $[d^n - 1/d^n]_{g_n}(z)$ Padé approximations to $g_n(z)$ when $\mu_1^0 = 0$, as we shall see later on).

Therefore the $\mu_k^{(n)}$ are independent of $g_0(z)$ for $k < d^n$, (or $k < 2d^n$, if g_0 is chosen in the restricted class $\mu_1^0 = 0$). If we choose $g_0(z)$ to be $g_\infty(z)$ itself, then, $g_n(z)$ will be $g_\infty(z)$ and

$$\mu_k^{(n)} = \mu_k^{(\infty)} \quad \text{for} \quad k < d^n. \tag{IV.41}$$

[We remark that we can always suppose, without loss of generality that $\mu_1^\infty = 0$, by redefining a new "centered measure".]

We can therefore compute unambiguously the $\mu_k^{(\infty)}$ up to $k < d^n$ by using the fact that from (IV.41)

$$g_\infty(z) - g_n(z) = O\left(z^{-[1+d^n]}\right) \tag{IV.42}$$

and from (IV.40)

$$g_n(z) - R_n(z) = O\left(z^{-[1+d^n]}\right) \tag{IV.43}$$

which gives

$$g_\infty(z) - R_n(z) = O\left(z^{-[1+d^n]}\right). \tag{IV.44}$$

To end the proof one should prove that the series generated in this way converges in a neighbourhood of ∞. In fact we have the following

Theorem 4. The Laurent series, obtained in Theorem 3 represents an holomorphic function in $A(\infty)$.

We recall that $A(\infty)$ is the basin of attraction of the point at infinity, that is, the open connected set containing ∞, made of the set of points such that $|T^{(n)}(z)|$ goes to infinity with n.

Proof: We first prove that $g_\infty(z)$ is analytic around ∞. One proves that there exists R such that for $|z| > R$

$$2|z|^d > |T(z)| > \frac{1}{2}|z|^d > R. \tag{IV.45}$$

We suppose $t_0 = 1$, here, to avoid cumbersome coefficients. It then results that for $|z| > R$

$$2^{\frac{d^j-1}{d-1}} > \left|T^{(j)}(z)\right| |z|^{-d^j} > 2^{-\frac{d^j-1}{d-1}}. \tag{IV.46}$$

Let a be an arbitrary point, not in $A(\infty)$, for instance, a fixed point of $T(z)$, i.e. $(T(a) = a)$, therefore $a \notin A(\infty)$, and let $g_0(z) = (z - a)^{-1}$. We want to study the convergence of the sequence $g_e(z) = \tau^e g_0(z)$ and we consider the infinite product

$$\pi(z) = g_0(z) \prod_{j=0}^{\infty} \frac{g_{j+1}(z)}{g_j(z)} = g_0(z) \prod_{j=0}^{\infty} \rho(u_j) \qquad (IV.47)$$

where $\rho(u) = \frac{(u-a)W(u)}{T(u)-a}$ and $u_j = T^{(j)}(z)$. (Using IV.35). Since $a \notin A(\infty)$, $\rho(u_j(z))$ has no pole for $z \in A(\infty)$. $[T^{j+1}(z) = a \Longrightarrow T^{j+1+k}(z) = a \ \forall k$, if we choose a to be a fixed point.]

By choosing R sufficiently large, we can also avoid any zero in $W(T^j(z))$. Therefore for $|u| > R$, $\exists\, C$ such that

$$|\ell n|\rho(u)|| < C|u|^{-1}. \qquad (IV.48)$$

Joining these estimates, we find that the infinite product is, for $|z|$ sufficiently large, uniformly convergent to an analytic function whose expansion is just the previous formal expansion we have obtained.

To end the proof we need to show that $g_\infty(z)$ which is holomorphic for $|z|$ sufficiently large, is in fact analytic everywhere in $A(\infty)$.

Given any $z \in A(\infty)$, $\exists N$ such that

$$\left|T^{(N)}(z)\right| > R \qquad (IV.49)$$

we now use the functional equation

$$g_\infty(z) = W(z)W(T(z))...W\left(T^{(N-1)}(z)\right) g_\infty\left(T^{(N)}(z)\right). \qquad (IV.50)$$

(IV.50) allows to continue analytically the analytic function $g_\infty(z)$ for $|z| > R$ everywhere inside $A(\infty)$.

Remark. The boundary of $A(\infty)$ is the Julia set of the polynomial $T(z)$[17]. J can be connected, or made of an infinite number of connected parts, or totally disconnected. Therefore $A(\infty)$ is not always simply connected. All rational functions $g_n(z)$ (corresponding to the special choice $g_0(z) = [z - a]^{-1}$) have their poles outside $A(\infty)$. a is fixed point of $T(z)$.

These rational functions are uniformly convergent rational approximations in any compact set included in $A(\infty)$, to the function $g_\infty(z)$. We shall see soon, that if we choose $a = \mu_1$ (the first moment), then $g_n(z)$ is the $[d^n - 1/d^n]$ Padé approximation to $g_\infty(z)$.

We end this section, by raising the following question: is $g_\infty(z)$ the generating function of the moments of an invariant measure μ_∞? We have seen that $g_k(z) = \tau^k g_0(z)$ has poles at $T^{(k)}(z) = a$, provided we choose $g_0(z) = (z - a)^{-1}$. If a is on J, the Julia set, the roots of this equation are also on J, because they are preimages of a point of J, J is totally invariant under $T(z)$ (that is backward and forward). In the same way, if a is strictly inside J (when J has interior point), all the preimages of such a point remain inside J. Furthermore, by choosing the complex number a outside the orbit of the iterates of the critical point of T (when they belong to J), all poles of $g_k(z)$ are simple (because a multiple zero of $T^{(k)}(z) - a = 0$, implies that $\left[T^{(k)}(z)\right]' = T'\left(T^{k-1}(z)\right) T'\left(T^{k-1}(z)\right)...T'(z) = 0$ and therefore z has to be the critical point or one of its preimages). Therefore $g_k(z)$ is the generating function of the moments of a discrete complex measure μ_k with d^k mass points on the preimages of a. We have to discuss existence and uniqueness of these limiting measures.

We list here some results[13] :

(i) When $W(z) = \frac{1}{d}T'(z)$, $g_\infty(z)$ is the generating function of the moments of the balanced measure which is unique and positive. This result is due to Brolin[2].

(ii) When $W(z)/T'(z)$ is real and positive on J, then one sees easily that the measures μ_k are probability measures (positive and normalized on J). The set of probability measures on the bounded closed set J being compact, limiting measures exist.

(iii) If the complement of $A(\infty)$ has an empty interior, a consequence of Mergelyan's theorem[27] is that if limiting measures exist, they all coincide with an invariant measure on J.

(iv) When the Julia set is real, and when $W(z)/T'(z)$ is positive on J, a consequence of (ii) and (iii) is that $g_\infty(z)$ is the generating function of an invariant measure on J.

Remarks. No general result is known yet. The main difficulty comes from the lack of positivity of the discrete measures μ_k. When the measure μ_∞ associated to $g_\infty(z)$ exists, the polynomials $g_\infty(z)$ are orthogonal with respect to μ_∞ in the sense of the scalar product $\langle f, g \rangle = \int_J f(z)g(z)d\mu_\infty$ which is *not* hermitian unless T, W, and the Julia set J are real.

IV.4. Construction of the Orthogonal Polynomials Associated with a Julia Set

We come back to the transformations

$$(\tau g)(z) = W(z)g(T(z)) \qquad (IV.51)$$

for which we have the following

Theorem 5. Provided they exist, the Padé approximants to $g(z)$ and $\tau g(z)$ are related by

$$[dm - 1/dm]_{\tau g}(z) = W(z)[m - 1/m]_g(T(z)) \qquad (IV.52)$$

and in particular for the fixed point g_∞, we get

$$[dm - 1/dm]_{g_\infty}(z) = W(z)[m - 1/m]_{g_\infty}(T(z)). \qquad (IV.53)$$

Proof: Starting from the definition

$$g(z) - [m - 1/m]_g[z] = O\left(z^{-(2m+1)}\right). \qquad (IV.54)$$

We substitute $T(z)$ for z in (IV.54) and get, after multiplication of both sides by $W(z)$

$$(\tau g)(z) - W(z)[m - 1/m]_g(T(z)) = W(z)O\left[(T(z))^{-(2m+1)}\right] = O\left(z^{-(2md+1)}\right). \quad (IV.55)$$

but $W(z)[m-1/m]_g(T(z))$ is a rational fraction of degree $dm-1/dm$, whose expansion coincides with the one of $(\tau g)(z)$ up to degree z^{-2md}, by uniqueness of the Padé approximation, it produces the desired result.

As a consequence of (IV.52) we get, iterating from, $m = d$ to $m = d^{n-1}$, first

$$[d - 1/d]_{g_\infty}(z) = W(z)[0/1]_{g_\infty}(T(z)) = \frac{W(z)}{T(z) - \mu_1^\infty}. \qquad (IV.56)$$

Noting that

$$[0/1]_{g_\infty}(z) = \frac{1}{z - \mu_1^\infty} \qquad \left(\mu_1^\infty = \frac{w_1 - t_1}{t_0}\right) \qquad (IV.57)$$

then

$$\left[d^2 - 1/d^2\right]_{g_\infty}(z) = W(z)W(T(z)) \Big/ T^{(2)}(z) - \mu_1^\infty \qquad (IV.58)$$

and in general

$$\left[d^n - 1/d^n\right]_{g_\infty}(z) = W(z)W(T(z))\cdots W\left(T^{(n-1)}(z)\right) \Big/ T^{(n)}(z) - \mu_1^\infty. \qquad (IV.59)$$

From this result, using Theorem 1 at the end of section III.7, we see that, choosing $w_1 = t_1$, the iterates $T^n(z)$ of any polynomials are subsequence of orthogonal polynomials.

If we now write the $[m - 1/m]_g(z)$ Padé approximant as

$$[m - 1/m]_g(z) = \frac{\rho_m^g(z)}{P_m^g(z)} \qquad (IV.60)$$

where $P_m^g(z)$ is the monic orthogonal polynomial of degree m associated to $g(z)$ (see section III.1), and $\rho_m^g(z)$ is a monic polynomials of degree $m - 1$, we have, as a consequence of (IV.52)

$$\begin{cases} P_{dm}^{\tau g}(z) &= t_0^{-m} P_m^g(T(z)) \\ \rho_{dm}^{\tau g}(z) &= t_0^{-m} \rho_m^g(T(z))W(z), \end{cases} \qquad (IV.61)$$

and for $g^\infty(z)$

$$\begin{cases} P_{dm}^{g_\infty}(z) &= t_0^{-m} P_m^{g_\infty}(T(z)) \\ \rho_{dm}^{g_\infty}(z) &= t_0^{-m} W(z) P_m^{g_\infty}(T(z)). \end{cases} \qquad (IV.62)$$

These results allow us to construct a recursive argument to compute the continued fraction expansion of $g_\infty(z)$. We know (see section III.8) that the numerator and denominator polynomials $\rho_m^{g_\infty}(z)$ and $P_m^{g_\infty}(z)$ fulfil the same three terms recursive relation, which for $P_m^{g_\infty}(z)$ reads

$$P_{m+1}^{g_\infty}(z) = (z - A_m^{g_\infty}) P_m^{g_\infty}(z) - R_m^{g_\infty} P_{m-1}^{g_\infty}(z). \qquad (IV.63)$$

To this three term recursive relations is associated the continued fraction expansion of $g^\infty(z)$

$$g^\infty(z) = 1 / (z - A_0^{g_\infty} - R_1^{g_\infty} / (z - A_1^{g_\infty} - R_2^{g_\infty} / \cdots / (z - A_n^{g_\infty} - R_n^{g_\infty} / (z - A_{n+1}^{g_\infty} \cdots) \cdots \qquad (IV.64)$$

and also the semi-infinite Jacobi matrix $\mathbf{J}$ (see section III.4), defined by its non vanishing elements

$$\mathbf{J}_{ii}^{g_\infty} = A_i^{g_\infty} \;;\; \mathbf{J}_{i+1,i}^{g_\infty} = R_{i+1}^{g_\infty} \text{ and } \mathbf{J}_{i,i+1}^{g_\infty} = 1 \quad i = 0,1,2... \qquad (IV.65)$$

It is also convenient to introduce a decimation matrix D_d, whose unique non zero element are

$$D_{d;ij} = t_0^i \text{ if } j = di, \quad 0 \text{ otherwise } i = 0,1,2,... \qquad (IV.66)$$

This matrix D_d is called a decimation matrix because acting on a column vector $\psi = [\psi_0, \psi_1, ..., \psi_k, ...]^T$ it kills most of the components

$$D_d \psi = \left[\psi_0, t_0 \psi_d, t_0^2 \psi_{2d}, ...\right]^T. \qquad (IV.67)$$

Furthermore, if we choose $t_0 = 1$, then $T(z)$ is a monic polynomial of degree d, and one checks readily that $D_d D_d^T = \mathbf{I}$, while $D_d^T D_d = \mathbf{P}_d$, where $\mathbf{P}_d$ is the projector on the infinite dimensional subspace generated by the set $\{e_0, e_d, e_{2d}, ...\}$. It is clear that, by a simple scaling can *always* reduce to this case. The reason why we want to explicitly keep a t_0, is because, in certain analysis, it appears to be convenient to let it go to infinity (Iterated Function Systems).

With this in mind we have the following

Theorem 6.

$$\mathbf{J}^g D_d = D_d T (\mathbf{J}^{\tau g}) \qquad (IV.68)$$

and for g_∞

$$\mathbf{J}^{g_\infty} D_d = D_d T (\mathbf{J}^{g_\infty}). \qquad (IV.69)$$

Proof: First we check that relation (IV.68) holds when applied to the column vector $\psi^{\tau g}(z) = [P_0^{\tau g}(z), P_1^{\tau g}(z), ...]^T$. [$P_e^g(z)$ will the monic orthogonal associated to a generic $g(z) \in H(\infty)$.]

Due to the particular structure of $\mathbf{J}^g$ (tridiagonal) and D_d, all matrix products involved contain only sums with a *finite* number of terms, therefore no specification on the growth of the components of a vector are needed here.

We have

$$\mathbf{J}^g \psi^g(z) = z \ \psi^g(z) \qquad (IV.70)$$

which expresses the three term recursive relation for $P_n^g(z)$ is compact form (see section III.4).

Introducing the matrix M, defined by

$$M \equiv \mathbf{J}^g D_d - D_d T (\mathbf{J}^{\tau g}), \qquad (IV.71)$$

we have

$$M \ \psi^{\tau g}(z) = \mathbf{J}^g D_d \psi^{\tau g}(z) - D_d T (\mathbf{J}^{\tau g}) \psi^{\tau g}(z). \qquad (IV.72)$$

From

$$\mathbf{J}^{\tau g} \psi^{\tau g}(z) = z \ \psi^{\tau g}(z) \qquad (IV.73)$$

we get

$$T (\mathbf{J}^{\tau g}) \psi^{\tau g}(z) = T(z) \psi^{\tau g}(z). \qquad (IV.74)$$

Furthermore,

$$D_d \psi^{\tau g}(z) = \psi^g(T(z)). \qquad (IV.75)$$

To prove this last result, we write

$$(e_i, D_d \psi^{\tau g}(z)) = \sum_j (e_i, D_d, e_j)(e_j, \psi^{\tau g}(z))$$
$$= t_0^i (e_{di}, \psi^{\tau g}(z)) = t_0^i P_{di}^{\tau g}(z) = P_i^g(T(z)) = (e_i, \psi^g(T(z))). \qquad (IV.76)$$

Combining (IV.74), (IV.75) with (IV.72), we find that

$$M \ \psi^{\tau g}(z) = 0. \qquad (IV.77)$$

Therefore
$$(e_i, M \; \psi^{\tau g}(z)) = 0 \quad \forall z \quad i = 0, 1, \dots \qquad (IV.78)$$

This is an infinite set of polynomials equations, because in each line of M appears only a *finite* number of matrix elements M_{ij}

$$\sum_{j=di-d}^{j=di+d} M_{ij} P_j^{\tau g}(z) = 0 \quad i = 0, 1, \dots \qquad (IV.79)$$

each $P_j^{\tau g}(z)$ being monic and of degree j those equations imply $M \equiv 0$.

This theorem allows to compute recursively the coefficients of the continued fraction expansion of (τg) knowing those of g. We have

$$
\begin{aligned}
(J^g D_d)_{kj} &= J^g_{k,k-1} t_0^{k-1} \delta_{dk-d,j} + J^g_{kk} t_0^k \delta_{dk,j} + J^g_{k,k+1} t_0^{k+1} \delta_{dk+d,j} \qquad (IV.80) \\
&= R_k^g t_0^{k-1} \delta_{dk-d,j} + A_k^g t_0^k \delta_{dk,j} + t_0^{k+1} \delta_{dk+d,j}
\end{aligned}
$$

and

$$[D_d T(J^{\tau g})]_{Kj} = t_0^K [T(J^{\tau g})]_{dK,j} \qquad (IV.81)$$

which is $\neq 0$ only if $j = dk - d, \; dk - d + 1, \dots, dk, \dots, dk + d - 1, dk + d$.

In particular, from

$$(J^d)_{i,i-d} = J_{i,i-1} J_{i-1,i-2} \dots J_{i-(d-1),i-d} = R_i R_{i-1} \dots R_{i-(d-1)}. \qquad (IV.82)$$

We have for $n \geq 1$

$$[D_d (T(J^{\tau g}))]_{n,(n-1)d} = [J^g D_d]_{n,(n-1)d} = t_0^{n-1} R_n^g, \qquad (IV.83)$$

$$t_0^n [T(J^{\tau g})]_{dn,dn-d} = t_0^n \left[t_0 (J^{\tau g})^d \right]_{dn,dn-d} = t_0^{n+1} R_{dn}^{\tau g} R_{dn-1}^{\tau g} \dots R_{dn-d+1}^{\tau g}. \qquad (IV.84)$$

and therefore for $n \geq 1$

$$t_0^{-2} R_n^g = R_{dn}^{\tau g} R_{dn-1}^{\tau g} \dots R_{dn-d+1}^{\tau g} \qquad (IV.85)$$

or

$$\begin{cases} t_0^{-2} R_1^g &= R_d^{\tau g} R_{d-1}^{\tau g} \dots R_1^{\tau g} \\ t_0^{-2} R_2^g &= R_{2d}^{\tau g} R_{2d-1}^{\tau g} \dots R_{d+1}^{\tau g} \end{cases} \qquad (IV.86)$$

expresses $R_d^{\tau g}$ knowing $R_{d-1}^{\tau g} \dots R_1^{\tau g}$ and $R_{dn}^{\tau g}$ knowing the R's of lower order. From the matrix element $(n, (n-1)d + 2k)$ one gets $R_{dn+k}^{\tau g}$ for $k = 1$ to $d - 1$ and from the matrix element $(n, (n-1)d + 2k + 1)$ one gets $A_{dn+k}^{\tau g}$ for $k = 0, 1, \dots, d - 1$. For $n = 0$ one needs to know $A_0^{\tau g}, \dots, A_{p-1}^{\tau g}, R_2^{\tau g}, \dots, R_{p-1}^{\tau g}$ for $d = 2p$ even, and the knowledge of $A_0^{\tau g}, \dots, A_{p-1}^{\tau g}, R_1^{\tau g}, R_2^{\tau g}, \dots, R_p^{\tau g}$ for $d = 2p + 1$ odd to start the recursion. The $(d-1)$ missing initial conditions do not depend on g, but are given by the $(d-1)$ first coefficients of the continued fraction expansion of $W(z)/T(z) - \mu_1^g$ as

$$[d - 1/d]_{\tau g}(z) = W(z)/T(z - \mu_1^g). \qquad (IV.87)$$

Let us illustrate this by an example. Take $d = 2$; $T(z) = z^2 - \lambda$; $W(z) = z - a$. That is for g_∞ the functional equation ($\mu_1^\infty = -a$)

$$g_\infty(z) = (z - a)g_\infty\left(z^2 - \lambda\right). \tag{IV.88}$$

We have

$$[1/2]_{g_\infty}(z) = \frac{z - a}{z^2 - \lambda + a} = \frac{1}{z + a - \dfrac{\left(\lambda - a - a^2\right)}{z - a}}. \tag{IV.89}$$

Therefore the initial missing conditions are

$$A_0 = -a, \quad R_1 = \lambda - a - a^2, \quad A_1 = a. \tag{IV.90}$$

Relation (IV.85) applied to g_∞, gives

$$R_{2n}R_{2n-1} = R_n \quad n \geq 1. \tag{IV.91}$$

Here from

$$D\left(\mathbf{J}^2 - \lambda\mathbf{I}\right) = \mathbf{J}D \tag{IV.92}$$

where we have dropped all non necessary indices ($D \equiv D_2$, $\mathbf{J} \equiv \mathbf{J}^{g_\infty}$), we get

$$(\mathbf{J}D)_{kj} = R_k\delta_{2k-2,j} + A_k\delta_{2k,j} + \delta_{2k+2,j}. \tag{IV.93}$$

Therefore the structure of $(\mathbf{J}D_2)$ is shown

$$\mathbf{J}D = \begin{bmatrix} A_0 & 0 & 1 & 0 & 0 & 0 & 0 & 0 & 0 & \cdots & \cdots & \cdots & \cdots & \cdots \\ R_1 & 0 & A_1 & 0 & 1 & 0 & 0 & 0 & 0 & \cdots & \cdots & \cdots & \cdots & \cdots \\ 0 & 0 & R_2 & 0 & A_2 & 0 & 1 & 0 & 0 & \cdots & \cdots & \cdots & \cdots & \cdots \\ 0 & 0 & 0 & 0 & R_3 & 0 & A_3 & 0 & 1 & \cdots & \cdots & \cdots & \cdots & \cdots \\ \cdots & \cdots & \cdots & \cdots & \cdots & \cdots & \cdots & \cdots & \cdots & \cdots & \cdots & \cdots & \cdots & \cdots \\ 0 & 0 & \cdots & \cdots & \cdots & \cdots & \cdots & R_K & 0 & A_K & 0 & 1 & 0 & \cdots \\ \cdots & \cdots & \cdots & \cdots & \cdots & \cdots & \cdots & \cdots & \cdots & \cdots & \cdots & \cdots & \cdots & \cdots \end{bmatrix} \tag{IV.94}$$

while

$$\left[D\left(\mathbf{J}^2 - \lambda\mathbf{I}\right)\right]_{kj} = \left[\mathbf{J}^2\right]_{2k,j} - \lambda\delta_{2k,j} \tag{IV.95}$$

and using

$$\mathbf{J}_{ij} = R_i\delta_{i-1,j} + A_i\delta_{ij} + \delta_{i+1,j} \tag{IV.96}$$

$$\left[\mathbf{J}^2\right]_{2k,j} = \sum_\ell \mathbf{J}_{2k,\ell}\mathbf{J}_{\ell,j} = \mathbf{J}_{2k,2k-1}\mathbf{J}_{2k-1,j} + \mathbf{J}_{2k,2k}\mathbf{J}_{2k,j} + \mathbf{J}_{2k,2k+1}\mathbf{J}_{2k+1,j}$$

$$= R_{2k}\mathbf{J}_{2k-1,j} + A_{2k}\mathbf{J}_{2k,j} + \mathbf{J}_{2k+1,j} = R_{2k}\left[R_{2k-1}\delta_{2k-2,j} + A_{2k-1}\delta_{2k-1,j} + \delta_{2k,j}\right]$$

$$+ A_{2k}\left[R_{2k}\delta_{2k-1,j} + A_{2k}\delta_{2k,j} + \delta_{2k+1,j}\right] + \left[R_{2k+1}\delta_{2k,j} + A_{2k+1}\delta_{2k+1,j} + \delta_{2k+2,j}\right] \tag{IV.97}$$

and

$$\left[D\left(\mathbf{J}^2 - \lambda\mathbf{I}\right)\right]_{kj} = R_{2k}R_{2k-1}\delta_{2k-2,j} + R_{2k}\left[A_{2k} + A_{2k-1}\right]\delta_{2k-1,j}$$

$$+ \left[R_{2k} + A_{2k}^2 + R_{2k+1} - \lambda\right]\delta_{2k,j} + \left(A_{2k} + A_{2k+1}\right)\delta_{2k+1,j} + \delta_{2k+2,j}. \tag{IV.98}$$

Equating the nonzero coefficients in (IV.92) gives

$$\begin{cases} R_{2k}R_{2k-1} = R_k & k \geq 1 \\ R_{2k}[A_{2k} + A_{2k-1}] = 0 & k \geq 1 \\ R_{2k} + R_{2k+1} + A_{2k}^2 - \lambda = A_k & (R_0 = 0) \\ A_{2k} + A_{2k+1} = 0 \end{cases} \qquad (IV.99)$$

with the initial conditions

$$\begin{cases} A_0 & = -a \\ A_1 & = a \\ R_1 & = \lambda - a - a^2. \end{cases} \qquad (IV.100)$$

Therefore

$$\begin{cases} A_{2k+1} & = a \\ A_{2k} & = -a \end{cases} \qquad (IV.101)$$

and

$$\begin{cases} R_{2k}R_{2k-1} & = R_k \quad k \geq 1 \\ R_{2k} + R_{2k+1} & = \lambda + A_k - A_{2k}^2 = \lambda - a^2 - (-)^k a. \end{cases} \qquad (IV.102)$$

In particular,

$$\begin{cases} R_2 R_1 & = R_1 \longrightarrow R_2 = 1 \\ R_1 & = \lambda - a^2 - a \\ R_3 & = \lambda - a^2 + a - 1 \\ R_4 & = \dfrac{R_2}{R_3} = \dfrac{1}{\lambda - a^2 + a - 1}. \end{cases} \qquad (IV.103)$$

IV.5. Quasiperiodic Properties
IV.5.1. An example of Hilbert space operator associated with a Julia set

We shall restrict ourselves to the case of a second degree polynomial, which, without loss of generality can be chosen to be (see II.10)

$$T(z) = z^2 - \lambda.$$

Any polynomial transformation $z' = t_0 z^d + t_1 z^{d-1} + \dots + t_d$ can be recast by a linear transformation, to the canonical form $z' = z^d + 0 \, z^{d-1} + \bar{t}_2 z^{d-2} + \dots + \bar{t}_d$ that is monic and centered as was explained in section III.2. Such linear transformations transform the Jacobi matrices in a trivial way, by recentering the A_m and rescaling the R_m. Furthermore, we shall contemplate the Jacobi matrix associated to the balanced measure, for simplicity. For such a case, the Jacobi matrix reads

$$J = \begin{bmatrix} 0 & 1 & 0 & 0 & 0 & \dots & \dots \\ R_1 & 0 & 1 & 0 & 0 & \dots & \dots \\ 0 & R_2 & 0 & 1 & 0 & \dots & \dots \\ 0 & 0 & R_3 & 1 & 0 & \dots & \dots \\ \dots & \dots & \dots & \dots & \dots & \dots & \dots \end{bmatrix} \qquad (IV.104)$$

and from (IV.102), we have

$$\begin{cases} R_{2k}R_{2k-1} & = R_k \quad k \geq 1 \\ R_{2k} + R_{2k+1} & = \lambda \\ R_1 & = \lambda. \end{cases} \qquad (IV.105)$$

One can compute easily the first few $R's$ $(\lambda \neq 0)$

$$R_1 = \lambda; \ R_2 = 1; \ R_3 = \lambda - 1; \ R_4 = \frac{1}{\lambda - 1}; \ R_5 = \frac{\lambda^2 - \lambda - 1}{\lambda - 1};$$

$$R_6 = \frac{(\lambda - 1)^2}{\lambda^2 - \lambda - 1}; \ R_7 = \frac{\lambda^3 - 2\lambda^2 + \lambda - 1}{\lambda^2 - \lambda - 1};$$

$$R_8 = \frac{\lambda^2 - \lambda - 1}{(\lambda - 1)(\lambda^3 - 2\lambda^2 + \lambda - 1)}; \ R_9 = \frac{\lambda^5 - 3\lambda^4 + 3\lambda^3 - 3\lambda^2 + 2\lambda + 1}{(\lambda - 1)(\lambda^3 - 2\lambda^2 + \lambda - 1)} \quad \text{etc.}$$

Before analyzing the properties of the system (IV.105), let us review some properties of $\mathbf{J}$, which are applications to this special case of the general theory developed previously.

(i) $\mathbf{J}$ has the renormalization group transformation property

$$\mathbf{J}D = D\left[\mathbf{J}^2 - \lambda\right] \tag{IV.106}$$

(ii) The spectrum of $\mathbf{J}$ is invariant under both transformations

$$T(x) = x^2 - \lambda \qquad \text{(Forward invariance)} \tag{IV.107}$$

and its two inverses

$$T_{\pm}^{-1}(x) = \pm\sqrt{x + \lambda} \qquad \text{(Backward invariance).} \tag{IV.108}$$

(iii) The "integrated density of states" is the equilibrium measure $\mu(x)$.
(iv) The spectrum K_N of $\mathbf{J}^{(2^N - 1)}$ is made of the zeros of

$$T^{(N)}(x) \tag{IV.109}$$

(see theorem 2).
    That is they are of the form

$$x\left(\vec{\sigma}^{(N)}\right) = \sigma_0^{(N)}\sqrt{\lambda + \sigma_1^{(N)}\sqrt{\lambda + \sigma_2^{(N)}\sqrt{\lambda + \ldots + \sigma_{N-1}^{N}\sqrt{\lambda}}}} \tag{IV.110}$$

where

$$\vec{\sigma}^{(N)} = \left(\sigma_0^{(N)}, \sigma_1^{(N)}, \ldots, \sigma_{N-1}^{(N)}\right) \quad \text{and} \quad \sigma_i^{(N)} = \pm 1. \tag{IV.111}$$

(v) In the special case λ real and $\lambda > 2$, the spectrum K of $\mathbf{J}$ is a Cantor set of the real axis [compact and perfect] of Lebesgue measure zero and the "continued square root expansion" (IV.110) converges $N \longrightarrow \infty$. In particular,

$$K \subset [-\xi, +\xi] \qquad \xi = \frac{1 + \sqrt{1 + 4\lambda}}{2}. \tag{IV.112}$$

(vi) The representation (IV.109) and (IV.110) is a well adapted coding of K_N and the actions of T is expressed on the sequences of signs $\vec{\sigma}^{(N)}$ as the usual shift S

$$S\left(\sigma_0^{(N)}, \sigma_1^{(N)}, \ldots, \sigma_{N-1}^{(N)}\right) = \left(\sigma_1^{(N)}, \sigma_2^{(N)}, \ldots, \sigma_{N-1}^{(N)}\right) \tag{IV.113}$$

which shows that

$$T\left(x\left(\vec{\sigma}^{(N)}\right)\right) = x\left[S\left(\vec{\sigma}^{(N)}\right)\right] \qquad (IV.114)$$

and therefore

$$T^{(N)}\left(x\left(\vec{\sigma}^{N}\right)\right) \equiv 0 \qquad (IV.115)$$

as it should be. Introducing

$$S_{\pm}^{-1}\left(\sigma_0^{(N)}, \sigma_1^{(N)}, ..., \sigma_{N-1}^{(N)}\right) = \left(\pm 1, \sigma_0^{(N)}, ..., \sigma_{N-1}^{(N)}\right) \qquad (IV.116)$$

we see that

$$T_{\pm}^{-1}\left(x\left(\vec{\sigma}^{(N)}\right)\right) = x\left(S_{\pm}^{-1}\left(\vec{\sigma}^{(N)}\right)\right). \qquad (IV.117)$$

Using this excellent coding, one can identify the measure $d\mu_N(x)$ associated to $g_N(z)$ as the probability measure

$$\int f(x)d\mu_N(x) = \underbrace{\int\int...\int}_{N \text{ times}} \prod_{n=0}^{n=N-1} \left\{d\sigma_n^{(N)} \frac{1}{2}\left[\delta\left(\sigma_{n+1}^{(N)} - 1\right) + \delta\left(\sigma_{n-1}^{(N)} + 1\right)\right]\right\} \qquad (IV.118)$$

$$f\left(x\left(\sigma_0^{(N)}, \sigma_1^{(N)}, ..., \sigma_{N-1}^{(N)}\right)\right)$$

where

$$d\mu_N(x) = 2^{-N} \sum_{i=0}^{2^N-1} \delta\left(x - \xi_i^{(N)}\right). \qquad (IV.119)$$

Here $\xi_i^{(N)}$ are the roots of $T^N(x) = 0$. The left hand side

$$\equiv \frac{1}{2^N} \sum_{\sigma_k^{(N)} = \pm 1} f\left(x\left(\sigma_0^N, \sigma_1^{(N)}, ..., \sigma_{N-1}^{(N)}\right)\right), \qquad 0 \le k \le N-1.$$

For $\lambda > 2$ real, one can take all limits $N \longrightarrow \infty$ in all the preceding formula. In particular,

$$\int f(x)d\mu(x) = \int...\int \prod_{n=0}^{\infty} \left\{d\sigma_n \frac{1}{2}[\delta(\sigma_n - 1) + \delta(\sigma_n + 1)]\right\} f(x(\sigma_0, \sigma_1, ..., \sigma_n, ...)). \qquad (IV.120)$$

There is no atomic part in $d\mu$ and the action of T in the spectrum has the ergodic properties of a Bernoulli shift.

(vii) The ergodicity and mixing properties of the invariant measure are all consequence of the following theorem which is proved for an arbitrary polynomial of degree d, and a more general measure than the equilibrium measure.

Theorem. [Quick mixing] [28] Let Q be a polynomial of degree less than d^n then

$$\int_J f\left(T^{(n)}(x)\right) Q(x)d\mu = \int_J f(x)d\mu \cdot \int_J Q(x)d\mu \qquad (IV.122)$$

where $f \in L^1(J, \mu)$, and T is a polynomial of degree d, for which $d\mu$ is an invariant measure with the choice (IV.20) for the place dependent weights.

Proof: The set of polynomials P of the form $\prod_{j=0}^{j=n-1} \left[T^{(j)}(x)\right]^{m_j}$, $m_j \in \{0, 1, ..., d-1\}$ contains a polynomial of each degree $0, 1, ..., d^n - 1$. So Q lies in the span of P. It is therefore sufficient to assume $Q(x) \in P$. Using invariance (see IV.15). We have

$$\int_J Q(x) f\left(T^{(n)}(x)\right) d\mu = \sum_{i=1}^{i=d} \int_J \gamma_i(x) \left[T_i^{-1}(x)\right]^{m_0} \prod_{j=1}^{n-1} \left[T^{(j-1)}(x)\right]^{m_j} f\left(T^{(n-1)}(x)\right) d\mu \tag{IV.123}$$

when $m_0 \leq d - 1$, we have

$$\sum_{i=1}^{i=d} \gamma_i(x) \left[T_i^{-1}(x)\right]^{m_0} = \text{constant independent of } x = C_{m_0}. \tag{IV.124}$$

To prove this last result, we only need to expand in $1/z$ the identity

$$\sum_{i=1}^{d} \frac{\gamma_i(x)}{z - T_i^{-1}(x)} = \frac{W(z)}{T(z) - x} \tag{IV.125}$$

and notice that the dependence on x appears in the right hand side expansion only for a power of $(1/z)^{d+1}$, that is for $m_0 \geq d$. Continuing the procedure, we find that

$$\int_J Q(x) f\left(T^{(n)}(x)\right) d\mu = \prod_{j=0}^{n-1} C_{m_j} \int_J f(x) d\mu = \int_J Q(x) d\mu \int_J f(x) d\mu. \tag{IV.126}$$

Note that the product $\prod_{j=0}^{n-1} C_{m_j}$ is independent of $f(x)$. We shall now end this section, by studying the properties of the coefficients R_n appearing in the Jacobi-matrix **J**.

IV.5.2. Almost periodicity of the coefficient R_n

We shall now list a certain number of properties of the system (IV.105).

i) Simple solution

$$\text{For } \lambda = 2 \quad \text{we have} \quad R_n = 1, \quad n \geq 2. \tag{IV.127}$$

ii) The $R_n(\lambda)$ are rational fractions. The union of the set of their zeros is identical to the union of the set of their poles[11]. We shall call this common set P, $\bar{P}$ its closure and P' the derived set. It is conjectured that the set P' is identical to the Mandelbrot set M.

We shall now restrict ourselves a little more, and consider the case of a real Julia set, λ real > 2. For a general discussion with λ complex see reference [11]. In this case we have

iii)

$$0 < R_{2n} < R_n \quad \text{and} \quad 0 < R_{2n} \leq 1 \tag{IV.128}$$

The proof is by induction, see Ref.[5], [6].

iv) A consequence of (iii) is

$$\lambda - 1 \leq R_{2n+1} < \lambda. \tag{IV.129}$$

v)

$$\lim_{k \to +\infty} R_{p2^k + s} = R_s, \quad p, k, s \in \mathbf{N}. \tag{IV.130}$$

Let us prove (IV.130) in the case $s = 0$. From

$$R_{2n} = \frac{R_n}{\lambda - R_{2n-2}} \quad \text{and} \quad 0 < R_{2n-2} \leq 1 \qquad (IV.131)$$

we see that

$$\frac{R_n}{\lambda} < R_{2n} \leq \frac{R_n}{\lambda - 1} \qquad (IV.132)$$

and therefore

$$\frac{R_p}{\lambda^k} < R_{p2^k} \leq \frac{R_p}{(\lambda - 1)^k} \qquad (IV.133)$$

which proves that

$$\lim_{k \to \infty} R_{p2^k} = 0 = R_0 \text{ by convention.} \qquad (IV.134)$$

Then, to prove (IV.130) in general, one proceeds by complete induction.
vi) The structure of the set $R_n(\lambda)$, $\lambda > 2$, is made more transparent by the following reordering.
  The set is made of two disjoint intervals

$$[0,1] \cup [\lambda - 1, \lambda], \text{ where } R_{2n} \in [0,1] \text{ and } R_{2n+1} \in [\lambda - 1, \lambda] \qquad (IV.135)$$

which can be rewritten

$$R_0 < R_{2n} \leq R_2 < R_3 \leq R_{2n+1} \leq R_1 \qquad (IV.136)$$

Then each of these sets separates again into disjoint sets

$$R_0 < R_{4n} \leq R_4 < R_6 \leq R_{4n+2} \leq R_2 \qquad (IV.137)$$

and

$$R_3 \leq R_{4n+3} \leq R_7 < R_5 \leq R_{4n+1} \leq R_1. \qquad (IV.138)$$

It is a *conjecture** that this splitting at each step continues, so that R_{p2^k+s} lies between R_s and R_{2^k+s}. If it is so, the closure of the set $\{R_n\}$ is nowhere dense and this set is perfect.
vii) For $\lambda \geq 3$

$$\left| R_{p2^k+s} - R_s \right| \leq \frac{\lambda}{(\lambda - 2)^k}. \qquad (IV.139)$$

The proof is by induction on s. It is true by (IV.133) for $s = 0$.
  This shows that the sequence R_n is almost periodic[29] because the bound is uniform in s. Therefore R_n, can be expanded (in n) in Fourier like series

$$R_n = \sum_{q=0}^{\infty} \sum_{p=0}^{2^q-1} r_{p,q} \exp\left\{ \frac{2i\pi n(2p + 1)}{2^q} \right\}. \qquad (IV.140)$$

---

* At least for $\lambda > \lambda^* > 2$ (λ real).

All those properties can be extended to complex values of λ, large enough and for λ real and slightly bigger than $2^{[11]}$. These quasi-periodic properties also extend to more general classes of polynomial $T(x)$ of degree $d^{[22]}$.

Let us end this section by mentioning some properties of the orthonormalized polynomial $\bar{P}_n(x)$ associated to the transformation $T(x) = x^2 - \lambda$, $\bar{P}_n(x) = h_n^{-1/2} P_n(x)$. The $\bar{P}_n(x)$ are the components of the eigenstate of $\mathbf{J}$ corresponding to the eigenvalue x.

i) Outside the spectrum of $\mathbf{J}$, that is when $x \notin k$ the $P_n(x)$ increase exponentially with n, because $P_{p2^k}(x) = P_p\left(T^{(k)}(x)\right)$ and $T^{(k)}(x)$ goes to infinity as x^{2^k} when x is outside the Julia set (called here K instead of J). Using the fact that $h_{p2^k} = h_p$, we see that this result applies as well to $\bar{P}_n(x)$.

ii) When $x \in K$, we have

$$|\bar{P}_n(x)| < \frac{1}{2}\left(\frac{4\sqrt{\lambda}}{\sqrt{1+4\lambda}-1}\right)^k \quad \text{for } 2^{k-1} \leq n \leq 2^k. \qquad (IV.141)$$

Therefore we have an explicit *Polynomial* bound (see Ref.[12]).

iii) The large n behaviour of $\bar{P}_n(x)$ is described by the Lyapounov index[30]

$$\gamma(x) = \lim_{n \to \infty} \frac{1}{n}\ell n\left(\frac{\bar{P}_n^2(x) + \bar{P}_{n+1}^2(x)}{\bar{P}_1^2(x) + \bar{P}_0^2(x)}\right). \qquad (IV.142)$$

It is easy to prove that

$$2\gamma(x) = \gamma(T(x)) \qquad x \in K. \qquad (IV.143)$$

Therefore

$$\gamma(x) = \operatorname{Re} G(x). \qquad (IV.144)$$

That is $\gamma(x)$ is the real part of the logarithm of the Böttcher function and it vanishes on K, which shows that these eigenstates are extended.

iv) Using

$$\bar{P}_{p2^k}\left(x\left(\vec{\sigma}\right)\right) = \vec{\sigma}_p\left(x\left(S^k\left(\vec{\sigma}\right)\right)\right), \qquad (IV.145)$$

we see that, because for almost every $x\left(\vec{\sigma}\right) \in K$, the sequence $\vec{\sigma}$ is chaotic, this implies that the behaviour of $\vec{\sigma}_{p2^k}\left(x\left(\vec{\sigma}\right)\right)$ will be itself chaotic with probability one. Exceptional sequences may be periodic, but this occurs with probability zero.

Besides its solvability, this matrix $\mathbf{J}$ has the most interesting property of exhibiting chaotic behaviour for its eigenstates. For large n, the eigenstates which are all extended, retain no memory of the almost periodicity and fluctuate at random.

V. CONCLUSION AND OUTLOOK

We have introduced the reader to the field of orthogonal polynomials on complex contour and more precisely we have discussed the theory of orthogonal polynomials induced by the iteration process of replacing a polynomial into itself.

The simplest results produced where studies of orthogonal polynomials on Cantor sets (for the iteration of the polynomial $x^2 - \lambda$, λ real greater than 2). The Jacobi matrices associated to these orthogonal polynomials have quasi periodic coefficients. The deep study of (IV.105) which generates those quasi periodic coefficients may produce unexpected and remarkable results. Also all the properties of the Hilbert space operator J associated to a Julia set are far from being exhausted.

We would like to end this paper by discussing a very important question which appears in image processing. Given a set of linear maps

$$\begin{cases} w_1(x) & = a_1 x + b_1 \\ w_2(x) & = a_2 x + b_2 \\ \\ w_d(x) & = a_d x + b_d \end{cases} \qquad (V.1)$$

the place dependent weight associated to these linear maps are constant and equal to $p_1, p_2, ..., p_d$, with

$$\sum_{i=1}^{i=d} p_i = 1 \qquad p_i > 0. \qquad (V.2)$$

The invariant measure associated to these maps and weights (This ensemble of map and weight is called an I.F.S. or Iterated Function System[31][32]) must satisfys, following (IV.15)

$$\int f(x) \mathrm{d}\mu = \sum_{i=1}^{i=d} p_i \int f(w_i(x)) \, \mathrm{d}\mu. \qquad (V.3)$$

If we choose for $f(x)$, the set $\{x^0, x^1, x^2, ..., x^4, ...\}$. We see, that the moments of μ

$$\mu_n = \int x^n \mathrm{d}\mu \qquad (V.4)$$

satisfy

$$\mu_n = \sum_{i=1}^{i=d} p_i \int (a_i x + b_i)^n \, \mathrm{d}\mu \qquad (V.5)$$

or

$$\mu_n = \left(1 - \sum_{i=1}^{i=d} p_i a_i^n\right)^{-1} \sum_{i=1}^{i=d} \sum_{j=0}^{j=n-1} \binom{n}{j} a_i^j b_i^{n-j} \mu_j p_i. \qquad (V.6)$$

Thus (V.6) allows to compute recursively the μ_n from the knowledge of the a_i and b_i.

Most interesting is the inverse problem[33][34]. Given μ, for instance, a digitalized image, or a T.V. image, find the linear map w_i and weight p_i which will reproduce it within a given error ε. The problem is to invert the equations (V.6): given the measured moments $\mu_0, \mu_1, ..., \mu_{3d-2}$ find the $a_1, ..., a_d, b_1, ..., b_d, p_1, ..., p_{d-1}$. A first attempt to solve these non linear equations is given in[35]. See for a constructive and general analysis[36].

Here we just want to point out that one can replace linear IFS (V.1) by a more sophisticated one, in which the w_i are the inverse branches of a polynomial of degree d. The linear maps being only the first truncated Taylor series expansion of such a system. In such a case we must find a polynomial T_d which produces a given Julia set. The answer is given by Theorem 5 specialized to $m = 1$. The $[d-1/d]$ Padé approximation to $g(z)$, constructed out of the moments $\mu_0, \mu_1, ..., \mu_{2d-1}$, has for denominator $T_d(z) - \mu$. This extremely simple and powerful result, will allow us to study the reconstruction of two dimensional maps using Julia-Padé approximations.

Acknowledgments

The author thanks Professors S. Demko, P. Moussa and J. Geronimo for their help and careful reading of the manuscript.

REFERENCES

[1] G. JULIA, Mémoire sur l'itération des fonctions rationnelles, J. de Math. Pures et Appliquées Ser. 8 1, 47-245 (1918).
P. FATOU, Sur les équations fonctionnelles, Bull. Soc. Math. France 47, 161-271 (1919).

[2] H. BROLIN, Invariant sets under iteration of rational functions, Arkiv för Matematik, Vol.6, p.103-144, 1965 (Stockholm).

[3] G. SZEGŐ, Orthogonal polynomials, Amer. Math. Soc. Colloquium publication 23 (1939).

[4] D. BESSIS and P. MOUSSA, Orthogonality properties of iterated polynomials mappings, Communication in Mathematical Physics 88, 503-529 (1983).
P. MOUSSA, Iteration des polynomes et propriétés d'orthogonalité, Annales IHP 44, 315-325 (1986).
T.S. PITCHER and J.R. KINNEY, Ark. Math. 8, 25-32 (1968).
R.L. ADLER and T.J. RIVLIN, Proc. Am. Math. Soc. 15, 749-796 (1964).

[5] D. BESSIS, M.L. MEHTA, P. MOUSSA, C.R. Acad. Sci. Paris 93, Ser. 1, 705-708 (1981).

[6] D. BESSIS, M.L. MEHTA, P. MOUSSA, Letters, Math. Phys. 6, 123-140 (1982).

[7] M.F. BARNSLEY, J.S. GERONIMO, A.N. HARRINGTON, Bull. Amer. Math. Soc. 7, 381-384 (1982).

[8] M.F. BARNSLEY, J.S. GERONIMO, A.N. HARRINGTON, On the invariant sets of a family of quadratic maps, Comm. Math. Phys. 88, 479-501 (1983).

[9] S. WALL, Continued fractions, Chelsea Publishing Co., New-York (1973).

[10] J. BELLISSARD, D. BESSIS and P. MOUSSA, Phys. Rev. Lett. 49, 701-704 (1982).

[11] G. BAKER, D. BESSIS and P. MOUSSA, Physica 124A, 61-78 (1984).
M.F. BARNSLEY, J.S. GERONIMO, A.N. HARRINGTON, Almost periodic Jacobi matrices associated with Julia sets, Comm. Math. Phys. 99, 303-317 (1988).

[12] P. MOUSSA and D. BESSIS, A solvable almost periodic Schrödinger operator in "Stochastic aspect of classical and quantum systems" p. 136-147, Lecture Notes in Mathematics 1109, Springer-Verlag 1985.
M.F. BARNSLEY, J.S. GERONIMO, A.N. HARRINGTON, Geometrical and electrical properties of some Julia sets in Anthology edited by G. and D. Chudnovsky.

[13] P. MOUSSA, Application of rational approximations to some functional equations, in Rational Approximation and its applications in Mathematical and Physics Lecture Notes in Mathematics 1237 Springer Verlag (1986).

[14] D. BESSIS, J.S. GERONIMO and P. MOUSSA, Function weighted measures and orthogonal polynomials on Julia sets, Constructive Approximation 4, 157-173 (1988).

[15] B. MANDELBROT, The fractal geometry of nature, Freeman and Co New-York (1983).

[16] D. BESSIS and P. MOUSSA, Physical applications of Julia sets, Fizika 17, 3, 345-360 (1985).

[17] For a review including the classical references, see Complex analytical dynamics on the Riemann sphere, P. BLANCHARD, Bull. Amer. Math. Soc. 11, 85-141 (1984).

[18] B.B. MANDELBROT, Ann. N.Y. Acad. Sci. 357, 249-259 (1980), Physica 7D, 224-239 (1983).

[19] A. DOUADY and J.H. HUBBARD, C.R. Acad. Sci. Paris 294, 123-126 (1982).
A. DOUADY, Séminaire Bourbaki 599, Asterisque 105-106 (1983).
H.O. PEITGEN and D. SAUPE, The science of fractal images, Springer-Verlag 1988.

[20] M.Y. LJUBICH, Ergod. Th. and Dynam. Sys. 3, 351-385 (1983).

[21] E. HILLE, Analytic function theory, Blaisdell Publ. Walthan, Mass 1962 Chapter 16.

[22] J. HERNDON Thesis: Limit periodicity of sequences defined by certain recurrence relation and Julia sets. School of Mathematics Georgia Institute of Technology Atlanta Georgia 1985.

[23] B. DERRIDA, J.P. ECKMANN, A. ERZAN, Renormalization groups with periodic and aperiodic orbits, J. Phys. A: Math. Gen. 16, 895-906 (1983).

[24] Th. NIEMEYER, J.M. VAN LEEUWEN, Renormalization theory for Ising like spin systems, Phase transition and critical phenomena, Vol.6, 425-505, C. Domb adn M.S. Green editors, Academic Press N.Y. (1976).

[25] D. BESSIS, J. GERONIMO and P. MOUSSA, Mellin transforms associated with Julia sets and physical applications, J. Stat. Phys. 34, 75-110 (1984).

[26] R. RAMMAL, Spectrum of harmonic excitations on fractals, J. Physique 45, 191-206 (1984).

[27] See for instance, W. RUDIN, Real and complex analysis p.386, Mc Graw Hill (1970).

[28] M.F. BARNSLEY, J.S. GERONIMO, A.N. HARRINGTON, Geometry, electrostatic measure and orthogonal polynomials on Julia sets for polynomials. Ergod. Th. and Dynam. Sys. 3, 509-520 (1983).

[29] H. BOHR, Almost periodic functions, Chelsea New-York (1951).

[30] D.J. THOULESS, J. Phys. C 5, 77-81 (1972).

[31] M.F. BARNSLEY and S. DEMKO, Iterated function systems, Proceedings of the Royal Society of London A, Vol399, 1985 p.245-275.

[32] J. HUTCHINSON, Fractals and self-similarity, Indiana University Journal of Mathematics 30, 713-747 (1981).

[33] M.F. BARNSLEY, V. ERVIN, D.HARDIN and J. LANCASTER, Solution of an inverse problems for fractals and other set, Proceedings of the National Academy of Science 83 (1986) p.1975-1977.

[34] P. DIACONIS and M. SHAHSHAHANI, Products of random matrices and computer image generation, Contemporary mathematics 50 (1986) 173-182.

[35] J. ELTON and Z. YAN, Approximation of measures by Markov processes and homogeneous affine iterated function systems. Constructive Approximation Vol.5 # 1 (1989).

[36] D. BESSIS and S. DEMKO, A constructive method using the theory of moments for signal processing.

THE THREE TERM RECURRENCE RELATION AND SPECTRAL PROPERTIES OF ORTHOGONAL POLYNOMIALS

T. S. Chihara
Department of Mathematics
Purdue University Calumet
Hammond, IN 46323
USA

ABSTRACT.   This paper surveys the theory of the three term recurrence relation for orthogonal polynomials and its relation to the spectral properties of the polynomials.

0. INTRODUCTION.   Arguably, the most characteristic feature of orthogonal polynomials (other than orthogonality itself) is the fact that every sequence of orthogonal polynomials satisfies a three term recurrence relation of a very special form. Written for monic polynomials $P_n(x)$, this takes the form

$$(0.1) \qquad P_n(x) = (x - c_n)P_{n-1}(x) - \lambda_n P_{n-2}(x),$$
$$P_{-1}(x) = 0, \quad P_0(x) = 1, \quad c_n \text{ real and } \lambda_{n+1} > 0, \quad n \geq 1.$$

The existence of (0.1) is of course an abstract consequence of the orthogonality and the fact that the first n+1 polynomials form a basis for the algebra of polynomials of degree at most n.   That the converse also holds is a celebrated fact that usually elicits surprise and interest among those not previously conversant with the theory of orthogonal polynomials.   This theorem, which says that any sequence $\{P_n(x)\}$ generated by a recurrence of the form (0.1) is an orthogonal polynomial sequence, is usually known as Favard's theorem.
   It has often been pointed out that this result was already known in many different contexts (e.g. [42], [43], [46], [58]) before Favard's paper [27] in 1935.   Indeed, one can argue that a form of the theorem is implicit in Stieltjes' great work [45] on the continued fractions and moment problem that are named after him, while the general form of the theorem is a consequence of Hamburger's extension [30] of Stieltjes' work to the general J-fractions and the Hamburger moment problem. Nevertheless, Favard's announcement was a major contribution because the theorem was mostly known (implicitly) to workers in continued fraction theory who were only peripherally interested in orthogonal polyomials while it was generally overlooked by--or unknown to--mathematicians who were involved with orthogonal polynomials.   Consider the solitary casual reference to Favard's theorem in Szegö's bible [47].

P. Nevai (ed.), Orthogonal Polynomials, 99–114.
© 1990 by Kluwer Academic Publishers.

During the past three decades, and especially during the current one, Favard's theorem and its consequences have become increasingly the focus of attention of many researchers in orthogonal polynomials. On the one hand, specialists in special functions have found (0.1) a fruitful source for discovering new specific classes of OPS. On the other hand, there is a natural challenge to search for conditions on the the coefficients in (0.1) which predict that the corresonding OPS will have specified desirable properties. It is the results of the second approach we plan to survey in this essay. We will summarize the principal theorems of this type which predict the distribution of zeros and related properties of orthogonal polynomials.

1. PRELIMINARIES AND NOTATION. We will be concerned with orthogonal polnomials in the classical sense. That is, we consider sequences $\{P_n(x)\}$ where $P_n(x)$ is a polynomial of degree n and for which there exists a distribution function ψ such that

$$(1.1) \qquad \int_{-\infty}^{\infty} P_m(x) \, P_n(x) \, d\psi(x) = K_n \, \delta_{mn}, \qquad (K_n > 0).$$

Here, by distribution function we will mean a bounded, nondecreasing function whose moments

$$(1.2) \qquad \mu_n = \int_{-\infty}^{\infty} t^n \, d\psi(t) \, , \qquad n = 0, 1, 2, \ldots$$

are all finite, and whose spectrum (= support of $d\psi$)

$$(1.3) \qquad S(\psi) = \{ \, t : \psi(t+\epsilon) - \psi(t-\epsilon) > 0 \quad \text{for all} \quad \epsilon > 0 \, \}$$

is an infinite set.

As noted in the introduction, it is an abstract consequence of the facts of orthogonality and the $P_n(x)$ being polynomials that three consecutive polynomials satisfy a simple recurrence relation. Assuming, as we do, that the polynomials are all monic, this recurrence relation has the form (0.1). Favard's theorem then says that the converse is true: given any real c_n and positive λ_{n+1} $(n > 1)$, let $\{P_n(x)\}$ be defined by (0.1). Then the $P_n(x)$ will satisfy (1.1) with $K_n = \lambda_1 \cdots \lambda_{n+1}$, $\lambda_1 = \psi(\infty) - \psi(-\infty)$. Some natural questions then present themselves. If $\{c_n\}$ and $\{\lambda_{n+1}\}$ have certain properties (e.g., both converge), what can one say about the resulting properties of $P_n(x)$ and/or ψ ? What conditions are sufficient in order that, for example, ψ is absolutely continuous or that $S(\psi)$ is discrete?

Many of the characteristic features one is interested in involve the distribution of the zeros of the $P_n(x)$. It is of course a most basic fact that $P_n(x)$ has only real, simple zeros. Denoting them by x_{ni},

$$x_{n1} < x_{n2} < \dots < x_{nn},$$

we recall the classical separation theorem that tells us that the zeros of $P_n(x)$ and $P_{n+1}(x)$ are interlaced:

$$x_{n+1,i} < x_{ni} < x_{n+1,i+1}, \qquad\qquad i = 1, 2, \dots, n.$$

From this follows the existence of the following limits in the extended real number system:

(1.4) $\qquad \xi_i = \lim_{n \to \infty} x_{ni}, \qquad \eta_j = \lim_{n \to \infty} x_{n,n-j+1},$

and we also have

(1.5) $\qquad -\infty \le \xi_1 \le \xi_2 \le \dots \le \dots \le \eta_2 \le \eta_1 \le \infty.$

The interval $[\xi_1, \eta_1]$ is the smallest closed interval that contains all of the zeros of the orthogonal polynomials and is called, following Shohat, the "true" interval of orthogonality. There always exists at least one distribution function ψ whose spectrum is a subset of $[\xi_1, \eta_1]$ for which (1.1) holds and this "true" interval of orthogonality is also the smallest closed interval for which the latter property holds. Thus it is also frequently referred to as the "spectral interval."

A lower bound for ξ_1 can usually be determined easily with the aid of the concept of a chain sequence [56], [9], [14].

DEFINITION. $\{a_n\}_{n=1}^{\infty}$ is a <u>chain sequence</u> if there exists a sequence $\{g_n\}_{n=0}^{\infty}$ such that

(i) $\qquad 0 \le g_0 < 1 ; \quad 0 < g_n < 1$ for $n \ge 1$;

(ii) $\qquad a_n = (1 - g_{n-1})g_n, \quad n \ge 1.$

THEOREM 1.1. $\xi_1 \ge c$ if and only if $c_n > c$ $(n \ge 1)$ and $\{\lambda_{n+1}/(c_n c_{n+1})\}_{n=1}^{\infty}$ is a chain sequence.

Wall's comparison theorem [56], [14] says that any sequence that is dominated by a chain sequence is itself a chain sequence. Since the constant sequence $\{1/4\}$ is a chain sequence $(1/4 = (1 - 1/2)1/2)$, we see, for example, that if $c_n > 0$ and $4\lambda_{n+1} \le c_n c_{n+1}$ $(n \ge 1)$, then $\xi_1 \ge 0$.

Of great significance also are the limits

(1.6) $\qquad \sigma = \lim_{i \to \infty} \xi_i, \qquad \tau = \lim_{j \to \infty} \eta_j.$

When the Hamburger moment problem given by (2.2) is determined , (i.e. when the distribution function ψ satisfying (1.2) is uniquely determined up to an equivalence class), then σ and τ are respectively the smallest and largest limit points of the spectrum. If the moment problem is indeterminate, then either $\sigma = \tau = \infty$ (or, essentially equivalently, $\sigma = \tau = -\infty$) or else $\sigma = -\tau = -\infty$. In the first case, the points ξ_i (η_j) will all be distinct and there will exist a unique solution of the moment problem whose spectrum consists of the points ξ_j (η_j) and thus is a denumerable set whose only limit point is ∞ ($-\infty$) [11]. In the latter case, there are relatively few general theorems that are applicable.

2. THE SPECTRAL INTERVAL IS BOUNDED. We first review results involving the situation when the spectral interval [ξ_1, η_1] is bounded. In this case the corresponding Hamburger moment problem is determined so the distribution function ψ is essentially unique. The following can be deduced from Theorem 1.1 but is essentially contained in Stieltjes' work [45]:

THEOREM 2.1. [ξ_1, η_1] is bounded if and only if both coefficient sequences $\{c_n\}$ and $\{\lambda_n\}$ are bounded.

The first spectral theorem of the type we are most interested in is the following which is also due to Stieltjes [45]. Let E' denote the derived set of E.

THEOREM 2.2. Let $c_n = 0$ for all n and let $\lim_{n \to \infty} \lambda_n = 0$. Then $S(\psi)' = \{0\}$.

Thus Stieltjes gives conditions for the spectrum to be a symmetric, denumerable set with 0 as its only accumulation point. Stieltjes proved his result using convergence properties of the corresponding S-fraction. Using the theory of completely continuous (compact) operators applied to the appropriate J-matrix, Krein in 1938 proved a remarkable generalization of Stieltjes' theorem. Let $E = \{\sigma_1, \ldots, \sigma_k\}$ where the σ_i are distinct real numbers.

THEOREM 2.3. Let $\{c_n\}$ and $\{\lambda_n\}$ be bounded. In order that $S(\psi)' = E$, it is necessary and sufficient that for all i,

$$(2.1) \qquad \lim_{n \to \infty} \int_{-\infty}^{\infty} f(x) P_n(x) P_{n+i}(x) d\psi(x) = 0,$$

when $f(x) = (x - \sigma_1) \cdots (x - \sigma_k)$, and that (2.3) fail for at least one i if $f(x)$ is a polynomial of degree less than k.

The terms in (2.3) are just the Fourier coefficients of the expansion of $f(x) P_n(x)$ and can be related to the coefficients c_n and λ_n by means of the corresponding Jacobi matrix. The precise relations are too involved to summarize here so we refer to [1], [14] for specifics in the general case and settle for noting the two simplest cases. In the special case, $E = \{a\}$, Krein's conditions become the natural generalization of Stieltjes' result:

$$\lim_{n \to \infty} c_n = a, \qquad \lim_{n \to \infty} \lambda_n = 0 ,$$

while for $E = \{a, b\}$, his conditions are

$$\lim_{n \to \infty} [\lambda_n + \lambda_{n+1} + (c_n - a)(c_{n+1} - b)] = 0 ,$$

$$\lim_{n \to \infty} \lambda_{n+1}(c_n + c_{n+1} - a - b) = 0 ,$$

$$\lim_{n \to \infty} \lambda_n \lambda_{n+1} = 0 .$$

Note that the latter does not require that either $\{c_n\}$ or $\{\lambda_n\}$ converge.

More recently, Maki [33] proved the following somewhat related result. Let

(2.2) $\mathcal{L} = \{ t : t$ is a subsequential limit point of $\{c_n\} \}$.

THEOREM 2.4. Let $\lim_{n \to \infty} \lambda_n = 0$. Then $\mathcal{L} \subseteq S(\psi)$.

Maki also used operator methods to prove his result which, it should be pointed out, does not require that $\{c_n\}$ be bounded. He also noted that by choosing the c_n to be, for example, an enumeration of the rationals, one would have an example whose spectrum is the real line. Maki also conjectured that, still with the hypothesis $\lim_{n \to \infty} \lambda_n = 0$, a point is a limit point of the sequence $\{c_n\}$ if and only if it is an accumulation point of the spectrum. We were able to verify Maki's conjecture [12] so we have the following.

THEOREM 2.5. Let $\lim_{n \to \infty} \lambda_n = 0$. Then $\mathcal{L} = S(\psi)'$.

There are many specific examples of the type covered by the preceding theorems. For example, there are Al-Salam and Carlitz q polynomials, the Meixner polynomials, the modified Lommel polynomials, the Wall polynomials (see [14]), and more recent examples such as the q-Askey-Wilson polynomials [7] and orthogonal polynomials associated with a certain queueing model [21], [53].

At the opposite extreme from the preceding results, there is the following theorem due to Blumenthal [7] which covers such familiar examples as the Jacobi and Pollaczek polynomials.

THEOREM 2.6. Let

(2.3) $\qquad \lim_{n \to \infty} c_n = c, \qquad \lim_{n \to \infty} \lambda_n = \lambda \qquad$ (both finite).

Then

$$\sigma = c - 2\sqrt{\lambda}, \qquad \tau = c + 2\sqrt{\lambda}$$

and the zeros of the $P_n(x)$ are dense in the interval $[\sigma, \tau]$.

The density of the zeros would suggest (but not imply) that the entire interval $[\sigma, \tau]$ belongs to the spectrum. This is in fact true and follows from an argument due to Nevai (oral communication in 1983) based on another theorem of Nevai himself [37, Th. 4.2.14]:

THEOREM 2.7. Under the conditions (2.3), $[\sigma, \tau] \subseteq S(\psi)$.

Recently, Máté, Nevai and van Assche [36] have shown that Theorem 2.6 can be proven from a theorem of Weil about perturbations of the spectra of self-adjoint operators.

The conclusion in Theorem 2.7 implies Blumenthal's conclusion that the zeros are dense in $[\sigma, \tau]$. With additional conditions on the rate of convergence in (2.2), more specific properties of the distribution function have been deduced. For example, Nevai [37] has proved:

THEOREM 2.8. Let

(2.4) $\qquad \sum (|c_n - c| + |\lambda_n - \lambda|) < \infty.$

Then ψ is absolutely continuous on (σ, τ) and ψ' is positive and continuous on (σ, τ).

Máté and Nevai [34] subsequently showed that (2.4) could be replaced by the weaker conditions: $c_n \to c$, $\lambda_{n+1} \to \lambda$ and

$$\sum (|c_n - c_{n+1}| + |\lambda_n - \lambda_{n+1}|) < \infty.$$

For additional results along these lines see for example [26], [29], [37], [38], [39].

The following result of Geronimo and Case [28] has significance for applications to scattering theory in physics.

THEOREM 2.9. Let

$$\sum n(|c_n - c| + |\lambda_n - \lambda|) < \infty.$$

Then $S(\psi)$ has at most finitely many points on the complement of (σ, τ) and ψ is continuous at σ and τ.

For alternate and and more direct proofs of Theorem 2.9, see [22].

Relative to the important conclusions in Theorem 2.9, the conditions

(2.5) $\qquad c_n - c = O(n^{-2})$, $\quad \lambda_n - \lambda = O(n^{-2})$

describe a borderline case. That is, it is possible to construct examples satisfying (2.3) for which the conclusions of Theorem 2.9 hold and others where these conclusions fail (see [15]). For other studies of (0.1) with bounded coefficients directed at determining various other properties of the polynomials (asymptotics, bounds for zeros, etc.), see for example [9], [13], [37], [39], [40], [41], [48], [52], [55].

3. $|\sigma| = \infty$. We turn next to the cases where the spectral interval is unbounded. Thus we assume at least one of the two sequences $\{c_n\}$ and $\{\lambda_n\}$ are unbounded. The question of whether σ is finite or not can be settled with the following modification of the concept of a chain sequence.

DEFINITION. $\{a_n\}_{n=1}^{\infty}$ is an _eventual chain sequence_ if there exists an index N such that $\{a_{N+n}\}_{n=1}^{\infty}$ is a chain sequence.

NOTATION. Let $\mathfrak{E}$ denote the class of all eventual chain sequences, and let

(3.1) $\qquad \alpha_n(x) = \dfrac{\lambda_{n+1}}{(c_n - x)(c_{n+1} - x)}$

We can then relate σ to the coefficients in (1.1) by the following [17]:

THEOREM 3.1. Let $\lim_{n \to \infty} c_n = \infty$.

(i) $\qquad$ If $\{\alpha_n(x)\} \in \mathfrak{E}$, then $\sigma \geq x$.

(ii) $\qquad$ If $\{\alpha_n(x)\} \notin \mathfrak{E}$, then $\sigma \leq x$.

Theorem 3.1 provides an easy means to obtain simple criteria for the conclusion $\sigma = \infty$ which is equivalent to the condition

$$\xi_1 < \xi_2 < \cdots < \xi_n, \qquad \xi_n \to \infty.$$

For we have seen that the constant sequence $\{1/4\}$ is a chain sequence and by Wall's comparison theorem, any sequence (weakly) dominated by a chain sequence is itself a chain sequence [14]. Thus if, for every real x, $\alpha_n(x) < 1/4$ for all sufficiently large n, it follows that $\{\alpha_n(x)\} \in \mathfrak{E}$ for every x, hence by Theorem 3.1(i), $\sigma = \infty$. Thus we have

THEOREM 3.2. Let

$$\lim_{n \to \infty} c_n = \infty, \qquad \limsup_{n \to \infty} \frac{\lambda_{n+1}}{c_n c_{n+1}} < \frac{1}{4}$$

Then $\sigma = \infty$.

Theorem 3.2 was originally proved [9] under the hypothesis that the Hamburger moment problem is determined and subsequently rediscovered [32] as a theorem about the convergence of continued fractions without any assumptions on the moment problem. This result is sufficiently general to cover the majority of examples found in the literature including the classical examples associated with the names of Charlier, Stieltjes-Wigert, Meixner, and Hahn as well as the more recently discovered examples of Al-Salam and Carlitz [3], Al-Salam and Chihara [4] (see also [5]), Askey and Ismail [5] (discussed below). However, certain special cases of the Al-Salam and Carlitz polynomials have the property that the ratio $\lambda_{n+1}/(c_n c_{n+1})$ converges to 1/4. The following theorem [17] handles the latter.

THEOREM 3.3. Let

$$\sum \frac{n}{c_n} < \infty, \qquad \sum n \left| \frac{\lambda_{n+1}}{c_n c_{n+1}} - \frac{1}{4} \right| < \infty$$

Then $\sigma = \infty$.

One can reduce the burden on $\lambda_{n+1}/(c_n c_{n+1})$ relative to its convergence to 1/4 by requiring a little more rapid growth of c_n. Thus, we can conclude $\sigma = \infty$ if $n^2/c_n \to 0$ and for some N,

$$\frac{\lambda_{n+N+1}}{c_{n+N} c_{n+N+1}} \leq \frac{1}{4} + \frac{s}{16n^2}, \qquad\qquad s < 1.$$

Askey and Ismail [15] have recently discovered an important new class of orthogonal polynomials which they denote by $v_n(x;q;a,b,c)$ ($|q| < 1$). If we consider the equivalent monic form:

$$P_n(x) = (-1)^n q^{-n(n-1)/2} (q;q)_n v_n(x),$$

then in the corresponding recurrence relation (1.1), we have

$$c_{n+1} = aq^{-n}, \qquad \lambda_{n+1} = q^{-2n+1}(1-q^n)(b-cq^{n-1}).$$

with a real and b and c restricted so that $\lambda_{n+1} > 0$. Askey and Ismail prove that the corresponding Hamburger moment problem is determined if and only if

$$\frac{b}{a^2} < 1 \quad \text{and} \quad |q| \geq \frac{|a| - \sqrt{a^2 - 4b}}{|a| + \sqrt{a^2 - 4b}}$$

When the moment problem is determined, they obtain the distribution function ψ explicitly. In particular, they show that the spectrum consists of the points x_n where

$$x_n = A(Bq^n) + (Bq^n)^{-1}, \qquad n = 0, 1, \ldots .$$

Here A and B are certain explictly given positive constants. In particular, this means that for $q > 0$ (and $a > 0$), $c_n \to \infty$ and $\lambda_{n+1}/(c_n c_{n+1}) \to b/a^2 < 1/4$ so this is another case covered by Theorem 3.2. However, if $q < 0$, then the spectral interval is $(-\infty, \infty)$ (as is also implied by the fact that $\inf_n c_n = -\sup_n c_n = -\infty$). We therefore will look briefly at this case next. However, before leaving the case $\sigma = \infty$, we mention that there are a number of criteria expressed in terms of the coefficients in (0.1) for predicting the determinacy or indeterminacy of the associated Hamburger moment problem. For example, if the conditions of Theorem 3.2 hold, the moment problem will be determined if $c_n = 0(n^p)$. However, if $c_n = f_n q^{-n}$, where $0 < q < 1$ and $\{f_n\}$ is both bounded and bounded away from 0, let L denote the limit superior that appears in the theorem. Then the moment problem is determined if $L < q(q+1)^{-2}$ and it is indeterminate if the opposite strict inequality holds. For these criteria, see [20].
   When $(\xi_1, \eta_1) = (-\infty, \infty)$, we also have $(\sigma, \tau) = (-\infty, \infty)$ (see Sherman [43]). Sufficient conditions for this to occur are, for example,

(3.1)

(i) $\quad \inf_n c_n = -\infty, \quad \sup_n c_n = \infty;$

(ii) $\quad \lim_{n \to \infty} c_n = \infty, \quad \liminf_{n \to \infty} \frac{\lambda_{n+1}}{c_n c_{n+1}} > \frac{1}{4};$

(iii) $\quad \lim_{n \to \infty} \frac{c_n}{n^2} = \infty, \quad \frac{\lambda_{n+1}}{c_n c_{n+1}} > \frac{1}{4} + \frac{t}{16n^2} \qquad (t > 1; n \geq N)$

see [9], [17]).

The Askey-Ismail polynomials with $q < 0$ are examples with a discrete spectrum satisfying (i). The Meixner polynomials of the second kind (Meixner-Pollaczek polynomials) (see [14]) provide examples satisfying (ii) where the spectrum is the entire real line . Note that for the latter polynomials, the Hamburger moment problem is determined.
   General spectral theorems dealing with the situation $(\sigma, \tau) = (-\infty, \infty)$ are about as rare as are specific examples. In the symmetric case $(c_n = 0)$, one can consider the related polynomials on $[0, \infty)$ (see (4.3) below) and translate results from theorems dealing with the case $(\xi_1, \eta_1) \subseteq (0, \infty)$ to obtain conclusions for the symmetric case. For the nonsymmetric case, one theorem which predicts a <u>discrete</u>

spectrum is the following [18]:

THEOREM 3.4. Let the Hamburger moment problem be determined and suppose

(i) $\lim\limits_{n \to \infty} |c_n| = \infty$;

(ii) $\inf_n c_n = -\infty$, $\sup_n c_n = \infty$;

(iii) $\limsup\limits_{n \to \infty} \left| \dfrac{\lambda_{n+1}}{c_n c_{n+1}} \right| < \dfrac{1}{4}$.

Then $\sigma = -\infty$, $\tau = \infty$ and $S(\psi)$ has no finite points of accumulation.

A second general result [18] is, to a certain extent, an analog of Theorem 2.5. Recall the notation (2.1) so that $\mathcal{L}$ denotes the set of subsequential limit points of $\{c_n\}$.

THEOREM 3.5. Let the Hamburger moment problem be determined and let

$$\lim\limits_{n \to \infty} \frac{\lambda_{n+1}}{c_n c_{n+1}} = 0.$$

Then $\sigma = \inf \mathcal{L}$, $\tau = \sup \mathcal{L}$ and $S(\psi)' \subseteq \mathcal{L}$.

Note that this theorem applies more generally than just to those cases where $(\sigma , \tau) = (-\infty, \infty)$. When we compare Theorem 3.5 with Theorem 2.5, we are led to the natural question: under what additional conditions will it be true that $S(\psi)' = \mathcal{L}$?

Criteria for deciding determinacy of the Hamburger moment problem in this case are rare. The simplest is Carleman's criterion (see [44]) which says that the moment problem is determined if $\sum \lambda_n^{-1/2} = \infty$.

However, this is rarely applicable when $\sigma = -\infty$, $\tau = \infty$. One point that suggests the limited applicability of Carleman's criterion is the fact that c_n is not involved. For a condition for determinacy that involves both c_n and λ_n, see Dennis and Wall [23].

4. $|\sigma| < \infty$, $\tau = \infty$.   Theorem 3.1 shows that when $c_n \to \infty$, the conclusion $|\sigma| < \infty$ is equivalent to the condition that $\{ \alpha_n(x) \}$ is an eventual chain sequence for at least one, but not every, finite value of x. In particular, when $\{ \alpha_n(0) \}$ converges, the limiting value 1/4 is the critical number (note the criterion (4.1)-(ii)). Keeping Theorem 3.1 in mind, we first cite the following analog of Blumenthal's Theorem 2.6 [10].

THEOREM 4.1. Let $\lim_{n \to \infty} c_n = \infty$ and let

$$\lim_{n \to \infty} \frac{\lambda_{n+1}}{c_n c_{n+1}} = \frac{1}{4}$$

If in addition σ is finite, then the zeros of the $P_n(x)$ are dense in (σ, ∞).

The hypothesis that σ is finite cannot be omitted. We noted in section 3 that certain special cases of the Al-Salam and Carlitz polynomials are examples whose recurrence coefficients satisfy the hypotheses in Theorem 4.1 but for which $\sigma = \infty$. The Hamburger moment problem associated with this case of the Al-Salam and Carlitz polynomials is actually indeterminate. In fact, the distribution function given by Al-Salam and Carlitz is an extremal solution (see [16]) which is in itself an interesting fact. The following example:

$$c_n = n^2 f_n \quad , \qquad \lambda_{n+1} = \frac{1}{4} n^2 (n+1)^2 f_n f_{n+1},$$

where $f_n \to \infty$, $\sum f_n^{-1} = \infty$, would provide another example with $\sigma = \infty$ and the moment problem would be determined [20, Th 4]. One can also construct examples with $\sigma = -\infty$. For example, if

$$c_n = 2n, \qquad \lambda_{n+1} = n^2 + n^\gamma \qquad (1 < \gamma < 2)$$

then $\sigma = -\infty$ [14, p. 125] and the corresponding moment problem is determined (by Carleman's criterion). However, we know of no examples in the literature where the orthogonal polynomials are explicitly known and the limit is 1/4 and $\sigma = -\tau = -\infty$.

There are of course a number of well known examples satisfying the hypotheses of Theorem 4.1. In additional to the classical Laguerre polynomials, there are two sequences of orthogonal polynomials related to the Meixner polynomials of the second kind which were studied by Al-Salam [2] (see also [14, p. 180]), and various hypergeometric orthogonal polynomials (such as the continuous dual Hahn polynomials) studied by Askey and Wilson [6], [7], [57].

To further emphasize the point that the additional hypothesis that σ is finite cannot be omitted in Theorem 4.1, we cite the following result [17]:

THEOREM 4.2. Given any sequence $\{c_n\}$ with $c_n \to \infty$, there exists a sequence $\{\lambda_{n+1}\}$ such that $\lambda_{n+1}/(c_n c_{n+1}) \to 1/4$ and $\sigma = \infty$ and there exists a $\{\lambda_{n+1}\}$ such that the above limit holds and $\sigma = -\infty$.

In this connection, it is also informative to consider the important special cases where c_n and λ_n are polynomials in n (see [17]). In this case, $\lim_{n \to \infty} c_n = \infty$ and $\lim_{n \to \infty} \lambda_{n+1}/(c_n c_{n+1}) = L$ always exist and $|\sigma| = \infty$ if $L \neq 1/4$. In the critical case $L = 1/4$, one can give conditions in terms of the coefficients c_n and λ_{n+1} for $|\sigma| = \infty$ or for σ to be finite. In the latter case, one can determine σ precisely in terms of the coefficients in c_n and λ_{n+1}. In particular, it can be shown that when the degree of c_n is at least three, σ is always infinite.

This borderline case, $\lim_{n \to \infty} \alpha_n(0) = 1/4$, calls out for further investigation. In connection with Theorem 3, the sequence $\{a_n\}$ with

$$a_n = \frac{1}{4} + \frac{1 + e_n}{16n(n+1)}$$

is of significance. For it can be shown [17] that

(i) if $e_n = O(1/n)$ or if $\sum e_n$ converges, then $\{a_n\} \in \xi$;

(ii) if $e_n \geq 0$ and $\sum e_n/n = \infty$, then $\{a_n\} \notin \xi$.

Recently, Jacobson and Masson [31] have improved (i) considerably. Closing the gap between (i) and (ii) will permit more precise determination of the nature of $S(\psi)$ and calculation of σ in many situations.

The next theorem is an analogue of Geronimo and Case's result, Theorem 2.9.

THEOREM 4.3. Let

$$\sum \frac{1}{c_n} = \infty \quad \text{and} \quad \sum n \left| \frac{\lambda_{n+1}}{c_n c_{n+1}} - \frac{1}{4} \right| < \infty$$

Then $\sigma = 0$, $S(\psi) \cap (-\infty, 0)$ is a finite set, and ψ is continuous at 0.

Theorem 4.3 and others that lead to similar conclusions can be found in [17].

Carleman's criterion for determinacy of the Hamburger moment problem is frequently applicable when the hypotheses of the preceding theorems are satisfied. A sufficient condition for indeterminacy which can sometimes be applied when $\lambda_{n+1}/(c_n c_{n+1})$ converges to 1/4 is the following [20]:

$$\lim_{n \to \infty} \inf c_n^{1/n} > 1.$$

We conclude by addressing the obvious question that arises upon seeing Theorem 4.1. "Where (or what) is the analog of Nevai's "half" of the Blumenthal-Nevai theorem (Theorem 2.7)?" An analogue of Theorem 2.8 would also be highly desirable. Some steps in this direction have been taken by Dombrowski. In [24], for example, she has considered the symmetric case. Let us write the recurrence relation for a symmetric OPS as

(4.1) $S_n(x) = x \, S_{n-1}(x) - \gamma_n S_{n-2}(x)$,

and denote the symmetric distribution with respect to which the $S_n(x)$ are orthogonal as $d\varphi(x)$. Dombrowski assumes that the coefficients in (4.1) satisfy:

(i) $\{\gamma_n\}$ increases monotonically to ∞;

(ii) $\sum \gamma_n^{-1/2} = \infty$.

Condition (ii) ensures that the Hamburger moment problem is determined (again by Carleman's criterion). Dombrowski then uses operator theory to prove:

THEOREM 4.4. Let $D_n = \sqrt{\gamma_{n+1}} - \sqrt{\gamma_n}$. If

(iii) $\{D_n\}$ is bounded and

(iv)   either (a) $D_{n+1} - D_n \le D_n - D_{n-1}$

       or   (b) $D_{n+1}^2 \le D_n D_{n+2}$

then φ is absolutely continuous on $(-\infty, \infty)$.

   Consider the corresponding polynomials, $P_n(x)$, whose symmetrization yields the polynomials $S_n(x)$:

$$P_n(x) = S_{2n}(\sqrt{x}).$$

Then $\{P_n(x)\}$ is orthogonal over $[0,\infty)$ with respect to $d\psi(x) = d\varphi(\sqrt{x})$ and satisfies (0.1) with

(4.2) $c_1 = \gamma_2$, $c_{n+1} = \gamma_{2n+1} + \gamma_{2n+2}$, $\lambda_{n+1} = \gamma_{2n}\gamma_{2n+1}$

(see [14]). The conditions (i) and (iii) above imply that $\lim_n \alpha_n(0) = 1/4$ so the polynomials of Dombrowski's theorem correspond to this borderline situation. However, it is not immediately clear how to state conditions in terms of $\{c_n\}$ and $\{\lambda_{n+1}\}$ which will be sufficient to guarantee that the conditions of Dombrowski's theorem are satisfied.

   For a related result dealing with (4.1) under conditions (i), (ii), see Dombrowski [26]. These and further results will be discussed by Dombrowski in the following talk so we conclude by citing a few additional studies of (0.1) which are applicable when the coefficient sequences are unbounded which deal with other properties of the orthogonal polynomials: [13], [35], [41], [49], [50], [54].

REFERENCES

1 Ahiezer, N. I. and Krein, M. G., 'Some Questions in the Theory of Moments,' Translations of Mathematical Monographs, vol. 2, Amer. Math. Soc., 1962.

2 Al-Salam, W. A., 'Characterization of a certain class of orthogonal polynomials related to elliptic functions,' Ann. Mat. Pura Appl., 67 (1965), 75-94.

3 Al-Salam, W. A. and Carlitz, L., 'Some orthogonal q-polyomials,' Math. Nachr., 30 (1965), 47-61.

4 Al-Salam, W. A. and Chihara, T. S., 'Convolutions of orthogonal poly-nomials,' SIAM J. Math. Anal., 7 (1976), 16-28.

5 Askey, R. A. and Ismail, M. E. H., Recurrence relations, continued fractions and orthogonal polynomials, Memoirs of the Amer. Math. Soc., No. 300, 1984.

6 Askey, R. A. and Wilson, J. A., 'A set of hypergeometric polynomi-als,' SIAM J. Math. Anal., 13 (1982), 651-655.

7 Askey, R. A. and Wilson, J., Some basic hypergeometric orthogonal polynomials that generalize Jacobi polynomials, Memoirs of the Amer. Math. Soc., No. 319, 1985..

8 Blumenthal, O., 'Über die Entwicklung einer willkurlichen Funktion nach den Nennern des Kettenbruches für $\int_{-\infty}^{0} [\emptyset(\xi)/(z-\xi)] d\xi$,' Dissertation, Göttingen, 1898.

9 Chihara, T. S., 'Chain sequences and orthogonal polynomials,' Trans. Amer. Math. Soc., 104 (1962), 1-16.

10 Chihara, T. S., 'Orthogonal polynomials whose zeros are dense in intervals,' J. Math. Anal. Appl., 24 (1968), 362-371.

11 Chihara, T. S., 'On indeterminate Hamburger moment problems,' Pacific J. Math., 27 (1968), 475-484.

12 Chihara, T. S., 'The derived set of the spectrum of a distribution function,' Pacific J. Math., 35 (1970), 571-574.

13 Chihara, T. S., 'Convergent sequences of orthogonal polynomials,' J. Math. Anal. Appl., 38 (1972), 335-347.

14 Chihara, T. S., An Introduction to Orthogonal Polynomials, Gordon and Breach Science Publishers, New York, 1978.

15 Chihara, T. S., 'Orthogonal polynomials whose distribution functions have finite point spectra,' SIAM J. Math. Anal., 11 (1980), 358-364.

16 Chihara, T. S., 'Indeterminate symmetric moment problems,' J. Math. Anal. Appl., 85 (1982), 331-346.

17 Chihara, T. S., 'Spectral properties of orthogonal polynomials on unbounded sets,' Trans. Amer. Math. Soc., 270 (1982), 623-639.

18 Chihara, T. S., 'Orthogonal polynomials with discrete spectra on the real line,' J. Approx. Theory, 42 (1984), 97-105.

19 Chihara, T. S., 'On the spectra of certain birth and death processes, SIAM J. Applied Math., 47 (1987), 662-669.

20 Chihara, T. S., 'Hamburger moment problems and orthogonal polynomi-als,' Trans. Amer. Math. Soc., to appear.

21 Chihara, T. S. and Ismail, M. E. H., 'Orthogonal polynomials sugges-ted by a queueing model,' Advances in Applied Math., 3 (1982), 441-462.

22 Chihara, T. S. and Nevai, P., 'Orthogonal polynomials and measures with finitely many point masses,' J. Approx. Theory, 35 (1982), 370-380.

23 Dennis, J. J. and Wall, H. S., 'The limit circle case for a positive-definite J-fraction,' Duke Math. J., 12 (1945), 255-273.

24 Dombrowski, J., 'Cyclic operators, commutators, and absolutely continuous measures,' Proc. Amer. Math. Soc., 100 (1987), 457-463.

25 Dombrowski, J., 'Spectral measures, orthogonal polynomials and absolute continuity,' SIAM J. Math. Anal., 19 (1988), 939-943.

26 Dombrowski, J. and Nevai, P., 'Orthogonal polynomials, measures and recurrence relations,' SIAM J. Math. Anal., 17 (1986), 752-759.

27 Favard, J., 'Sur les polynomes de Tchebicheff,' C. R. Acad. Sci. Paris, 200 (1935), 2052-2053.

28 Geronimo, J. S. and Case, K. M., 'Scattering theory and polynomials orthogonal on the real line,' Trans. Amer. Math. Soc., 258 (1980), 467-494.

29 Geronimo, J. S. and Nevai, P. G., 'Necessary and sufficient conditions relating the coefficients in the recurrence formula to the spectral function for orthogonal polynomials,' SIAM J. Math. Anal., 14 (1983), 622-637.

30 Hamburger, H., 'Uber eine Erweiterung des Stieltjesschen Momentenproblems,' Math. Anal., 81 (1920), 253-319; 82 (1921), 120-164; 168-187.

31 Jacobsen, L. and Masson, D. R., 'On the convergence of limit periodic continued fractions $K(a_n/1)$, where $a_n \to -1/4$. Part III., Constructive Approximation, to appear.

32 Maki, D. P., 'On constructing distribution functions: with applications to Lommel polynomials and Bessel functions,' Trans. Amer. Math. Soc., 130 (1968), 281-297.

33 Maki, D. P., 'A note on recursively defined orthogonal polynomials,' Pacific J. Math., 28 (1969), 611-613.

34 Máté, A. and Nevai, P. G., 'Orthogonal polynomials and absolutely continuous measures,' in Approximation Theory, IV, C. K. Chui et al., eds., Academic Press, New York, 1983, 611-617.

35 Máté, A., Nevai, P. G. and Totik, V., 'Asymptotics for the zeros of orthogonal polynomials associated with infinite intervals, J. London Math. Soc., (2) 33 (1986), 303-310.

36 Máté, A., Nevai, P. G. and van Assche, W., 'The supports of measures associated with orthogonal polynomials and the spectra of the related self adjoint operators,' Rocky Mtn. J. Math., to appear.

37 Nevai, P. G., Orthogonal Polyomials, Memoirs of Amer. Math. Soc., No. 213, 1979.

38 Nevai, P. G., 'Orthogonal polynomials defined by a recurrence relation,' Trans. Amer. Math. Soc., 250 (1979), 369-384.

39 Nevai, P. G., 'On orthogonal polynomials,' J. Approx. Theory, 25 (1979), 34-37.

40 Nevai, P. G., 'Two of my favorite ways of obtaining asymptotics for orthogonal polynomials,' International Series of Numerical Mathematics, 65 (1984), 417-436.

41 Nevai, P. G. and Dehesa, J. S., 'On asymptotic average properties of zeros of orthogonal polynomials,' SIAM J. Math. Anal., 10 (1979), 1184-1192.

42 Perron, O., Die Lehre von den Kettenbrüchen, 3rd ed., Leipzig, 1957.

43 Sherman, J., 'On the numerators of the convergents of the Stieltjes continued fractions,' Trans. Amer. Math. Soc., 35 (1933), 64-87.

44 Shohat, J. A. and Tamarkin, J. D., The Problem of Moments, Mathematical Surveys No. 1, Amer. Math. Soc., 1943.

45 Stieltjes, T. J., 'Recherches sur les fractions continues,' Annales de la Faculté des Sciences de Toulouse, 8 (1984), J 1-122; 9 (1895), A 1-47; Oeuvres, Vol. 2, 398-566.

46 Stone, M. H., Linear Transformations in Hilbert Spaces and Their Applications to Analysis, Colloquium Publications no. 15, Amer. Math. Soc., N.Y., 1932.

47 Szegö, G. Orthogonal Polynomials, Colloquium Publications no. 23, Amer. Math. Soc., Providence, 4th edition, 1975.

48 Van Assche, W., 'Asymptotic properties of orthogonal polynomials from their recurrence formula, I,' J. Approx. Theory, 44 (1985), 258-276.

49 Van Assche, W., 'Asymptotic properties of orthogonal polynomials from their recurrence formula, II,' J. Approx. Theory, 52 (1988), 322-338.

50 Van Assche, W., 'The ratio of q-like orthogonal polynomials,' J. Math. Anal. Appl., to appear.

52 Van Assche, W., Asymptotics for Orthogonal Polynomials, Lecture Notes in Mathematics, 1265, Springer, Berlin, 1987.

51 Van Assche, W. and Geronimo, J.S., 'Asymptotics for orthogonal polynomials on and off the essential spectrum,' J. Approx. Theory, 55 (1988), 220-231.

53 Van Doorn, E., 'The transient state probabilities for a queueing model where potential customers are discouraged by queue length,' J. Appl. Prob., 18 (1981), 499-506.

54 Van Doorn, E. A., 'On orthogonal polynomials with positive zeros and the associated kernel polynomials,' J. Math. Anal. Appl., 113 (1986), 441-450.

55 Van Doorn, E. A., 'Representations and bounds for zeros of orthogonal polynomials and eigenvalues of sign-symmetric tri-diagonal matrices,' J. Approx. Theory, 51 (1987), 254-266.

56 Wall, H. S., Continued Fractions, van Nostrand, N. Y., 1948.

57 Wilson, J. A., 'Some hypergeometric orthogonal polynomials,' SIAM J. Math. Anal., 11 (1980), 690-701.

58 Wintner, A., Spektraltheorie der unendlichen Matrizen, 2nd ed., Leipzig, 1929.

ON THE ROLE OF ORTHOGONAL POLYNOMIALS ON THE UNIT CIRCLE IN DIGITAL SIGNAL PROCESSING APPLICATIONS

P. DELSARTE and Y. GENIN
Philips Research Laboratory Brussels
Av. Van Becelaere 2 - Box 8
B-1170 Brussels - Belgium

ABSTRACT. The aim of this contribution is first to show how the Szegö theory of orthogonal polynomials on the unit circle is intimately related to the celebrated Levinson algorithm, which is commonly used in digital signal processing (DSP) applications to solve various least-squares problems. A computationally more efficient substitute for the Levinson algorithm, termed the split Levinson algorithm, has recently been proposed in the DSP literature. In the case of real data, this new algorithm can be interpreted naturally in the framework of a well-defined one-to-one correspondence between the families of real Szegö polynomials and the families of polynomials orthogonal on the interval [−1, 1] with respect to a symmetric measure. More generally, the philosophy underlying the split Levinson algorithm opens the door to an interesting "tridiagonal approach" to the theory of complex Szegö polynomials, nonnegative definite Hermitian Toeplitz matrices, and related algebraic and function theoretic questions. Some of the main topics of this new mathematical framework are briefly reviewed and are shown on specific examples to be of particular interest in DSP applications.

1 Introduction

Orthogonal polynomial theory is known to have numerous applications in a vast collection of engineering problems. This is especially true in the field of digital signal processing (DSP) applications [1], [2], [3], [4]. In particular, the Szegö theory of orthogonal polynomials on the unit circle [5], [6] provides a mathematical framework into which a variety of DSP problems fit naturally. The present paper concentrates on this part of orthogonal polynomial theory, and attempts to show why and how it is of special importance in DSP applications, both from a theoretical viewpoint and from a practical viewpoint. For general information about DSP, the reader is referred to [7]; specific information on speech processing can be found in [8].

By definition, digital signal processing deals with sampled signals. In other words, it is concerned with sequences of numbers that contain some useful information. Typical DSP problems include e.g. extracting, compressing, transmitting, or filtering this information. In all these examples, the DSP designer is confronted with the question of selecting appropriate methods which will act on the given sequences of numbers so as to produce a desirable effect in some optimal way. The linear least-squares method arises naturally in this context [9]; it provides not only a mathematically sound approach to a collection of such DSP problems, but also yields solution techniques that are "feasible" in the sense that they can be implemented at a reasonable cost.

The Szegö theory of orthogonal polynomials on the unit circle comes naturally into the picture at this very point. Indeed, a family of such orthogonal polynomials is classically associated with a well-defined nonnegative definite Hermitian Toeplitz matrix. It turns out

P. Nevai (ed.), Orthogonal Polynomials, 115–133.
© *1990 by Kluwer Academic Publishers.*

that the Toeplitz structure is of special interest from a DSP viewpoint, since a nonnegative definite Toeplitz matrix can be interpreted as the covariance matrix of a discrete-time stationary stochastic process, or as the autocorrelation matrix of a finite sampled signal record (in the positive definite case). It is therefore not surprising that the system matrix of the normal equations that define the optimal linear least-squares solution exhibits the Toeplitz structure in many concrete DSP situations. This explains why the problem of numerically solving a Toeplitz system of linear equations is an important issue in DSP applications. The Levinson algorithm [10], [11] is the numerical procedure that is classically used in DSP to compute the solution of a positive definite Toeplitz system in a fast and robust way. Although originally derived independently of the Szegö theory, this algorithm actually relies on the well-known recurrence relation that links successive members of a family of orthogonal polynomials on the unit circle.

The extensive investigation of the Toeplitz structure within the DSP research community in turn has had some impact on the Szegö theory itself. Recently, and rather unexpectedly, the Levinson algorithm has been shown to be inherently redundant in complexity, and a parsimonious reformulation of the algorithm, termed the split Levinson algorithm, has been proposed [12]. The reduction of complexity stems from the fact that the split Levinson algorithm recursively computes a family of symmetric polynomials instead of the classical Szegö family. This recursive calculation is based on a three-term recurrence relation, of unusual type, which is satisfied by the symmetric polynomials in question. It turns out that the general philosophy underlying this new algorithm produces an attractive new approach to the theory of Szegö polynomials and related questions.

In addition to the algebraic type considerations above, there exists a significant reason of a function theoretic nature that explains why Szegö orthogonal polynomial theory is of specific interest in DSP problems. Szegö's theory is known to be intimately related to analytic function theory and, more precisely, to the theory of Carathéodory (positive) functions and Schur (bounded) functions [13]. Such functions are introduced naturally in the DSP context due to the fact that their properties correspond to physical constraints such as system passivity or losslessness [14], which are inherent in a variety of popular cascade filtering structures, e.g. the wave digital filter structure [15] and the lattice filter structure [16]. Furthermore, it is known since the work of Kolmogorov, Carathéodory and Toeplitz, that stationary stochastic processes can be characterized in terms of Carathéodory functions; this explains why the study of that class of functions is of great significance in stochastic process theory [6].

The aim of the present contribution is to provide some basic elements that put into light the important rôle of Szegö orthogonal polynomial theory in digital signal processing applications. This rôle can be demonstrated in so many different ways that it is not at all evident how to select the particular topics to be presented in an introductory paper of moderate size. The authors have made the choice to focus on basic ideas and concepts rather than on more advanced techniques and developments. This explains why the scalar (one-dimensional) case is only considered, and why no attempt is made to discuss some elaborate DSP problems which can be approached with the help of generalized Szegö theory arguments; in other words, generality is traded for simplicity. It should also be stressed that the selection of questions discussed in this paper is biased by the authors own interests in the field.

In section 2, we first consider two simple but typical DSP problems: the autoregressive modelling problem for a finite sampled signal record, and the linear least-squares prediction problem for a stationary stochastic process. These two problems are shown to give rise to a Toeplitz system of linear equations [1], [2]. It is explained how the numerical solution of such a system can be obtained with the help of the classical Levinson algorithm. This algorithm is then revisited at the light of Szegö's theory of orthogonal polynomials on the unit circle; it is shown to be nothing but an appropriate implementation of the recurrence relation satisfied by a family of such orthogonal polynomials. Finally, the function theoretic aspects of the subject and their relevance in DSP problems are briefly mentioned but not elaborated upon.

In section 3, we restrict our attention to the case of real (positive definite) Toeplitz matrices. The split Levinson algorithm is first derived from the classical Levinson algorithm by replacing the family of Szegö polynomials by an appropriate family of symmetric polynomials that can be obtained recursively at a lower computational cost [12]. The main properties of the symmetric polynomials in question are then reviewed. In particular, it is shown how these properties can be discussed by associating with our polynomials a well-defined Jacobi matrix, which is seen to be positive definite together with the given Toeplitz matrix. A simple change of variable is then shown to transform a family of such symmetric polynomials into a family of even/odd polynomials orthogonal on the real interval $[-1, 1]$, so that the recurrence relation of the split Levinson algorithm can be viewed as a reformulation of the classical three-term formula relative to these orthogonal polynomials [12], [17]. This one-to-one correspondence between real Szegö polynomials and even/odd orthogonal polynomials on $[-1, 1]$ is illustrated by deriving the Szegö counterparts of the ultraspherical polynomials [17].

In section 4, the ideas that underly the split Levinson algorithm are extended to the complex case [18], [19], [20]. This leads to a novel approach to the general theory of orthogonal polynomials on the unit circle [19], [21], [22]. This approach involves a tridiagonal polynomial matrix of a special form, together with an associated family of complex symmetric polynomials. This matrix is further assumed to be positive definite for a specified value of the variable, with unit modulus. A one-to-one correspondence is then established between such families of complex symmetric polynomials (which are "quasi orthogonal" on the unit circle) and the families of complex Szegö polynomials. From this result it follows that the split Levinson algorithm can be generalized to accommodate Hermitian Toeplitz matrices, and that the resulting algorithm achieves a reduction of computational complexity with respect to the complex Levinson algorithm just as in the real case. As an illustration of this new approach, the problem of determining the mass points and the mass coefficients of the discrete measure underlying a singular family of Szegö polynomials [6], [13] is considered in some detail and is shown to be equivalent to a generalized eigen problem for a tridiagonal Hermitian-definite matrix pencil [22]. It turns out that this particular problem is important in DSP applications for it is closely related to the Pisarenko modelling technique [23], which realizes a stationary stochastic process from a given covariance matrix in some well-defined optimal manner. It is briefly mentioned that the tridiagonal approach also leads to an interesting new characterization theorem for the class of Carathéodory functions [19], [24], a subject outside the scope of the present contribution.

Let us again stress that the segments of orthogonal polynomial theory and its DSP applications discussed in this paper provide only an introduction to the field. They have been selected in order to whet the reader curiosity; they should not be viewed as an exact reflection of the state of the art and especially not of the most advanced developments of the subject. In that respect, it should be emphasized that most of the results presented herein can be extended to more general settings (see especially [25], [26], [27], [28]).

2   Levinson's algorithm and Szegö's orthogonal polynomials

To begin with, let us briefly consider two typical DSP applications, which illustrate how positive definite Toeplitz systems of linear equations arise naturally in the context of least-squares approximation or estimation problems.

118

2.1 Autoregressive modelling problem

Let there be given a sequence of complex signal samples $s(0)$, $s(1),\ldots, s(N-1)$, of finite
length N, with $s(0) \neq 0$. By convention, we set $s(t) = 0$ for $t \leq -1$ and for $t \geq N$. The
question is then to find an *autoregressive (AR) filter*, together with an impulse input, which
will best "explain" the given samples $s(t)$, considered as output data, in the least-squares
sense. An AR filter of degree n is defined to be a recursive computational scheme that
transforms an input sequence of complex numbers $u(0)$, $u(1),\ldots$ into an output sequence
of complex numbers $v(0)$, $v(1),\ldots$ according to the rule

$$v(t) + \sum_{i=1}^{n} a_{n,i}\, v(t-i) = u(t), \qquad t = 0, 1, \ldots \tag{2.1}$$

In the present case, as we have assumed $u(t) = 0$ for $t \geq 1$, the problem reduces to
determining the complex numbers $u(0)$ (the input pulse value) and $a_{n,1}, a_{n,2}, \ldots, a_{n,n}$ (the
filter coefficients) so as to minimize the squared Euclidean norm

$$E_n = \sum_{t=0}^{N+n-1} |e_n(t)|^2 \tag{2.2}$$

of the vector whose entries are the modelling errors $e_n(t) = s(t) + a_{n,1}\, s(t-1) + \cdots + a_{n,n}\, s(t-n) - u(0)\, \delta_{t,0}$. By elementary manipulations we can show that the normal equations of this
least-squares problem are given by $u(0) = s(0)$ and

$$C_n [1, a_{n,1}, \ldots, a_{n,n}]^T = [\sigma_n, 0, \ldots, 0]^T, \tag{2.3}$$

where C_n is the *positive definite Hermitian Toeplitz matrix*

$$C_n = \begin{bmatrix} c_0 & \bar{c}_1 & \cdots & \bar{c}_n \\ c_1 & c_0 & \cdots & \bar{c}_{n-1} \\ \vdots & \vdots & & \vdots \\ c_n & c_{n-1} & \cdots & c_0 \end{bmatrix} \tag{2.4}$$

whose entries are the autocorrelation lags of the signal samples, i.e., $c_k = \sum_{t=k}^{N-1} s(t)\,\bar{s}(t-k)$.
Note that the number σ_n is the right hand side of (2.3) is positive and is uniquely determined
from the data. Thus, it is seen that the AR filter vector $[1, a_{n,1}, \ldots, a_{n,n}]^T$ is proportional
to the first column of the inverse of C_n. Let us also point out that the minimum value of
the squared error norm (2.2) is given by $E_n = \sigma_n - s^2(0)$.

2.2 Linear least-squares prediction of a stationary stochastic process

Let $(x(t) : t = \ldots -2, -1, 0, 1, 2, \ldots)$ be a discrete-time complex-valued stationary stochastic process, with zero mean $(E[x(t)] = 0)$ and with given covariances $E[x(t)\,\bar{x}(t-k)] = c_k$
(for all k). We consider the *linear least-squares forward prediction problem* of order n relative to such a process: to find a prediction $\xi_n(t)$ of the random variable $x(t)$ as a linear
combination of the n past random variables $x(t-1),\ldots, x(t-n)$ such that the variance
$E[|e_n(t)|^2]$ of the prediction error $e_n(t) = x(t) - \xi_n(t)$ be minimum. Thus, we have

$$\xi_n(t) = -a_{n,1}\, x(t-1) - a_{n,2}\, x(t-2) - \cdots - a_{n,n}\, x(t-n), \tag{2.5}$$

and we select the complex coefficients $a_{n,i}$ so as to minimize the quadratic function $E[|x(t)+ a_{n,1}\ x(t-1) + \cdots + a_{n,n}\ x(t-n)|^2]$.

The normal equations of the problem are easily shown to take exactly the form (2.3), where the entries c_k of the Toeplitz matrix C_n are the given covariances. In this situation, C_n is nonnegative definite (and may be singular). The minimum value of the prediction error variance is given by $E[|e_n(t)|^2] = \sigma_n$ (and is independent of t). Note that C_n is singular if and only if σ_n vanishes. This possibility is ruled out in the sequel, which entails no real loss of generality.

2.3   The Levinson algorithm

The Levinson algorithm is the standard technique used in DSP applications to solve Toeplitz systems of the form (2.3). It computes the solution vector of such a system, in a recursive manner, by considering the Toeplitz subsystems of increasing dimension formed with the nested sequence $(C_k : k = 0, 1, \ldots, n)$ of the top principal submatrices C_k of the given positive definite Toeplitz matrix C_n. Let us examine the mechanism in some detail.

Thus, we assume the solution σ_{k-1}, $a_{k-1,1}, \ldots, a_{k-1,k-1}$ to be known, for the C_{k-1} matrix. In view of the Toeplitz structure, we can then write the matrix identity

$$
C_k \begin{bmatrix} 1 & a_{k,1} & \cdots & a_{k,k-1} & a_{k,k} \\ 1 & a_{k-1,1} & \cdots & a_{k-1,k-1} & 0 \\ 0 & \overline{a}_{k-1,k-1} & \cdots & \overline{a}_{k-1,1} & 1 \end{bmatrix}^T = \begin{bmatrix} \sigma_k & 0 & \cdots & 0 & 0 \\ \sigma_{k-1} & 0 & \cdots & 0 & -\rho_k\sigma_{k-1} \\ -\overline{\rho}_k\sigma_{k-1} & 0 & \cdots & 0 & \sigma_{k-1} \end{bmatrix}^T , \quad (2.6)
$$

where ρ_k is an appropriate complex number, defined by

$$
\sigma_{k-1}\, \rho_k = -\sum_{i=0}^{k-1} c_{k-i}\, a_{k-1,i}, \tag{2.7}
$$

with $a_{k-1,0} = 1$. In the DSP literature, the numbers ρ_k are usually referred to as *reflection coefficients*, or as *partial correlation (PARCOR) coefficients*; the numbers σ_k are the *prediction error variances*. From (2.6) and (2.7) it is clear that the solution relative to C_k can be expressed in terms of the numbers $a_{k-1,i}$ and σ_{k-1} by

$$
a_{k,i} = a_{k-1,i} + \rho_k\, \overline{a}_{k-1,k-i}, \qquad i = 1, 2, \ldots, k, \tag{2.8}
$$

$$
\sigma_k = \sigma_{k-1}(1 - |\rho_k|^2). \tag{2.9}
$$

The *Levinson algorithm* [10], [11] results from applying equations (2.7)–(2.9) recursively, for $k = 1, 2, \ldots, n$, with the obvious initialization $a_{0,0} = 1$ and $\sigma_0 = c_0$. It provides the numerical solution of the problem (2.3) in $O(n^2)$ flops, instead of $O(n^3)$ flops as in the case of classical algorithms to solve general (positive definite) systems of linear equations. Furthermore, it is known to be numerically stable when applied to positive definite Toeplitz matrices [29]. Note that we have $\sigma_k = \det C_k/\det C_{k-1}$, by definition, whence

$$
\det C_k = c_0(1 - |\rho_1|^2)(1 - |\rho_2|^2) \cdots (1 - |\rho_k|^2), \tag{2.10}
$$

in view of (2.9). As a consequence, the positive definiteness of C_n is equivalent to the set of inequalities $c_0 > 0$ and $|\rho_k| < 1$ for $k = 1, 2, \ldots, n$.

The Levinson algorithm gives rise to several interesting "companion algorithms", such as the *Schur algorithm* [30] and the *lattice algorithm* [31], which are of frequent use in DSP applications. Moreover, it can be extended to Toeplitz systems with arbitrary right hand sides [10], [11], and it can be generalized to non-Toeplitz ("close to Toeplitz") systems so as to accommodate more elaborate DSP least-squares problems [32].

2.4 Szegö's orthogonal polynomials on the unit circle

The Hermitian Toeplitz matrix that has $[c_0, c_1, \ldots, c_n]^T$ as its first column is known to be nonnegative definite if and only if there exists a *positive measure* $d\omega$, defined on the interval $[0, 2\pi)$, that admits the numbers $c_0, c_1, \ldots, c_n$ as its first trigonometric moments, i.e., that satisfies

$$c_k = \int_0^{2\pi} e^{-ik\theta} \, d\omega(\theta), \qquad k = 0, 1, \ldots, n. \tag{2.11}$$

Thus, $d\omega$ is a solution of the *partial trigonometric moment problem* [13] relative to C_n. Here, we assume that C_n is positive definite, which means exactly that the function ω has more than n points of increase.

With respect to any suitable measure $d\omega$ let us define the *positive definite inner product* $< \cdot, \cdot >$ over the space of complex polynomials of degree not exceeding n by setting

$$<p(z), q(z)> = \int_0^{2\pi} p(e^{i\theta}) \, \overline{q}(e^{i\theta}) \, d\omega(\theta), \tag{2.12}$$

with $p(z) = \sum_{j=0}^n p_j z^j$ and $q(z) = \sum_{j=0}^n q_j z^j$ for arbitrary complex numbers p_j and q_j. If $\mathbf{p} = [p_0, \ldots, p_n]^T$ and $\mathbf{q} = [q_0, \ldots, q_n]^T$ denote the coefficient vectors of $p(z)$ and $q(z)$, then, by use of (2.11), we can rewrite the inner product (2.12) in the form

$$<p(z), q(z)> = \mathbf{q}^* C_n \, \mathbf{p}. \tag{2.13}$$

The polynomials $a_k(z) = 1 + a_{k,1} z + \cdots + a_{k,k} z^k$ built from the variables processed by the Levinson algorithm (2.7)–(2.9), for $k = 0, 1, \ldots, n$, are known in the DSP literature as the *forward predictor polynomials*. In view of (2.6) and (2.11), they are orthogonal to each of their nonconstant monomials, i.e.,

$$<a_k(z), z^j> = 0, \qquad j = 1, 2, \ldots, k, \tag{2.14}$$

and to satisfy the relation $<a_k(z), a_k(z)> = \sigma_k$. If, instead of the $a_k(z)$, we consider the associated reciprocal polynomials $b_k(z) = \hat{a}_k(z) = z^k \overline{a}_k(1/\overline{z})$, termed *backward predictor polynomials* in the DSP environment, then we can translate the orthogonality relations above into

$$<b_k(z), b_\ell(z)> = \sigma_k \, \delta_{k,\ell}, \qquad k, \ell = 0, 1, \ldots, n. \tag{2.15}$$

Therefore, the monic polynomials $b_0(z), b_1(z), \ldots, b_n(z)$ constitute a family of *Szegö polynomials orthogonal on the unit circle*, with respect to the measure $d\omega$ [5], [6]. Such polynomials are known to satisfy a recurrence relation of the form

$$b_k(z) = z \, b_{k-1}(z) + \overline{\rho}_k \, \hat{b}_{k-1}(z), \tag{2.16}$$

where the coefficients ρ_k, called *Szegö parameters* in this context, obey the inequality $|\rho_k| < 1$, as a consequence of $\sigma_k = \sigma_{k-1}(1 - |\rho_k|^2)$. A comparison of (2.8) with (2.16) reveals that the Levinson algorithm can be viewed as a numerical implementation of the Szegö recurrence relation, supplemented with formula (2.7) so as to allow the recursive computation of the predictor polynomials $a_k(z)$.

In the same context let us also briefly mention the *Schur-Cohn stability criterion* [33] to check whether a given polynomial has all its zeros in the open unit disc $|z| < 1$, which is a crucial question in digital filter design. The descending version (from degree k to degree $k - 1$) of the Szegö recurrence relation (2.16) can be recognized in the numerical recursion involved in this criterion [4], [34].

2.5 Function theoretic applications of Szegö's theory

The Levinson algorithm provides a good illustration of the relevance of the algebraic properties of the Szegö orthogonal polynomials in DSP applications. There exists however a more fundamental reason that explains the primary importance of the general theory of Szegö's polynomials in digital signal processing: this theory constitutes a natural framework for the study of *Carathéodory* (or *positive*) *functions* and of *Schur* (or *bounded*) *functions* [13]. These two classes of functions (which are equivalent under a simple bilinear transformation) are often considered in the technical literature because their specific properties correspond to some physical constraints relative to various engineering problems, such as spectral analysis, digital filter design, or inverse scattering synthesis, just to mention a few [4], [15], [35], [36]. Roughly speaking, the properties of these functions reflect the fact that the DSP process under consideration is stable and passive. Explaining the rôle played by the theory of orthogonal polynomials on the unit circle in this function theoretic context goes beyond the scope of the present contribution. The interested reader is especially referred to [4] and to the references therein for details on the question.

Remark The Levinson algorithm can be generalized so as to accommodate positive definite *block Toeplitz matrices*. The resulting algorithm can also be viewed as a numerical implementation of suitable Szegö type recurrence relations satisfied by families of matrix polynomials orthogonal on the unit circle [25]. Compared to the scalar situation, the only noticeable difference lies in the fact that *two* families of matrix polynomials have to be introduced; they are respectively left orthogonal and right orthogonal with respect to the matrix measure associated with the given block Toeplitz matrix.

3 Even/odd polynomials orthogonal on the interval $[-1, 1]$, and the split Levinson algorithm

For reasons to be explained later, we shall restrict our attention in the present section to the case of *real data*. Thus, the Toeplitz matrix C_n is assumed to be real symmetric and positive definite, which implies the predictor polynomials $a_k(z)$ and the Szegö parameters ρ_k to be real.

3.1 The split Levinson algorithm

Rather surprisingly, the Levinson algorithm has been shown recently to be redundant in computational complexity [12]. It turns out that about half the number of multiplications and half the number of variables in the Levinson algorithm can be saved by suitably exploiting the inherent redundancy hidden in the family of predictor polynomials $a_k(z)$. This result is achieved in what is called the split Levinson algorithm by substituting an appropriate family of symmetric polynomials for the original family.

Let us now explain the construction, and the mechanism of the algorithm. From the family of predictor polynomials $a_k(z)$ we define a new family of polynomials, denoted $p_k(z)$, sometimes called *singular predictor polynomials* in the DSP literature, by the formula

$$p_k(z) = p_k(0)[a_{k-1}(z) + z\,\hat{a}_{k-1}(z)], \qquad k = 1, 2, \ldots, n+1, \qquad (3.1)$$

together with $p_0(z) = 2$ (by convention). The coefficients $p_k(0)$ in (3.1) are appropriate positive numbers, chosen as indicated below. Note the resulting *symmetry relation* $\hat{p}_k(z) = p_k(z)$, i.e., $p_{k,j} = p_{k,k-j}$, for all j and k.

The counterpart of the scalar product formula (2.7) in terms of the coefficients $p_{k,j}$ of $p_k(z)$ is found, with the help of (2.6), to assume the form

$$\tau_k = \sum_{j=0}^{k} c_j\, p_{k,j}, \tag{3.2}$$

for $k = 1, 2, \ldots, n$. Note that $\tau_k = p_k(0)\sigma_{k-1}(1 - \rho_k)$. Relation (3.2) defines the positive number τ_k from $p_k(z)$ just as σ_k is defined from $a_k(z)$. Let us then recursively freeze the normalization coefficients $p_k(0)$ by imposing the constraints

$$p_k(0) = c_0/\tau_{k-1}, \qquad k = 1, 2, \ldots, n + 1, \tag{3.3}$$

with the initialization $\tau_0 = c_0$. By use of (3.1) and (3.2), together with the identity $a_k(z) + \hat{a}_k(z) = (1 + \rho_k)\left[a_{k-1}(z) + z\,\hat{a}_{k-1}(z)\right]$ deduced from (2.8), it is shown, after elementary manipulations, that the predictor polynomial $a_k(z)$ and the prediction error variance σ_k can be recovered from $p_k(z)$ and $p_{k+1}(z)$ by the simple expressions

$$a_k(z) = \frac{p_{k+1}(z) - \lambda_{k+1}\, z\, p_k(z)}{p_{k+1}(0)(1 - z)}, \qquad \sigma_k = \frac{c_0\,\lambda_{k+1}}{p_{k+1}^2(0)}, \tag{3.4}$$

with $\lambda_{k+1} = p_{k+1}(0)\,\sigma_k/\tau_k$, actually evaluated as $\lambda_{k+1} = p_{k+1}(1)/p_k(1)$. The positive numbers λ_k thus defined play an important rôle in the present approach; they are referred to as the *Jacobi parameters* of the problem.

Inserting the expression (3.4) of $a_k(z)$ into the Szegö-Levinson formula $a_k(z) = a_{k-1}(z) + \rho_k\, z\, \hat{a}_{k-1}(z)$ yields the remarkable three-term recurrence relation

$$p_{k+1}(z) = \alpha_k(1 + z)\, p_k(z) - z\, p_{k-1}(z), \tag{3.5}$$

where the coefficients α_k are given by

$$\alpha_k = \tau_{k-1}/\tau_k, \tag{3.6}$$

as can easily be verified. Formula (3.5) constitutes the corner-stone of (one of the possible versions of) the *split Levinson algorithm* [12], which consists of the following three stages:

1. set the initializations $p_0(z) = 2$, $p_1(z) = 1 + z$, and $\tau_0 = c_0$;

2. compute the family of polynomials $p_k(z)$ by recursive use of (3.2), (3.5) and (3.6) for $k = 1, 2, \ldots, n$;

3. determine $a_n(z)$ and σ_n by use of (3.4) with $k = n$.

The complexity reduction of the split Levinson algorithm with respect to its standard counterpart is achieved by taking benefit of the symmetry property $\hat{p}_k(z) = p_k(z)$; the reader is referred to [12] for more details.

It can be shown that the companions of the Levinson algorithm mentioned in section 2 also admit parsimonious reformulations, based on the "split technique" [37], [38], [39], [40]. Furthermore, let us point out that there exist essentially four different versions of the split Levinson algorithm; roughly speaking, the other three versions are obtained by slightly altering the definition (3.1) of $p_k(z)$ from $a_{k-1}(z)$ so as to produce the relations $\hat{p}_k(z) = \varepsilon_k\, p_k(z)$ with $\varepsilon_k = -1$, $(-1)^k$, or $(-1)^{k+1}$, respectively, instead of $\varepsilon_k = 1$. These four versions have equivalent properties (see details in [20]).

3.2 Properties of the singular predictor polynomials

From (2.6), (3.1) and (3.2), the coefficient vector p_k of $p_k(z)$ is seen to obey the system of linear equations $C_k p_k = [\tau_k, 0, \ldots, 0, \tau_k]^T$, for $1 \leq k \leq n$. Therefore, the family of singular predictor polynomials $p_k(z)$ can be viewed as "*quasi orthogonal*" on the unit circle (with respect to the measure $d\omega$ that underlies the Toeplitz matrix C_n), in the sense that it satisfies the inner product relations

$$<p_k(z), z^j> = 0, \qquad j = 1, 2, \ldots, k-1, \tag{3.7}$$

for $k = 2, \ldots, n$. This should be compared with (2.14).

From Szegö's orthogonal polynomial theory (or from the Schur-Cohn criterion), the predictor polynomials $a_k(z)$ are known to be devoid of zeros in the closed unit disc $|z| \leq 1$. A standard argument of polynomial stability theory applied to definition (3.1) then shows that the zeros of the singular predictor polynomials $p_k(z)$ are simple and are located on the unit circle $|z| = 1$. Furthermore, the rational function $p_k(z)/(1-z) p_{k-1}(z)$ can be proved to be a Carathéodory function of lossless type and of degree exactly k (see [19]); this implies that the zeros of $p_k(z)$ interlace those of $(1-z) p_{k-1}(z)$ (on the unit circle).

Let us stress that the Jacobi parameters $\lambda_k = p_k(1)/p_{k-1}(1)$ are well-defined *positive numbers*, for $k = 1, 2, \ldots, n+1$, as a consequence of the positive definiteness of C_n. Moreover, it is seen from (3.5) that the λ_k's can alternatively be determined from the recursion

$$\lambda_{k+1} = 2\alpha_k - \lambda_k^{-1}, \qquad k = 0, 1, \ldots, n, \tag{3.8}$$

initialized with $\alpha_0 = 1/2$ and $\lambda_0 = \infty$, which yields $\lambda_1 = 1$. There exists an interesting reformulation of (3.8) in terms of the Jacobi (or tridiagonal) matrices

$$J_k = \begin{bmatrix} 2\alpha_0 & -1 & & \\ -1 & 2\alpha_1 & \ddots & \\ & \ddots & \ddots & -1 \\ & & -1 & 2\alpha_k \end{bmatrix}. \tag{3.9}$$

A comparison of (3.9) with (3.8) reveals the identity

$$\lambda_k = \det J_{k-1}/\det J_{k-2}, \qquad k = 2, 3, \ldots, n+1. \tag{3.10}$$

Thus the Jacobi matrix J_n and the Toeplitz matrix C_n are positive definite simultaneously (see details in [19]).

3.3 Orthogonal polynomials on $[-1, 1]$

Let $[\phi_k(x) : k = 0, 1, \ldots, n+1]$ be a family of polynomials orthogonal on the real interval $[-1, 1]$ relatively to a positive measure $d\mu$ that is *symmetric* with respect to 0. The orthogonality relations are of the form

$$\int_{-1}^{1} \phi_k(x) \phi_\ell(x) \, d\mu(x) = \gamma_k \delta_{k,\ell}, \qquad k, \ell = 0, 1, \ldots, n+1, \tag{3.11}$$

where the γ_k's are some positive numbers. Our symmetry assumption implies that $\phi_k(x)$ is an *even* or *odd* function of x, depending on the parity of the degree k; thus, we have

$\phi_k(-x) = (-1)^k \phi_k(x)$. For a suitable normalization, such a family is known to satisfy a three-term recurrence relation of the form

$$\phi_{k+1}(x) = 2\alpha_k \, x \, \phi_k(x) - \phi_{k-1}(x), \tag{3.12}$$

for well-defined positive numbers α_k. Conversely, it can be shown that, given a real sequence $(\alpha_0, \alpha_1, \ldots, \alpha_n)$ with $\alpha_0 > 0$, the polynomials $\phi_k(x)$ generated by formula (3.12) from the initial values $\phi_0(x) = \phi_{0,0} > 0$ and $\phi_{-1}(x) = 0$ are orthogonal on the interval $[-1, 1]$ relatively to a symmetric measure $d\mu$ under the necessary and sufficient positivity condition

$$\phi_k(1) > 0, \qquad k = 1, 2, \ldots, n+1. \tag{3.13}$$

With such a family of orthogonal polynomials let us associate the family of functions $p_k(z)$ defined by

$$p_k(z) = z^{k/2} \, \phi_k(x), \qquad \text{with } x = \frac{1}{2}(z^{1/2} + z^{-1/2}). \tag{3.14}$$

The even/odd property of the $\phi_k(x)$ polynomials is easily seen to imply that $p_k(z)$ is a symmetric polynomial of degree k in the variable z. Moreover, when expressed in terms of these $p_k(z)$ polynomials, the classical recurrence relation (3.12) is found to read

$$p_{k+1}(z) = \alpha_k(1 + z) \, p_k(z) - z \, p_{k-1}(z). \tag{3.15}$$

This is identical with the formula (3.5) that underlies the split Levinson algorithm. Note also that the condition (3.13) can be written as $p_k(1) > 0$, since we have $p_k(1) = \phi_k(1)$.

Conversely, let $[p_k(z) : k = 0, 1, \ldots, n+1]$ be the family of singular predictor polynomials produced by the split Levinson algorithm, from a given positive definite Toeplitz matrix C_n. The transformation (3.14) is readily verified to define a family of even/odd polynomials $\phi_k(x)$ that obeys the recurrence relation (3.12), as a consequence of (3.15). Furthermore, the identity $\phi_k(1) = p_k(1)$ implies that the constraint (3.13) is satisfied, as a consequence of the positivity of the Jacobi parameters λ_k. Therefore, the resulting polynomials $\phi_k(x)$ are orthogonal on the interval $[-1, 1]$ with respect to a symmetric measure $d\mu$. In fact, this measure can be expressed in terms of the measure $d\omega$ of Szegö's theory by the simple relation $d\mu(x) = d\omega(\theta)$ with $x = \cos \theta/2$.

To sum up, formulas (3.4) and (3.14) establish a one-to-one correspondence between the families of real polynomials $b_k(z) = \hat{a}_k(z)$ orthogonal on the unit circle and the families of even/odd polynomials $\phi_k(x)$ orthogonal on the interval $[-1, 1]$ of the real line.

Let us illustrate this correspondence by an example [17]. Consider the family of *ultraspherical polynomials* $C_k^\nu(x)$, defined by the recurrence relation

$$C_{k+1}^\nu(x) = x \, C_k^\nu(x) - \frac{k(k + 2\nu - 1)}{4(k + \nu)(k + \nu - 1)} \, C_{k-1}^\nu(x), \tag{3.16}$$

for any value of the parameter ν subject to the constraint $\nu > -1/2$. These polynomials are known to be orthogonal on the interval $[-1, 1]$ relatively to a well-defined symmetric measure. Renormalizing $C_k^\nu(x)$ so as to put (3.16) in the desired form (3.12), we find that the corresponding coefficients α_k are determined by the identity

$$\frac{1}{\alpha_k \, \alpha_{k-1}} = \left(1 + \frac{\nu}{k + \nu - 1}\right) \left(1 - \frac{\nu}{k + \nu}\right). \tag{3.17}$$

On the other hand, an elementary manipulation involving (3.2), (3.3) and (3.6) provides the equality $(\alpha_k \, \alpha_{k-1})^{-1} = (1 + \rho_{k-1})(1 - \rho_k)$. Consequently, the family of Szegö polynomials on the unit circle that is associated with the ultraspherical family is generated by the recurrence (2.15) where the Szegö parameters ρ_k are given by

$$\rho_k = \nu/(k+\nu), \qquad k = 1, 2, \ldots \tag{3.18}$$

Note that the inequality $\nu > -1/2$ precisely amounts to $|\rho_k| < 1$ for all k. Let us also observe that the Chebyshev polynomials, which are the ultraspherical polynomials with $\nu = 0$, have $\rho_k = 0$ and thus admit the monomials $b_k(z) = z^k$ as their associated Szegö polynomial (the Toeplitz matrix C_n reduces to a multiple of the unit matrix).

Remark In the next section, the general philosophy underlying the split Levinson algorithm will be shown to produce a new efficient approach to the theory of Szegö polynomials that applies not only to the real case but also to the complex case. In fact, it applies even to the matrix case. However, it must be emphasized that the correspondence described above between the theories of orthogonal polynomials on the unit circle and on the real line is restricted to the case of real scalar data exclusively.

4 Tridiagonal approach to Szegö orthogonal polynomials and related questions

This section is devoted to a study of tridiagonal matrix binomials of a special form, which are intimately related to the theory of Szegö polynomials, Toeplitz matrices and their mathematical environment. An essential feature of this recently proposed approach [19], [21], [22] is that it involves symmetric polynomials only, and thus yields attractive substitutes for the computational techniques that are classically used in the field. Part of the material below is an extension of the contents of subsections 3.1 and 3.2 above to the case of complex (scalar) data.

4.1 Basic definitions

Let there be given a sequence $(\alpha_0, \alpha_1, \ldots, \alpha_n)$ of nonzero complex numbers α_k. For the sake of consistency with the results of the preceding section, we assume $|\alpha_0| \geq 1/2$, which entails no loss of generality. Then, there exists a complex number ζ_0 that satisfies $|\zeta_0| = 1$ and $\mathrm{Re}(\zeta_0^{-1/2}\alpha_0) = 1/2$. (Here and in the sequel, $\zeta_0^{1/2}$ denotes either of the square roots of ζ_0.) From the sequence $(\alpha_k)_{k=0}^n$ let us construct the nested set of tridiagonal matrix binomials

$$J_k(z) = \begin{bmatrix} \alpha_0 + \overline{\alpha}_0 z & -1 & & \\ -z & \alpha_1 + \overline{\alpha}_1 z & \ddots & \\ & \ddots & \ddots & -1 \\ & & -z & \alpha_k + \overline{\alpha}_k z \end{bmatrix}, \tag{4.1}$$

for $k = 0, 1, \ldots, n$. Then, we define the family of complex polynomials

$$p_k(z) = p_0(z)\det J_{k-1}(z), \qquad k = 1, 2, \ldots, n+1, \tag{4.2}$$

with $p_0(z) = \pm|\alpha_0|^{-1}$. Thus, $p_k(z)$ has degree k and exhibits the *symmetry property*

$$\hat{p}_k(z) = p_k(z). \tag{4.3}$$

(As usual, $\hat{v}_k(z)$ denotes the reciprocal of a polynomial $v_k(z)$ of formal degree k; it is defined by $\hat{v}_k(z) = z^k \overline{v}_k(1/\overline{z})$.) Moreover, the $p_k(z)$ family obeys the recurrence relation

$$p_{k+1}(z) = (\alpha_k + \overline{\alpha}_k z)\, p_k(z) - z\, p_{k-1}(z)\,, \tag{4.4}$$

for $k = 0, 1, \ldots, n$, with the initial values $p_0(z) = \pm|\alpha_0|^{-1}$ and $p_{-1}(z) = 0$. This follows from (4.2), by applying the Laplace expansion rule to the determinant of (4.1).

Henceforth, let us *assume the Jacobi matrix* $\zeta_0^{-1/2} J_n(\zeta_0)$ *to be positive definite*. In view of (4.2), this assumption is equivalent to the positivity condition

$$\lambda_k > 0\,, \qquad k = 1, 2, \ldots, n+1\,, \tag{4.5}$$

where the λ_k's, termed *Jacobi parameters*, are defined by

$$\lambda_k = \zeta_0^{-1/2}\, p_k(\zeta_0)/p_{k-1}(\zeta_0)\,. \tag{4.6}$$

Note the identity $\lambda_1 = 1$, which results from our particular choice of ζ_0. In view of (4.4), the Jacobi parameters can alternatively be computed by means of the recursive formula

$$\lambda_{k+1} = 2\mathrm{Re}\,(\alpha_k\, \zeta_0^{-1/2}) - \lambda_k^{-1}\,, \tag{4.7}$$

for $k = 0, 1, \ldots, n$, with the initialization $\lambda_0 = \infty$.

4.2 Associated Szegö polynomials

From the family of symmetric polynomials $p_k(z)$ let us construct the family of polynomials $a_k(z)$, for $0 \le k \le n+1$, as follows:

$$a_k(z) = \begin{cases} p_{k+1}^{-1}(0)[p_{k+1}(z) - \lambda_{k+1}\, \zeta_0^{-1/2}\, z\, p_k(z)]/(1 - \zeta_0^{-1} z)\,, & k = 0, \ldots, n\,, \\[2mm] p_{n+1}^{-1}(0)\, p_{n+1}(z)\,, & k = n+1\,. \end{cases} \tag{4.8}$$

Note that $a_k(z)$ is actually a polynomial of formal degree k, in view of (4.6). For future use, let us define the parameters $\rho_k = a_{k,k}$ (the coefficient of z^k in $a_k(z)$), and $\omega_k = \overline{p}_k(0)/p_k(0)$. Thus, we have $|\omega_k| = 1$. Taking into account the symmetry property (4.3), we deduce from (4.8) the formulas

$$\rho_k = \begin{cases} \zeta_0\, \omega_{k+1}(\lambda_{k+1}\, \zeta_0^{-1/2}\, \overline{\alpha}_k^{-1} - 1)\,, & k = 1, \ldots, n\,, \\[2mm] \omega_{n+1}\,, & k = n+1\,. \end{cases} \tag{4.9}$$

By tedious but elementary manipulations it can be shown that the recurrence relation (4.4) translates into the Szegö-Levinson type formula

$$a_k(z) = a_{k-1}(z) + \rho_k\, z\, \hat{a}_{k-1}(z)\,, \qquad k = 1, 2, \ldots, n+1\,. \tag{4.10}$$

Furthermore, we have $|\alpha_k|^2\,(1 - |\rho_k|^2) = \lambda_{k+1}/\lambda_k > 0$, whence $|\rho_k| < 1$, for $k = 1, \ldots, n$, due to (4.5), (4.7) and (4.9), and $|\rho_{n+1}| = 1$. Therefore, the reciprocals $b_k(z) = \hat{a}_k(z)$ of the polynomials $a_k(z)$ constitute a family of monic Szegö polynomials orthogonal on the unit circle with respect to a suitable positive measure. This measure is unique (within a constant factor); it is discrete and has $n+1$ mass points (see details below). Equivalently, there is a unique nonnegative definite Toeplitz matrix C_{n+1}, of order $n+2$ and rank $n+1$, associated with this family of Szegö polynomials. (More precisely, C_{n+1} is unique within an arbitrary choice of its diagonal entry c_0, subject to $c_0 > 0$.)

The Toeplitz matrix $C_{n+1} = [c_{i-j} : 0 \le i, j \le n+1]$, with $c_{-k} = \bar{c}_k$, can be determined from the data α_k in various ways. Let us now explain one of the possible derivations. The entries c_k can be proved to satisfy the relations

$$\tau_k = \sum_{j=0}^{k} c_{k-j}\, p_{k,j}\,, \qquad k = 1, 2, \ldots, n+1\,, \tag{4.11}$$

where the numbers τ_k are defined by

$$\tau_k = \begin{cases} c_0\, \zeta_0^{1/2}\, p_{k+1}^{-1}(0)\,, & k = 1, \ldots, n\,, \\ 0\,, & k = n+1\,. \end{cases} \tag{4.12}$$

Therefore, the entries of C_{n+1} can be computed from the diagonal entry c_0 (any positive number) with the help of the recursive formula

$$c_k = p_k^{-1}(0) \left[\tau_k - \sum_{j=1}^{k} c_{k-j}\, p_{k,j} \right], \qquad k = 1, \ldots, n+1\,, \tag{4.13}$$

Without going into detail, let us emphasize that the symmetric polynomials $p_k(z)$ enjoy the *quasi orthogonality property* $<p_k(z), z^j> = 0$ for $1 \le j \le k-1$ and $2 \le k \le n+1$, with respect to the positive measure underlying the Toeplitz matrix C_{n+1} (see [19], [20]).

Conversely, let us consider a family of monic polynomials $b_k(z) = \hat{a}_k(z)$, for $k = 0, 1, \ldots, n+1$, that satisfies the Szegö recurrence relation (2.16), (4.10), with the "singular property" $|\rho_{n+1}| = 1$ and the "regular property" $|\rho_k| < 1$ for $k = 1, \ldots, n$. We shall briefly indicate how a suitable complex $(n+1)$-tuple $(\alpha_0, \alpha_1, \ldots, \alpha_n)$ can be associated with this family of Szegö polynomials $b_k(z)$. To that end, we first select an arbitrary complex number ζ_0 of unit modulus, with the sole restriction $a_{n+1}(\zeta_0) \ne 0$. It turns out that the numbers α_k can be determined by recursive use of the formula

$$\alpha_k = \zeta_0 [(1 + \zeta_0\, \omega_k \bar{\rho}_{k-1})(1 - \bar{\omega}_k\, \rho_k)\alpha_{k-1}]^{-1}\,, \tag{4.14}$$

from the initial value $\alpha_0 = (\zeta_0^{-1/2} + \omega_1 \zeta_0^{1/2})^{-1}$. The parameters ω_k in (4.14) are those mentioned above; they can be obtained with the help of the recurrence relation

$$\omega_k = (\rho_k + \zeta_0\, \omega_{k+1})/(1 + \zeta_0\, \omega_{k+1}\, \bar{\rho}_k)\,, \tag{4.15}$$

for $k = n, n-1, \ldots, 1$, with the initialization $\omega_{n+1} = \rho_{n+1}$. As for the symmetric polynomials $p_k(z)$, they can be computed by means of the three-term recurrence (4.4) from the initial values $p_{-1}(z) = 0$ and

$$p_0(z) = 2\mathrm{Re}(\zeta_0^{1/2}\, \omega_1^{1/2})\,. \tag{4.16}$$

A *split Levinson algorithm* for complex data [18], [19], [20], [21] can be derived immediately from the results above. Just as the classical Levinson algorithm, it computes the predictor polynomial $a_n(z)$ and the prediction error variance σ_n relative to a given positive definite Toeplitz matrix C_n. It consists of the following three stages:

1. select two complex numbers ζ_0 and ω_1 satisfying $|\rho_0| = |\omega_1| = 1$ and $\zeta_0\, \omega_1 \ne -1$; set the initial values $\tau_0 = c_0$, $p_0(z)$ as in (4.16) and $p_1(z) = \omega_1^{-1/2} + \omega^{1/2} z$;

2. compute the family of symmetric polynomials $p_k(z)$ by recursive use of (4.4) with $\alpha_k = \tau_{k-1}/\tau_k$, where τ_k is obtained from (4.11);

3. determine $a_n(z)$ by use of (4.8), with $k = n$, where λ_{n+1} is given by (4.6), and determine σ_n by means of $\sigma_n = c_0 \lambda_{n+1} |p_{n+1}(0)|^{-2}$.

In case of a real Toeplitz matrix C_n, this algorithm reduces to the simple formulation given in subsection 3.1 provided the parameters ζ_0 and ω_1 are set equal to 1.

It is interesting to point out that, for a fixed choice of ζ_0, the parameter ω_1 produces a well-defined singular (nonnegative definite) extension C_{n+1} of the given positive definite Hermitian Toeplitz matrix C_n. Indeed, ω_1 is in one-to-one correspondence with the unit modulus Szegö parameter $\rho_{n+1} = \overline{p}_{n+1}(0)/p_{n+1}(0)$ that defines C_{n+1}. Note that all singular extensions C_{n+1} are obtainable in this way, except the one that corresponds to the choice $\omega_1 = \zeta_0^{-1}$ ruled out above. In fact, this choice is perfectly acceptable (although it yields the property $p_k(\zeta_0) = 0$ for all $k \geq 0$) for a slightly modified version of the split Levinson algorithm [20], [21].

4.3 Lossless rational functions

The results above show that a family of symmetric polynomials $[p_k(z) : k = 0, 1, \ldots, n+1]$ and a singular family of Szegö polynomials $[\hat{a}_k(z) : k = 0, 1, \ldots, n + 1]$ carry the same information. (Here, "singular" refers to the property $|\rho_{n+1}| = 1$.) Note however that, due to the symmetry relation (4.3), polynomial $p_k(z)$ is entirely determined by its $\lceil (k+1)/2 \rceil$ first coefficients, which is a significant advantage from a computational complexity viewpoint. Besides, it turns out that considering the $p_k(z)$ family instead of the $a_k(z)$ family often leads to interesting new approaches to standard questions.

To illustrate this, let us examine the problem of determining the $n + 1$ *mass points* $\zeta_1, \zeta_2, \ldots, \zeta_{n+1}$ (with $|\zeta_j| = 1$) and the corresponding *mass coefficients* $h_1, h_2, \ldots, h_{n+1}$ (with $h_j > 0$) of the discrete measure $d\omega$ that underlies a singular family of Szegö polynomials $[\hat{a}_k(z) : k = 0, 1, \ldots, n + 1]$. Thus, we can write the expressions

$$c_k = \int_0^{2\pi} e^{-ik\theta} \, d\omega(\theta) = \sum_{j=1}^{n+1} h_j \zeta_j^{-k}, \qquad (4.17)$$

for all integers k, where the numbers c_k with $0 \leq |k| \leq n+1$ are the entries of the nonnegative definite Hermitian Toeplitz matrix C_{n+1} (of nullity 1) that is associated with the given singular Szegö family [6]. From the positive measure $d\omega$ let us define the *Carathéodory function* [13]

$$f(z) = \int_0^{2\pi} \frac{e^{i\theta} + z}{e^{i\theta} - z} \, d\omega(\theta), \qquad (4.18)$$

i.e., $f(z) = c_0 + 2 \sum_{k=1}^{\infty} c_k z^k$. In view of (4.17) it appears that $f(z)$ is a *rational lossless function of degree* $n + 1$, since it can be written in the form

$$f(z) = \sum_{j=1}^{n+1} h_j \frac{\zeta_j + z}{\zeta_j - z}, \qquad (4.19)$$

with $|\zeta_j| = 1$ and $h_j > 0$ for $j = 1, 2, \ldots, n + 1$.

A classical result of Szegö polynomial theory [6] states that *the mass points ζ_j are the zeros of* $a_{n+1}(z)$, and that the mass coefficients h_j are given by the Christoffel formula

$$h_j = \left[\sum_{k=0}^{n} \sigma_k^{-1} |a_k(\zeta_j)|^2 \right]^{-1}. \qquad (4.20)$$

Thus, the second expression (4.8) allows us to write

$$p_{n+1}(z) = p_{n+1}(0) \prod_{j=1}^{n+1} (1 - \zeta_j^{-1} z).$$ (4.21)

Moreover, it can be shown that there exists a *Christoffel type formula* that expresses the mass coefficients in terms of the symmetric polynomials $p_k(z)$; it reads

$$h_j = c_0 |\zeta_0 - \zeta_j|^2 [\mathbf{p}^*(\zeta_j)\zeta_0^{-1/2} J_n(\zeta_0) \mathbf{p}(\zeta_j)]^{-1},$$ (4.22)

with the vector polynomial $\mathbf{p}(z) = [p_0(z), p_1(z), \ldots, p_n(z)]^T$, evaluated at the point $z = \zeta_j$. This result is proved in [22].

By use of (4.1) we can write the set of identities (4.4), with $0 \leq k \leq n$, in the form of the linear system

$$J_n(z)\,\mathbf{p}(z) = [0, \ldots, 0, p_{n+1}(z)]^T.$$ (4.23)

Consequently, the ingredients requisite to setting up the lossless function $f(z)$ through (4.21) and (4.22) can be defined by the values ζ_j that satisfy $\det J_n(\zeta_j) = 0$ and by the corresponding (suitably normalized) solution vectors $\mathbf{p}(\zeta_j)$ of the homogeneous linear system $J_n(\zeta_j)\,\mathbf{p}(\zeta_j) = 0$. Let us see how this problem can be stated as a *generalized eigen problem* [22]. To that end, we introduce the lower bidiagonal matrix

$$A = \zeta_0^{1/2} \begin{bmatrix} \bar{\alpha}_0 & & & \\ -1 & \bar{\alpha}_1 & & \\ & \ddots & \ddots & \\ & & -1 & \bar{\alpha}_n \end{bmatrix}.$$ (4.24)

This allows us to rewrite (4.23) in the form

$$(\zeta_0 A^* + z\,A)\,\mathbf{p}(z) = [0, \ldots, 0, \zeta_0^{1/2} p_{n+1}(z)]^T,$$ (4.25)

which shows that the problem above amounts to the generalized eigen problem for the matrix pencil $(\zeta_0 A^*, -A)$.

With the mass points $\zeta_j = \exp(i\theta_j)$ let us associate the real numbers $t_j = \cot(\theta_j - \theta_0)/2$. Then the system (4.25) with $z = \zeta_j$ can further be transformed into

$$(\mathrm{Im}\,A - t_j\,\mathrm{Re}\,A)\,\mathbf{p}(\zeta_j) = 0,$$ (4.26)

with $\mathrm{Re}\,A = (A + A^*)/2 = \zeta_0^{-1/2} J_n(\zeta_0)/2$ and $\mathrm{Im}\,A = (A - A^*)/2i$. Recall that $\mathrm{Re}\,A$ is positive definite. It is seen from (4.26) that the numbers t_j and the vectors $\mathbf{p}(\zeta_j)$ are the generalized eigenvalues and eigenvectors of the *tridiagonal Hermitian-definite pencil* ($\mathrm{Im}\,A$, $\mathrm{Re}\,A$). This result leads to an attractive procedure for computing the zeros ζ_j of $a_{n+1}(z)$ and the corresponding mass coefficients h_j, since there exist efficient algorithms to solve that type of generalized eigen problems [11].

The question of computing the mass points ζ_j (and possibly the mass coefficients h_j) from the underlying Szegö parameters ρ_k or from the entries c_k of the Toeplitz matrix C_{n+1} occurs in various interesting applications. Let us first mention that the *eigenvalue problem for unitary Hessenberg matrices* is of that type, and can thus be translated into the form (4.26). This stems from the fact that a unitary Hessenberg matrix of order $n + 1$ can be specified by means of a well-defined sequence $(\rho_k)_{k=1}^{n+1}$ of Szegő parameters ρ_k satisfying $|\rho_k| < 1$ for $k = 1, \ldots, n$, and $|\rho_{n+1}| = 1$. We shall not go into any detail about that subject; the reader is referred to [41], [42], [43], and [22]. Next, let us point out that the same basic problem occurs in some DSP applications and, more particularly, in a popular stationary stochastic process modelling technique known as the *Pisarenko technique* [23]. We shall briefly outline this significant DSP method.

Let there be given a positive definite Hermitian Toeplitz matrix C'_{n+1}, of order $n + 2$, and consider the problem of designing a discrete-time zero-mean stochastic process that admits the entries $c'_0, c'_1, \ldots, c'_{n+1}$ of C'_{n+1} as its first $n + 1$ covariances, i.e., that satisfies $E[x(t)\,\overline{x}(t - k)] = c'_k$ for $k = 0, 1, \ldots, n + 1$. Stated in such general terms, the problem has infinitely many solutions. Let us henceforth restrict our attention to the particular situation where the process $x(t)$ decomposes as the sum of two uncorrelated stationary stochastic processes, $y(t)$ and $v(t)$, where $v(t)$ is white noise of maximum variance λ. Thus, we have

$$c'_k = c_k + \delta_{k,0}\lambda, \qquad k = 0, 1, \ldots, n + 1, \tag{4.27}$$

with $c_k = E[y(t)\,\overline{y}(t - k)]$, and we require that λ be as large as possible. Consider the covariance matrix $C_{n+1} = C'_{n+1} - \lambda I$ of the $y(t)$ process. Since C_{n+1} has to be nonnegative definite, the maximum value of λ is identified at once to be the smallest eigenvalue, $\lambda_{\min}$, of C'_{n+1}. As a consequence, the Toeplitz matrix C_{n+1} is singular.

To avoid technicalities, we assume that the eigenvalue $\lambda_{\min}$ is simple, which implies that C_{n+1} has nullity 1. Therefore, as explained above, there exist $n + 1$ distinct numbers $\zeta_j = \exp(i\theta_j)$ of unit modulus, together with $n + 1$ positive numbers h_j, with $j = 1, 2, \ldots, n + 1$, such that the entries c_k of C_{n+1} can be written in the form (4.17). Interpreted in terms of random variables, this shows that the $y(t)$ process can be modelled as the output sum of $n + 1$ sinusoidal wave generators of frequencies $\theta_j/2\pi$ and amplitudes $h_j^{1/2}$ whose phases ϕ_j are uncorrelated zero-mean random variables. Thus, we can write

$$y(t) = \sum_{j=1}^{n+1} h_j^{1/2}\, e^{-i(t\theta_j + \phi_j)}. \tag{4.28}$$

To sum up, the Pisarenko modelling technique realizes a stationary stochastic process of given covariance matrix C'_{n+1} as the output signal that results from the parallel connection of a white noise generator of variance $\lambda_{\min}$ and of $n + 1$ sinusoidal wave generators, which can be constructed directly from the representation (4.17) of the entries c_k of the Toeplitz matrix $C_{n+1} = C'_{n+1} - \lambda_{\min} I$.

It turns out that the results of this section can be simplified considerably in case of *real data* [12], [22], [38], [42], [44], [45]. On the basis of the correspondence (explained in subsection 3.3) between families of real Szegő polynomials and families of even/odd polynomials orthogonal on the $[-1, 1]$ interval, the generalized eigen problem (4.26) can be reduced, with the help of the three-term formula (3.12), to the ordinary eigen problem for a well-defined zero-diagonal Jacobi matrix.

4.4 Function theoretic aspects of the tridiagonal approach

The material of this paper gives only a partial account of the algebraic results that can be obtained from the tridiagonal approach to Szegő polynomials. The reader is referred to the literature and, more particularly, to recent papers by the authors [21], [22], for a more extensive treatment of the subject and of its applications.

Our tridiagonal approach has not only an algebraic facet, but also a quite interesting function theoretic facet. It has been shown to bring about a new characterization of

Carathéodory functions (or positive functions). The new criterion is (of course) equivalent to the celebrated Carathéodory-Fejér and Schur criteria [13], but it relies on a different kind of "parametrization". It involves as infinite sequence $[J_k(z) \,:\, 0 \leq k < \infty]$ of nested tridiagonal matrices $J_k(z)$ of form (4.1), whose elements α_k are computed recursively from the Maclaurin coefficients of the function under study. The criterion states that the given function is a Carathéodory function if and only if all the Jacobi parameters relative to the $J_k(z)$ matrices are nonnegative (roughly speaking). Furthermore, this characterization theorem leads to attractive new algorithms for the recursive solution of the Carathéodory-Fejér and Nevanlinna-Pick interpolation problems, which are known to play an important rôle in DSP applications (among other areas). A detailed treatment of this subject can be found in [19].

References

[1] T. Kailath, 'A view of three decades of linear filtering theory', IEEE Trans. Inform. Theory, **IT-20**, pp. 145–181, 1974.

[2] J. Makhoul, 'Linear prediction: a tutorial review', Proc. IEEE, **63**, pp. 561–580, 1975.

[3] W.B. Jones and A. Steinhardt, 'Applications of Schur fractions to digital filtering and signal processing', in *Rational Approximation and Interpolation*, R.P. Graves-Morris, E.B. Saff and R.S. Varga eds, Springer Lecture Notes Math. no 1105, Springer-Verlag, Berlin, pp. 210–226, 1984.

[4] Y. Genin, 'An introduction to the modern theory of positive functions and some of its today applications to signal processing, circuits and systems problems', ECCTD-87 Course, Proc. Europ. Conf. Circuit Theory and Design, **3**, pp. 195–234, 1987.

[5] G. Szegö, *Orthogonal Polynomials*, American Mathematical Society, New York, 1959.

[6] U. Grenander and G. Szegö, *Toeplitz Forms and their Applications*, University of California Press, Berkeley, 1958.

[7] A.V. Oppenheim and R.W. Schafer, *Digital Signal Processing*, Prentice-Hall, Englewood Cliffs, 1975.

[8] J.D. Markel and A.H. Gray, Jr., *Linear Prediction of Speech*, Springer-Verlag, Berlin, 1976.

[9] T. Kailath, ed., *Linear Least-Squares Estimation*, Benchmark Papers in Electrical Engineering and Computer Science, Dowden, Hutchinson & Ross, **17**, 1977.

[10] N. Levinson, 'The Wiener rms (root mean square) error criterion in filter design and prediction', J. Math. Phys., **25**, pp. 261–278, 1946.

[11] G.H. Golub and C.F. Van Loan, *Matrix Computations*, North Oxford Academic, Oxford, 1983.

[12] P. Delsarte and Y. Genin, 'The split Levinson algorithm', IEEE Trans. Acoust. Speech Signal Process., **ASSP-34**, pp. 470–478, 1986.

[13] N.I. Akhiezer, *The Classical Moment Problem*, Oliver and Boyd, London, 1965.

[14] V. Belevitch, *Classical Network Theory*, Holden-Day, San Francisco, 1968.

[15] A. Fettweis, 'Pseudopassivity, sensitivity, and stability of wave digital filters', IEEE Trans. Circuit Theory, **CT-19**, pp. 668–673, 1972.

[16] H. Lev-Ari and T. Kailath, 'Lattice filter parametrization and modeling of nonstationary processes', IEEE Trans. Inform. Theory, **IT-30**, pp. 2–16, 1984.

[17] Y. Genin, 'On a duality relation in the theory of orthogonal polynomials and its application in signal processing', Proc. 1st Internat. Conf. Industrial and Applied Mathematics, Paris, pp. 102–113, 1987.

[18] H. Krishna and S.D. Morgera, 'The Levinson recurrence and fast algorithms for solving Toeplitz systems of linear equations', IEEE Trans. Acoust. Speech Signal Process., **ASSP-35**, pp. 839–848, 1987.

[19] P. Delsarte and Y. Genin, 'The tridiagonal approach to Szegö's orthogonal polynomials, Toeplitz linear systems and related interpolation problems', SIAM J. Math. Anal., **19**, pp. 718–735, 1988.

[20] P. Delsarte and Y. Genin, 'An introduction to the class of split Levinson algorithms', in *Numerical Linear Algebra, Digital Signal Processing and Parallel Algorithms*, G.H. Golub and P. Van Dooren eds, Springer NATO-ASI Series (to appear).

[21] P. Delsarte and Y. Genin, 'Tridiagonal approach to the algebraic environment of Toeplitz matrices, part I: basic results' (submitted for publication).

[22] P. Delsarte and Y. Genin, 'Tridiagonal approach to the algebraic environment of Toeplitz matrices, part II: zero and eigenvalue problems' (submitted for publication).

[23] V.P. Pisarenko, 'The retrieval of harmonics from a covariance function', Geophys. J. R. Astr. Soc., **33**, pp. 347–366, 1973.

[24] P. Delsarte and Y. Genin, 'The tridiagonal approach to inverse scattering problems', Proc. IEEE Internat. Symp. Circuits and Systems, Philadelphia, **1**, pp. 140–142, 1987.

[25] P. Delsarte, Y. Genin, and Y. Kamp, 'Orthogonal polynomial matrices on the unit circle', IEEE Trans. Circuits and Systems, **CAS-25**, pp. 149–160, 1978.

[26] Y. Bistritz, H. Lev-Ari, and T. Kailath, 'Immittance-domain Levinson algorithms', Proc. IEEE Internat. Conf. Acoustics Speech Signal Processing, pp. 253–256, 1986.

[27] P. Delsarte and Y. Genin, 'The multichannel split Levinson algorithm' in *Linear Circuits, Systems and Signal Processing Theory and Applications*, C.I. Byrnes, C.F. Martin and R.E. Saeks eds, North-Holland, Amsterdam, pp. 183–190, 1988.

[28] P. Delsarte and Y. Genin, 'Multichannel singular predictor polynomials', IEEE Trans. Circuits and Systems, **CAS-35**, pp. 190–200, 1988.

[29] G. Cybenko, 'The numerical stability of the Levinson-Durbin algorithm for Toeplitz systems of equations', SIAM J. Sci. Statist. Comput., **1**, pp. 303–320, 1980.

[30] J. Le Roux and C. Gueguen, 'A fixed point computation of partial correlation coefficients', IEEE Trans. Acoust. Speech Signal Process., **ASSP-25**, pp. 257–259, 1977.

[31] F. Itakura and S. Saito, 'Digital filtering techniques for speech analysis and synthesis', Proc. 7th Internat. Congr. Acoust., Budapest, pp. 261–264, 1971.

[32] T. Kailath, S.-Y. Kung, and M. Morf, 'Displacament ranks of matrices and linear equations', J. Math. Anal. Appl., 68, pp. 395–407, 1979.

[33] M. Marden, *Geometry of Polynomials*, American Mathematical Society, Providence, RI, 1966.

[34] A. Vieira and T. Kailath, 'Another approach to the Schur-Cohn criterion', IEEE Trans. Circuits and Systems, CAS-24, pp. 218–220, 1977.

[35] A.H. Gray, Jr. and J.D. Markel, 'Digital lattice and ladder filter synthesis', IEEE Trans. Audio Electroacoust., AU-21, pp. 491–500, 1973.

[36] P. Dewilde and H. Dym, 'Lossless inverse scattering, digital filters, and estimation theory', IEEE Trans. Inform. Theory, IT-30, pp. 644–662, 1984.

[37] P. Delsarte and Y. Genin, 'On the splitting of classical algorithms in linear prediction theory', IEEE Trans. Acoust. Speech Signal Process., ASSP-35, pp 645–653, 1987.

[38] P. Delsarte and Y. Genin, 'A survey of the split approach based techniques in digital signal processing applications', Philips J. Res., 43, pp. 346–374, 1988.

[39] P. Delsarte and Y. Genin, 'An extension of the split Levinson algorithm and its relatives to the joint process estimation problem', IEEE Trans. Inform. Theory (to appear).

[40] P. Delsarte and Y. Genin, 'On the split approach based algorithms for DSP problems', in *Numerical Linear Algebra, Digital Signal Processing and Parallel Algorithms*, G.H. Golub and P. Van Dooren eds, Springer NATO-ASI Series (to appear).

[41] H. Kimura, 'Generalized Schwarz form and lattice-ladder realizations of digital filters', IEEE Trans. Circuits and Systems, CAS-32, pp. 1130–1139, 1985.

[42] G.S. Ammar, W.B. Gragg, and L. Reichel, 'On the eigenproblem for orthogonal matrices', Proc. 25th IEEE Conf. Decision and Control, Athens, 1986.

[43] W.B. Gragg, 'The QR algorithm for unitary Hessenberg matrices', J. Comp. Appl. Math., 16, pp. 1–8, 1986.

[44] G. Cybenko, 'Computing Pisarenko frequency estimates', Proc. Internat. Conf. Systems and Sciences, Princeton University, 1984.

[45] L. Reichel and G. Ammar, 'Fast approximation of dominant harmonics by solving an orthogonal eigenvalue problem', Proc. 2nd IMA Conf. Mathematics in Signal Processing, University of Warwick, Coventry, 1988.

A SURVEY ON THE THEORY OF ORTHOGONAL SYSTEMS AND SOME OPEN PROBLEMS

M. M. Djrbashian

Institute of Mathematics
Marshall Bagramian St. 24 B
375019 - Yerevan, Armenian SSR, USSR

The present report is a brief survey of the major results established in a series of works by the author. The information we have makes it possible to conclude that these results have been overlooked by the overwhelming majority of mathematicians working in general orthogonal systems and, in particular, in orthogonal polynomials. We did not include in this report a number of the author's results on the theory of bi-orthogonal systems, which are far-reaching generalizations of the theory of Fourier orthogonal systems in the complex domain.

§1. Orthogonal systems of rational functions on the circle with a given set of poles, generalization of the Szegő polynomials.

The results formulated below were first announced in [1], and subsequently they were proved in [2].

1. Let $\{\alpha_k\}_0^\infty$ where $|\alpha_k| < 1$ be an arbitrary sequence of complex numbers and let some of them have finite or even infinite multiplicity (it is not necessary for them to appear successively). We associate with the sequence $\{\alpha_k\}_0^\infty$ the sequence of rational functions

$$\left\{ \frac{z^{s_k-1}}{(1 - \overline{\alpha_k} \cdot z)^{s_k}} \right\}_0^\infty , \tag{1.1}$$

where $s_k \geq 1$ is the multiplicity of occurence of α_k in the interval $\{\alpha_0, \alpha_1, \ldots, \alpha_k\}$ of our sequence. A somewhat different, less precise definition of the same system was given in [1] and [2]. Given $k \geq 0$, if $\alpha_k \neq 0$ then the function $z^{s_k-1}/(1 - \overline{\alpha}_k \cdot z)^{s_k}$ has a pole at $z = 1/\overline{\alpha}_k$ (here $|1/\overline{\alpha}_k| > 1$) of order s_k, and if $\alpha_k = 0$ then it has a pole at $z = \infty$ of the order $s_k - 1$. Observe also that in the extreme case $\alpha_k = 0$ $(k = 0, 1, 2, \ldots)$ we have $s_k = k+1$ $(k \geq 0)$, hence to the sequence $\{0\}_0^\infty$ there corresponds the sequence $\{z^k\}_0^\infty$ of natural degrees of the argument z. The so-called Malmquist system of rational functions (cf. [3]) is defined in the following way

$$\varphi_n(z) = \frac{(1 - |\alpha_n|^2)^{1/2}}{1 - \overline{\alpha}_n \cdot z} \prod_{k=0}^{n-1} \frac{\alpha_k - z}{1 - \overline{\alpha}_k \cdot z} \cdot \frac{|\alpha_k|}{\alpha_k} \qquad (n = 0, 1, 2, \ldots) \tag{1.2}$$

135

P. Nevai (ed.), Orthogonal Polynomials, 135–146.
© *1990 by Kluwer Academic Publishers.*

where for $\alpha_k = 0$ we put $|\alpha_k|/\alpha_k = \overline{\alpha}_k/|\alpha_k| = -1$. It is easy to show that this Malmquist system is orthonormal on the unit circle in the sense that

$$\frac{1}{2\pi}\int_{-\pi}^{\pi}\varphi_n(z)\cdot\overline{\varphi_m(z)}dx = \delta_{n,m} \qquad (n,m = 0,1,2,\ldots; z = e^{ix}). \tag{1.3}$$

This system is the result of orthogonalization of the ordered sequence of functions of the system (1.1) on the unit circle $z = e^{ix}$ $(-\pi \le x \le \pi)$ with the weight function $dx/2\pi$.

2. Let $\alpha(x)$ be a bounded nondecreasing function in the interval $[-\pi,\pi]$ with infinitely many points of increase. Then denote by $\{\phi_k(z)\}_0^\infty$ the orthogonalization of the ordered system of functions (1.1) or, which is the same, the Malmquist system (1.2), on the unit circle $z = e^{ix}$ $(-\pi \le x \le \pi)$ with the same weight function $1/2\pi\, d\alpha(x)$. Thus, we shall come to the sequence of rational functions $\{\phi_n(z)\}_0^\infty$, which satisfies the following conditions and is uniquely determined by them: $\phi_n(z)$ is "a polynomial of degree n" of the first $n+1$ Malmquist functions:

$$\phi_n(z) = \sum_{k=0}^{n} c_{k,n}\cdot\varphi_k(z), \qquad c_{n,n} = \kappa_n > 0,$$

and

$$\frac{1}{2\pi}\int_{-\pi}^{\pi}\phi_n(z)\cdot\overline{\phi_m(z)}d\alpha(x) = \delta_{n,m} \qquad (z = e^{ix}; \qquad n,m = 0,1,\ldots).$$

Notice that in the extreme case, when $\alpha_k = 0$ $(0 \le k < +\infty)$ the system of functions $\{\phi_n(z)\}_0^\infty$ turns into the system of Szegö [4] polynomials $\{P_n(z)\}_0^\infty$, which are orthogonal on the circle $z = e^{ix}$ with the same weight function $(1/2\pi)d\alpha(x)$.

3. The algebraic properties of the system $\{\phi_n(z)\}_0^\infty$ compared with the corresponding properties and relations of the Szegö polynomials, have more complicated nature and are established by more complicated constructions. These properties will be formulated in the theorems below. First consider the functions of two variables

$$S_n(\xi;z) = \sum_{k=0}^{n}\overline{\phi_k(\xi)}\cdot\phi_k(z) \qquad (n = 0,1,2,\ldots) \tag{1.4}$$

and call them again kernels of the distribution $(1/2\pi)d\alpha(x)$.

Theorem 1.1. *For every $n \ge 0$ and $\xi, z \in \mathbb{C}$ the kernel $S_n(\xi;z)$ satisfies the functional equation*

$$S_n(\xi;z) = \frac{\overline{B_{n+1}(\xi)}\cdot B_{n+1}(z)}{\overline{\xi}\cdot z}\cdot S_n\left(\frac{1}{\overline{z}};\frac{1}{\overline{\xi}}\right), \tag{1.5}$$

where

$$B_{n+1}(z) = \prod_{k=0}^{n} \frac{\alpha_k - z}{1 - \overline{\alpha}_k \cdot z} \cdot \frac{|\alpha_k|}{\alpha_k}. \tag{1.6}$$

Theorem 1.2. *The formulas*

$$S_n(\alpha_n; z) = -\frac{|\alpha_n|}{\alpha_n} \cdot \frac{\kappa_n}{(1 - |\alpha_n|^2)^{1/2}} \cdot \frac{B_{n+1}(z)}{z} \cdot \overline{\phi}_n\left(\frac{1}{z}\right) \tag{1.7}$$

and

$$S_n(\alpha_n; \alpha_n) = \frac{\kappa_n^2}{1 - |\alpha_n|^2}.$$

hold for every $n \geq 0$.

Theorem 1.3. *For every* $z, \xi \in \mathbb{C}$ *and* $n \geq 0$, *for the kernel* $S_n(\xi; z)$ *the Christoffel type formula holds:*

$$S_n(\xi; z) =$$
$$\frac{(1 - \alpha_{n+1}\overline{\xi})(1 - \overline{\alpha_{n+1}}z)}{1 - |\alpha_{n+1}|^2} \cdot \frac{\overline{\phi^*_{n+1}(\xi)} \cdot \phi^*_{n+1}(z) - \overline{\phi_{n+1}(\xi)} \cdot \phi_{n+1}(z)}{1 - \overline{\xi} \cdot z} \tag{1.8}$$

where

$$\phi^*_{n+1}(z) = \frac{B_{n+2}(z)}{z} \cdot \overline{\phi}_{n+1}\left(\frac{1}{z}\right).$$

Theorem 1.4. *There is a recurrence relation of the form*

$$-\frac{|\alpha_{n+1}|}{\alpha_{n+1}} \cdot \frac{\kappa_{n+1}}{(1 - |\alpha_{n+1}|^2)^{1/2}} \cdot \phi_{n+1}(z)$$

$$= \frac{B_{n+2}(z)}{z} \cdot \left\{ \phi_{n+1}(\alpha_{n+1}) \cdot \overline{\phi}_{n+1}\left(\frac{1}{z}\right) + \sum_{k=0}^{n} \phi_k(\alpha_{n+1}) \cdot \overline{\phi}_k\left(\frac{1}{z}\right) \right\} \tag{1.9}$$

which holds for every $n \geq 0$.

Note that the analogues of the Szegő recurrences for the polynomials the $\{P_n(z)\}_0^{\infty}$ can be deduced from (1.9) and (1.8).

4. Here we will give explicit formulas for the functions of orthogonal system $\{\phi_k(z)\}_0^{\infty}$ and for corresponding kernels in the case of special, but important classes of distribution functions. For given integer $p \geq 0$ fix $A_p > 0$ and $\{\gamma_k\}_0^p$ with $(0 < \gamma_k < 1)$, and consider the functions

$$\omega_p(z) = A_p^{-1/2} \cdot \prod_{k=0}^{p} \frac{z - \gamma_k}{1 - \overline{\alpha}_k \cdot z}, \quad D_p(z) = A_p^{1/2} \cdot \prod_{k=0}^{p} \frac{1 - \overline{\alpha}_k \cdot z}{1 - \overline{\gamma}_k \cdot z}, \tag{1.10}$$

and also the distribution function $\alpha(x) \equiv \alpha_p(x)$ of the form $d\alpha_p(x) = W_p(x)dx$, where

$$W_p(x) = |D_p(e^{ix})|^2 \qquad (-\pi \leq x \leq \pi). \tag{1.11}$$

Theorem 1.5. *The system of rational functions orthonormal on the circle* $z = e^{ix}$ ($-\pi \leq x \leq \pi$) *associated with the distribution* $\alpha_p(x)$ *admits the representation*

$$\phi_n(z) = e^{i\mathcal{X}_n} \cdot \omega_p(z)\frac{(1 - |\alpha_n|^2)^{1/2}}{1 - \overline{\alpha}_n \cdot z} \cdot \prod_{k=p+1}^{n} \frac{\alpha_k - z}{1 - \overline{\alpha}_k \cdot z} \cdot \frac{|\alpha_k|}{\alpha_k} \tag{1.12}$$

for $n = p+1, p+2, \ldots$, *where* $\mathcal{X}_n$ ($Im\mathcal{X}_n = 0$) *is a constant. In addition, the formula*

$$S_n(\xi; z) \equiv \sum_{k=0}^{n} \overline{\phi_k(\xi)} \cdot \phi_k(z) =$$

$$= \frac{1}{(1 - \overline{\xi} \cdot z) \cdot \overline{D_p(\xi)} \cdot D_p(z)} - \frac{\overline{B_{n+1}(\xi)} \cdot B_{n+1}(z)}{(1 - \overline{\xi} \cdot z) \cdot \overline{D}_p(1/\xi) \cdot \overline{D}_p(1/z)} \tag{1.13}$$

holds for $n = p+1, p+2, \ldots$.

From the Theorem 1.5 one obtains

Theorem 1.6. *If* $d\alpha(x) = W_p(x)dx$ *and*

$$B = \sum_{k=0}^{\infty}(1 - |\alpha_k|) = +\infty,$$

then for arbitrary z *and* ξ *we have*

$$S_\infty(\xi; z) = \sum_{k=0}^{\infty} \overline{\phi_k(\xi)} \cdot \phi_k(z) = \frac{1}{(1 - \overline{\xi} \cdot z) \cdot \overline{D_p(\xi)} \cdot D_p(z)}. \tag{1.14}$$

If $\alpha(x)$ *is arbitrary and satisfies*

$$\int_{-\pi}^{\pi} \log \alpha'(x)dx > -\infty, \tag{1.15}$$

as in the case of Szegő polynomials, then the function $D_p(z)$ *in the formula (1.14) can be replaced by the Szegő function*

$$D(z) = \exp\left\{\frac{1}{4\pi} \int_{-\pi}^{\pi} \log \alpha'(t) \cdot \frac{e^{it} + z}{e^{it} - z} dt\right\}. \tag{1.16}$$

5. Two open problems. (a) From the example of the distribution $W_p(x)dx/2\pi$ considered above it is evident that the convergence set and the sum of the bilinear series

$$S_\infty(\xi; z) = \sum_{k=0}^{\infty} \overline{\phi_k(\xi)} \cdot \phi_k(z)$$

depends on divergence or convergence of the series B.

Problem I. Investigate the nature of convergence and the value of the sum of the series $S_\infty(\xi; z)$ depending on various combinations of finiteness or infiniteness of the quantities

$$A = \int_{-\pi}^{\pi} \log \alpha'(x)dx \quad \text{and} \quad B = \sum_{k=0}^{\infty}(1 - |\alpha_k|). \tag{1.17}$$

(b) Let us consider a union of two systems of rational functions

$$\{\tau_k(z)\}_{-\infty}^{+\infty} \equiv \left\{ \frac{z^{s_k-1}}{(1 - \overline{\alpha}_k \cdot z)^{s_k}} \right\}_0^{\infty} \cup \left\{ \frac{z}{(z - d_k)^{s_k}} \right\}_1^{\infty}. \tag{1.18}$$

In the extreme case, when $\alpha_k = 0 \; (k \geq 0)$, this system coincides with Fourier system $\{z^k\}_{-\infty}^{+\infty}$ on the circle $z = e^{ix} \; (-\pi \leq x \leq \pi)$. Denote by $C_\infty\{\alpha_k\}$ the closure of the system $\{\tau_k(z)\}_{-\infty}^{+\infty}$ on the circle $z = e^{ix} \; (-\pi \leq x \leq \pi)$ in uniform metric. It's well known that if C_∞ is the set of all continuous functions on the circle $z = e^{ix} \; (-\pi \leq x \leq \pi)$, then the identity

$$C_\infty\{\alpha_k\} \equiv C_\infty$$

holds if and only if $B = +\infty$. Hence, under the condition $B < +\infty$ we have the strict inclusion

$$C_\infty\{\alpha_k\} \subset C_\infty.$$

For the solution of problem I the following problem has a principal significance.

Problem II. Under the condition $B < +\infty$ give complete inner characterization of the subspace $C_\infty\{\alpha_k\} \subset C_\infty$ of the functions continuous on the circle $z = e^{ix} \; (-\pi \leq x \leq \pi)$.

§2. Orthonormal systems of rational functions on $[-1, +1]$; a generalization of Chebyshev polynomials.

1. Let $\{\omega_k\}_0^{\infty}$ be an arbitrary sequence of complex numbers outside the interval $K = [-1, +1]$. Then we form another sequence of numbers $\{\alpha_k\}_0^{\infty} \; (|\alpha_k| < 1)$ by

$$\overline{\alpha}_k = [\omega_k + \sqrt{\omega_k^2 - 1}]^{-1} \quad (k = 0, 1, 2, \dots)$$

and associate with it the Malmquist system of rational functions $\{\varphi_n(z)\}_0^{\infty}$, which is orthonormal on the unit (cf. (1.2)).

The following main theorems were established in [6].

Theorem 2.1. If $\omega_0 = \infty$, then the system of functions $\{M_n^{(0)}(x)\}_0^\infty$, where

$$M_0^{(0)}(x) \equiv 1, \quad M_n^{(0)}(x) = \varphi_n(x + \sqrt{x^2 - 1}) + \varphi_n(x - \sqrt{x^2 - 1}), \qquad (2.1)$$

$n = 1, 2, \ldots$, is orthogonal in $[-1, +1]$ with the weight $(1 - x^2)^{-1/2}$ in the sense that

$$\int_{-1}^{+1} \frac{M_n^{(0)}(x) \cdot \overline{M_m^{(0)}(x)}}{\sqrt{1 - x^2}} dx = \begin{cases} 0, & \text{when } m \neq n \ (n, m = 0, 1, 2, \ldots) \\ 2\pi, & \text{when } m = n \ (n, m = 1, 2, \ldots) \\ \pi, & \text{when } m = n = 0. \end{cases}$$

Theorem 2.2. The system of functions $\{M_n^{(1)}(x)\}_0^\infty$, where

$$M_0^{(1)}(x) = \varphi_0(x + \sqrt{x^2 - 1}) + \varphi_0(x - \sqrt{x^2 - 1}) - \varphi_0(0), \qquad (2.3)$$

and

$$M_n^{(1)}(x) = \frac{(x + \sqrt{x^2 - 1}) \cdot \varphi_n(x + \sqrt{x^2 - 1}) - (x - \sqrt{x^2 - 1}) \cdot \varphi_n(x - \sqrt{x^2 - 1})}{\sqrt{x^2 - 1}},$$

$$(2.4)$$

$n = 1, 2, \ldots$, is orthogonal in $[-1, +1]$ with the weight $(1 - x^2)^{1/2}$ in the sense that

$$\int_{-1}^{+1} M_n^{(1)}(x) \cdot \overline{M_m^{(1)}(x)} \cdot \sqrt{1 - x^2} dx = \begin{cases} 2\pi, & \text{when } m = n \\ 0, & \text{when } m \neq n, \end{cases} \qquad (2.5)$$

$n, m = 0, 1, 2, \ldots$.

If we choose the sequence of complex numbers $\{\omega_k\}_0^\infty$ in a special way, the functions $M_n^{(0)}(x)/2$ and $M_n^{(1)}(x)/2$ become the classical Chebyshev polynomials of first and second kinds, respectively. Indeed, if we put $\omega_k = \infty \ (k \geq 0)$, then $\alpha_k = 0 \ (k \geq 0)$, and so

$$\varphi_n(z) = z^n \ (n \geq 0) \quad \text{and} \quad \varphi_n(0) = 0 \ (n \geq 1).$$

Consequently, in the case under consideration we have

$$M_0^{(0)}(x) \equiv 1, \quad M_n^{(0)}(x) = (x + \sqrt{x^2 - 1})^n + (x - \sqrt{x^2 - 1})^n =$$
$$= 2 \cdot T_n(x) = 2 \cdot \cos(n \cdot \arccos x), \qquad (n = 1, 2, \ldots)$$

and

$$M_n^{(1)}(x) = \frac{(x + \sqrt{x^2 - 1})^{n+1} - (x - \sqrt{x^2 - 1})^{n+1}}{\sqrt{x^2 - 1}} =$$
$$= 2 \cdot U_n(x) = \frac{\sin((n + 1) \cdot \arccos x)}{\sqrt{1 - x^2}}, \qquad (n = 0, 1, 2, \ldots).$$

2. In the general case, when the sequence $\{\omega_k\}_0^\infty$ is arbitrary, we denote again by $s_k \geq 1$ $(k \geq 0)$ the multiplicity of occurrence of the number ω_k in $\{\omega_0, \omega_1, \ldots, \omega_k\}$ and consider the system of rational functions

$$\left\{\frac{1}{(x - \omega_k)^{s_k}}\right\}_0^\infty . \tag{2.6}$$

Then both the systems $\{M_n^{(0)}(x)\}_0^\infty$ and $\{M_n^{(1)}(x)\}_0^\infty$ are, essentially, the result of orthogonalization of the system (2.6) in the interval $[-1, +1]$ with the weights $(1 - x^2)^{-1/2}$ and $(1 - x^2)^{1/2}$, respectively. The orthogonal systems introduced above also have integral representations. For given $n \geq 0$ let $\rho_n > 1$ be chosen so that all the points $\{\omega_k\}_0^n$ lie outside the closed ellipse

$$|z + \sqrt{z^2 - 1}| \leq \rho_n$$

with foci at ± 1.

Theorem 2.3. *The integral representations*

$$M_n^{(0)}(x) = \frac{+1}{2\pi i} \int\limits_{|t| = \rho_n} \frac{\varphi_n(t)}{t^2 - 2tx + 1} \cdot \frac{t^2 - 1}{t} dt, \tag{2.7}$$

and

$$M_n^{(1)}(x) = \frac{+1}{\pi i} \int\limits_{|t| = \rho_n} \frac{t \cdot \varphi_n(t)}{t^2 - 2tx + 1} dt \tag{2.8}$$

hold for $n = 0, 1, 2, \ldots$.

Problem III. (a) Compute the Fourier coefficients

$$a_k = \frac{1}{\sqrt{2\pi}} \int\limits_{-1}^{+1} |t| \cdot \overline{M_k^{(0)}(t)} \frac{dt}{\sqrt{1 - t^2}} \qquad (k = 1, 2, \ldots)$$

of the function $|x|$ with respect to the system $\{M_n^{(0)}(x)\}_0^\infty$.
    (b) Obtain an optimal estimate for the sum of the series

$$\rho_n = \sum_{k=n+1}^\infty |a_k|^2, \qquad n \to \infty, \tag{2.9}$$

by varying the arrangements of the points $\{\omega_k\}_0^\infty$ outside the interval $[-1, +1]$. Is it possible to establish in such manner that

$$\rho_n = O(\exp(-c\sqrt{n})) \qquad (c > 0)? \tag{2.10}$$

Such a statement will immediately follow from the well-known theorem on the best approximation on $[-1,+1]$ of the function $|x|$ by rational functions with free poles out of $[-1,+1]$.

§3. Integral representations of some orthogonal systems.

Now we will briefly describe some of the results in [7].

1. Orthogonalization of generalized Müntz-Szász systems.

Let $\{\mu_j\}_0^\infty$ $(Re\mu_j > 0)$ be an arbitrary sequence of complex numbers in right-hand half-plane $D^{(+)} = \{z : Re\,z > 0\}$. For given $k \geq 1$ we denote by $s_k \geq 1$ the multiplicity of the occurence of number μ_k in $\{\mu_1,\ldots,\mu_k\}$ of this sequence. Consider the system of functions

$$\{e^{-\mu_j x} \cdot x^{s_j-1}\}_1^\infty \in L_2(0,+\infty) \tag{3.1}$$

and denote by $\{\gamma_n(x)\}_1^\infty$ its orthogonalization on $(0,+\infty)$ by the well-known Gram-Schmidt method.

Theorem 3.1. [7] *For functions of the system* $\{\gamma_n(x)\}_1^\infty$ *the integral representation*

$$\frac{1}{\sqrt{2\pi}} \cdot \int_{-\infty}^{+\infty} \psi_n(t) \cdot e^{-ixt} dt = \begin{cases} \gamma_n(x), & x \in (0,+\infty) \\ 0, & x \in (-\infty,0) \end{cases} \tag{3.2}$$

holds where

$$\psi_n(z) = \sqrt{\frac{Re\mu_n}{\pi}} \cdot \frac{i}{z+i\mu_n} \cdot \prod_{j=1}^{n-1} \frac{z - i\overline{\mu}_j}{z+i\mu_j} \quad (n = 1, 2, \ldots). \tag{3.3}$$

Theorem 3.2. [7] *The condition*

$$\sum_{n=1}^{\infty} \frac{Re\mu_n}{1+|\mu_n|^2} = +\infty \tag{3.4}$$

is necessary and sufficient for the $L_2(0,+\infty)$ *completeness of the system* $\{\gamma_n(x)\}_1^\infty$, *or, what is the same, of the system (3.1).*

As a conclusion of this section, observe that if we put $\mu_j = 1/2$ and $s_j = j$ for $1 \leq j < \infty$, then Theorems 3.1 and 3.2 yield the formula for Laguerre polynomials

$$L_n(x) = e^{x/2} \cdot \gamma_{n+1}(x) \quad (n = 0, 1, 2, \ldots)$$

and the $L_2(0,+\infty)$ completeness of the Laguerre system $\{e^{-x/2} \cdot L_n(x)\}_0^\infty$.

2. Orthogonalization and completeness of generalized Wiener-Paley systems.

Let $\{\sigma_j\}_0^\infty$ satisfy $|\operatorname{Im}\sigma_j| < \pi/2$, and let $p_k \geq 1$ denote the multiplicity of the occurrence of σ_k in $\{\sigma_j\}_0^k$. Consider the sequence of polynomials

$$\kappa_0(t) \equiv 1, \quad \kappa_j(t) = \left(\frac{1}{2} + it\right) \cdot \left(\frac{3}{2} + it\right) \cdots \left(\frac{2j-1}{2} + it\right), \tag{3.5}$$

$1 \leq j < \infty$, and the system of functions

$$\{\kappa_{p_j-1}(t) \cdot e^{-i\sigma_j t}\}_0^\infty \in L_2(\mathbf{R}). \tag{3.6}$$

Denote by $\{\rho_j(t)\}_0^\infty$ the orthogonalization of the system (3.6) on the real line with respect to the weight

$$W(t) = \frac{1}{\pi} \cdot |\Gamma(1/2 + it)|^2 = \frac{1}{ch(\pi t)}. \tag{3.7}$$

Theorem 3.3. *The functions of the system* $\{\rho_n(t)\}_0^\infty$ *have the representations of the form*

$$\rho_n(t) = \{\sqrt{2} \cdot \Gamma(1/2 + it)\}^{-1} \cdot \int_0^{+\infty} \omega_{n+1}(x) \cdot x^{-1/2+it} dx, \tag{3.8}$$

and

$$\rho_n(t) = \frac{\sqrt{Re\,(e^{\sigma_n})}}{2\pi} \int_{-\infty}^{+\infty} \frac{i}{u + ie^{\sigma_n}} \prod_{j=0}^{n=1} \frac{u - ie^{\overline{\sigma}_j}}{u + ie^{\sigma_j}} \cdot (e^{i\frac{\pi}{2}sgn u} \cdot |u|)^{-\frac{1}{2} - it} du, \tag{3.9}$$

where

$$\omega_{n+1}(x) = \frac{1}{\sqrt{2\pi}} \int_{-\infty}^{+\infty} \sqrt{\frac{Re\,(e^{\sigma_n})}{\pi}} \cdot \frac{i}{u + ie^{\sigma_n}} \cdots \prod_{j=0}^{n-1} \frac{u - ie^{\overline{\sigma}_j}}{u + ie^{\sigma_j}} \cdot e^{-ixu} du, \tag{3.10}$$

$n \geq 0$, *where for* $n = 0$ *we set* $\prod_{j=0}^{n-1} \equiv 1$.

Notice that if each element of $\{\sigma_j\}_0^\infty$ different, and consequently

$$p_j = 1, \quad \kappa_0(t) = \kappa_{p_j-1}(t) \equiv 1, \quad (0 \leq j < +\infty),$$

then our system (3.5) turns into the system

$$\{e^{-\pi/2 \cdot |t|} \cdot e^{-i\sigma_j t}\}_0^\infty, \quad (|Im\,\sigma_j| < \frac{\pi}{2}), \tag{3.11}$$

introduced by Wiener and Paley.

The following theorem represents an essential generalization of Wiener-Paley's result.

Theorem 3.4. *The condition*

$$\sum_{j=0}^{\infty} \frac{\cos(Im\,\sigma_j)}{ch(Re\,\sigma_j)} = +\infty \tag{3.12}$$

is necessary and sufficient for the in $L_2(\mathbf{R})$ *completeness of the orthonormal system* $\{\rho_j(t)/\sqrt{ch(\pi t)}\}_0^{\infty}$ *, or what is the same, of* $\{e^{-i\sigma_j t} \cdot \kappa_{p_j-1}(t)/\sqrt{ch(\pi t)}\}_0^{\infty}$ *.*

3. Pollaczek's orthogonal polynomials on $(-\infty, +\infty)$ and related polynomials on the half-line $(0, +\infty)$.

Let us consider the function

$$W^{(\alpha)}(t;\varphi) = \frac{2^{\alpha}}{\pi} \cdot (\sin\varphi)^{1+\alpha} \cdot |\Gamma\left(\frac{1+\alpha}{2} + it\right)|^2 \cdot e^{-(\pi-2\varphi)t} \tag{3.13}$$

where $t \in \mathbf{R}$, and suppose that the parameters satisfy $\varphi \in (0,\pi)$ and $\alpha \in (-1, +\infty)$. Using Stirling's formula it can be easily shown that for $t \to \pm\infty$

$$\log W^{(\alpha)}(t;\varphi) \sim -c_{\pm}(\varphi;\alpha) \cdot |t|, \tag{3.14}$$

where $c_{\pm}(\varphi;\alpha) > 0$ are certain constants. Following Pollaczek, we denote by $\{P_n^{(\alpha)}(t;\varphi)\}_0^{\infty}$ the system of polynomials, which represents the result of orthogonalization of the system $\{t^n\}_0^{\infty}$ on $(-\infty, +\infty)$ with respect to the weight $W^{(\alpha)}(t;\varphi)$ normalized by

$$\int_{-\infty}^{+\infty} W^{(\alpha)}(t;\varphi) \cdot P_n^{(\alpha)}(t;\varphi) \cdot P_m^{(\alpha)}(t;\varphi)dt = \frac{\Gamma(1+\alpha+n)}{\Gamma(1+n)}\delta_{nm}. \tag{3.15}$$

It should be mentioned that Pollaczek constructed the theory of his polynomials by a method quite different from that we have used in our work. His method enabled him to establish only their completeness, their generating functions and recurrence relations. The proofs of these results are found neither in the Pollaczek's paper nor in the book by Szegő [4]. The method developed by us makes it possible to give complete proofs of all Pollaczek's results mentioned above and, in addition, to discover a new important fact: the integral representations of these polynomials.

Theorem 3.5. *The polynomials* $P_n^{(\alpha)}(t;\varphi)$ *have an integral representation of the form*

$$P_n^{(\alpha)}(t;\varphi) =$$

$$e^{in\varphi} \cdot \frac{[2^{-1} \cdot (1 - i\cot\varphi)]^{\frac{1+\alpha}{2} + it}}{\Gamma\left(\frac{1+\alpha}{2} + it\right)} \cdot \int_0^{+\infty} e^{-\frac{1}{2}(1-i\cot\varphi)x} \cdot L_n^{(\alpha)}(x) \cdot x^{\frac{\alpha-1}{2} + it}dx \tag{3.16}$$

where $L_n^{(\alpha)}(x)$ is the generalized Laguerre polynomial, that is

$$L_0^{(\alpha)}(x) \equiv 1, \quad L_n^{(\alpha)}(x) = \frac{x^{-\alpha} \cdot e^x}{n!} \cdot \frac{d^n}{dx^n}\{x^{n+\alpha} \cdot e^{-x}\} \tag{3.17}$$

for $n = 1, 2, \cdots$. In addition, the representation

$$P_n^{(\alpha)}(t; \varphi) = e^{in\varphi} \cdot [2^{-1} \cdot (1 - i \cot \varphi)]^{\frac{1+\alpha}{2}+it}.$$
$$\cdot \frac{1}{2\pi i} \int\limits_{|w-\frac{1}{2}|=\rho} \frac{(w+1/2)^{n+\alpha} \cdot (w - i/2 \cdot \cot \varphi)^{-\frac{1+\alpha}{2}-it}}{(w-1/2)^{n+1}} dw, \tag{3.18}$$

holds where ρ is arbitrary number such that $\rho \in (0, 1/2)$.

The Pollaczek's results mentioned above follow from the representation (3.18). In [7] we have introduced polynomials for the half-line $(0, +\infty)$ related to the system $\{P_n^{(\alpha)}(t; \varphi)\}_0^\infty$ as follows.

Theorem 3.6. *Let the systems of polynomials* $\{A_n^{(\alpha)}(t)\}_0^\infty$ *and* $\{B_n^{(\alpha)}(t)\}_0^\infty$ *be orthonormal on the half-line* $(0, +\infty)$ *in the sense that*

$$\int\limits_0^{+\infty} W_-^{(\alpha)}(t) \cdot A_n^{(\alpha)}(t) \cdot A_m^{(\alpha)}(t) dt = \frac{\Gamma(1+\alpha+2n)}{\Gamma(1+2n)} \delta_{nm} \qquad (n, m = 0, 1, 2, \ldots)$$

and

$$\int\limits_0^{+\infty} W_+^{(\alpha)}(t) B_m^{(\alpha)}(t) B_n^{(\alpha)}(t) dt = \frac{\Gamma(2+\alpha+2n)}{\Gamma(2+2n)} \delta_{nm} \qquad (n, m = 0, 1, 2, \ldots),$$

respectively, where

$$W_\mp^{(\alpha)}(t) = \frac{2^\alpha}{\pi} \cdot |\Gamma\left(\frac{1+\alpha}{2} + i\sqrt{t}\right)|^2 \cdot t^{\mp\frac{1}{2}}.$$

Then the integral representations

$$A_n^{(\alpha)}(t) = e^{i\pi n} \cdot \frac{2^{-\frac{1+\alpha}{2}+i\sqrt{t}}}{\Gamma\left(\frac{1+\alpha}{2} + i\sqrt{t}\right)} \cdot \int\limits_0^{+\infty} e^{-\frac{x}{2}} \cdot L_{2n}^{(\alpha)}(x) \cdot x^{\frac{\alpha-1}{2}+i\sqrt{t}} dx$$

and

$$B_n^{(\alpha)}(t) = e^{i\pi(n+\frac{1}{2})} \cdot \frac{2^{-\frac{1+\alpha}{2}+i\sqrt{t}}}{\Gamma\left(\frac{1+\alpha}{2} + i\sqrt{t}\right)} \cdot \int\limits_0^{+\infty} e^{-\frac{x}{2}} \cdot L_{2n+1}^{(\alpha)}(x) \cdot x^{\frac{\alpha-1}{2}+i\sqrt{t}} dx$$

hold for $n = 0, 1, \ldots$.

Theorem 3.7. *Our systems have the generating functions*

$$G_A^{(\alpha)}(t; z) = \sum_{n=0}^{\infty} z^n \cdot A_n^{(\alpha)}(t) =$$

$$= \frac{1}{2}(1+z)^{-\frac{1+\alpha}{2}-i\sqrt{t}} \cdot \{(1+i\sqrt{z})^{2i\sqrt{t}} + (1-i\sqrt{z})^{2i\sqrt{t}}\}$$

and

$$G_B^{(\alpha)}(t; z) = \sum_{r=0}^{\infty} z^r \cdot B_r^{(\alpha)}(t) =$$

$$= \frac{1}{2\sqrt{tz}} \cdot (1+z)^{-\frac{1+\alpha}{2}-i\sqrt{t}} \cdot \{(1+i\sqrt{z})^{2i\sqrt{t}} - (1-i\sqrt{z})^{2i\sqrt{t}}\}.$$

The explicit recurrence relations for both systems of polynomials have been also established.

Problem IV. Investigate asymptotic properties of Pollaczek polynomials $\{P_n^{(\alpha)}(t)\}_0^{\infty}$ and the polynomials $\{A_n^{(\alpha)}(t)\}_0^{\infty}$, $\{B_n^{(\alpha)}(t)\}_0^{\infty}$, when $|t| \to +\infty$ or $n \to +\infty$, and investigate the distribution of the zeros of these polynomials.

Similar problems for Hermite and Laguerre polynomials were investigated in detail by many authors.

REFERENCES

1. M. M. Djrbashian, Orthogonal systems of rational functions on the circle with given set of poles, Doklady SSSR **147**(6)(1961), 1278–1281.

2. M. M. Djrbashian, Orthogonal systems of rational functions on the circle, Izvestija AN Arm. SSR, ser. Matematika **1**(1) (1966), 3-24 and **1**(2) (1966) 106–125.

3. J. L. Walsh, Interpolation and Approximation in the Complex Domain; Amer. Math. Soc. Publ., 1950.

4. G. Szegő, "Orthogonal Polynomials", Amer. Math. Soc. Publ., 1959.

5. U. Grenander and G. Szegő, "Toeplitz Forms and Their Applications", Univ. Calif. Press, 1959.

6. M. M. Djrbashian and A.A. Kitbalyan, On a generalization of the Chebyshev polynomials, Doklady AN Arm. SSR **38**(5)(1964), 263–270.

7. M. M. Djrbashian, Representation and completeness of some orthogonal systems, Izvestija AN Arm. SSR, ser. Matematika **14**(6)(1979), 446–493.

ORTHOGONAL POLYNOMIALS AND FUNCTIONAL ANALYSIS

J. Dombrowski
Department of Mathematics
Wright State University
Dayton, Ohio 45435, U.S.A.

ABSTRACT. This paper studies the measure of orthogonality for a system of polynomials defined by a three term recursion formula, using the techniques of operator theory and functional analysis. Spectral properties of self-adjoint operators and compact operators, perturbation theorems, and commutator equations are used in the development of the ideas.

1. Introduction

Let μ be a measure with infinite support, defined on the Borel subsets of the real line with the property that $\int t^n d\mu < \infty$ for $n = 0,1,2,\ldots$ Since the Hilbert space $L^2(\mu)$ contains the set of polynomials, the Gram-Schmidt orthogonalization process can be applied to the set $\{1, t, t^2, \ldots\}$ to obtain a system of polynomials $\{P_n\}$ which are orthonormal with respect to the measure μ. Assuming that $\mu(R) = 1$, it is easily shown that these polynomials $\{P_n\}$ satisfy the following conditions:

$$(1.1) \qquad P_1(t) = 1 \qquad P_2(t) = \frac{t - b_1}{a_1} \qquad \text{and} \qquad P_n(t) = \frac{(t - b_{n-1})P_{n-1}(t) - a_{n-2}P_{n-2}(t)}{a_{n-1}}$$

for $n > 2$, with $a_n > 0$ and b_n real for all n.

The characteristic properties of the sequence $\{P_n\}$ are, of course, determined by the nature of the given measure μ. Extracting that information from the given measure is one of the interesting (and challenging) problems in the study of

147

P. Nevai (ed.), Orthogonal Polynomials, 147–161.

systems of orthogonal polynomials.

Conversely, suppose that the sequence of polynomials $\{P_n\}$ in (1.1) is given, and that the existence, uniqueness and nature of a measure of orthogonality are to be determined. Questions on existence and uniqueness are related to the Hamburger moment problem, which asks when a sequence of constants is the moment sequence of a positive Borel measure on the real line. Necessary and sufficient conditions are known. (See [16] and [17].) Favard's Theorem states that the conditions $a_n>0$ and b_n real in (1.1) imply the existence of a measure μ with respect to which the polynomials $\{P_n\}$ are orthonormal. The measure is unique if and only if the associated moment problem has a unique solution. One sufficient condition for uniqueness, due to Carleman, is given by $\sum \frac{1}{a_n} = \infty$. If a unique measure of orthogonality exists, it then is of interest to study the relationship between the nature and support of that measure and properties of the sequence $\{P_n\}$. This is the focus of the discussion that follows.

Throughout this paper it will be assumed that the system (1.1) is given and that the measure of orthogonality is to be determined. Questions concerning the existence, uniqueness and nature of this measure will be explored from the viewpoint of operator theory. While operator theory is obviously only one possible approach to these complex issues, it does offer an interesting perspective and some surprisingly useful techniques.

To establish the theoretical setting for the discussion that follows, it is first necessary to relate the measure of orthogonality for a system of orthonormal polynomials of type (1.1) to the spectral measure of a self-adjoint operator. To do so, a few basic facts about self-adjoint operators are needed.

Let A be a self-adjoint operator defined on a dense subset D of a Hilbert space h. Recall that the vector ϕ is said to be a cyclic vector for the operator A if the collection of all finite linear combinations of vectors of the form $A^n\phi$ is dense in h. If A has a cyclic vector then one version of the spectral theorem asserts that the operator A is unitarily equivalent to a multiplication operator on an appropriate L^2 space. In particular, there exists a Borel measure μ such that $A:h \to h$ is unitarily equivalent to the operator $M_t: L^2(\mu) \to L^2(\mu)$ defined as follows:

$$(M_t\, f)(t) = t\, f(t) \quad \text{for} \quad f \in L^2(\mu).$$

The support of the measure μ is the spectrum of the operator A, which consists of all real numbers λ for which $A - \lambda I$ is not invertible. Furthermore, the correspondence between h and the space $L^2(\mu)$ associates the cyclic vector ϕ with the function $f(t)=1$. It then follows that vectors of the form $A^n\phi$ correspond to the functions t^n.

A second version of the spectral theorem asserts that the operator A has an integral representation of the form $A = \int \lambda dE_\lambda$ where $E(\beta)$ is a projection valued measure defined on the Borel subsets of the real line. Note that if the operator A is represented as a multiplication operator on the function space $L^2(\mu)$, then the spectral projection $E(\beta)$ corresponds to multiplication by the characteristic function of the Borel set β. That is, for f in $L^2(\mu)$, $E(\beta)f = \chi_\beta f$.

The aim now is to find a self-adjoint operator, related to the system (1.1), to which these ideas can be applied. The self-adjoint operator will provide, (via the spectral theorem), a measure of orthogonality for the given system of polynomials. The definition of this operator is the subject of the next section.

2. The Definition of the Operator

Given a system of polynomials $\{P_n\}$ which satisfy (1.1), the coefficient sequen-ces $\{a_n\}$ and $\{b_n\}$ can be used define a matrix operator which acts (via matrix multiplication) on sequences in l^2. Let

(2.1)
$$
C = \begin{bmatrix}
b_1 & a_1 & 0 & . & . & . \\
a_1 & b_2 & a_2 & 0 & . & . \\
0 & a_2 & b_3 & a_3 & 0 & . \\
0 & 0 & a_3 & b_4 & a_4 & . \\
. & . & . & . & . & . \\
. & . & . & . & . & .
\end{bmatrix}
$$

with domain D_C consisting of all sequences in l^2 for which matrix multiplication yields a sequence in l^2. This operator with the indicated domain plays a key role in the discussion that follows. To show that the system (1.1) has an essentially unique measure of orthogonality it is, in fact, enough to show that the operator in (2.1) is self-adjoint. For if the matrix operator is self-adjoint then it is possible to apply the spectral theorem and represent this operator as a multi-plication operator on an appropriate L^2 space. As noted above, the correspon-dence (i.e. the unitary equivalence) associates the cyclic vector $\phi_1 = (1, 0, 0, \ldots)$ with the function $f(t) = 1$, and $C^n\phi_1$ with t^n. An induction argument establishes that $\phi_n= (0, 0, ..., 1, 0, 0, ...)$ corresponds to $P_n(t)$ as defined in (1.1). Thus the spectral theorem establishes the existence of a measure μ with respect to which the polynomials $\{P_n\}$ are orthogonal. From the second version of the spectral theorem it follows that if $C=\int \lambda dE_\lambda$ then $\mu(\beta) = \| E(\beta) \phi_1 \|^2$. The nature and support of that measure now depend on the spectral properties of the operator C.

The results of this paper require that the matrix operator defined in (2.1) be self-adjoint (so that the spectral theorem applies). However, the problem of determining self-adjointness is itself fairly intriguing. To appreciate the difficulties, let Q denote the restriction of the matrix operator defined in (2.1) to the set D_Q consisting of all finite linear combinations of the standard basis vectors in l^2. Assuming only that the terms of the sequence $\{a_n\}$ are positive and that those of $\{b_n\}$ are real it is easily verified that $<Qx,y> = <x,Qy>$ for all x and y in D_Q. If the sequences $\{a_n\}$ and $\{b_n\}$ are also bounded, then the operator Q has a unique extension to a bounded self-adjoint operator on l^2. That extension is, of course, the matrix operator C with domain D_C described above. Thus $<Cx,y>=<x,Cy>$ for all x and y in l^2. If both sequences $\{a_n\}$ and $\{b_n\}$ are not bounded, it is nevertheless true that Q with domain D_Q is symmetric (i.e. $<Qx,y> = <x,Qy>$ for all x and y in D_Q) and that $Q^* = C$ (i.e. $<Qx,y> = <x,Cy>$ for all x in D_Q and y in D_C). The same statements hold for the closure of Q. However, it is not in general true that the closure of the operator Q equals C (which would imply that C is self-adjoint). In fact, without additional conditions on the coefficient sequences $\{a_n\}$ and $\{b_n\}$ the operator C may not even be symmetric.

Fortunately, as shown in [7], Carleman's condition $\Sigma \frac{1}{a_n} = \infty$ is sufficient to show that C is selfadjoint with domain D_C dense in l^2 and cyclic vector ϕ_1.

It should be noted that the theoretical issue of finding a self-adjoint extension of the symmetric operator Q has other practical implications for the given sequence $\{P_n\}$. If C is a self-adjoint extension of Q then there exists a unique measure of orthogonality for the sequence $\{P_n\}$ in (1.1) and, as remains to be shown, the operator C provides a convenient means for studying that measure. But even if this is not the case the symmetric operator Q does have a family of self-adjoint extensions which yield a corresponding family of Borel measures related to (1.1). A discussion of the latter case from the perspective of the moment problem can be found in the excellent expository article by Sarason [17]. The results that follow require that C be self-adjoint. Sufficient conditions for self-adjointness are discussed in [3 , Chapter VII]. The sufficiency of $\Sigma \frac{1}{a_n} =\infty$

is established and the following result is presented.

Theorem Assume (1.1) and (2.1). If either $a_{j-1} + a_j + b_j \leq M$ or $a_{j-1} + a_j - b_j \leq M$ (for $j = 2,3, \ldots$) then C is self-adjoint.

It is also shown in [3] that if $\{b_n\}$ is bounded, $\Sigma\frac{1}{a_n} < \infty$, and $a_{j-1}a_{j+1} \leq a_j^2$ for all $j \geq N$, then C is not self-adjoint. The proof presented shows that in this case every complex, non-real value of z is an eigenvalue for C. Thus C cannot be symmetric.

The aim now is to survey some of the results in the literature about systems of orthogonal polynomials from this operator theoretic perspective. To do so, it will henceforth be assumed that the sequences $\{a_n\}$ and $\{b_n\}$ in the given system (1.1) satisfy the following conditions:

(2.2)   i) $a_n > 0$ and b_n is real for each n

ii) $\Sigma\frac{1}{a_n} = \infty.$

Under these conditions the operator defined in (2.1) is self-adjoint. The measure μ, obtained from the spectral theorem, will, without further ado, be called both the spectral measure of the operator (2.1) and the measure of orthogonality for the system (1.1). The discussion will first focus on the bounded case and eventually arrive at the unbounded case. The following section deals with the singular component of the measure μ which is supported on the point spectrum of the operator C.

3. Eigenvalues

Basic structural properties of self-adjoint operators defined by tridiagonal (or Jacobi) matrices, are discussed in Stone [18] . It is shown there, for example, that the eigenvalues of the nxn truncated matrix, obtained by letting $b_i = 0$ for $i > n$ and $a_i = 0$ for $i > n-1$, correspond to the zeros of the polynomial P_{n+1} and that λ is an eigenvalue for the operator C if and only if $x = \{x_n\}$, where $x_n = P_n(\lambda)$, is in l^2. In this case the sequence $\{x_n\}$ is the corresponding eigenvector.

If $\lim a_n = 0$ and $\lim b_n = 0$ then (2.1) defines a compact self-adjoint operator. In this case the spectrum consists entirely of eigenvalues, and the support of the measure μ, which is the spectrum of C, is a countable set with zero as its only accumulation point. (That C is compact follows from the fact that $C = VA + AV^* + B$ where $V\phi_n = \phi_{n+1}$, $A\phi_n = a_n\phi_n$ and $B\phi_n = b_n\phi_n$. Since A and B are compact diagonal operators and since the compact operators form a closed two sided ideal in the set of bounded operators, C is compact .)

It is natural to ask under what other conditions the support of the spectral measure is a countable set with a finite number of limit points. A complete answer is provided by Krein's Theorem. To understand Krein's Theorem within the context of this presentation it should be noted that, if the self-adjoint matrix operator C is represented as the multiplication operator M_t on the Hilbert space $L^2(\mu)$, then for any polynomial $\pi(x)$, the operator $\pi(C)$ corresponds to the multiplication operator $M_{\pi(t)}$ on $L^2(\mu)$.

<u>Krein's Theorem</u> Let $\pi(x)$ be a polynomial of degree p. Every limit point of the spectrum of C is a zero of the polynomial $\pi(x)$ if and only if

$$\lim \int \pi(t) P_n(t) P_{n+i}(t) d\mu = 0 , \qquad i = 0, 1, 2, \ldots$$

In terms of matrices, the theorem states that every limit point of the spectrum of C is a zero of $\pi(x)$ if and only if the diagonals in the matrix representation of $\pi(C)$ converge to zero (which implies that C is compact). It remains then to investigate the elements which appear on the diagonals in the matrix representation of $\pi(C)$ to obtain specific conditions on the sequences $\{a_n\}$ and $\{b_n\}$.

It follows from Krein's theorem that if the support of the measure μ is a countable set with a finite number of limit points then $\lim \inf a_n = 0$. (If $\lambda_1, \lambda_2, \ldots, \lambda_k$ are the k limit points, let $\pi(\lambda) = (\lambda - \lambda_1)(\lambda - \lambda_2) \cdots (\lambda - \lambda_k)$, and look at the diagonals of $\pi(C)$. See the argument given in [5].) Thus if $\lim \inf a_n > 0$, the support of μ must contain an infinite number of limit points. It will be shown below that if $\lim a_n = a$, $a > 0$, and if $\lim b_n = b$ then the support of μ must contain the interval $[-2a+b, 2a+b]$, and may contain isolated eigenvalues outside of this interval. Further information is needed about the sequences $\{a_n\}$ and $\{b_n\}$ to determine the nature of the measure on the interval $[-2a+b, 2a+b]$.

A necessary condition for λ to be an eigenvalue is given in [5]. If the sequences $\{a_n\}$ and $\{b_n\}$ are bounded (so that C is a bounded operator on l^2), and if λ is an eigenvalue for the operator C, then

(3.1) $a_1^2 P_1^2(\lambda) + \sum_{n=2}^{\infty} [(a_n^2-a_{n-1}^2)P_n^2(\lambda)+a_{n-1}(b_n-b_{n-1})P_{n-1}(\lambda)P_n(\lambda)]=0.$

Furthermore, the partial sums of this infinite series satisfy the following:

$S_N(\lambda) = a_1^2 P_1^2(\lambda) + \sum_{n=2}^{N} [(a_n^2-a_{n-1}^2)P_n^2(\lambda)+a_{n-1}(b_n-b_{n-1})P_{n-1}(\lambda)P_n(\lambda)]$

$= a_{N-1}^2 P_{N-1}^2(\lambda)-a_{N-1}(\lambda-b_N)P_{N-1}(\lambda)P_N(\lambda)+a_N^2 P_N^2(\lambda).$

These facts are useful for special cases. It follows immediately, for example, that there can be no eigenvalues if $\{b_n\}$ is the zero sequence and $\{a_n\}$ is monotone increasing with $\lim a_n = a$, and that there are at most a finite number of eigenvalues if the coefficient sequences are eventually constant. More generally, if $\lim a_n=a$, $a>0$, and $\lim b_n=b$, it is relatively easy to impose conditions (see [5]) on the coefficient sequences which imply that the related operator C has no eigenvalues in the interval $(-2a+b, 2a+b)$, which must be in the spectrum.

Further observations about the singular component of the measure μ will be made below, after the introduction of several needed perturbation theorems.

4. Perturbation Theorems

Perturbation theorems provide a useful technique in the analysis of spectral measures. Several definitions and theorems are needed to begin this phase of the discussion. Recall that a compact self-adjoint operator on a separable Hilbert space is a self-adjoint operator whose spectrum is a sequence of eigenvalues with zero as the only accumulation point. A compact, self-adjoint operator is of trace class if the corresponding sequence of eigenvalues is absolutely summable. Weyl's Theorem states that if A is a bounded self-adjoint operator and if B=A+D where D is a compact self-adjoint operator, then the spectrum of A and the spectrum of B will have the same set of limit points. A compact perturbation may, however, change the nature of the spectrum (or, equivalently, the spectral measure) unless additional conditions are imposed. The Kato-Rosenblum Theorem states that if the self-adjoint compact perturbation D is of trace class then the absolutely continuous parts of the operators A and B will be unitarily equivalent. These theorems can be applied to obtain information about small (i.e. compact) perturbations of well understood operators. Examples follow.

Assume that the polynomials $\{P_n\}$ in (1.1) are defined by $a_n = \frac{1}{2}$ and $b_n = 0$ for all n. It follows that the related self-adjoint operator C in (2.1) is the real part of the unilateral shift operator V defined on l^2 by $V\phi_n = \phi_{n+1}$. That is, $C = \frac{1}{2}(V + V^*)$, where $V^*\phi_n = \phi_{n-1}$ for $n > 1$ and $V^*\phi_1 = 0$. The spectrum of the unilateral shift operator V is known to be the closed unit disk and since V is hyponormal (i.e. $V^*V - VV^* \geq 0$) the spectrum of its real part is the projection onto the real axis of the spectrum of V. Thus the spectrum of C, which is the support of the spectral measure μ, is the closed interval [-1, 1]. It is also known (and will be shown below) that in this case, the spectral measure μ is absolutely continuous with respect to Lebesgue measure. In fact, $\mu(\beta) = \int_\beta \sqrt{1 - t^2}\, dt$ for every Borel subset β of [-1, 1]. The above theorems can then be applied to obtain information about compact perturbations of this operator.

It follows from Weyl's Theorem, for example, that if (1.1) holds with $\lim a_n = \frac{1}{2}$ and $\lim b_n = 0$ then the spectrum of the related self-adjoint operator contains the interval [-1, 1]. Points in the spectrum which lie outside of this closed interval must be isolated points in the support of μ, and hence must be eigenvalues for the operator. (Note that if $a_n = \frac{1}{2}$ for $n \geq N$, then it follows from (3.1) that there can be at most a finite number of eigenvalues.) Furthermore, if $\Sigma |a_n - \frac{1}{2}| < \infty$ and $\Sigma |b_n| < \infty$ then the Kato-Rosenblum Theorem says that the spectral measure of the corresponding self-adjoint operator has an absolutely continuous part.

The Kato-Rosenblum Theorem is also used in [4], with a perturbation type argument, to show that if $\lim \inf a_n = 0$ then the spectral measure μ cannot have an absolutely continuous part. For if $\lim \inf a_n = 0$ then it is possible to write C as the sum of two operators A and B such that the operator A has a pure point spectrum and the operator B is of trace class. Since C - B = A, the Kato-Rosenblum Theorem implies that C has an absolutely continuous part if and only if A does. But A, of course, does not.

To further discuss the absolutely continuous part of the measure μ, it is first necessary to explore the role of commutator equations in the study of spectral measures.

5. Commutator Equations

It was shown in Putnam [15] that if A and B are bounded self-adjoint operators defined on a Hilbert space h and if AB-BA = -iK where K≥0 then the range of the operator K can provide information about the spectral measure of the operator A. If A= $\int \lambda dE_\lambda$ then for any x in h, $v_x(\beta) = ||E(\beta)x||^2$ defines a measure on the Borel subsets of the real line. Let h_a denote the set of vectors x in h for which the measure v_x is absolutely continuous with respect to Lebesque measure. Putnam's theorem asserts that the range of the operator K is a subset of h_a.

To apply this result to the problem under discussion, consider again the real part of the unilateral shift operator (for which $a_n=\frac{1}{2}$ and $b_n=0$). Recall that C= $\frac{1}{2}$(V + V*), with $V\phi_n=\phi_{n+1}$. Let J= $\frac{1}{2i}$ (V-V*). Then CJ-JC = -iK where K = [k_{ij}] is a diagonal operator with k_{11} being the only non-zero entry in the matrix. It follows from Putnam's theorem that the measure $||E(\beta) \phi_1||2$ is absolutely continuous with respect to Lebesque measure. Since ϕ_1 is a cyclic vector for C, this is precisely the spectral measure μ which is the measure of orthogonality for the system (1.1) with $a_n=\frac{1}{2}$ and $b_n=0$.

This argument also applies if the sequence {a_n} monotonically increases to $\frac{1}{2}$ and {b_n} is the zero sequence. In this case define T by $T\phi_n=2a_n\phi_{n+1}$. Then C is the real part of T, and if J is the corresponding imaginary part, it follows that CJ - JC = -2iK with K≥0. In fact, K= [k_{ij}] is a diagonal matrix with $k_{11}=a_1^2$. An application of Putnam's Theorem then shows that the measure $\mu(\beta)= ||E(\beta)\phi_1||^2$ is absolutely continuous with respect to Lebesgue measure.

Assume now that the constants a_n form a decreasing sequence and that b_n is 0 for all n. For this example the above construction leads to a diagonal operator K which does not satisfy the condition K≥ 0. Thus Putnam's theorem cannot be applied. However it is shown in [6] that a commutator equation can still be used to obtain information about the spectral measure of C. The following theorem is established:

<u>Theorem</u> If lim a_n = a (a>0), lim b_n = b, $\Sigma|a_n-a_{n-1}|<\infty$, and $\Sigma|(b_n - b_{n-1})| < \infty$, then the spectral measure of C has a non-trivial absolutely continuous part with support (-2a+b, 2a+b).

156

The key idea in the proof of this theorem is the commutator equation CJ - JC = −2iK, where C is the real part of the operator T defined by $T\phi_n=2a_n\phi_{n+1}+b_n\phi_n$ and J is the corresponding imaginary part. The operator K, whose matrix with respect to the standard basis in l^2 can be computed by the usual matrix operations, has tridiagonal form, and the hypotheses imply that the diagonals are absolutely summable. That is, if K = [k_{ij}] then $\Sigma|k_{ii}| < \infty$ and $\Sigma|k_{i,i+1}| < \infty$. It follows that K is a trace class operator. The trace class properties of K are then used in conjunction with the commutator equation to establish the existence of an absolutely continuous part for the spectral measure of C.

Before moving on to the unbounded case it is perhaps worth noting that the sufficient condition for λ to be an eigenvalue for the bounded self-adjoint operator C stated in (3.1) can be restated in terms of the operator K as follows:

<u>Lemma</u> If the sequences $\{a_n\}$ and $\{b_n\}$ are bounded and if λ is an eigenvalue for the corresponding operator C with eigenvector x = $\{P_n(\lambda)\}$ then <Kx,x> = 0.

6. The Unbounded Case

The ideas presented thus far have applied to systems (1.1) for which the coefficient sequences $\{a_n\}$ and $\{b_n\}$ are bounded. Information about the measure of orthogonality μ has obtained by analyzing the corresponding bounded self-adjoint matrix operator (2.1) defined on l^2. This section considers the much more difficult problem in which the coefficient sequences are not bounded but (2.1) is satisfied. In this case the domain of the corresponding matrix operator (2.1) is a proper dense subset of l^2 (which includes the standard basis vectors in l^2). The condition $\Sigma\frac{1}{a_n}=\infty$ is sufficient to show that <Cx, y>=<x, Cy> for all x, y in this dense domain. (See [7].) Thus C is symmetric, and it can then be shown that C=C*. As before, the spectral theorem (for unbounded) self-adjoint operators provides a measure of orthogonality μ for the system (1.1). In this case the support of the measure is an unbounded subset of the real line. Initial results will be presented under the following assumptions:

(6.1) i) The sequence $\{a_n\}$ is a monotonically increasing sequence of positive real numbers with lim $a_n= \infty$.

ii) $\Sigma \frac{1}{a_n} = \infty$.

Additional conditions will now be imposed on the sequences $\{a_n\}$ and $\{b_n\}$ which will insure that the measure μ is absolutely continuous. The following theorems are established in [7].

__Theorem__ Assume (6.1). Let $d_n = a_n-a_{n-1}$ and assume $\{b_n\}$ is the zero sequence. If $\{d_n\}$ is bounded and $d_{n+1} + d_{n-1} \le 2d_n$ for $n \ge N \ge 2$ then the spectral measure μ is absolutely continuous.

__Theorem__ Assume (6.1). Let $d_n = a_n-a_{n-1}$ and assume $\{b_n\}$ is the zero sequence. If $\{d_n\}$ is bounded and $d_{n+1}^2 \le d_{n+2}d_n$ for $n \ge N \ge 2$ then C is absolutely continuous.

Obviously the conditions of both theorems are satisfied if the difference sequence $\{d_n\}$ is a constant sequence. Further examples for the first theorem can be constructed by choosing the difference sequence $\{d_n\}$ to be monotone increasing and bounded above with $d_{n+1} - d_n \le d_n - d_{n-1}$. The normalized Hermite polynomials provide an interesting example for the second theorem. These polynomials are orthonormal on $(-\infty,\infty)$ with respect to $d\mu=w(x)dx$ where $w(x) = e^{-x^2}dx$. The polynomials satisfy a recursion formula of the form (1.1) with $b_n=0$ and $a_n=\sqrt{\frac{n}{2}}$. Thus $d_n=(1/\sqrt{2})(\sqrt{n} - \sqrt{n-1})$ and it can be shown that $d_{n+1}^2 \le d_n d_{n+2}$. (Since $\sqrt{n}-\sqrt{n-1} =1/(\sqrt{n} + \sqrt{n-1})$ the required inequality is equivalent to $(\sqrt{n+1} + \sqrt{n})^2 \ge (\sqrt{n} +\sqrt{n-1})(\sqrt{n+2} + \sqrt{n+1})$ which is easily verified.)

The proofs of both theorems rely on a commutator equation to establish a relationship between the spectral measure μ and Lebesgue measure. The operators for this commutator equation are obtained by adapting the natural generalization of the technique used in the bounded case to circumvent the difficulties encountered in working with commutator equations in which all participating operators are unbounded. As before, the operator C is viewed as the real part of the operator T defined by $T\phi_n =2a_n\phi_{n+1}$. That is, $C=1/2(T+T^*)$ and $J=1/2i (T-T^*)$. The bounded operator J_N is obtained from J by letting $a_n=a_N$ for $n \ge N$. The operator K_N, which is also bounded, is defined by the equation $CJ_N-J_NC = -2iK_N$, which is valid on a dense subset of l^2. The operator K_N then plays a key role in establishing the needed connection between μ and Lebesgue measure on a bounded subset of R. Results on absolute continuity on R are achieved by using a sequence of such commutator equations. (See [7] for details.)

The study of the unbounded case is continued in [8], where the assumption that the sequence $\{a_n\}$ be a monotonically increasing sequence of positive numbers is replaced by a weaker condition. The following theorem is established:

Theorem Assume $\{a_n\}$ is a sequence of positive numbers which satisfies the conditions $\lim a_n = \infty$, $\Sigma \frac{1}{a_n} = \infty$, and $\Sigma[a_n^2 - a_{n-1}^2]^- < \infty$. If $d_n = |a_n - a_{n-1}|$ and if $\Sigma|d_n - d_{n-1}| < \infty$, then the measure μ has an absolutely continuous part with support equal to the spectrum of C.

In particular, the theorem applies if the sequence $\{a_n\}$ is eventually monotone increasing and the difference sequence $\{d_n\}$ is such that the differences of the differences are absolutely summable. To obtain an interesting class of examples, choose α to be an even positive integer and let $\{P_n\}$ be the sequence of polynomials obtained by orthonormalizing the sequence $\{t^n\}_{n=0}^{\infty}$ with respect to the measure $\mu(\beta) = \int_\beta e^{-x^\alpha/\alpha} dx$. These polynomials satisfy a recursion formula of the form (1.1) with $b_n = 0$, and as shown in [13], $a_n = n^{1/\alpha}[c_0 + \frac{c_2}{n^2} + O(\frac{1}{n^4})]$. It can then be shown (see[8]) that the conditions of the theorem are satisfied.

These results can, in turn, be extended to allow the possibility of a non-zero diagonal sequence in (2.1). The following result is established in [9]:

Theorem Assume $\{a_n\}$ is a sequence of positive numbers such that $\lim a_n = \infty$ and $\Sigma \frac{1}{a_n} = \infty$, that $\{b_n\}$ is a sequence of real numbers, and that $d_n = |a_n - a_{n-1}|$. If $\Sigma[a_n^2 - a_{n-1}^2]^- < \infty$, $\Sigma|d_n - d_{n-1}| < \infty$, and $\Sigma a_{n-1}|b_n - b_{n-1}| < \infty$ then μ has an absolutely continuous part with support equal to the spectrum of C.

The conditions of this theorem were motivated in part by the analysis in [2], which also provides an interesting set of examples. Let $\mu(\beta) = \int_\beta e^{-Q(x)} dx$ where $Q(x) = x^4/4 + q_3 x^3/3 + q_2 x^2/2 + q_1 x$. It is shown in [2] that in this case

$$a_n = (\tfrac{n}{3})^{1/4}\left[c_0 + c_1(\tfrac{n}{3})^{-1/2} + c_2(\tfrac{n}{3})^{-1} + c_3(\tfrac{n}{3})^{-3/2} + o((\tfrac{n}{3})^{-3/2})\right],$$

$$b_n = d_0 + d_1\left(\tfrac{n}{3}\right)^{-1/2} + d_2\left(\tfrac{n}{3}\right)^{-1} + d_3\left(\tfrac{n}{3}\right)^{-3/2} + o\left(\left(\tfrac{n}{3}\right)^{-3/2}\right),$$

where the coefficients c_i and d_i are determined by q_1, q_2 and q_3. It is shown in [9] that for this choice of coefficient sequences the conditions of the last theorem are indeed satsfied.

Two stronger results presented in [9] require more stringent conditions on the coefficient sequences $\{a_n\}$ and $\{b_n\}$.

Theorem Assume $\{a_n\}$ is a monotonically increasing sequence of positive numbers such that $\lim a_n = \infty$ and $\Sigma \tfrac{1}{a_n} = \infty$, that $\{b_n\}$ is a monotonically increasing sequence of real numbers, and that $\{d_n\}$, where $d_n = a_n - a_{n-1}$, is bounded. Suppose there exists $N \geq 2$ such that the following conditions are satisfied:

i) $(b_2 - b_1) \leq a_1$

ii) For $n = 2, \ldots, N-1$, $\tfrac{1}{2}(b_{n+1} - b_{n-1}) \leq d_n$

iii) $\tfrac{1}{2}(b_{N+1} - b_{N-1}) + \tfrac{1}{2}d_{N+1} \leq d_N$

iv) For $n > N$, $\tfrac{1}{2}[d_{n-1} + d_{n+1} + (b_{n+1} - b_{n-1})] \leq d_n$.

Then the measure μ is absolutely continuous.

In the next result, conditions iii) and iv) above are replaced by new conditions which also lead to the conclusion that the measure is absolutely continuous.

Theorem Assume $\{a_n\}$ is a monotonically increasing sequence of positive numbers such that $\lim a_n = \infty$ and $\Sigma \tfrac{1}{a_n} = \infty$, that $\{b_n\}$ is a monotonically increasing sequence of real numbers, and that $\{d_n\}$, where $d_n = a_n - a_{n-1}$, is bounded. Suppose there exists $N \geq 2$ such that the following conditions are satisfied:

i) $(b_2 - b_1) \leq a_1$

ii) For $n = 2, \ldots, N-1$, $\tfrac{1}{2}(b_{n+1} - b_{n-1}) \leq d_n$

iii) $b_{N+1} - b_{N-1} \leq d_N$

iv) For $n > N$, $d_{n-1}^2/d_{n-2} + (b_{n+1} - b_{n-1}) \leq d_n$.

Then the measure μ is absolutely continuous.

7. Final Remarks

This article has been an attempt to convey (in a non-technical fashion) some of the interesting applications of operator theory to the study of systems of orthogonal polynomials. For a more complete study of this topic the interested reader should probably begin with Stone [18] for a discussion of Jacobi matrices, Sarason [17] for background on the moment problem, and Berezanskii [3] for a discussion of self-adjointness. More technical information on the moment problem can be found in Shohat and Tamarkin [16]. Perturbation theorems are presented in Kato [11] and commutator equations are discussed in Putnam [15].

References

[1] N. I. Akhiezer and I. M. Glazman, *Theory of Linear Operators in Hilbert Space, Vol. I and II*, Monographs and Studies in Mathematics, **9, 10**, Pitman, London, 1981.

[2] W. Bauldry, A. Máté, and P. Nevai, *Asymptotic Expansions of Recurrence Coefficients of Asymmetric Freud Polynomials*, Approximation Theory V, New York: Academic Press, 251-254.

[3] J. M. Berezanskii, *Expansions in Eigenfunctions of Selfadjoint Operators*, Translations of Mathematical Monographs, **17**, Amer. Math. Soc., Providence, 1968.

[4] J. Dombrowski, *Quasitriangular Matrices*, Proc. Amer. Math. Soc. **69**(1978), 95-96.

[5] J. Dombrowski, *Tridiagonal Matrix Representations of Cyclic Self-adjoint Operators*, Pacific J. Math., **114**(1984), 325-334.

[6] J. Dombrowski, *Tridiagonal Matrix Representations of Cyclic Self-adjoint Operators II*, Pacific J. Math. **120**(1985), 47-53.

[7] J. Dombrowski, *Cyclic Operators, Commutators, and Absolutely Continuous Measures*, Proc. Amer. Math. Soc. **100**(1987), 457-463.

[8] J. Dombrowski, *Spectral Measures, Orthogonal Polynomials and Absolute Continuity*, SIAM J. Math. Anal.,**19**(1988), 939-943.

[9] J. Dombrowski, *Spectral Measures Corresponding to Orthogonal Polynomials with Unbounded Recurrence Coefficients*, Constr. Approx. **5**(1989), 371-381.

[10] J. Dombrowski and P. Nevai, *Orthogonal Polynomials, Measures and Recurrence Relations,* SIAM J. Math. Anal. **17**(1986), 752-759.

[11] T. Kato, *Perturbation of Continuous Spectra by Trace Class Operators,*Proc. Japan Acad., **33**(1957), 260-264.

[12] A. Máté and P. Nevai, *Orthogonal Polynomials and Absolutely Continuous Measures,* Approximation Theory IV (C. Chui, L. L. Schumaker, J. D. Ward, eds.) New York: Academic Press (1983), 611-617.

[13] A. Máté, P. Nevai and T. Zaslavsky, *Asymptotic Expansion of Ratios of Coefficients of Orthogonal Polynomials With Exponential Weights,* Trans. Amer. Math. Soc., 287 (1985), 495-505.

[14] P. Nevai, *Orthogonal Polynomials,* Mem. Amer. Math. Soc., **213**(1979).

[15] C. R. Putnam, *Commutation Properties of Hilbert Space Operators and Related Topics,* Ergebnesse der Mathematik, **36**, Springer, Berlin, 1967.

[16] J. Shohat and J. Tamarkin, *The Problem of Moments,* Mathematical Surveys, No. 1, Amer. Math. Soc., Providence, R. I., 1943/1950.

[17] D. Sarason, Moment Problems and Operators in Hilbert Space, '*Moments in Mathematics',* Proceedings of Symposia in Applied Mathematics, **37**, Amer. Math. Soc. 1987, 54-70.

[18] M. Stone, *Linear Transformations in Hilbert Space,* Amer. Math. Soc., New York, 1932.

[19] G. Szegö, *Orthogonal Polynomials,* Amer. Math. Soc., Providence, R.I., 1939, Fourth edition, 1975.

USING SYMBOLIC COMPUTER ALGEBRAIC SYSTEMS TO DERIVE FORMULAS INVOLVING ORTHOGONAL POLYNOMIALS AND OTHER SPECIAL FUNCTIONS

GEORGE GASPER*
Department of Mathematics
Northwestern University
Evanston, IL 60208, U.S.A.

ABSTRACT. It is shown how symbolic computer algebraic systems such as Mathematica, Macsyma, SMP, etc., can be used to derive transformation and expansion formulas for orthogonal polynomials that are expressible in terms of either hypergeometric or basic hypergeometric series. In particular, we demonstrate how Mathematica can be used to apply transformation formulas to the Racah and q-Racah polynomials, to derive an indefinite bibasic summation formula, an expansion formula for Laguerre polynomials, Clausen's formula for the square of hypergeometric series, a q-analogue of a Fields and Wimp expansion formula, and to prove the Askey-Gasper inequality which de Branges used in his proof of the Bieberbach conjecture. We also make some observations and conjectures related to Jensen's necessary and sufficient conditions for the Riemann Hypothesis to hold.

1. Introduction

Now that several symbolic computer algebraic systems such as Derive, Reduce, Scratchpad, SMP, and the three M's "Macsyma, Maple, and Mathematica" are available for various computers, it is natural for persons having access to such a system to try to have it perform the tedious symbolic manipulations needed to derive certain formulas involving orthogonal polynomials and other special functions. Here, for definiteness, we will use Mathematica to illustrate how the author has been employing it (and Macsyma, SMP, etc.) to derive transformation and expansion formulas for orthogonal polynomials that are expressible in terms of either hypergeometric series

$$(1.1) \qquad {}_rF_s\left[\begin{matrix} a_1,\ldots,a_r \\ b_1,\ldots,b_s \end{matrix};z\right] = \sum_{n=0}^{\infty} \frac{(a_1)_n\cdots(a_r)_n}{n!(b_1)_n\cdots(b_s)_n}z^n$$

*This material is based upon research supported in part by the National Science Foundation under grant number DMS-8601901.

P. Nevai (ed.), Orthogonal Polynomials, 163–179.
© *1990 by Kluwer Academic Publishers.*

or q-(basic) hypergeometric series

$$(1.2) \qquad {}_r\phi_s \left[\begin{matrix} a_1, \ldots, a_r \\ b_1, \ldots, b_s \end{matrix} ; q, z \right]$$

$$= \sum_{n=0}^{\infty} \frac{(a_1; q)_n \cdots (a_r; q)_n}{(q; q)_n (b_1; q)_n \cdots (b_s; q)_n} [(-1)^n q^{n(n-1)/2}]^{1+s-r} z^n$$

where $(a)_n = \prod_{k=0}^{n-1}(a + k)$ is the shifted factorial (Pochhammer symbol) and $(a; q)_n = \prod_{k=0}^{n-1}(1 - aq^k)$ is the q-shifted factorial. Unless stated otherwise we will assume that $0 < |q| < 1$. For such orthogonal polynomials, which include the classical orthogonal polynomials of Gegenbauer, Hahn, Hermite, Jacobi, Krawtchouk, Laguerre, Meixner, Tchebichef, and their q-analogues, see Askey and Wilson [5], Chihara [10], Erdélyi [14], Gasper and Rahman [27], and Szegö [30].

Although our methods are applicable to all of the above mentioned orthogonal polynomials, in order to demonstrate them in this paper we will only consider certain formulas involving the Laguerre polynomials

$$(1.3) \qquad L_n^a(x) = \frac{(a + 1)_n}{n!} \, {}_1F_1 \left[\begin{matrix} -n \\ a + 1 \end{matrix} ; x \right]$$

the Gegenbauer (ultraspherical) polynomials

$$(1.4) \qquad C_n^a(x) = \frac{(2a)_n}{n!} \, {}_2F_1 \left[\begin{matrix} -n, n + 2a \\ a + 1/2 \end{matrix} ; \frac{1 - x}{2} \right]$$

the Racah polynomials

$$(1.5) \qquad W_n(x; a, b, c, N) = {}_4F_3 \left[\begin{matrix} -n, n + a + b + 1, -x, c + x - N \\ a + 1, b + c + 1, -N \end{matrix} ; 1 \right]$$

and the q-Racah polynomials

$$(1.6) \qquad W_n(x; a, b, c, N; q) = {}_4\phi_3 \left[\begin{matrix} q^{-n}, abq^{n+1}, q^{-x}, cq^{x-N} \\ aq, bcq, q^{-N} \end{matrix} ; q, q \right].$$

In §2 we illustrate how the symbolic manipulation capabilities of Mathematica can be utilized to apply certain transformation formulas to the Racah and q-Racah polynomials. Symbolic factorization is employed in §3 to derive an indefinite bibasic summation formula. A technique for manipulating multiple series is demonstrated in §4 by using it to derive the expansion formula

$$(1.7) \qquad L_n^b(x) = \sum_{k=0}^{n} \frac{(b - a)_{n-k}}{(n - k)!} L_k^a(x).$$

The coefficients in (1.7) are called the connection coefficients between the sequences $\{L_n^b(x)\}$ and $\{L_n^a(x)\}$. Notice that the above connection coefficients are clearly nonnegative when $a \leq b$. For applications of the nonnegativity of the connection coefficients between two sequences of orthogonal polynomials to positive definite functions, isometric embeddings of metric spaces, the derivation of inequalities, etc., see Askey [1], Askey and Gasper [2], Gangolli [16], Gasper [17], [18], and Gasper and Rahman [27].

Multiple series manipulations are also used in §5 to derive Clausen's [11] formula

$$(1.8) \qquad \left\{ {}_2F_1 \left[\begin{matrix} a,b \\ a+b+1/2 \end{matrix} ; x \right] \right\}^2 = {}_3F_2 \left[\begin{matrix} 2a, 2b, a+b \\ 2a+2b, a+b+1/2 \end{matrix} ; x \right]$$

and, in §6, to derive a q-extension (q-analogue) of the Fields and Wimp [15] expansion formula

$$(1.9) \qquad {}_{r+t}F_{s+u} \left[\begin{matrix} a_R, c_T \\ b_S, d_U \end{matrix} ; xw \right] = \sum_{n=0}^{\infty} \frac{(c_T)_n (e_K)_n (-x)^n}{(d_U)_n (f_M)_n (n+\gamma)_n n!}$$

$$\cdot {}_{k+t}F_{m+u+1} \left[\begin{matrix} n+c_T, n+e_K \\ 2n+1+\gamma, n+d_U, n+f_M \end{matrix} ; x \right]$$

$$\cdot {}_{m+r+2}F_{k+s} \left[\begin{matrix} -n, n+\gamma, a_R, f_M \\ b_S, e_K \end{matrix} ; w \right]$$

where we employed the contracted notation of representing $a_1, a_2, \ldots, a_r$ by a_R, $(a_1)_n (a_2)_n \cdots (a_r)_n$ by $(a_R)_n$, and $n+a_1, n+a_2, \ldots, n+a_r$ by $n+a_R$. In (1.9), as elsewhere, either the parameters and variables are assumed to be such that the (multiple) series converge absolutely or the series are considered to be formal power series in the variables x and w. As an application of (1.8) and (1.9) we point out how they can be used to prove the Askey-Gasper [3], [4] inequality

$$(1.10) \qquad {}_3F_2 \left[\begin{matrix} -n, n+a+2, (a+1)/2 \\ a+1, (a+3)/2 \end{matrix} ; x \right] \geq 0, \qquad 0 \leq x \leq 1,$$

for $a > -2$ and $n = 0, 1, \ldots$, which was used by de Branges [7], [8] in his proof of the Bieberbach conjecture.

In §7 we make some observations and conjectures related to Jensen's necessary and sufficient conditions for the Riemann Hypothesis to hold.

2. Transformation formulas

In order to use Mathematica to apply Whipple's ${}_4F_3$ transformation formula [6, 7.2(1)]

$$(2.1) \qquad {}_4F_3 \left[\begin{matrix} -n, a, b, c \\ d, e, f \end{matrix} ; 1 \right]$$

$$= \frac{(e-a)_n (f-a)_n}{(e)_n (f)_n} {}_4F_3 \left[\begin{matrix} -n, a, d-b, d-c \\ d, a+1-n-e, a+1-n-f \end{matrix} ; 1 \right]$$

where $d+e+f = a+b+c+1-n$, to the Racah polynomials we let $p[a, n]$ symbolically denote the Pochhammer symbol $(a)_n$ (without defining it as a product) and enter the following function definition into a Mathematica session

$In[1] :=$ **transform4F3[mn_,a_,b_,c_,d_,e_,f_] := p[e-a,-mn] p[f-a,-mn]***
  fourF3[mn,a,d-b,d-c,d,a+1+mn-e,a+1+mn-f]/(p[e,-mn] p[f,-mn])

which represents the right side of (2.1). The a_, b_, etc., on the left side refer to any expressions, to be named a, b, etc., and the := defines the transformation rule to be used automatically each time the left side is requested. Either a * or a space may be used between variables and functions to denote multiplication in Mathematica, and, as above, the * has to be used when the next factor is continued on the next line. Notice that we used the symbol mn to denote the $-n$ argument on the left side of (2.1). Since Mathematica's built-in functions begin with capital letters (function names cannot begin with numbers and all function arguments must be enclosed in square brackets), we chose function names that begin with lower case letters to prevent any possible confusion with Mathematica's built-in functions.

In view of the $_4F_3$ series representation for the Racah polynomials in (1.5) we enter

$In[2] :=$ **transform4F3[-n,n+a+b+1,-x,c+x-N,a+1,b+c+1,-N]**

and then Mathematica responds with

$Out[2]=$ (fourF3[$-n, 1 + a + b + n, 1 + a + x, 1 + N + a - c - x, 1 + a, 1 + a - c,$
  $2 + N + a + b$] p[$-a + c - n, n$] p[$-1 - N - a - b - n, n$])/
  (p[$-N, n$] p[$1 + b + c, n$])

which is the $_4F_3$ series representation for the Racah polynomials that results when Whipple's transformation formula (2.1) is applied to the $_4F_3$ on the right side of (1.5).

Analogously, to apply Sears' [29] q-analogue of Whipple's formula

$$(2.2) \qquad {}_4\phi_3\left[\begin{matrix} q^{-n}, a, b, c \\ d, e, f \end{matrix}; q, q\right]$$

$$= \frac{(e/a; q)_n (f/a; q)_n}{(e; q)_n (f; q)_n} a^n {}_4\phi_3\left[\begin{matrix} q^{-n}, a, d/b, d/c \\ d, aq^{1-n}/e, aq^{1-n}/f \end{matrix}; q, q\right]$$

where $def = abcq^{1-n}$, to the q-Racah polynomials we let $o[a, q, n]$ symbolically denote the q-shifted factorial $(a; q)_n$ and enter the function definition

$In[3] :=$ **transform4phi3[mn_,a_,b_,c_,d_,e_,f_,q_] := o[e/a,q,-mn] o[f/a,q,-mn]***
  a^(-mn) fourphi3[q^mn,a,d/b,d/c,d,a q^(1+mn)/e,a q^(1+mn) /f,q,q]/
  (o[e,q,-mn] o[f,q,-mn])

Then, entering

$In[4] :=$ **transform4phi3[-n,a b q^(n+1),q^(-x),c q^(x-N),a q,b c q,q^(-N),q]**

yields the $_4\phi_3$ series representation

$Out[4] = (a^n \, b^n \, q^{n \, (1 + n)}$ fourphi3[$q^{-n}, a \, b \, q^{1 + n}, \, a \, q^{1 + x}, \dfrac{a \, q^{1 + N - x}}{c}, a \, q, \dfrac{a \, q}{c},$
  $a \, b \, q^{2 + N}, \, q, \, q$] $o[\dfrac{c}{a \, q^n}, \, q, \, n]$ $o[\dfrac{q^{-1 - N - n}}{a \, b}, \, q, \, n])/$
  $(o[b \, c \, q, \, q, \, n] \, o[q^{-N}, \, q, \, n])$

for the q-Racah polynomials.

Several other formulas for the Racah and q-Racah polynomials may be obtained by applying these transformations to the corresponding series on the right sides of (1.5) and (1.6) with their second, third and fourth numerator arguments interchanged or their denominator arguments interchanged, and by iterating these transformations. On a Macintosh II with five megabytes of RAM, it only took Mathematica about one-half second and one second, respectively, of CPU time to compute Out[2] and Out[4]. Application of Watson's transformation formula [30, (3.4.1.5)]

$$(2.3) \qquad {}_4\phi_3 \left[\begin{matrix} q^{-n}, a, b, c \\ d, e, f \end{matrix} ; q, q \right] = \frac{(d/b; q)_n (d/c; q)_n}{(d; q)_n (d/bc; q)_n}$$

$$\cdot {}_8\phi_7 \left[\begin{matrix} \sigma, q\sqrt{\sigma}, -q\sqrt{\sigma}, f/a, e/a, b, c, q^{-n} \\ \sqrt{\sigma}, -\sqrt{\sigma}, e, f, ef/ab, ef/ac, efq^n/a \end{matrix} ; q, \frac{efq^n}{bc} \right]$$

where $def = abcq^{1-n}$ and $\sigma = ef/aq$, to the q-Racah polynomials took about two seconds. One advantage of using a symbolic computer algebraic system is that once a function is defined its definition can be stored in a file (called a Notebook in Mathematica) from which it can be quickly read into any other session whenever it is needed. Another advantage is that one can also have the computer automatically check that the required "balanced" conditions $d + e + f = a + b + c + 1 - n$ in (2.1) and $def = abcq^{1-n}$ in (2.2) or (2.3) are satisfied before it computes the transformations. For example, rather than using the definition given in In[3] it is preferable to replace it by

```
transform4phi3[mn_,a_,b_,c_,d_,e_,f_,q_] :=
    If [d e f == a b c q^(1+mn), o[e/a,q,-mn] o[f/a,q,-mn] a^(-mn)*
    fourphi3[q^mn,a,d/b,d/c,d,a q^(1+mn)/e,a q^(1+mn)/ f,q,q]/
    (o[e,q,-mn] o[f,q,-mn]),
    Print["ERROR — Nonbalanced Series"],
    Print["ERROR — Nonbalanced Series"]]
```

which will immediately print "ERROR — Nonbalanced Series" whenever the required "balanced" condition is not satisfied. In the above display the logical operator == tests whether the expressions on the left and right sides of it are equal. For a limited time persons who wish to obtain copies of my Mathematica input files (which can easily be converted to work with Macsyma) containing symbolic forms of most of the identities and the summation and transformation formulas in the three Appendices of the book [27] may obtain them via email by contacting me at either gasper@nuacc.bitnet or george@math.nwu.edu, or by mailing me a formatted Macintosh 3.5" disk along with a self-addressed stamped envelope.

3. An indefinite bibasic summation formula

As in Gasper [21], let

$$(3.1) \qquad s_k = \frac{(ap; p)_k (bp; p)_k (cq; q)_k (aq/bc; q)_k}{(q; q)_k (aq/b; q)_k (ap/c; p)_k (bcp; p)_k}$$

for $k = 0, 1, 2, \ldots, s_{-1} = 0$, and define the difference operator Δ by $\Delta s_k = s_k - s_{k-1}$. If we enter the Product definition for s_k into Mathematica and apply the Factor command, we only get the difference of the two products that we started with. Therefore, we first observe that

$$(3.2) \qquad \Delta s_k = \frac{(ap;p)_{k-1}(bp;p)_{k-1}(cq;q)_{k-1}(aq/bc;q)_{k-1}}{(q;q)_k(aq/b;q)_k(ap/c;p)_k(bcp;p)_k}$$
$$\cdot \left[(1-ap^k)(1-bp^k)(1-cq^k)(1-aq^k/bc)\right.$$
$$\left. - (1-q^k)(1-aq^k/b)(1-ap^k/c)(1-bcp^k)\right]$$

and then ask Mathematica to factor the above term in square brackets by entering

$In[1] :=$ **Factor[(1-a p^k)(1-b p^k)(1-c q^k)(1-a q^k/b/c)-**
(1-q^k)(1-a q^k/b)(1-a p^k/c)(1-b c p^k)]

to obtain

$$Out[1] = \frac{(-1+c)\,(1-a\,p^k\,q^k)\,(-a+b\,c)\,(b\,p^k-q^k)}{b\,c}$$

which combined with (3.2) shows that

$$(3.3) \qquad \Delta s_k = \frac{(1-ap^kq^k)(1-bp^kq^{-k})}{(1-a)(1-b)} \, \frac{(a;p)_k(b;p)_k(c;q)_k(a/bc;q)_k}{(q;q)_k(aq/b;q)_k(ap/c;p)_k(bcp;p)_k}\, q^k.$$

Since $\Delta s_0 = s_0 = 1$ and

$$(3.4) \qquad \sum_{k=0}^{n} \Delta s_k = s_n$$

for $n \geq 0$, it follows from (3.3) that we have the indefinite bibasic summation formula [21]

$$(3.5) \qquad \sum_{k=0}^{n} \frac{(1-ap^kq^k)(1-bp^kq^{-k})}{(1-a)(1-b)} \, \frac{(a;p)_k(b;p)_k(c;q)_k(a/bc;q)_k}{(q;q)_k(aq/b;q)_k(ap/c;p)_k(bcp;p)_k}\, q^k$$
$$= \frac{(ap;p)_n(bp;p)_n(cq;q)_n(aq/bc;q)_n}{(q;q)_n(aq/b;q)_n(ap/c;p)_n(bcp;p)_n}.$$

Observing that $(q^{1-n};q)_n = 0$ for $n \geq 1$, we find that when $c = q^{-n}$, $n = 0,1,2,\ldots$, formula (3.5) reduces to the summation formula

$$(3.6) \qquad \sum_{k=0}^{n} \frac{(1-ap^kq^k)(1-bp^kq^{-k})}{(1-a)(1-b)} \, \frac{(a;p)_k(b;p)_k(q^{-n};q)_k(aq^n/b;q)_k}{(q;q)_k(aq/b;q)_k(apq^n;p)_k(bpq^{-n};p)_k}\, q^k = \delta_{n,0}$$

for $n = 0,1,2,\ldots$, where $\delta_{n,m}$ is the Kronecker delta function. Formula (3.6) was independently derived in an equivalent form by Bressoud [9, §4]. This formula will be employed in our derivation in §6 of the q-analogue of the Fields and Wimp formula (1.9). The above derivation of (3.5) can be extended to give the generalization derived in Gasper and Rahman [25, (1.7)].

4. Derivation of the Laguerre polynomial expansion formula (1.7)

Although Mathematica's Sum[c[i, j, ...], {i, imin, imax}, {j, jmin, jmax}, ...] can be used to represent a multiple series such as those on the right sides of (1.7) and (1.9), one runs into difficulties in trying to perform the series manipulations needed to derive formulas involving multiple series such as those in [18], [19], [20], [21]. Here we will use the expansion formula (1.7) to demonstrate how one can derive formulas involving multiple series by working directly with the terms in the series.

Since the use of an identity such as

$$(4.1) \qquad\qquad (a)_{n+k} = (a)_n(a + n)_k$$

to replace $(a)_{n+k}$ by $(a)_n(a + n)_k$ corresponds to multiplying by 1 in the form

$$(4.2) \qquad\qquad (a)_n(a + n)_k/(a)_{n+k}$$

and cancelling the two $(a)_{n+k}$ products, we start by entering the following defined function into Mathematica

   $In[1] := \text{asnpk}[a\_,n\_,k\_] := p[a,n]\ p[a+n,k]/p[a,n+k]$

Then asnpk[a, n, k] equals 1 for all choices of its arguments, and multiplication of an expression such as a term in a series by this function followed by symbolic cancellation corresponds to using the identity (4.1). Sometimes the Simplify command has to be applied so that the desired symbolic cancellations occur and the simplest expressions are formed. The function name asnpk is a mnemonic for "a sub n plus k," which makes it easy to remember that multiplication by this function corresponds to applying the identity (4.1). Similarly, to apply the identity

$$(4.3) \qquad\qquad (-n)_k = (-1)^k(1)_n/(1)_{n-k}$$

we enter the definition

   $In[2] := \text{mnsk}[mn\_,k\_] := (-1)\text{\textasciicircum}k\ p[1,-mn]/(p[1,-mn-k]\ p[mn,k])$

where mnsk stands for "minus n sub k." Then mnsk[-n, k] equals 1 for $k = 0, 1, \ldots, n$. Notice that even if we multiply an expression by one of these functions after making a mistake in typing the desired choice of arguments, we will still get a correct answer (i.e., an equal expression) because we only multiplied by 1.

Analogously, to apply a summation formula

$$(4.4) \qquad\qquad \sum_k a_k = A$$

with $A \neq 0$, we first rewrite it in the form

$$(4.5) \qquad\qquad \sum_k A^{-1}a_k = 1$$

and then represent this series by its kth term $A^{-1}a_k$. For example, with Vandermonde's summation formula [6, p. 3]

$$(4.6) \qquad {}_2F_1\left[\begin{matrix} -n,a \\ c \end{matrix};1\right] = \frac{(c-a)_n}{(c)_n}$$

rewritten in the form

$$(4.7) \qquad \sum_{k=0}^{n} \frac{(c)_n(-n)_k(a)_k}{(c-a)_n(1)_k(c)_k} = 1$$

we define

$In[3] := $ **vand2F1[mn_,a_,c_,k_] := p[c,-mn] p[mn,k] p[a,k]/**
  (p[c-a,-mn] p[1,k] p[c,k])

so that the sum of vand2F1[-n,a,c,k] from $k = 0$ to n equals 1 for all values of a and c when $n = 0, 1, 2, \ldots$. Similarly, to be able to apply the expansion

$$(4.8) \qquad x^j = \sum_{k=0}^{j} \frac{(a+1)_j(-j)_k}{(a+1)_k} L_k^a(x)$$

we define the function

$In[4] := $ **jthpowerofx[j_,x_,a_,k_] := x^(-j) p[a+1,j] p[-j,k] laguerre[k,a,x] /**
  p[a+1,k]

whose sum over all k equals 1.

Now we are ready to derive formula (1.7). Enter

$In[5] := $ **p[b+1,n] p[-n,j] x^j/(p[1,n] p[1,j] p[b+1,j])**

to get

$$Out[5] = \frac{x^j\, p[-n,\ j]\, p[1+b,\ n]}{p[1,\ j]\, p[1,\ n]\, p[1+b,\ j]}$$

which is the jth term of the series representation for the Laguerre polynomial $L_n^b(x)$ on the left side of (1.7). Then

$In[6] := $ **% jthpowerofx[j,x,a,k]**

where % stands for the last result (Out[5] in this case which could have been used instead of the %), gives

$$Out[6] = \frac{\text{laguerre}[k,\ a,\ x]\, p[-j,\ k]\, p[-n,\ j]\, p[1+a,\ j]\, p[1+b,\ n]}{p[1,\ j]\, p[1,\ n]\, p[1+a,\ k]\, p[1+b,\ j]}$$

which is the j, k^{th} term in the double sum obtained by using the expansion (4.8) in the series for $L_n^b(x)$. Apply the identity (4.3) by using

$In[7] := $ **% mnsk[-j,k]**

to obtain

$$Out[7] = \frac{(-1)^k \, \text{laguerre}[k, \, a, \, x] \, p[-n, \, j] \, p[1+a, \, j] \, p[1+b, \, n]}{p[1, \, n] \, p[1, \, j-k] \, p[1+a, \, k] \, p[1+b, \, j]}$$

and then use

$In[8] := \%/.\ \text{j->j+k}$

to replace j by $j + k$ and get

$$Out[8] = \frac{(-1)^k \, \text{laguerre}[k, \, a, \, x] \, p[-n, \, j+k] \, p[1+a, \, j+k] \, p[1+b, \, n]}{p[1, \, j] \, p[1, \, n] \, p[1+a, \, k] \, p[1+b, \, j+k]}$$

In view of the above $j + k$'s we can use

$In[9] := \%\ \text{asnpk[-n,k,j] asnpk[a+1,k,j]/asnpk[b+1,k,j]}$

to obtain

$$Out[9] = \frac{(-1)^k \, \text{laguerre}[k, \, a, \, x] \, p[-n, \, k] \, p[1+b, \, n] \, p[k-n, \, j] \, p[1+a+k, \, j]}{p[1, \, j] \, p[1, \, n] \, p[1+b, \, k] \, p[1+b+k, \, j]}$$

Noticing that Vandermonde's summation formula may be applied to evaluate the sum over j, we enter

$In[10] := \%/\text{vand2F1[k-n,1+a+k,1+b+k,j]}$

to get

$$Out[10] = \frac{(-1)^k \, \text{laguerre}[k, \, a, \, x] \, p[-n, \, k] \, p[1+b, \, n] \, p[-a+b, \, -(k-n)]}{p[1, \, n] \, p[1+b, \, k] \, p[1+b+k, \, -(k-n)]}$$

Finally, by entering

$In[11] := \text{Simplify}[\%\ \text{asnpk[b+1,k,n-k] mnsk[-n,k]}\ (-1)\hat{\ }(-2k)]$

we obtain

$$Out[11] = \frac{\text{laguerre}[k, \, a, \, x] \, p[-a+b, \, -k+n]}{p[1, \, -k+n]}$$

whose sum over k is the right side of (1.7), which concludes our derivation of (1.7).

5. Derivation of Clausen's formula

Observing that

$$(5.1) \qquad \left\{ {}_2F_1 \left[\begin{matrix} a,b \\ a+b+1/2 \end{matrix} ; x \right] \right\}^2 = \sum_{j=0}^{\infty} \sum_{k=0}^{\infty} \frac{(a)_j (b)_j (a)_k (b)_k x^{j+k}}{(1)_j (a+b+1/2)_j (1)_k (a+b+1/2)_k}$$

we start by entering

$In[1] := $ p[a,j] p[b,j] p[a,k] p[b,k] x^(j+k)/
(p[1,j] p[a+b+1/2,j] p[1,k] p[a+b+1/2,k])

and obtaining

$$Out[1] = \frac{x^{j\,+\,k}\, p[a,\,j]\, p[a,\,k]\, p[b,\,j]\, p[b,\,k]}{p[1,\,j]\, p[1,\,k]\, p\,[\frac{1}{2}+a+b,\,j]\, p[\frac{1}{2}+a+b,\,k]}$$

which is the j, k^{th} term of the double series on the right side of (5.1). In view of the $j + k^{\text{th}}$ power of x, we apply the substitution

$In[2] :$ %/. j->j-k

to obtain

$$Out[2] = \frac{x^{j}\, p[a,\,k]\, p[a,\,j-k]\, p[b,\,k]\, p[b,\,j-k]}{p[1,\,k]\, p[1,\,j-k]\, p[\frac{1}{2}+a+b,\,k]\, p[\frac{1}{2}+a+b,\,j-k]}$$

To be able to apply the identity

(5.2) $$(a)_{n-k} = (-1)^k (a)_n /(1 - n - a)_k$$

enter the definition

$In[3] := $ asnmk[a_,n_,k_] := (-1)^k p[a,n]/(p[1-n-a,k] p[a,n-k])

where asnmk stands for a sub n minus k. Then

$In[4] := $ Out[2] asnmk[a,j,k] asnmk[b,j,k]/
(asnmk[1,j,k] asnmk[a+b+1/2,j,k])

gives

$$Out[4] = \frac{x^{j}\, p[a,\,j]\, p[a,\,k]\, p[b,\,j]\, p[b,\,k]\, p[-j,\,k]\, p[\frac{1}{2}-a-b-j,\,k]}{p[1,\,j]\, p[1,\,k]\, p[\frac{1}{2}+a+b,\,j]\, p[\frac{1}{2}+a+b,\,k]\, p[1-a-j,\,k]\, p[1-b-j,\,k]}$$

To sum over k it suffices to apply the summation formula

(5.3) $$\,_4F_3\left[\begin{array}{c} -n, a, b, 1/2 - a - b - n \\ a + b + 1/2, 1 - a - n, 1 - b - n \end{array}; 1\right] = \frac{(2a)_n (2b)_n (a+b)_n}{(a)_n (b)_n (2a + 2b)_n}$$

which follows easily from Dougall's $\,_7F_6$ summation formula [6, 4.3.(5)]. Enter this formula into Mathematica by defining

$In[5] := $ specialsum[n_,a_,b_,k_] := p[-n,k] p[a,k] p[b,k] p[1/2-a-b-n,k]*
p[a,n] p[b,n] p[2a+2b,n]/(p[1,k] p[a+b+1/2,k] p[1-a-n,k]*
p[1-b-n,k] p[2a,n] p[2b,n] p[a+b,n])

whose sum over k equals 1. Then,

$In[6] := $ Out[4]/specialsum[j,a,b,k]

yields

$$Out[6] = \frac{x^j \, p[2\,a,\,j] \, p[2\,b,\,j] \, p[a+b,\,j]}{p\,[1,\,j] \, p[2\,a+2\,b,\,j] \, p[\frac{1}{2}+a+b,\,j]}$$

which is the jth term of the $_3F_2$ series on the right side of (1.8). This completes our derivation of Clausen's formula.

One particularly important special case of Clausen's formula is that for the ultraspherical polynomials it gives the formula

$$(5.4) \qquad \{C_n^a(x)\}^2 = \left(\frac{(2a)_n}{n!}\right)^2 \, _3F_2\left[\begin{array}{c} -n, n+2a, a \\ 2a, a+1/2 \end{array} ; 1 - x^2\right]$$

by using the series representation

$$(5.5) \qquad C_n^a(x) = \frac{(2a)_n}{n!} \, _2F_1\left[\begin{array}{c} -n/2, a+n/2 \\ a+1/2 \end{array} ; 1 - x^2\right].$$

An extension of (5.4) to the continuous q-ultraspherical polynomials

$$(5.6) \qquad C_n(\cos\theta; \beta | q) = \frac{(\beta^2; q)_n}{\beta^{n/2}(q; q)_n} \, _4\phi_3\left[\begin{array}{c} q^{-n}, \beta^2 q^n, \beta^{1/2}e^{i\theta}, \beta^{1/2}e^{-i\theta} \\ \beta q^{1/2}, -\beta q^{1/2}, -\beta \end{array} ; q, q\right]$$

is derived in Gasper [22], and a nonterminating q-analogue of Clausen's formula is derived in Gasper and Rahman [26].

6. q-Extensions of the Fields and Wimp expansion formula (1.9)

Verma [32] showed that the Fields and Wimp expansion formula (1.9) is a special case of the expansion

$$(6.1) \qquad \sum_{j=0}^{n} A_j B_j \frac{(xw)^j}{j!} = \sum_{n=0}^{\infty} \frac{(-x)^n}{n!(c+n)_n} \sum_{k=0}^{\infty} \frac{A_{n+k} x^k}{k!(c+2n+1)_k}$$
$$\cdot \sum_{j=0}^{n} \frac{(-n)_j (c+n)_j}{j!} B_j w^j$$

and derived the q-analogue

$$(6.2) \qquad \sum_{j=0}^{\infty} A_j B_j \frac{(xw)^j}{(q;q)_j} = \sum_{n=0}^{\infty} \frac{(-x)^n q^{n(n-1)/2}}{(q;q)_n(cq^n;q)_n}$$
$$\cdot \sum_{k=0}^{\infty} \frac{A_{n+k} x^k}{(q;q)_k(cq^{2n+1};q)_k} \sum_{j=0}^{n} \frac{(q^{-n};q)_j(cq^n;q)_j}{(q;q)_j} B_j (wq)^j.$$

From (6.2) it follows that (1.9) has a q-analogue of the form

(6.3)
$$_{r+t}\phi_{s+u}\left[\begin{matrix}a_R,c_T\\b_S,d_U\end{matrix};q,xw\right]$$

$$=\sum_{n=0}^{\infty}\frac{(c_T;q)_n(e_K;q)_n x^n[(-1)^n q^{n(n-1)/2}]^{2+m+u-k-t}}{(q;q)_n(d_U;q)_n(f_M;q)_n(\gamma q^n;q)_n}$$

$$\cdot {}_{k+t}\phi_{m+u+1}\left[\begin{matrix}c_T q^n,e_K q^n\\\gamma q^{2n+1},d_U q^n,f_M q^n\end{matrix};q,xq^{n(2+m+u-k-t)}\right]$$

$$\cdot {}_{m+r+2}\phi_{k+s}\left[\begin{matrix}q^{-n},\gamma q^n,a_R,f_M\\b_S,e_K\end{matrix};q,wq\right]$$

where we used a contracted notation analogous to that used in (1.9).

To derive (6.2) first observe that since the identities (4.1) and (4.3) have the q-analogues

(6.4) $(a;q)_{n+k}=(a;q)_n(aq^n;q)_k$

(6.5) $(q^{-n};q)_k=(-1)^k q^{k(k-1)/2-nk}(q;q)_n/(q;q)_{n-k}$

and since

(6.6) $(a;q)_n=(1-a)(aq;q)_{n-1}$

we can enter the definitions

$In[1] := $ **asnpk[a_,q_,n_,k_]** := o[a,q,n] o[a q^n,q,k]/o[a,q,n+k]

$In[2] := $ **mnsk[mn_,q_,k_]** := (-1)^k q^(k(k-1)/2 +mn k) o[q,q,-mn]/
(o[q,q,-mn-k] o[q^mn,q,k])

$In[3] := $ **shift[a_,q_,n_]** := (1-a) o[a q,q,n-1]/o[a,q,n]

It should be noted that both of the functions asnpk[a,q,n,k] and asnpk[a,n,k] can be used in the same session because Mathematica will distinguish them by their different number of arguments. Similarly, both of the functions mnsk[mn,q,k] and mnsk[mn,k] can be used in the same Mathematica session.

So that we can also employ the $b \to 0$ limit case of the $p = q$ case of the summation formula (3.5), let's enter its kth term in the form

$In[4] := $ **deltasum[n_,a_,q_,k_]** := (1-a q^(2k)) o[a,q,k] o[q^(-n),q,k] q^(n k)/
((1-a) o[q,q,k] o[a q^(n+1),q,k])

Then the product $A[j]x^j$ can be represented by entering

$In[5] := $ **deltasum[m,c q^(2j),q,i] A[j+m] x^(j+m)/(o[q,q,m] o[c q^(2j+1),q,m])**

whose sum over all $m \geq 0$ equals $A[j]x^j$, to get

$$Out[5] = \frac{q^{i\,m}\,x^{j+m}\,A[j+m]\,(1-c\,q^{2\,i+2\,j})\,o[c\,q^{2\,j},\,q,\,i]\,o[q^{-m},\,q,\,i]}{(1-c\,q^{2\,j})\,o[q,\,q,\,i]\,o[q,\,q,\,m]\,o[c\,q^{1+2\,j},\,q,\,m]\,o[c\,q^{1+2\,j+m},\,q,\,i]}$$

where i and m are the indexes of summation. Next use

$In[6] := \%$ asnpk[c q^(2j+1),q,m,i] shift[c q^(2j),q,i] mnsk[-m,q,i]/
asnpk[c q^(2j+1),q,i-1,m+1]

to obtain

$$Out[6] = \frac{(-1)^i \; q^{\frac{i\,(-\frac{1}{2} + i)}{2}} \; x^{j \,+\, m} \; A[j+m] \; (1 - c\, q^{2\,i\,+\,2\,j})}{o[q,\,q,\,i]\; o[q,\,q,\,-i+m]\; o[c\, q^{i\,+\,2\,j},\,q,\,1+m]}$$

Applying the change in indexes of summation

$In[7] := $ Simplify[%/. {i->n-j, m->n+k-j}]

we have

$$Out[7] = \frac{(-1)^{-j\,+\,n} \; q^{\frac{(-j\,+\,n)\,(-\frac{1}{2}\,-\,j\,+\,n)}{2}} \; x^{k\,+\,n} \; A[k+n] \; (1 - c\, q^{2\,n})}{o[q,\,q,\,k]\; o[q,\,q,\,-j+n]\; o[c\, q^{j\,+\,n},\,q,\,1-j+k+n]}$$

where n and k are the indexes of summation. Then

$In[8] := $ Simplify[% asnpk[c q^n,q,j,n+k+1-j]/
(mnsk[-n,q,j] asnpk[c q^n,q,n,k+1] shift[c q^(2n),q,k+1])]

gives

$$Out[8] = \frac{(-1)^{-2\,j\,+\,n} \; q^{j\,-\,\frac{n}{2}\,+\,\frac{n^2}{2}} \; x^{k\,+\,n} \; A[k+n] \; o[q^{-n},\,q,\,j] \; o[c\, q^n,\,q,\,j]}{o[q,\,q,\,k]\; o[q,\,q,\,n]\; o[c\, q^n,\,q,\,n]\; o[c\, q^{1\,+\,2\,n},\,q,\,k]}$$

Hence if we multiply by B[j] w^j /o[q,q,j] and sum from $j = 0$ to ∞ by using

$In[9] := \%$ (-1)^(2j) B[j] w^j/o[q,q,j]

we finally obtain

$$Out[9] = \frac{(-1)^n \; q^{j\,-\,\frac{n}{2}\,+\,\frac{n^2}{2}} \; w^j \, x^{k\,+\,n} \; A[k+n] \; B[j] \; o[q^{-n},\,q,\,j] \; o[c\, q^n,\,q,\,j]}{o[q,\,q,\,j]\; o[q,\,q,\,k]\; o[q,\,q,\,n]\; o[c\, q^n,\,q,\,n]\; o[c\, q^{1\,+\,2\,n},\,q,\,k]}$$

which gives the n, j, k^{th} term of the triple sum on the right side of (6.2) and so concludes our derivation of (6.2). This technique can also be employed to derive the extensions of (6.2) and (6.3) in [21,§4] and the bibasic extension in [21,§3] of Euler's transformation formula

(6.7)
$$\sum_{n=0}^{\infty} a_n b_n x^n = \sum_{k=0}^{\infty} (-1)^k \frac{x^k}{k!} f^{(k)}(x) \Delta^k a_0$$

where

$$f(x) = b_0 + b_1 x + b_2 x^2 + \cdots$$

and

$$\Delta^k a_0 = \sum_{j=0}^{k} (-1)^j \binom{k}{j} a_{k-j}.$$

The Fields and Wimp expansion formula (1.9) follows from (6.3) by replacing each parameter in (6.3) by a power of q and letting $q \to 1$. In [18,23] it was pointed out that by using (5.4) in (1.9) we obtain the sum of squares expansion

$$(6.8) \qquad {}_3F_2 \left[\begin{matrix} -n, n+a+2, (a+1)/2 \\ a+1, (a+3)/2 \end{matrix} ; (1-x^2)(1-y^2) \right]$$

$$= \sum_{k=0}^{n} \frac{n!(n+a+2)_k((a+2)/2)_k}{k!(n-k)!((a+3)/2)_k(k+a+1)_k} (1-y^2)^k$$

$$\cdot \left\{ \frac{k!(n-k)!}{(a+1)_k(2k+a+2)_{n-k}} C_k^{(a+1)/2}(x) C_{n-k}^{k+(a+2)/2}(y) \right\}^2$$

which immediately gives the Askey-Gasper inequality (1.10) since each term on the right side of (6.8) is clearly nonnegative. The special case $y = 0$ of (6.8) gives the expansion in [3, (1.16)]. For a q-extension of (6.8), see [22].

7. Jensen's necessary and sufficient conditions for the Riemann Hypothesis

Among Jensen's necessary and sufficient conditions for the Riemann Hypothesis given in Pólya [28] is the condition that

$$(7.1) \qquad \int_{-\infty}^{\infty} \int_{-\infty}^{\infty} \Phi(s)\Phi(t) e^{i(s+t)x}(s-t)^{2n} \, ds \, dt \geq 0$$

for all real x when $n = 0, 1, 2, \ldots$, where

$$(7.2) \qquad \Phi(t) = 4 \sum_{k=1}^{\infty} (2k^4 \pi^2 e^{9t} - 3k^2 \pi e^{5t}) e^{-k^2 \pi e^{4t}}$$

is an even function of t which is positive for all real t. Fourteen years ago, I pointed out in a survey paper [18, §9] on positivity and special functions that, since the above integral is a square when $n = 0$, the method of sums of squares (discussed earlier in the paper) is suggested for proving (7.1). I also stated that a computer analysis of (7.1) and of the other necessary and sufficient conditions for the Riemann Hypothesis in [28] might lead to some interesting observations.

Recently, Csordas and Varga [12] (also see [13]) considered the inequalities [28, (18)]

$$(7.3) \qquad \int_{-\infty}^{\infty} \int_{-\infty}^{\infty} \Phi(s)\Phi(t) e^{i(s+t)x} e^{(s-t)y}(s-t)^2 \, ds \, dt \geq 0$$

for real x and y, which is one of Jensen's necessary and sufficient conditions for the Riemann Hypotheses to hold, and showed that it suffices to prove (7.3) for $0 \leq x < \infty$ when y is in the bounded interval $0 \leq y < 1$. In view of the maximum principles that are known to hold for certain kernels involving orthogonal polynomials and other special functions [18, §5], this suggests the conjecture that if the inequalities in (7.3) hold for $0 \leq x < \infty$ when $y = 0$ then they hold for all real x and y. A proof of this conjecture would reduce proving

(7.3) for $0 \leq x < \infty$, $0 \leq y < 1$ to just proving the more tractable single variable special case $0 \leq x < \infty$, $y = 0$.

When $n = 1$ in (7.1) or, equivalently, $y = 0$ in (7.3), the evenness of $\Phi(t)$ and the identity $e^{i\theta} = \cos\theta + i\sin\theta$ can be used to show that

$$(7.4) \quad \frac{1}{8} \int_{-\infty}^{\infty} \int_{-\infty}^{\infty} \Phi(s)\Phi(t)e^{i(s+t)x}(s-t)^2\,ds\,dt$$

$$= \left(\int_0^{\infty} \Phi(t)\cos(xt)dt \right) \left(\int_0^{\infty} \Phi(t)t^2\cos(xt)dt \right) + \left(\int_0^{\infty} \Phi(t)t\sin(xt)dt \right)^2.$$

An extension of (7.4) to arbitrary y is derived in [12]. The main advantage of (7.4) is that it reduces the computation of the double integral to that of the single integrals on the right side, which takes a lot less time when using numerical integration to approximate the integrals for particular values of x. Since, as expected, the use of Mathematica's numerical integration NIntegrate function (with appropriate truncations of the series (7.2) and range of integration and suitable settings of the WorkingPrecision, AccuracyGoal, MinRecursion, MaxRecursion, and Points options) only gave positive values for the right side of (7.4) for each chosen value of x, I decided to investigate what happens when the function $\Phi(t)$ in the integrals on the right side of (7.4) is replaced by the nth partial sum (call it $\Phi_n(t)$) of the series representation for $\Phi(t)$ in (7.2). Letting $f_n(x)$ denote the right side of (7.4) with $\Phi(t)$ replaced by $\Phi_n(t)$, it was found that $f_1(x)$ changes sign from positive to negative in the interval $(37, 38)$, and that when $n = 2, 3, 4, 5, 6, 7, 8$, and 9 the functions $f_n(x)$ change sign from positive to negative in the intervals $(85, 86)$, $(134, 135)$, $(210, 211)$, $(302, 303)$, $(401, 402)$, $(519, 520)$, $(657, 658)$, and $(817, 818)$, respectively. I conjecture that for each natural number n there is an $x_n > 0$ such that $x_n \to \infty$ and $f_n(x) \geq 0$ for $0 \leq x \leq x_n$, which would imply that (7.1) and (7.3) hold for real x when $n = 1$ and $y = 0$.

REFERENCES

[1] R. Askey, 'Orthogonal polynomials and positivity,' *Studies in Applied Mathematics* **6**, Special Functions and Wave Propagation (D. Ludwig and F. W. J. Olver, eds.), SIAM, Philadelphia, 1970, pp. 64–85.

[2] R. Askey and G. Gasper, 'Jacobi polynomial expansions of Jacobi polynomials with non-negative coefficients,' *Proc. Camb. Phil. Soc.* **70** (1971), 243–255.

[3] R. Askey and G. Gasper, 'Positive Jacobi polynomial sums II,' *Amer. J. Math.* **98** (1976), 709–737.

[4] R. Askey and G. Gasper, 'Inequalities for polynomials,' *The Bieberbach Conjecture: Proc. of the Symposium on the Occasion of the Proof* (A. Baernstein, D. Drasin, P. Duren, and A. Marden, eds.), Math. Surveys and Monographs **21**, Amer. Math. Soc., Providence, R. I., 1986, pp. 7–32.

[5] R. Askey and J. Wilson, 'Some basic hypergeometric polynomials that generalize Jacobi polynomials,' *Memoirs Amer. Math. Soc.* **319**, Amer. Math. Soc., Providence, R. I., 1985.

[6] W. N. Bailey, *Generalized Hypergeometric Series*, Cambridge University Press, Cambridge, 1935; reprinted by Stechert-Hafner, New York, 1964.

[7] L. de Branges, 'A proof of the Bieberbach conjecture,' *Acta Math.* **154** (1985), 137–152.

[8] L. de Branges, 'Powers of Riemann mapping functions,' *The Bieberbach Conjecture: Proc. of the Symposium on the Occasion of the Proof* (A. Baernstein, et al., eds.), Math. Surveys and Monographs **21**, Amer. Math. Soc., Providence, R. I., 1986, pp. 51–67.

[9] D. M. Bressoud, 'The Bailey Lattice: an introduction,' *Ramanujan Revisited* (G. E. Andrews et al., eds.), Academic Press, New York, 1988, pp. 57–67.

[10] T. S. Chihara, *An Introduction to Orthogonal Polynomials*, Gordon and Breach, New York, 1978.

[11] T. Clausen, 'Ueber die Fälle, wenn die Reihe von der Form ... ein Quadrat von der Form ... hat,' *J. reine angew. Math.* **3** (1828), 89–91.

[12] G. Csordas and R. S. Varga, 'Fourier transforms and the Hermite-Biehler theorem,' to appear.

[13] G. Csordas and R. S. Varga, 'Necessary and sufficient conditions and the Riemann Hypothesis,' to appear.

[14] A. Erdélyi, *Higher Transcendental Functions, Vols. I & II*, McGraw-Hill, New York, 1953.

[15] J. L. Fields and J. Wimp, 'Expansions of hypergeometric functions in hypergeometric functions,' *Math. Comp.* **15** (1961), 390–395.

[16] R. Gangolli, 'Positive definite kernels on homogeneous spaces and certain stochastic processes related to Lévy's Brownian motion of several parameters,' *Ann. Inst. H. Poincaré, Sect. B Vol. III*, 1967, pp. 121–226.

[17] G. Gasper, 'Projection formulas for orthogonal polynomials of a discrete variable,' *J. Math. Anal. Appl.* **45** (1974), 176–198.

[18] G. Gasper, 'Positivity and special functions,' *Theory and Applications of Special Functions* (R. Askey, ed.), Academic Press, New York, 1975, pp. 375–433.

[19] G. Gasper, 'Positive sums of the classical orthogonal polynomials,' *SIAM J. Math. Anal.* **8** (1977), 423–447.

[20] G. Gasper, 'A short proof of an inequality used by de Branges in his proof of the Bieberbach, Robertson and Milin conjectures,' *Complex Variables* **7** (1986), 45–50.

[21] G. Gasper, 'Summation, transformation, and expansion formulas for bibasic series,' *Trans. Amer. Math. Soc.* **312** (1989), 257–277.

[22] G. Gasper, 'q-Extensions of Clausen's formula and of the inequalities used by de Branges in his proof of the Bieberbach, Robertson, and Millin conjectures,' *SIAM J. Math. Anal.* **20** (1989), 1019–1034.

[23] G. Gasper, 'Bibasic summation, transformation and expansion formulas, q-analogues of Clausen's formula, and nonnegative basic hypergeometric series,' *Workshop on q-Series and Partitions* (D. Stanton, ed.), IMA Volumes in Mathematics and Its Applications, Springer, Berlin and New York, to appear.

[24] G. Gasper and M. Rahman, 'Product formulas of Watson, Bailey and Bateman types and positivity of the Poisson kernel for q-Racah polynomials,' *SIAM J. Math. Anal.* **15** (1984), 768–789.

[25] G. Gasper and M. Rahman, 'An indefinite bibasic summation formula and some quadratic, cubic, and quartic summation and transformation formulas,' *Canad. J. Math.*, to appear.

[26] G. Gasper and M. Rahman, 'A nonterminating q-Clausen formula and some related product formulas,' *SIAM J. Math. Anal.*, to appear.

[27] G. Gasper and M. Rahman, *Basic Hypergeometric Series*, Cambridge University Press, Cambridge, to appear.

[28] G. Pólya, 'Über die algebraisch-funktionentheoretischen Untersuchungen von J. L. W. V. Jensen,' *Kgl. Danske Videnskabernes Selskab. Math.-Fys. Medd.* **7** (17) (1927), pp. 3–33; reprinted in his Collected Papers, Vol. II, pp. 278–308.

[29] D. B. Sears, 'On the transformation theory of basic hypergeometric functions,' *Proc. London Math. Soc.* (2) **53** (1951), 158–180.

[30] L. J. Slater, *Generalized Hypergeometric Functions*, Cambridge University Press, Cambridge, 1966.

[31] G. Szegö, *Orthogonal Polynomials*, 4th edition, Amer. Math. Soc. Colloq. Publ. 23, 1975.

[32] A. Verma, 'Certain expansions of the basic hypergeometric functions,' *Math. Comp.* **20** (1966), 151–157.

COMPUTATIONAL ASPECTS OF ORTHOGONAL POLYNOMIALS[†]

WALTER GAUTSCHI
Department of Computer Sciences
Purdue University
West Lafayette, IN 47907
USA

ABSTRACT. Our concern here is with computational methods for generating orthogonal polynomials and related quantities. We focus on the case where the underlying measure of integration is nonclassical. The main problem, then, is that of computing the coefficients in the basic recurrence relation satisfied by orthogonal polynomials. Two principal methods are considered, one based on modified moments, the other on inner product representations of the coefficients. The first method is the more economical one, but may be subject to ill-conditioning. A study is made of the underlying reasons for instability. The second method, suitably implemented, is more widely applicable, but less economical. A number of problem areas in the physical sciences and in applied mathematics are described where these methods find useful applications.

1. Recurrence Relation: The Key Towards the Computational Use of Orthogonal Polynomials

1.1 THE RECURRENCE RELATION

Compared to other systems of orthogonal functions, orthogonal polynomials have the distinguishing feature of enjoying a simple algebraic structure. It derives from the fact that each member of the system has degree exactly one larger than the previous member. This means that in order to proceed from one member to the next, a multiplication by the independent variable t must take place to raise the degree; at the same time, orthogonality is maintained by adding a suitable linear combination of members already at hand. If the inner product $(\,\cdot\,,\,\cdot\,)$ with respect to which the polynomials are orthogonal is symmetric with respect to multiplication by t, i.e., if

$$(Mu,v) = (u,Mv), \quad \text{all} \quad u,v \in \mathbf{P}, \tag{1.1}$$

where $(Mu)(t) = tu(t)$ and $\mathbf{P}$ is the class of polynomials, then, remarkably enough, only the two most recent members of the orthogonal system are needed to form the linear combination to be added for orthogonality. This is the content of the well-known *three-term recurrence*

[†] Work supported, in part, by the National Science Foundation under grant CCR-8704404.

P. Nevai (ed.), Orthogonal Polynomials, 181–216.
© 1990 *by Kluwer Academic Publishers.*

relation (for monic orthogonal polynomials),

$$\pi_{k+1}(t) = (t - \alpha_k)\pi_k(t) - \beta_k\pi_{k-1}(t), \quad k = 0,1,\dots,$$
$$\pi_{-1}(t) = 0, \quad \pi_0(t) = 1. \tag{1.2}$$

The coefficients α_k, β_k depend on the particular inner product under consideration.

A widely used inner product satisfying (1.1) – and one to which we will restrict ourselves in the following – is given by

$$(u,v) = \int_{\mathbb{R}} u(t)v(t)d\sigma(t), \quad u,v \in \mathbf{P}, \tag{1.3}$$

where $d\sigma$ is a measure supported on the real line $\mathbf{R}$ and such that the integral in (1.3) is meaningful for polynomials u,v. In other words, we assume that all moments

$$\mu_k = \mu_k(d\sigma) = \int_{\mathbb{R}} t^k d\sigma(t), \quad k = 0,1,2,\dots, \tag{1.4}$$

exist and are finite. It is then well known (cf., e.g., Chihara [1978, Ch.1, §3]) that the inner product (1.3) defines a unique (infinite) system of monic orthogonal polynomials provided all Hankel determinants in the moments (1.4) are nonsingular,

$$\det H_n \neq 0, \quad n = 1,2,3,\dots; \quad H_n = [\mu_{i+j}]_{i,j=0}^{n-1}. \tag{1.5}$$

If (1.5) holds only for a finite number N of consecutive integers n, then the system of orthogonal polynomials consists only of N members. (This is the case, e.g., when $d\sigma$ in (1.3) is a discrete measure supported on exactly N distinct points. For the sake of definiteness, we shall in the following disregard such measures, even though they are not devoid of interesting applications and computational problems.) We denote the uniquely defined orthogonal polynomials by

$$\pi_n(\cdot) = \pi_n(\cdot;d\sigma), \quad n = 0,1,2,\dots, \tag{1.6}$$

and the recursion coefficients in (1.2) by

$$\alpha_k = \alpha_k(d\sigma), \quad \beta_k = \beta_k(d\sigma), \quad k = 0,1,2,\dots, \tag{1.7}$$

where $d\sigma$ is included in the notation when we wish to emphasize dependence on the measure $d\sigma$. While the coefficient β_0 in (1.2) is arbitrary, we find it convenient to define

$$\beta_0 = \beta_0(d\sigma) = \int_{\mathbb{R}} d\sigma(t). \tag{1.8}$$

We recall that (1.5) is satisfied in the stronger form $\det H_n > 0$ if $d\sigma$ is a *positive measure*, i.e., $d\sigma(t) \geq 0$ and the support of $d\sigma$, supp $d\sigma = \{t \in \mathbf{R}: d\sigma(t) > 0\}$, contains infinitely many points. In this case,

$$\beta_k(d\sigma) > 0, \quad \text{all } k \geq 0 \quad (d\sigma \geq 0). \tag{1.9}$$

In most applications to be considered, the measure $d\sigma$ is indeed positive, but there are occasions when more general (real-valued) measures on $\mathbf{R}$ need to be considered.

The smallest closed interval containing the support of $d\sigma$ will from now on be briefly (and somewhat imprecisely) referred to as the *support interval* of $d\sigma$.

1.2 THE BASIC PROBLEM

It is our view that the recursion coefficients α_k, β_k in (1.2) are the fundamental quantities in the constructive theory of orthogonal polynomials. For positive measures $d\sigma$ it makes sense (cf. (1.9)) to define the real symmetric tridiagonal matrix

$$
J = J(d\sigma) =
\begin{bmatrix}
\alpha_0 & \sqrt{\beta_1} & & & & 0 \\
\sqrt{\beta_1} & \alpha_1 & \sqrt{\beta_2} & & & \\
& \sqrt{\beta_2} & \alpha_2 & \sqrt{\beta_3} & & \\
& & \cdot & \cdot & \cdot & \\
& & & \cdot & \cdot & \cdot \\
0 & & & & \cdot & \cdot
\end{bmatrix},
\tag{1.10}
$$

called the *Jacobi matrix* associated with the measure $d\sigma$. The leading principal $(n \times n)$ submatrix of J will be denoted by

$$
J_n = J_n(d\sigma) = [J(d\sigma)]_{n \times n}, \quad n = 1,2,3, \dots .
\tag{1.11}
$$

We can now formulate what we consider to be the basic computational problem for orthogonal polynomials.

Basic problem: Given a measure $d\sigma$ on $\mathbf{R}$ and a positive integer n, compute

$$
\alpha_k(d\sigma), \quad \beta_k(d\sigma) \quad \text{for} \quad k = 0,1, \dots, n-1.
\tag{1.12}
$$

In particular, if $d\sigma \geq 0$, compute the Jacobi matrix $J_n(d\sigma)$.

In the next subsection, we indicate how the ability to solve this problem enables us to compute many other quantities of interest.

1.3 APPLICATIONS

(i) *Computation of orthogonal polynomials*. For fixed t (real or complex) we can compute $\{\pi_k(t;d\sigma)\}_{k=0}^{n}$ by a straightforward application of (1.2) for $k = 0,1, \dots, n-1$, with the coefficients α_k, β_k given in (1.12). Likewise, any derivative of $\pi_k(t;d\sigma)$ can be computed from the recurrence relation obtained by differentiating (1.2) the appropriate number of times with respect to t. The procedure is quite robust with regard to the effects of rounding errors.

(ii) *Norms*. The normalization constant $||\pi_n||_{d\sigma}^2 = \int_{\mathbf{R}} \pi_n^2(t;d\sigma)d\sigma(t)$ is easily computed from

$$
||\pi_n||_{d\sigma}^2 = \beta_0\beta_1\beta_2 \cdots \beta_{n-1}\beta_n.
\tag{1.13}
$$

This is an instance where the definition (1.8) of β_0 pays off. Also note that (1.13) requires one additional coefficient, β_n, beyond those given in (1.12).

(iii) *Evaluation of orthogonal sums.* Expansions in orthogonal polynomials require the evaluation of partial sums

$$s_n(t) = \sum_{k=0}^{n} c_k \pi_k(t; d\sigma).$$ (1.14)

For fixed n and t, an economical way to compute $s_n(t)$ is *Clenshaw's algorithm* (Clenshaw [1955])

$$u_n = c_n, \quad u_{n+1} = 0,$$
$$u_k = (t - \alpha_k)u_{k+1} - \beta_{k+1}u_{k+2} + c_k,$$
$$k = n-1, n-2, \ldots, 0,$$ (1.15)
$$s_n(t) = u_0.$$

Here, $\alpha_k = \alpha_k(d\sigma)$, $k = 0, 1, \ldots, n-1$, and $\beta_k = \beta_k(d\sigma)$, $k = 1, 2, \ldots, n-1$, while β_0 is not required and β_n is arbitrary.

(iv) *Gaussian quadrature.* Any measure $d\sigma$ (satisfying (1.5)) gives rise to a sequence of Gaussian quadrature rules

$$\int_{\mathbb{R}} f(t)d\sigma(t) = \sum_{v=1}^{n} \sigma_v f(\tau_v) + R_n(f), \quad n = 1, 2, 3, \ldots,$$ (1.16)

where

$$R_n(f) = 0, \quad \text{all } f \in \mathbf{P}_{2n-1},$$ (1.17)

and $\mathbf{P}_k$ denotes the class of polynomials of degree $\leq k$. The *Gaussian nodes* $\tau_v = \tau_v^{(n)}(d\sigma)$ are precisely the zeros of $\pi_n(\cdot; d\sigma)$, whereas the weights $\sigma_v = \sigma_v^{(n)}(d\sigma)$ – also called the *Christoffel numbers* for $d\sigma$ – can be expressed in various ways in terms of the orthogonal polynomials. If $d\sigma$ is not a positive measure, the zeros of $\pi_k(\cdot; d\sigma)$ may or may not lie in the support interval of $d\sigma$; they could be complex and might conceivably be multiple. For positive measures, however, it is well known that all τ_v are simple and contained in the support interval of $d\sigma$; in addition, all weights σ_v are positive.

For computational purposes it is useful to observe that τ_v are also the eigenvalues of J_n (cf. (1.11)),

$$\det(J_n - \tau_v I_n) = 0, \quad v = 1, 2, \ldots, n,$$ (1.18)

whereas the Christoffel numbers σ_v are expressible in terms of the first components $u_{v,1}$ of the corresponding normalized eigenvectors u_v,

$$J_n u_v = \tau_v u_v, \quad u_v^T u_v = 1,$$ (1.19)

namely as

$$\sigma_v = \beta_0 u_{v,1}^2, \quad v = 1, 2, \ldots, n.$$ (1.20)

Using the QR (or QL) algorithm (for example, the procedure *imtql 2* in Wilkinson & Reinsch [1971] or in Smith et al. [1976]) for solving the eigensystem problem (1.19) (simplifying it so that only the first component of the eigenvectors is computed) leads to an efficient and accurate method for computing the Gauss formula (1.16); see Golub & Welsch [1969], Gautschi [1979a].

The ability to generate Gaussian quadrature formulae lies at the heart of many applications involving orthogonal polynomials; cf. Subsection 2.2.

(v) *Padé approximation.* Padé approximants $f[m,n](z)$ are rational functions with numerator degree m and denominator degree n associated with a formal power series

$$f(z) = \mu_0 + \mu_1 z + \mu_2 z^2 + \cdots \qquad (1.21)$$

in the sense that the series expansion (at zero) of $f[m,n]$ agrees with the given series (1.21) to as many terms as possible. If the coefficients μ_k in (1.21) are the moments $\mu_k = \mu_k(d\sigma)$ of a (positive) measure $d\sigma$ (cf. (1.4)), then the theory of Padé approximation becomes closely tied up with the theory of orthogonal polynomials and Gaussian quadrature. For example, if $m = n - 1$, then

$$f[n-1,n](z) = \sum_{v=1}^{n} \frac{\sigma_v}{1 - \tau_v z}, \qquad n = 1,2,3,\ldots, \qquad (1.22)$$

where $\tau_v = \tau_v^{(n)}(d\sigma)$, $\sigma_v = \sigma_v^{(n)}(d\sigma)$ are the nodes and weights of the Gaussian quadrature formula (1.16). More generally,

$$f[n-1+j,n](z) = \mu_0 + \cdots + \mu_{j-1} z^{j-1} + z^j \sum_{v=1}^{n} \frac{\sigma_{v,j}}{1 - \tau_{v,j} z},$$
$$n = 1,2,3,\ldots ; j \text{ even}, \qquad (1.23)$$

where $\tau_{v,j} = \tau_v^{(n)}(d\sigma_j)$, $\sigma_{v,j} = \sigma_v^{(n)}(d\sigma_j)$ are the nodes and weights of the Gauss formula for $d\sigma_j(t) = t^j d\sigma(t)$. For j even, $d\sigma_j$ is a positive measure, so that the orthogonal polynomial $\pi_n(\cdot; d\sigma_j)$ indeed exists. The formula (1.23) holds true also for odd j if $\pi_n(\cdot; d\sigma_j)$ exists and has simple zeros.

(vi) *Functions of the second kind.* For any complex z not in the support of $d\sigma$, the functions of the second kind are defined by

$$\rho_r(z; d\sigma) = \int_{\mathbb{R}} \frac{\pi_r(t; d\sigma)}{z - t} \, d\sigma(t), \quad r = 0,1,2,\ldots ; \ z \in \mathbb{C} \setminus \text{supp } d\sigma. \qquad (1.24)$$

It is elementary to show that $\{\rho_r(\cdot; d\sigma)\}$ is also a solution of the recurrence relation (1.2), (1.7). More interestingly, if $d\sigma$ satisfies a mild additional condition, namely to give rise to a *determined moment problem*, then $\{\rho_r(\cdot; d\sigma)\}$ is the *minimal solution* of (1.2),

$$\lim_{r \to \infty} \frac{\rho_r(z; d\sigma)}{\pi_r(z; d\sigma)} = 0, \quad z \in \mathbb{C} \setminus \text{supp } d\sigma, \qquad (1.25)$$

normalized by

$$\rho_{-1}(z; d\sigma) = 1 \qquad (1.26)$$

(cf. Gautschi [1981b]). Therefore, known algorithms (see, e.g., Gautschi [1967a], Wimp [1984]) can be applied to compute $\rho_n(z \, ; d\sigma)$ for fixed n and z, assuming that the coefficients $\alpha_k(d\sigma)$, $\beta_k(d\sigma)$ are known for $k > n$ sufficiently large.

If $d\sigma$ has bounded support, the moment problem for $d\sigma$ is known to be determined. In this case, the ratio in (1.25),

$$K_n(z \, ; d\sigma) = \frac{\rho_n(z \, ; d\sigma)}{\pi_n(z \, ; d\sigma)} , \tag{1.27}$$

is just the *kernel* in the remainder term of the Gauss formula (1.16), when f is assumed analytic in a domain $\mathcal{D}$ of the complex plane containing the support interval of $d\sigma$ in its interior. More precisely, for any contour $\Gamma \subset \mathcal{D}$ surrounding that interval, one has

$$R_n(f) = \frac{1}{2\pi i} \oint_\Gamma K_n(z \, ; d\sigma) f(z) dz. \tag{1.28}$$

This often allows the error of Gaussian quadrature rules to be estimated realistically as

$$|R_n(f)| \le \frac{1}{2\pi} \text{ length } (\Gamma) \cdot \max_{z \in \Gamma} |K_n(z \, ; d\sigma)| \cdot \max_{z \in \Gamma} |f(z)|, \tag{1.29}$$

provided Γ is chosen appropriately, and precise information about $\max_{z \in \Gamma} |K_n(z \, ; d\sigma)|$ can be obtained. For the latter, see Gautschi & Varga [1983], Gautschi, Tychopoulos & Varga [to appear], Gautschi [to appear].

2. Standard and Nonstandard Measures

A large body of applications, especially in the physical sciences, deals with classical orthogonal polynomials, usually in connection with orthogonal expansions, Padé approximation, and Gaussian quadrature. Here the computational problems are rather straightforward, since the recursion coefficients are all explicitly known and easy to compute. Some of the more frequently used measures and their recurrence relations will be recorded below in Subsection 2.1. Occasions, nevertheless, arise – both in the physical sciences and in applied mathematics – when one has to deal with nonstandard measures whose orthogonal polynomials are not explicitly known and must be generated by computational methods. In Subsection 2.2 we mention a number of such instances; in all of them, Gaussian quadrature rules play a prominent role. References to tables of nonstandard recursion coefficients will be provided in Subsection 2.3.

2.1 SOME CLASSICAL MEASURES

(i) *Legendre measure.* This is the equally weighted measure $d\sigma(t) = dt$ on a finite interval (i.e., $d\sigma = 0$ outside the interval), for which we take, without restricting generality, the canonical interval $[-1,1]$. The corresponding orthogonal polynomials are the *Legendre polynomials*; their recursion coefficients in (1.2), (1.7) are given by

$$\alpha_k = 0, \quad k \ge 0,$$

$$\beta_0 = 2, \quad \beta_k = \frac{1}{4 - k^{-2}}, \quad k \ge 1 \qquad (d\sigma(t) = dt \text{ on } [-1,1]). \tag{2.1}$$

The vanishing of all the α's is characteristic for a measure which is symmetric with respect to the origin. Note also that $\beta_k \to {}^1\!/_4$ as $k \to \infty$, which, together with $\alpha_k \to 0$, is shared by a large class of measures on $[-1,1]$; cf. the class $M(0,1)$ in Nevai [1979].

(ii) *Chebyshev measure.* This may be thought of as the uniform measure on the upper unit semicircle, projected down to the interval $[-1,1]$, that is, $d\sigma(t) = (1 - t^2)^{-1/2} dt$. It generates the important *Chebyshev polynomials* of the first kind, whose recursion coefficients are particularly simple,

$$\alpha_k = 0, \quad k \geq 0,$$
$$\beta_0 = \pi, \; \beta_1 = {}^1\!/_2, \; \beta_k = {}^1\!/_4, \; k \geq 2 \quad (d\sigma(t) = (1 - t^2)^{-1/2} dt \text{ on } [-1,1]). \tag{2.2}$$

The interest in Chebyshev polynomials stems from the extremum properties they satisfy, in particular the fact that the nth degree (monic) Chebyshev polynomial has the smallest ∞-norm on $[-1,1]$ among all monic polynomials of the same degree:

$$||p||_\infty \geq ||\pi_n(\cdot \,; d\sigma)||_\infty = \frac{1}{2^{n-1}}, \text{ all } p(\text{monic}) \in \mathbf{P}_n, \; n \geq 1$$
$$(d\sigma \text{ as in (2.2)}); \tag{2.3}$$

here, $||u||_\infty = \max\limits_{-1 \leq t \leq 1} |u(t)|$. The extreme values of π_n are attained at $n + 1$ consecutive points on $[-1,1]$ with alternating signs; indeed $\pi_n(t \,; d\sigma) = 2^{-(n-1)} \cos(n\theta)$, $\cos \theta = t$. For the many interesting properties and applications of Chebyshev polynomials, we refer to Fox & Parker [1968], Rivlin [1974], Paškovskii [1983].

(iii) *Jacobi polynomials.* These generalize the Legendre and Chebyshev polynomials, and are orthogonal relative to the measure $d\sigma(t) = (1 - t)^\alpha (1 + t)^\beta dt$ on $[-1,1]$, where $\alpha > -1$, $\beta > -1$. Here, too, the recursion coefficients are explicitly known; for their precise expressions we refer to Chihara [1978, p. 220].

(iv) *Generalized Laguerre measure.* This is the best-known measure on the half-infinite interval $[0, \infty]$, given by $d\sigma(t) = t^\alpha e^{-t} dt$, $\alpha > -1$. It defines the (generalized) *Laguerre polynomials*, whose recursion coefficients are

$$\alpha_k = 2k + \alpha + 1, \quad k \geq 0,$$
$$\beta_0 = \Gamma(\alpha + 1), \; \beta_k = k(\alpha + k), \; k \geq 1 \quad (d\sigma(t) = t^\alpha e^{-t} dt \text{ on } \mathbb{R}_+). \tag{2.4}$$

(v) *Hermite measure* $d\sigma(t) = e^{-t^2} dt$ on $[-\infty,\infty]$. A measure widely used in quantum physics, it gives rise to the *Hermite polynomials* with the very simple recursion coefficients

$$\alpha_k = 0, \quad k \geq 0,$$
$$\beta_0 = {}^1\!/_2 \sqrt{\pi}, \; \beta_k = {}^1\!/_2 k, \; k \geq 1 \quad (d\sigma(t) = e^{-t^2} dt \text{ on } \mathbb{R}). \tag{2.5}$$

The same measure, supported however on the half-infinite interval $[0,\infty]$, is nonstandard, and its recursion coefficients, to our knowledge, are not known explicitly.

(vi) *Discrete equally weighted measure.* This is the discrete analogue of the Legendre measure, being supported at $N + 1$ uniformly distributed points in $[-1,1]$, say, $t_r = -1 + 2r/N$, $r = 0,1,2,\ldots, N$, and all weights being the same, say $2/(N+1)$; thus (in terms of the Dirac delta function), $d\sigma(t) = (2/(N+1)) \sum_{r=0}^{N} \delta(t - t_r)$. The corresponding (discrete) orthogonal polynomials are useful in polynomial least squares approximation and have already been obtained (for precisely this purpose) by Chebyshev [1859]. The recursion coefficients are

$$\alpha_k = 0, \quad 0 \leq k \leq N,$$

$$\beta_0 = 2, \quad \beta_k = \left[1 + \frac{1}{N}\right]^2 \left[1 - \left[\frac{k}{N+1}\right]^2\right] \left[4 - \frac{1}{k^2}\right]^{-1}, \quad 1 \leq k \leq N.$$

(2.6)

Not surprisingly, they converge to the Legendre coefficients in (2.1) when $N \to \infty$ for fixed k.

2.2 EXAMPLES OF NONSTANDARD MEASURES OCCURRING IN APPLICATIONS

In most applications, the need for orthogonal polynomials arises in connection with integration problems: to evaluate integrals with difficult behavior (usually packaged into the integration measure) by means of Gaussian quadrature. All applications below are of this type.

(i) *Reciprocal gamma function.* Fransén [1979] suggests that the probability density function $d\sigma(t) = c \, dt / \Gamma(t)$ on $[0, \infty]$, where $c^{-1} = \int_0^\infty dx / \Gamma(x)$, could be useful in reliability theory, owing to its rapid decay as $t \to \infty$. The computation of expected values $Ef = c \int_0^\infty f(t) dt / \Gamma(t)$ then calls for Gaussian integration relative to this measure.

(ii) *Inhomogeneous Airy functions.* These are solutions $Hi(x)$ and $Gi(x)$ to the inhomogeneous differential equation $dy^2/dx^2 - xy = \pm 1/\pi$ satisfying appropriate initial conditions. They arise, e.g., in harmonic oscillator models for large quantum numbers. In view of the integral representations (Lee [1980])

$$Hi(x) = \frac{1}{\pi} \int_0^\infty e^{-\frac{1}{3} t^3 + t x} \, dt,$$

$$Gi(x) = -\frac{1}{\pi} \int_0^\infty e^{-\frac{1}{3} t^3 - \frac{1}{2} t x} \cos \left[\frac{\sqrt{3}}{2} t x + \frac{2\pi}{3} \right] dt,$$

they can be effectively evaluated by Gaussian quadrature relative to the "Airy measure" $d\sigma(t) = \exp(-t^3/3) dt$ on $[0, \infty]$.

(iii) *One-sided Hermite measures.* Integrals involving the measure $d\sigma(t) = t^p e^{-t^2} dt$ on the half-infinite interval $[0, \infty]$, for $p = 0, 1, 2$, are encountered in the solution of the Boltzmann equation in kinetic theory (see, e.g., Shizgal [1981]). The case $p = 0$ also has statistical applications (Kahaner, Tietjen & Beckman [1982]).

(iv) *Finite Hermite and Laguerre measures.* These are the measures with $d\sigma(t) = e^{-t^2} dt$ and $d\sigma(t) = e^{-t} dt$, respectively, supported on a finite interval, $[0, c]$ or $[-c, c]$, where $c > 0$. They abound in quantum mechanical computations; see, e.g., King & Dupuis [1976] and Mach [1984].

(v) *Bessel transforms.* Gaussian quadrature relative to the measure $d\sigma(t) = t^\mu K_\nu(t) dt$, $\mu \pm \nu > -1$, with K_ν the modified Bessel function, are used by Wong [1982] to construct asymptotic approximations (for large $|x|$) to Bessel transforms $(H_i f)(x) = \int_0^\infty t^\mu H_\nu^{(i)}(xt)$ $\cdot f(t)dt$, $i = 1,2$, where $H_\nu^{(i)}$ are the Hankel functions and f is analytic in the right half plane.

(vi) *Singular integrals with* coth-*kernel.* Cauchy principal value integrals of the type

$$(I_a f)(x) = f_{-1}^1 f(t) \coth(a(t-x))dt, \quad -1 < x < 1, \quad a > 0,$$

are difficult to evaluate, not so much because of the pole at $t = x$, but because of the infinitely many additional poles in the complex plane, especially those closest to the real line. Writing $I_a f$ as a conventional Cauchy integral,

$$(I_a f)(x) = \frac{1}{a} f_{-1}^1 \frac{f(t)}{t-x} d\sigma(t;a,x), \quad d\sigma(t;a,x) = \omega(a(t-x))dt,$$

where $\omega(u) = u \coth u$, however, allows the use of standard procedures (cf. Gautschi, Kovačević & Milovanović [1987]) to reduce the problem to one of Gaussian quadrature relative to the measure $d\sigma(\cdot;a,x)$.

(vii) *Slowly convergent series involving Laplace transforms.* With $F(s) = \int_0^\infty e^{-st} f(t)dt$, Re $s > 0$, denoting the Laplace transform of f, infinite series such as $\sum_{k=1}^\infty (-1)^{k-1} F(k)$ are apt to converge very slowly. If the original function f is known, however, one can make use of

$$\sum_{k=1}^\infty (-1)^k F(k) = \int_0^\infty f(t) \frac{dt}{e^t + 1}$$

and evaluate the integral, often quite effectively (see Gautschi & Milovanović [1985]), by Gaussian quadrature for the measure $d\sigma(t) = (e^t + 1)^{-1} dt$ on $[0,\infty]$. Likewise, sums of the form $-\sum_{k=1}^\infty F'(k)$ and $-\sum_{k=1}^\infty (-1)^{k-1} F'(k)$ lead to integrals respectively with measures $d\sigma(t) = t\, dt/(e^t - 1)$ and $d\sigma(t) = t\, dt/(e^t + 1)$ on $[0,\infty]$. The same measures are of interest in solid state physics.

(viii) *Moment-preserving spline approximation.* Given a function $f \in C^{m+1}[0,\infty]$ satisfying certain growth conditions at ∞, the n-point Gauss formula for the measure

$$d\sigma_m(t) = \frac{(-1)^{m+1}}{m!} t^{m+1} f^{(m+1)}(t)dt \quad \text{on } [0,\infty]$$

yields the solution to the following approximation problem: find the spline function $s_{n,m}(t) = \sum_{\nu=1}^n a_\nu (\tau_\nu - t)_+^m$ of degree m with (unknown) knots $\tau_1 > \tau_2 > \cdots > \tau_n > 0$ and coefficients $a_\nu \in \mathbb{R}$ such that the first $2n$ moments of $s_{n,m}$ are the same as those of f,

$$\int_0^\infty t^j s_{n,m}(t)dt = \int_0^\infty t^j f(t)dt, \quad j = 0,1,\ldots,2n-1.$$

Indeed, the knots must be the Gaussian points $\tau_\nu = \tau_\nu^{(n)}(d\sigma_m)$ (assumed all positive), and the coefficients are given by $a_\nu = \tau_\nu^{-(m+1)} \sigma_\nu$, where $\sigma_\nu = \sigma_\nu^{(n)}(d\sigma_m)$ are the Christoffel numbers for $d\sigma_m$ (cf. Gautschi & Milovanović [1986]). The solution is guaranteed to exist uniquely if $d\sigma_m$ is a positive measure, for example, if f is completely monotonic on $[0,\infty]$. In this case, $d\sigma_m \geq 0$ for all $m \geq 0$. Analogous problems can be posed on finite intervals and lead to generalized Gauss-Radau and Gauss-Lobatto formulae (Frontini, Gautschi & Milovanović [1987]); see also Micchelli [1988] for still other variations, and Milovanović & Kovačević [1988] for approximation by defective splines.

2.3 TABLES OF RECURSION COEFFICIENTS

Below in Table 2.1 we list nonstandard measures $d\sigma(t) = \omega(t)dt$ for which recursion coefficients $\alpha_k(d\sigma)$, $\beta_k(d\sigma)$, $k = 0,1,\ldots,n-1$, have been tabulated in the literature. In the first four columns we show respectively the weight function ω, the associated interval $[a,b]$, the value of n, and the given precision (S = significant decimal digits; D = decimal digits after the decimal point). The last column lists the sources. Tables that have been superseded by more accurate and/or more extensive tables are not included. We are not claiming that Table 2.1 is complete, as we relied more on our recollection than on systematic data retrieval. Ultimately, it is not numerical tables that are needed anyway, but good and reliable computer routines. The outline of a software package, developed and used by the author for some time now (but not yet published), is given in the appendix at the end of this paper.

TABLE 2.1 *Recursion coefficients* α_k, β_k, $0 \leq k \leq n-1$ *for measures* $d\sigma(t) = \omega(t)dt$ *on* $[a,b]$.

$\omega(t)$	$[a,b]$	n	Accuracy	Reference
$\ln(1/t)$	$[0,1]$	20	12 S	Danloy [1973].
$\ln(1/t)$	$[0,1]$	16	30 S	Stroud & Secrest [1966]
e^{-t^2}	$[0,\infty]$	20	20 S	Galant [1969]
$e^{-\frac{1}{3}t^3}$	$[0,\infty]$	15	16 D	Gautschi [1983]
$E_1(t)$ (exponential integral)	$[0,\infty]$	20	12 S	Danloy [1973]
$1/\Gamma(t)$	$[0,\infty]$	40	20 S	Gautschi [1982b]
$[t/(e^t - 1)]^r$, $r = 1,2$	$[0,\infty]$	40	25 S	Gautschi & Milovanović [1985]
$[1/(e^t + 1]^r$, $r = 1,2$	$[0,\infty]$	40	25 S	"
$(2\pi^{-d/2})t^{d+1}e^{-t^2}$, $d = 1,2,3$	$[0,\infty]$	50	25 S	Gautschi [1984b]
$t[t/(1 - e^{-t})]^2 e^{-t}$	$[0,\infty]$	50	25 S	"
$2[t^2/(1 + e^{-t^2})]^2 e^{-t^2}$	$[0,\infty]$	50	25 S	"
$|t|^{\alpha}e^{-|t|}$, $\alpha = 1,2,3$	$[-\infty,\infty]$	20	30 S	Stroud & Secrest [1966]

3. Moment-Related Methods

In principle, the first $2n$ moments $\mu_k(d\sigma)$, $k = 0,1,2,\ldots, 2n-1$ (cf. (1.4)), determine uniquely the $2n$ recursion coefficients $\alpha_k(d\sigma)$, $\beta_k(d\sigma)$, $k = 0,1,\ldots, n-1$ (cf. (1.2), (1.7)), and there are well-known methods to implement the underlying map. The difficulty is that the map becomes ill-conditioned at an exponential rate as n increases; see, e.g., Gautschi [1982a, §3.2]. Rounding errors present in the moments, and those committed during the computation, thus conspire to undermine any algorithm attempting to compute the recursion coefficients from the moments.

An alternative approach is to start with *modified moments*. These are defined by

$$m_k = m_k(d\sigma) = \int_{\mathbb{R}} p_k(t)d\sigma(t), \quad k = 0,1,2,\ldots, \tag{3.1}$$

where $\{p_k\}$ is a given system of polynomials. We assume that they, too, satisfy a three-term recurrence relation,

$$p_{k+1}(t) = (t - a_k)p_k(t) - b_k p_{k-1}(t), \quad k = 0,1,2,\ldots,$$
$$p_{-1}(t) = 0, \quad p_0(t) = 1, \tag{3.2}$$

with coefficients a_k, b_k assumed to be known. [The special case $a_k = b_k = 0$, all k, yields $p_k(t) = t^k$, hence $m_k = \mu_k(d\sigma)$.] Typically, the p_k are themselves orthogonal polynomials,

$$p_k(\cdot) = p_k(\cdot\,; ds), \tag{3.3}$$

with respect to some given (usually classical) measure ds. We will demonstrate in Subsection 3.2 how the desired recursion coefficients $\alpha_k(d\sigma)$, $\beta_k(d\sigma)$, $k = 0,1,\ldots, n-1$, can be obtained from the first $2n$ modified moments (3.1). Before we do this, however, we analyze in Subsection 3.1 the condition of the underlying map in the case where p_k is given by (3.3) for some (positive) measure ds. (This, of course, excludes the case of ordinary moments.) Examples in Subsection 3.3 then illustrate both the theoretical and algorithmic aspects of the problem.

3.1 THE MAP FROM MODIFIED MOMENTS TO RECURSION COEFFICIENTS: SENSITIVITY ANALYSIS

We denote by $m \in \mathbb{R}^{2n}$ the vector of the first $2n$ modified moments, $m = [m_0, m_1, \ldots, m_{2n-1}]^T$ (cf. (3.1)), and by ρ the vector of the desired recursion coefficients, $\rho = [\alpha_0, \ldots, \alpha_{n-1}, \beta_0, \ldots, \beta_{n-1}]^T$ (cf. (1.2), (1.7)). The map of interest, then, is

$$K_n: \mathbb{R}^{2n} \to \mathbb{R}^{2n} \quad m \to \rho. \tag{3.4}$$

Somewhat more convenient (for analysis) is the map

$$G_n: \mathbb{R}^{2n} \to \mathbb{R}^{2n} \quad m \to \gamma, \tag{3.5}$$

where γ is the "Gauss vector", $\gamma = [\tau_1, \ldots, \tau_n, \sigma_1, \ldots, \sigma_n]^T$ (cf. (1.16)). More precisely, normalizing the modified moments by $\tilde{m}_k = m_k/\|p_k\|_{ds}$, we consider

$$\tilde{G}_n: \mathbb{R}^{2n} \to \mathbb{R}^{2n} \quad \tilde{m} \to \gamma. \tag{3.6}$$

The original map then is the composition

$$K_n = H_n \circ \tilde{G}_n \circ M_n, \tag{3.7}$$

where $M_n: m \to \tilde{m}$ is a trivial diagonal map, and $H_n: \gamma \to \rho$ is expected to be a well-conditioned map.

We shall measure the sensitivity (i.e., the "condition") of the map $\tilde{G}_n$ by the magnitude of its (Fréchet) derivative $\tilde{G}_n'$ (i.e., the Jacobian matrix of $\tilde{G}_n$), taking as matrix norm the Frobenius norm $||\cdot||_F$.

Since the map $\tilde{G}_n$ amounts to solving the nonlinear system

$$\Phi(\gamma) = \tilde{m}, \quad \Phi_r(\gamma) = s_r^{-1} \sum_{\nu=1}^{n} \sigma_\nu p_r(\tau_\nu), \quad r = 0, 1, \dots, 2n - 1, \tag{3.8}$$

where $s_r = ||p_r||_{ds}$, the Jacobian of $\tilde{G}_n$ is the inverse of the Jacobian $\partial\Phi/\partial\gamma$,

$$\tilde{G}_n' = \left[\frac{\partial\Phi}{\partial\gamma}(\gamma) \right]^{-1}.$$

It is elementary to show that $\partial\Phi/\partial\gamma = D_s^{-1} P D_\sigma$, where D_s, D_σ are diagonal matrices, $D_s = \text{diag}(s_0, s_1, \dots, s_{2n-1})$, $D_\sigma = \text{diag}(1, \dots, 1, \sigma_1, \dots, \sigma_n)$, and $P \in \mathbb{R}^{2n \times 2n}$ is a confluent Vandermonde matrix in the polynomials $\{p_k\}$, i.e.,

$$\text{row}_k P = [p_k(\tau_1), \dots, p_k(\tau_n), \ p_k'(\tau_1), \dots, p_k'(\tau_n)], \quad k = 0, 1, \dots, 2n - 1. \tag{3.9}$$

Therefore,

$$\tilde{G}_n' = D_\sigma^{-1} P^{-1} D_s. \tag{3.10}$$

In order to invert the matrix P in (3.9), we let h_ν, k_ν be the fundamental Hermite interpolation polynomials of degree $2n - 1$ belonging to the Gaussian abscissae $\tau_1, \tau_2, \dots, \tau_n$,

$$h_\nu(\tau_\mu) = \delta_{\nu\mu}, \quad h_\nu'(\tau_\mu) = 0;$$
$$k_\nu(\tau_\mu) = 0, \quad k_\nu'(\tau_\mu) = \delta_{\nu\mu}, \tag{3.11}$$

and expand them in the polynomials $\{p_k\}$,

$$h_\nu(t) = \sum_{\mu=1}^{2n} a_{\nu\mu} p_{\mu-1}(t), \quad k_\nu(t) = \sum_{\mu=1}^{2n} b_{\nu\mu} p_{\mu-1}(t). \tag{3.12}$$

By virtue of the interpolation conditions (3.11), one then finds that

$$P^{-1} = \begin{bmatrix} A \\ B \end{bmatrix}, \quad A = [a_{\nu\mu}], \quad B = [b_{\nu\mu}].$$

Since

$$(D_\sigma^{-1}P^{-1}D_s)_{\nu\mu} = s_{\mu-1}a_{\nu\mu}, \quad (D_\sigma^{-1}P^{-1}D_s)_{\nu+n,\mu} = \frac{1}{\sigma_\nu} s_{\mu-1}b_{\nu\mu},$$

$$\nu = 1,2,\dots, n; \quad \mu = 1,2,\dots, 2n,$$

one obtains from (3.10) that

$$||\widetilde{G}_n'||_F^2 = \sum_{\nu=1}^{n} \sum_{\mu=1}^{2n} s_{\mu-1}^2 \left[a_{\nu\mu}^2 + \frac{1}{\sigma_\nu^2} b_{\nu\mu}^2 \right]. \tag{3.13}$$

On the other hand, by (3.12),

$$\int_{\mathbb{R}} h_\nu^2(t)ds(t) = \sum_{\mu,\kappa=1}^{2n} a_{\nu\mu}a_{\nu\kappa} \int_{\mathbb{R}} p_{\mu-1}(t)p_{\kappa-1}(t)ds(t) = \sum_{\mu=1}^{2n} s_{\mu-1}^2 a_{\nu\mu}^2,$$

where the last equation follows from the orthogonality of the p_k. Similarly,

$$\int_{\mathbb{R}} k_\nu^2(t)ds(t) = \sum_{\mu=1}^{2n} s_{\mu-1}^2 b_{\nu\mu}^2.$$

Hence, (3.13) finally yields

$$||\widetilde{G}_n'||_F = \left\{ \int_{\mathbb{R}} g_n(t)ds(t) \right\}^{1/2}, \tag{3.14}$$

where

$$g_n(t) = \sum_{\nu=1}^{n} \left[h_\nu^2(t) + \frac{1}{\sigma_\nu^2} k_\nu^2(t) \right]. \tag{3.15}$$

The result (3.14), (3.15) clearly identifies the factors responsible for the magnitude of $||\widetilde{G}_n'||_F$, i.e., for the degree of sensitivity of the map $\widetilde{G}_n: \bar{m} \to \gamma$. On the one hand, we have the polynomial g_n of degree $\leq 4n - 2$ appearing as integrand in (3.14), which depends only on the measure $d\sigma$ (through the Gaussian nodes $\tau_\nu = \tau_\nu^{(n)}(d\sigma)$). On the other hand, there is integration with respect to the measure ds. It is a combination of both, namely the *magnitude of g_n on the support of ds*, which determines the magnitude of $||\widetilde{G}_n'||_F$.

We note from (3.11) and (3.15) that g_n satisfies

$$g_n(t) > 0 \text{ on } \mathbb{R},$$
$$g_n(\tau_\nu) = 1, \quad g_n'(\tau_\nu) = 0, \quad \nu = 1,2,\dots, n. \tag{3.16}$$

(By themselves, these conditions of course do not yet determine g_n.) Ideally, one would like g_n to remain less than or equal to one throughout the support of ds. In that case, $||\widetilde{G}_n'||_F$ would be bounded by s_0, *uniformly* in n. Unfortunately, this is not normally the case.

3.2 MODIFIED CHEBYSHEV ALGORITHM

There is a simple algorithm to carry out the map in (3.4) from the modified moments to the recursion coefficients. It has complexity $O(n^2)$, and in the case of ordinary moments (and discrete measure $d\sigma$) it was already given by Chebyshev [1859]. In a different form, it has been rediscovered by Sack & Donovan [1971/72], and in the form given here by Wheeler [1974]. We call it the *modified Chebyshev algorithm*.

To derive the algorithm, let

$$\sigma_{k\ell} = \int_{\mathbb{R}} \pi_k(t)p_\ell(t)d\sigma(t), \quad k \geq -1, \quad \ell \geq -1, \tag{3.17}$$

denote the "mixed moments" of $d\sigma$, where $\pi_k(\cdot) = \pi_k(\cdot\,; d\sigma)$ are the target polynomials and p_ℓ the auxiliary polynomials defining the modified moments (3.1) and satisfying (3.2) with known coefficients a_k, b_k. By orthogonality, clearly $\sigma_{k\ell} = 0$ if $k > \ell$, while for $k = \ell \geq 1$, using (3.2),

$$\begin{aligned}
\sigma_{kk} &= \int_{\mathbb{R}} \pi_k(t)[(t - a_{k-1})p_{k-1}(t) - b_{k-1}p_{k-2}(t)]d\sigma(t) \\
&= \int_{\mathbb{R}} \pi_k(t) \cdot tp_{k-1}(t)d\sigma(t) = \int_{\mathbb{R}} \pi_k^2(t)d\sigma(t).
\end{aligned} \tag{3.18}$$

The relation $\sigma_{k+1,k-1} = 0$, in combination with (1.2), (3.2), gives

$$\begin{aligned}
0 &= \int_{\mathbb{R}} \pi_{k+1}(t)p_{k-1}(t)d\sigma(t) = \int_{\mathbb{R}} [(t - \alpha_k)\pi_k(t) - \beta_k\pi_{k-1}(t)]p_{k-1}(t)d\sigma(t) \\
&= \int_{\mathbb{R}} \{\pi_k(t)[p_k(t) + a_{k-1}p_{k-1}(t) + b_{k-1}p_{k-2}(t)] - \beta_k\pi_{k-1}(t)p_{k-1}(t)\}d\sigma(t) \\
&= \sigma_{kk} - \beta_k\sigma_{k-1,k-1}.
\end{aligned}$$

Thus,

$$\beta_k = \frac{\sigma_{kk}}{\sigma_{k-1,k-1}}, \quad k = 1,2,3,\ldots. \tag{3.19}$$

(Recall that β_0 has been defined by $\beta_0 = \int_{\mathbb{R}} d\sigma(t) = m_0$.) Similarly, $\sigma_{k+1,k} = 0$ gives

$$\begin{aligned}
0 &= \int_{\mathbb{R}} \pi_{k+1}(t)p_k(t)d\sigma(t) = \int_{\mathbb{R}} [(t - \alpha_k)\pi_k(t) - \beta_k\pi_{k-1}(t)]p_k(t)d\sigma(t) \\
&= \int_{\mathbb{R}} \pi_k(t)[p_{k+1}(t) + a_kp_k(t) + b_kp_{k-1}(t)]d\sigma(t) - \alpha_k\sigma_{kk} - \beta_k\sigma_{k-1,k} \\
&= \sigma_{k,k+1} + (a_k - \alpha_k)\sigma_{kk} - \beta_k\sigma_{k-1,k}.
\end{aligned}$$

Combining this with (3.19), and using that $\sigma_{-1,\ell} = 0$, $\sigma_{0\ell} = m_\ell$ for all $\ell \geq 0$, we find

$$\begin{aligned}
\alpha_0 &= a_0 + \frac{m_1}{m_0}, \\
\alpha_k &= a_k + \frac{\sigma_{k,k+1}}{\sigma_{kk}} - \frac{\sigma_{k-1,k}}{\sigma_{k-1,k-1}}, \quad k = 1,2,3,\ldots.
\end{aligned} \tag{3.20}$$

The relations (3.19), (3.20) express the desired recursion coefficients in terms of the mixed moments $\sigma_{k\ell}$. These latter can be updated by means of the relation

$$\sigma_{k\ell} = \sigma_{k-1,\ell+1} - (\alpha_{k-1} - a_\ell)\sigma_{k-1,\ell} - \beta_{k-1}\sigma_{k-2,\ell} + b_\ell \sigma_{k-1,\ell-1}, \qquad (3.21)$$

which follows from (1.2) and (3.2) (with k replaced by ℓ). The algorithm is now complete: to compute α_k, β_k for $k = 0, 1, \ldots, n-1$, first initialize

$$\sigma_{-1,\ell} = 0, \quad \ell = 1, 2, \ldots, 2n-2,$$
$$\sigma_{0\ell} = m_\ell, \quad \ell = 0, 1, \ldots, 2n-1,$$
$$\alpha_0 = a_0 + \frac{m_1}{m_0}, \qquad (3.22_0)$$
$$\beta_0 = m_0,$$

and then continue for $k = 1, 2, \ldots, n-1$ with

$$\sigma_{k\ell} = \sigma_{k-1,\ell+1} - (\alpha_{k-1} - a_\ell)\sigma_{k-1,\ell} - \beta_{k-1}\sigma_{k-2,\ell} + b_\ell \sigma_{k-1,\ell-1},$$
$$\ell = k, k+1, \ldots, 2n-k-1,$$
$$\alpha_k = a_k + \frac{\sigma_{k,k+1}}{\sigma_{kk}} - \frac{\sigma_{k-1,k}}{\sigma_{k-1,k-1}}, \qquad (3.22_k)$$
$$\beta_k = \frac{\sigma_{kk}}{\sigma_{k-1,k-1}}.$$

Note that the algorithm requires as input $2n$ modified moments $\{m_r\}_{r=0}^{2n-1}$ and $2n-1$ recursion coefficients, $\{a_s\}_{s=0}^{2n-2}$ and $\{b_s\}_{s=0}^{2n-2}$, in order to produce the first n recursion coefficients $\{\alpha_k\}_{k=0}^{n-1}$, $\{\beta_k\}_{k=0}^{n-1}$ (and, at the same time, the first n normalization constants σ_{kk}; cf. (3.18)).

If the underlying map $G_n\colon m \to \gamma$ (or, equivalently, $\tilde{G}_n$) is well-conditioned, the modified Chebyshev algorithm is the method of choice for generating orthogonal polynomials. The required modified moments can sometimes be computed in closed form, at other times from a judicious application of recurrence relations, or else they can be approximated by suitable discretizations similar to those used in Subsection 4.2 below in another context.

3.3 EXAMPLES

(i) $d\sigma(t) = [(1 - k^2 t^2)(1 - t^2)]^{-1/2} dt$ on $[-1,1]$, $0 < k < 1$ (an example of Christoffel).

For small k, the measure $d\sigma$ is close to the Chebyshev measure, which suggests taking the Chebyshev moments

$$m_r = \int_{-1}^1 p_r(t;ds)d\sigma(t)dt, \quad ds(t) = (1 - t^2)^{-1/2} dt, \qquad (3.23)$$

and the recursion coefficients for Chebyshev polynomials (cf. 2.1 (ii)), as input to the modified Chebyshev algorithm. The modified moments (3.23) can be computed very accurately (see Gautschi [1982a, pp. 310–311]). Moreover, for all n as large as $n = 80$, and for various values of k as close to 1 as $k^2 = .99$, it was found by computation that the polynomial g_n (cf. (3.15)), rather remarkably, remained less than or equal to 1 on the whole interval $[-1,1]$. For these values, and very likely in general, this means, by (3.14), that the map G_n in (3.5) is uniformly (in n) well-conditioned. Our computations indeed produced results accurate to machine precision.

(ii) $d\sigma(t) = t^{\alpha}\ln(1/t)dt$ on $[0,1]$, $\alpha > -1$.

Measures of this type, having a branch point and logarithmic singularity at the same point, would be difficult to use without the help of the relevant orthogonal polynomials. These, fortunately, can be generated quite accurately by means of the modified Chebyshev algorithm. The simplest choice of modified moments is the one based on shifted (monic) Legendre polynomials $p_k(\cdot\,;ds)$, with $ds(t) = dt$ on $[0,1]$. The modified moments can then be written down explicitly (Gautschi [1979b]); for example, if α is not an integer,

$$\frac{(2k)!}{k!^2} m_k = \frac{1}{\alpha+1}\left\{\frac{1}{\alpha+1} + \sum_{r=1}^{k}\left[\frac{1}{\alpha+1+r} - \frac{1}{\alpha+1-r}\right]\right\}$$
$$\times \prod_{r=1}^{k}\frac{\alpha+1-r}{\alpha+1+r}, \qquad k = 0,1,2,\dots . \tag{3.24}$$

Furthermore, the recursion coefficients required for the algorithm (3.22) are simply $a_k = \frac{1}{2}$ $(k \geq 0)$, $b_0 = 1$, $b_k = [4(4 - k^{-2})]^{-1}$ $(k \geq 1)$.

Our experience with algorithm (3.22) has been quite satisfactory for values of α such as $\alpha = 0$, $\alpha = \pm\frac{1}{2}$. (Others have not been tried.) For example, when $\alpha = -\frac{1}{2}$, we were able to obtain results with an accuracy close to machine precision for n as large as $n = 80$. Further inspection revealed, nevertheless, that the polynomial g_n (cf. (3.15)), in this case, develops sharp peaks in the last third of the interval $[0,1]$. Fortunately, integration in (3.14) sufficiently smooths them out that $||\tilde{G}_n'||_F$ remains less than about 10^3 for $n \leq 80$.

(iii) *Finite one- and two-sided Hermite measures:* $d\sigma(t) = e^{-t^2}dt$ on $[0, c]$ or $[-c, c]$, $0 < c \leq \infty$.

Consider first the one-sided case. If we use the (monic) Hermite polynomials $p_k(t) = 2^{-k}H_k(t)$ in (3.1) (for which $a_k = 0$, $b_k = \frac{1}{2}k$ in (3.2)), the modified moments $m_k = \int_0^c p_k(t)e^{-t^2}dt$ become

$$m_0 = \frac{\sqrt{\pi}}{2}\,\text{erf}\,c, \quad m_k = \frac{1}{2}[p_{k-1}(0) - e^{-c^2}p_{k-1}(c)], \quad k \geq 1. \tag{3.25}$$

This is easily obtained from the recurrence formula for Hermite polynomials and the fact that $H_n'(t) = 2nH_{n-1}(t)$. The theory in Subsection 4.1, however, warns us that the modified Chebyshev algorithm based on these moments might be subject to severe loss of accuracy, more so the smaller c. Indeed, the polynomial $g_n(t)$ in (3.15), once the argument leaves the support $[0, c]$ of $d\sigma$, is going to increase rapidly on either side of $[0, c]$. Since $ds(t) = e^{-t^2}dt$ in (3.14) is supported on the whole real line, the large values of g_n on the left of $[0, c]$, and to a lesser degree on the right, contribute to making the value of the integral in (3.14) rather large, notwithstanding the damping power of $ds(t)$ for large t.

We have computed $||\tilde{G}_n'||_F$ for $n = 3(3)15$, using a more accurate procedure (cf. Example (ii) in Subsection 4.3) to generate $\alpha_k(d\sigma)$, $\beta_k(d\sigma)$ [hence, to compute by the method in Subsection 1.3 (iv) the Gauss nodes τ_ν and weights σ_ν needed in (3.14)]. The results are reported for $c = .2, 1., 5., \infty$ in Table 4.1. Actual computation with the modified Chebyshev algorithm showed that in single precision (ca. 15 decimal digits) on the Cyber 205, all accuracy was lost by the time n became 4, 7, 11, 12 for the four values of c, respectively.

TABLE 4.1 $||\tilde{G}_n'||_F$ for Hermite measure on $[0, c]$, $c = .2, 1., 5., \infty$
(Integers in parentheses denote decimal exponents.)

n	$c = .2$	$c = 1.$	$c = 5.$	$c = \infty$
3	4.00(6)	3.08(3)	3.60(2)	3.60(2)
6	7.20(16)	6.01(9)	2.59(6)	2.59(6)
9	9.94(27)	7.68(16)	2.42(10)	2.41(10)
12	4.45(39)	2.96(24)	2.86(14)	2.40(14)
15	4.59(51)	2.52(32)	5.69(18)	2.48(18)

For symmetric support $[-c,c]$ of $d\sigma$ one expects the conditioning of the map G_n to improve, since the effect of large g_n is now mitigated by the dampening of ds on *both* sides of the interval $[-c,c]$.

TABLE 4.2 $||\tilde{G}_n'||_F$ for Hermite measure on $[-c,c]$, $c = .2, 1., 5.$

n	$c = .2$	$c = 1.$	$c = 5.$	n	$c = .2$	$c = 1.$	$c = 5.$
3	1.19(5)	4.80(1)	3.86(0)	18	1.97(53)	8.61(28)	7.84(7)
6	3.17(13)	8.23(5)	6.96(1)	21	9.94(63)	2.78(35)	1.13(10)
9	6.50(22)	1.08(11)	1.54(3)	24	7.88(74)	1.41(42)	2.92(12)
12	4.33(32)	4.62(16)	3.71(4)	27	9.25(85)	1.06(49)	1.19(15)
15	6.67(42)	4.55(22)	1.18(6)	30	1.54(97)	1.12(56)	7.11(17)

The magnitudes of $||\tilde{G}_n'||_F$ indeed are as shown in Table 4.2. This time, single precision was observed to be "exhausted" as n becomes equal to 6, 11, 31 for the three values of c, respectively. When $c = \infty$, of course, since $p_k(\cdot; ds) = \pi_k(\cdot; ds)$, the modified Chebyshev algorithm implements essentially the identity map, and therefore is perfectly stable.

4. "Bootstrap" Procedures

4.1 STIELTJES' PROCEDURE

Explicit formulae for the recursion coefficients $\alpha_k(d\sigma)$, $\beta_k(d\sigma)$ in the basic recurrence relation

$$\pi_{k+1}(t) = (t - \alpha_k)\pi_k(t) - \beta_k\pi_{k-1}(t), \quad k = 0,1,2,\ldots,$$
$$\pi_{-1}(t) = 0, \quad \pi_0(t) = 1 \tag{4.1}$$

for the orthogonal polynomials $\pi_k(\cdot) = \pi_k(\cdot; d\sigma)$ can be obtained if we take the inner product (1.3) of both sides of (4.1) with π_k and π_{k-1}, respectively:

$$\alpha_k = \frac{(M\pi_k, \pi_k)}{(\pi_k, \pi_k)}, \quad k \geq 0; \quad \beta_k = \frac{(\pi_k, \pi_k)}{(\pi_{k-1}, \pi_{k-1})}, \quad k \geq 1, \tag{4.2}$$

where $(Mu)(t) = t\, u\,(t)$. While the expressions in (4.2) do not lend themselves directly to compute α_k, β_k (they require knowledge of π_k and π_{k-1}!), we can combine them with (4.1) to successively build up the coefficients – one by one – starting with the known polynomial $\pi_0(t) = 1$. Thus, α_0 can be computed (in principle!) from the first relation in (4.2) with $k = 0$, whereas β_0 is computable from $\beta_0 = (\pi_0, \pi_0)$. We are now ready to use (4.1) with $k = 0$ to generate π_1. Knowing π_1, we can go into (4.2) with $k = 1$ and compute α_1, β_1, which then in turn allows us to use (4.1) with $k = 1$ to get π_2. Clearly, we can proceed in this manner to generate as many coefficients α_k, β_k as desired. The algorithm (to our knowledge) was first suggested by Stieltjes [1884], and we have thus called it the *Stieltjes procedure* (Gautschi [1981a], [1982a]).

The problem with Stieltjes' procedure, of course, is the difficulty of evaluating the inner products in (4.2). The Gauss formula for $d\sigma$, which would be the most natural approach, is simply not available, since to generate it (cf. 1.3 (iv)), we would need precisely the quantities α_k, β_k we are in the process of computing! Neither is it feasible to use the simple-minded idea of expressing the polynomials π_k explicitly in power form, evaluating the inner products explicitly in terms of the coefficients of the polynomials involved and the moments of $d\sigma$, and finally updating the coefficients by using (4.1). In effect, this procedure would simply implement the map $G_n: \mu \to \rho$ from ordinary moments to the recursion coefficients and thus would suffer from the exponential growth of condition noted at the beginning of Section 3. To make the Stieltjes procedure a viable computational tool, one has to work harder! Possible improvements will be discussed in the next subsection.

It must be noted, nevertheless, that the Stieltjes procedure is readily applicable in the case of discrete measures $d\sigma$, since the inner products in (4.2) then only require finite summations. Forsythe [1957] indeed used and studied this procedure in connection with least squares data-fitting.

4.2 DISCRETIZED STIELTJES PROCEDURE

The basic idea, here, is to approximate the inner product (1.3) by a *discrete* inner product

$$(u,v)_N = \sum_{k=1}^{N} w_k u(t_k) v(t_k) \tag{4.3}$$

involving $N > n$ distinct points $t_k = t_k^{(N)}$ and suitable weights $w_k = w_k^{(N)}$, $k = 1, 2, \ldots, N$. Thereafter, one applies Stieltjes' procedure (in its discrete form) with the inner product $(\,\cdot\,,\,\cdot\,)$ in (4.2) replaced by $(\,\cdot\,,\,\cdot\,)_N$. The coefficients $\alpha_k^{(N)}$, $\beta_k^{(N)}$, $k = 0, 1, \ldots, n-1$, so computed are taken to approximate α_k, β_k. There is some skill required in selecting the discretization (4.3) so as to induce reasonably fast convergence of the process as $N \to \infty$. In general, this requires a careful adaptation of the discretization to the particular measure $d\sigma$ on hand.

To describe how this might be done, suppose that the measure has the form

$$d\sigma(t) = \omega(t)dt \quad \text{on } [a,b], \tag{4.4}$$

where $[a,b]$ is a finite or infinite interval and ω an appropriate weight function. The first step, in general, is to decompose $[a,b]$ into a finite number of (possibly overlapping) subintervals,

$$[a,b] = \bigcup_{i=1}^{m} [a_i,b_i] \quad (m \geq 1), \tag{4.5}$$

and to rewrite integrals with respect to the measure $d\sigma$ as

$$\int_a^b f(t)\omega(t)dt = \sum_{i=1}^{m} \int_{a_i}^{b_i} f(t)\omega_i(t)dt. \tag{4.6}$$

In specific instances, such decompositions are often suggested in a natural way by the properties and structure of the given weight function ω (cf. the examples in Subsection 4.3). In the next step, each integral on the right of (4.6) is approximated by an appropriate quadrature rule,

$$\int_{a_i}^{b_i} f(t)\omega_i(t)dt \approx Q_i f, \quad Q_i f = \sum_{r=1}^{N_i} w_{r,i}f(t_{r,i}). \tag{4.7}$$

The relations (4.6), (4.7), in which f is replaced by uv, then give rise to an approximation of the inner product $(u,v) = \int_a^b u(t)v(t)\omega(t)dt$ by a discrete inner product of the type (4.3),

$$(u,v)_N = \sum_{i=1}^{m} \sum_{r=1}^{N_i} w_{r,i}u(t_{r,i})v(t_{r,i}), \quad N = \sum_{i=1}^{m} N_i. \tag{4.8}$$

There is enough flexibility in this approach – choosing the subdivision (4.5), the local "weight functions" ω_i in (4.6), and the quadrature rules Q_i in (4.7) – to come up with an effective scheme of approximating the true inner product $(\,\cdot\,,\,\cdot\,)$. Indeed, it is straightforward to adapt the approach to deal with measures supported on disjoint intervals (which suggests an obvious decomposition (4.5) of the support interval) and to measures containing, in addition to an absolutely continuous component (4.4), a discrete point spectrum,

$$d\sigma(t) = \omega(t)dt + \sum_j \omega_j \delta(t - \tau_j). \tag{4.9}$$

All one has to do is to add on to (4.8) the term $\sum_j \omega_j u(\tau_j)v(\tau_j)$.

Strictly for use as a *general purpose procedure*, ignoring specific properties or structural characteristics of $d\sigma$, we recommend the following choices: Take (4.5) to be a decomposition in nonoverlapping intervals, and on each interval let $\omega_i(t) = \omega(t)$ and do the integration in (4.7) by the Fejér quadrature rule applied to $f\omega$ and properly transformed to the interval $[a_i,b_i]$, using equal numbers of nodes, $N_i = N^F$, all i. More precisely, recall that the N^F-point Fejér rule on the canonical interval $[-1,1]$ is the interpolatory N^F-point quadrature rule based on the Chebyshev abscissae $t_r^F = \cos((2r - 1)\pi/2N^F)$,

$$Q_{N^F}^F g = \sum_{r=1}^{N^F} w_r^F g(t_r^F). \tag{4.10}$$

The weights w_r^F can be computed explicitly in terms of trigonometric functions and are all positive (see, e.g., Gautschi [1967b]). To apply (4.10) in (4.7), where now $\omega_i = \omega$ and $g = f\omega$, and subsequently in (4.8), select monotone functions ϕ_i (a linear function if $[a_i,b_i]$ is finite) which map $[-1,1]$ onto $[a_i,b_i]$, and then use

$$\int_{a_i}^{b_i} f(t)\omega(t)dt = \int_{-1}^{1} f(\phi_i(\tau))\omega(\phi_i(\tau))\phi_i'(\tau)d\tau$$

$$\approx \sum_{r=1}^{N^F} w_r^F f(\phi_i(t_r^F))\omega(\phi_i(t_r^F))\phi_i'(t_r^F) =: Q_i f. \tag{4.11}$$

In other words, in (4.7), take $t_{r,i} = \phi_i(t_r^F)$ and $w_{r,i} = w_r^F \omega(\phi_i(t_r^F))\phi_i'(t_r^F)$, $i = 1,2,\ldots$, $N^F(= N_i)$. This results in a discrete inner product (4.8) with $N = mN^F$ points – the Chebyshev points in the respective subintervals.

In sufficiently simple cases, one can dispense with the subdivision (4.5) (i.e., take $m = 1$) and apply the procedure directly to the entire interval $[a,b]$.

There are several reasons why we favor Fejér's quadrature rule over other possible rules: (i) As already mentioned, the Fejér rule can be computed from explicit formulae, as opposed, for example, to the Gauss-Legendre rule, and can therefore be generated more economically. (For a comparison with the Gauss-Legendre rule, see Gautschi [1982a, Table 2.1].) (ii) Convergence of (4.10), as $N^F \to \infty$, is assured not only for continuous integrands g, but also for integrands having singularities at the endpoints of $[-1,1]$ (owing, e.g., to a singular factor ω in $g = f\omega$), provided the singularities are monotonic and integrable (cf. Gautschi [1967b]). (iii) If ω happens to be a polynomial of degree s (for example, $\omega \equiv 1$), if all intervals $[a_i,b_i]$ are finite, and if $f \in \mathbf{P}_{2n-1}$, then (4.11) is exact for each i whenever $N^F \geq 2n + s$. With this choice of N^F, our implementation of Stieltjes' procedure thus produces exact answers, except for rounding errors.

4.3 EXAMPLES

(i) *Chebyshev measure with a constant added to it:* $d\sigma(t\,;a) = [(1 - t^2)^{-1/2} + a]dt$ on $[-1,1]$, $a > 0$.

It would be difficult, in this example, to find a discretization that works well when applied to the entire interval $[-1,1]$. Using $m = 2$ in (4.5), and $[a_1,b_1] = [a_2,b_2] = [-1,1]$, $\omega_1(t) = (1 - t^2)^{-1/2}$, $\omega_2(t) = a$, however, makes it easy to get exact results: simply take for Q_1 in (4.7) the n-point Gauss-Chebyshev rule and for Q_2 the $(2n)$-point Fejér rule – both explicitly computable – where n is the number of recursion coefficients α_k, β_k desired. Actual computations for $0 \leq a \leq 1000$, $1 \leq n \leq 80$ indeed produced results accurate to essentially machine precision. As one would expect, there is a continuous transition from the Chebyshev case to the Legendre case as a increases from 0 to ∞.

While it is true that the discretized Stieltjes procedure is capable of producing exact results in this example, it does so at a price! Much more economical (by almost a factor of 10, when $n = 80$) is the modified Chebyshev algorithm, which – it so happens – can also be applied, with Chebyshev modified moments $m_k = m_k(d\sigma(\cdot\,;a))$,

$$m_0 = \pi + 2a,$$

$$m_k = \frac{1}{2^{k-1}} \int_{-1}^{1} T_k(t)d\sigma(t\,;a) = -\frac{a}{2^{k-2}(k^2 - 1)}, \quad k \text{ even} > 0,$$

$$m_k = 0, \quad k \text{ odd}.$$

Moreover, the condition of the underlying map G_n, as measured by $||\tilde{G}'_n||_F$ in (3.14), is almost perfect for a and n in the ranges indicated above. The numerical results, accordingly, have been observed to be as accurate as with the previous method.

(ii) *Finite Hermite measures:* $d\sigma(t) = e^{-t^2}dt$ on $[0, c]$ or $[-c, c]$, $c > 0$ (cf. 2.2 (iv)).

Applying our general purpose discretization, with Fejér's rule used on the whole interval, works reasonably well for $c < \infty$, but has difficulty converging when $c = \infty$. We found that for $n = 40$ and $c = .2, 1., 5.$, we can achieve an accuracy of 12 significant decimal digits with $N^F = 161$ in (4.10). Decomposition of the interval $[0, c]$ into four subintervals of equal length, when $c < \infty$, and into $[0,3] \cup [3,6] \cup [6,9] \cup [9,\infty]$ when $c = \infty$, gave the same accuracy already with $N^F = 81$ in (4.11). Analogous results were obtained for the symmetric interval $[-c, c]$, $c < \infty$, except that the case $c = 5$ required $N^F = 201$ without decomposition.

Alternatively, the recursion coefficients for the symmetric interval can be obtained from those for the one-sided interval by well-known recursive algorithms (see, e.g., Chihara [Ch. I, §§8-9]).

(iii) *A measure involving the modified Bessel function:* $d\sigma(t) = t^\mu K_0(t)dt$ on $[0,\infty]$, $\mu > -1$ (cf. 2.2 (v)).

Here it is indispensable that one find a discretization that does justice to the special properties of the weight function, in particular its behavior for small and large t. This can be described, for the factor K_0, by

$$K_0(t) = \begin{cases} R(t) + I_0(t)\ln(1/t), & 0 < t \le 1, \\ t^{-1/2}e^{-t}S(t), & 1 \le t < \infty, \end{cases} \qquad (4.12)$$

where R, S are well-behaved smooth functions, and I_0 is the "regular" modified Bessel function. All three functions can be easily and accurately evaluated on their respective intervals by rational approximations (Russon & Blair [1969]). Therefore,

$$\int_0^\infty f(t)d\sigma(t) = \int_0^1 t^\mu[R(t)f(t)]dt + \int_0^1 t^\mu\ln(1/t)[I_0(t)f(t)]dt$$
$$+ \int_1^\infty e^{-t}[t^{\mu-1/2}S(t)f(t)]dt.$$

This suggests the decomposition $[0,\infty] = \bigcup_{i=1}^{3} [a_i, b_i]$ with $a_1 = a_2 = 0$, $b_1 = b_2 = 1$, $a_3 = 1$, $b_3 = \infty$, and the discretizations (where for simplicity we take $N_i = N$, $i = 1,2,3$)

$$\int_0^1 t^\mu[R(t)f(t)]dt \approx \sum_{r=1}^{N} w_{r,1}f(t_{r,1}), \quad t_{r,1} = \tau_{r,1}^{(N)}, \quad w_{r,1} = \sigma_{r,1}^{(N)}R(t_{r,1}),$$

$$\int_0^1 t^\mu\ln(1/t)[I_0(t)f(t)]dt \approx \sum_{r=1}^{N} w_{r,2}f(t_{r,2}), \quad t_{r,2} = \tau_{r,2}^{(N)}, \quad w_{r,2} = \sigma_{r,2}^{(N)}I_0(t_{r,2}),$$

$$\int_1^\infty e^{-t}[t^{\mu-1/2}S(t)f(t)]dt = e^{-1}\int_0^\infty e^{-\tau}[(1+\tau)^{\mu-1/2}S(1+\tau)f(1+\tau)]d\tau$$

$$\approx \sum_{r=1}^{N} w_{r,3}f(t_{r,3}), \quad t_{r,3} = 1 + \tau_{r,3}^{(N)}, \quad w_{r,3} = e^{-1}\sigma_{r,3}^{(N)}t_{r,3}^{\mu-1/2}S(t_{r,3}),$$

where $\{\tau_{r,i}^{(N)}, \sigma_{r,i}^{(N)}\}$ $(i = 1, 2, 3)$ are the nodes and weights of the N-point Gauss formula for, respectively, the measures $d\sigma_1(t) = t^\mu dt$ on $[0,1]$, $d\sigma_2(t) = t^\mu \ln(1/t)dt$ on $[0,1]$, and $d\sigma_3(t) = e^{-t}dt$ on $[0,\infty]$. The first and last are classical measures – the Jacobi measure shifted to the interval $[0,1]$ with parameters $\alpha = 0$, $\beta = \mu$ (cf. 2.1 (iii)), and the Laguerre measure (cf. 2.1 (iv)) – and the Gauss formulae for them are easily obtained by the method of 1.3 (iv). The latter is true also for the second measure, once its recursion coefficients have been obtained as described in Example (ii) of Subsection 3.3.

5. Modification Algorithms

It is sometimes necessary to multiply a given (positive) measure $d\sigma$ by a polynomial, or by a rational function, and to construct the orthogonal polynomials for the new measure $d\hat{\sigma}$ in terms of those for the old measure, $d\sigma$. The problem for polynomials is solved by a classical formula of Christoffel (Szegö [1975, Thm. 2.5]), the one for rational functions more recently by similar (but more complicated) formulae of Uvarov [1959], [1969]. In both cases, the result is expressed in determinantal form and requires knowledge of the zeros of the polynomial, or the zeros and poles of the rational function.

For computational purposes it is more convenient to try obtaining the recursion coefficients $\hat{\alpha}_k = \alpha_k(d\hat{\sigma})$, $\hat{\beta}_k = \beta_k(d\hat{\sigma})$ for the new orthogonal polynomials in terms of those for the old ones. This is the subject of the present section. In Subsection 5.1 we consider the case of multiplication, in Subsection 5.2 the case of division by a polynomial. Applications are given in Subsection 5.3.

5.1 MODIFICATION BY LINEAR AND QUADRATIC FACTORS

It clearly suffices to consider the case of multiplying $d\sigma(t)$ by a linear factor $t - x$, or by a quadratic factor $(t - x)^2 + y^2$, since the general case can then be resolved by a repeated application of these special cases.

Assuming that the support of $d\sigma$ is a strict subset of $\mathbf{R}$, the case

$$d\hat{\sigma}(t) = (t - x)d\sigma(t), \quad x \in \mathbf{R} \setminus \text{supp } d\sigma, \tag{5.1}$$

is closely related to "kernel polynomials" (cf. Chihara [1978, Ch.I, §7]). From work of Galant [1971] and Gautschi [1982c] one obtains the following algorithm for computing the desired recursion coefficients $\hat{\alpha}_k$, $\hat{\beta}_k$:

$$e_{-1} = 0,$$

$$\left.\begin{aligned}
q_k &= \alpha_k - e_{k-1} - x, \\
\hat{\beta}_k &= q_k \cdot \begin{cases} \beta_0 & \text{if } k = 0 \\ e_{k-1} & \text{if } k > 0 \end{cases} \\
e_k &= \beta_{k+1}/q_k, \\
\hat{\alpha}_k &= x + q_k + e_k
\end{aligned}\right\} \quad k = 0, 1, 2, \ldots, n - 1. \tag{5.2}$$

Here, $\alpha_k = \alpha_k(d\sigma)$. $\beta_k = \beta_k(d\sigma)$. Note that algorithm (5.2) requires β_n in addition to the first n recursion coefficients α_k and β_k, $k = 0, 1, \ldots, n - 1$. The measure (5.1) is positive if x is to the left of the support of $d\sigma$, but negative otherwise. In the latter case, $q_0 < 0$, and $\hat{\beta}_0$ in (5.2) is negative (as it should be). Replacing $t - x$ in (5.1) by $|t - x|$ requires only one change in (5.2): $\hat{\beta}_0 = |q_0|\beta_0$ instead of $\hat{\beta}_0 = q_0\beta_0$.

The case of quadratic factors

$$d\hat{\sigma}(t) = [(t - x)^2 + y^2]d\sigma(t), \quad x \in \mathbf{R}, \quad y > 0, \tag{5.3}$$

can be dealt with by applying a complex version of (5.2) twice, first with the linear factor $t - z$, then with $t - \bar{z}$, where $z = x + iy$. The result can be rewritten in real form, and simplified, and produces the following algorithm (Gautschi [1982c, Eq. (4.7)]).

Initialization:

$$s_{-1} = x - \alpha_0, \quad t_{-1} = y, \quad e''_{-1} = \beta_0 y. \tag{5.4_0}$$

Continuation for $k = 0, 1, 2, \ldots, n-1$:

$$e'_k = -\beta_{k+1}\frac{s_{k-1}}{s_{k-1}^2 + t_{k-1}^2}, \quad e''_k = \beta_{k+1}\frac{t_{k-1}}{s_{k-1}^2 + t_{k-1}^2},$$

$$s_k = x + e'_k - \alpha_{k+1},$$

$$t_k = y + e''_k,$$

$$\hat{\alpha}_k = x + t_k\frac{e'_k}{e''_k} - \frac{s_k}{t_k}e''_k, \tag{5.4_k}$$

$$\hat{\beta}_k = t_k e''_{k-1}\left\{1 + \left[\frac{e'_k}{e''_k}\right]^2\right\}.$$

This algorithm requires both α_n and β_n, in addition to the α_k, β_k for $0 \le k \le n - 1$. In the case of *symmetric* measures $d\sigma$, i.e., those having support symmetric with respect to the origin and satisfying $d\sigma(-t) = d\sigma(t)$, and if $x = 0$, that is, for

$$d\hat{\sigma}(t) = (t^2 + y^2)d\sigma(t), \quad y > 0, \quad d\sigma \text{ symmetric}, \tag{5.5}$$

one has $\alpha_k = \hat{\alpha}_k = 0$, all $k \ge 0$, and the algorithm (5.4) simplifies considerably:

$$t_{-1} = y, \quad e_{-1} = \beta_0 y,$$

$$\left.\begin{aligned} e_k &= \beta_{k+1}/t_{k-1} \\ t_k &= y + e_k \\ \hat{\beta}_k &= t_k e_{k-1} \end{aligned}\right\} \quad k = 0, 1, \ldots, n - 1. \tag{5.6}$$

Limited experience with the algorithms (5.2), (5.4), (5.6) suggests that they are all numerically stable.

In (5.3) we assumed $y > 0$, since the case $y = 0$ can be reduced to two applications of (5.1). It is interesting, however, to note (Kautsky & Golub [1983]) that the Jacobi matrix $\hat{J}_n = J_n(d\hat{\sigma})$ (cf. (1.11)), when $d\hat{\sigma}(t) = (t - x)^2 d\sigma(t)$, can be obtained from $J_{n+1} = J_{n+1}(d\sigma)$ by one step of the *QR algorithm* with shift x; that is, if

$$J_{n+1} - xI_{n+1} = QR \quad (Q \text{ orthogonal, } R \text{ upper triangular}) \tag{5.7}$$

is the *QR* decomposition of $J_{n+1} - xI_{n+1}$, then

$$\hat{J}_n = (RQ + xI_{n+1})_{n \times n}, \tag{5.8}$$

i.e., $\hat{J}_n$ is the $n \times n$ leading principal submatrix of $RQ + xI_{n+1}$.

5.2 MODIFICATION BY LINEAR AND QUADRATIC DIVISORS

Algorithms for dividing a given (positive) measure $d\sigma$ by linear and quadratic polynomials can be obtained, essentially, by interchanging the roles of $d\sigma$ and $d\hat{\sigma}$ in Subsection 5.1 and "inverting" the algorithms given there.

For

$$d\hat{\sigma}(t) = \frac{d\sigma(t)}{t - x}, \quad x \in \mathbf{R} \setminus \text{supp } d\sigma, \tag{5.9}$$

one so obtains (Gautschi [1982c])

$$q_0 = 1/H(x), \quad \hat{\alpha}_0 = x + q_0,$$

$$\left.\begin{array}{l} e_{k-1} = \alpha_{k-1} - x - q_{k-1}, \\ \hat{\beta}_k = q_{k-1}e_{k-1}, \\ q_k = \beta_k/e_{k-1}, \\ \hat{\alpha}_k = q_k + e_{k-1} + x \end{array}\right\} \quad k = 1, 2, \ldots, n - 1, \tag{5.10}$$

where

$$H(x) = \frac{1}{\beta_0} \int_{\mathbf{R}} \frac{d\sigma(t)}{t - x} \tag{5.11}$$

is assumed to be known. Naturally, $\hat{\beta}_0 = \beta_0 H(x)$, and $\hat{\beta}_0 = \beta_0 |H(x)|$ if $t - x$ in (5.9) is replaced by $|t - x|$.

A similar algorithm can be derived for

$$d\hat{\sigma}(t) = \frac{d\sigma(t)}{(t-x)^2 + y^2}, \quad x \in \mathbf{R}, \quad y > 0, \tag{5.12}$$

but it is somewhat involved (cf. Gautschi [1982c, Eq. (5.8)]). We state here only the special case of a symmetric measure $d\sigma$, and $x = 0$ (for which, of course, $\alpha_k = \hat{\alpha}_k = 0$):

$$e_0 = h - y, \quad e_1 = \frac{\beta_1}{e_0} - y,$$

$$\hat{\beta}_1 = y e_0, \quad \hat{\beta}_2 = h e_1,$$

$$\left.\begin{array}{l} e_k = \dfrac{\beta_k}{e_{k-1}} - y \\[3mm] \hat{\beta}_{k+1} = \beta_{k-1} \dfrac{e_k}{e_{k-2}} \end{array}\right\} \quad k = 2, 3, \ldots, n-2, \tag{5.13}$$

where

$$h = \frac{\beta_0}{y \displaystyle\int_{\mathbf{R}} \frac{d\sigma(t)}{t^2 + y^2}} \tag{5.14}$$

is assumed known. Clearly, $\hat{\beta}_0 = \beta_0/(yh)$.

While both algorithms (5.10), (5.13) become rapidly unstable as x, resp. iy, moves away from the support of $d\sigma$, it is precisely the case where x, or iy, is very close to the support of $d\sigma$ which is often of considerable practical interest. In this case, the algorithms may still be used to advantage.

5.3 APPLICATIONS

(i) *Constrained least squares approximation.* Given a function f on the support of some (positive, possibly discrete) measure $d\sigma$, the problem here is to find a polynomial $\hat{p}_n \in \mathbf{P}_n$ which solves

$$\text{minimize} \int_{\mathbf{R}} [p(t) - f(t)]^2 d\sigma(t): \quad p \in \mathbf{P}_n$$

$$\text{subject to} \tag{5.15}$$

$$p(s_j) = f(s_j), \quad j = 0, 1, \ldots, m; \quad m < n.$$

Here, s_j are given distinct points on the support of $d\sigma$. Writing

$$p(t) = p_m(t;f) + s_m(t)q(t), \tag{5.16}$$

where $p_m(\cdot;f)$ is the interpolation polynomial of degree $\leq m$ corresponding to the constraints in (5.15), $s_m(t) = \prod_{j=0}^{m}(t - s_j)$, and $q \in \mathbf{P}_{n-m-1}$ is a polynomial of degree $\leq n - m - 1$ that can be varied freely, one has

$$\int_{\mathbb{R}} [p(t) - f(t)]^2 d\sigma(t) = \int_{\mathbb{R}} [f(t) - p_m(t;f) - s_m(t)q(t)]^2 d\sigma(t)$$

$$= \int_{\mathbb{R}} \left[\frac{f(t) - p_m(t;f)}{s_m(t)} - q(t) \right]^2 s_m^2(t) d\sigma(t).$$

Hence, we are led to minimize the last integral over all $q \in \mathbf{P}_{n-m-1}$, which is an *unconstrained* least squares problem for the new function

$$g(t) = \frac{f(t) - p_m(t;f)}{s_m(t)} = [s_0, s_1, \ldots, s_m, t]f \tag{5.17}$$

relative to the new measure $d\hat\sigma(t) = s_m^2(t)d\sigma(t)$. The required orthogonal polynomials $\pi_k(\cdot; d\hat\sigma)$ can be obtained by a repeated application of the QR steps (5.7), (5.8) with shifts $x = s_j$, $j = 0, 1, \ldots, m$.

Similar techniques can be used to deal with preassigned poles (cf. Lin [1988]).

(ii) *Integration in the presence of nearby poles.* Integrals in which the integrand has a simple pole very close to the interval of integration can be treated effectively by absorbing the principle part of the pole into a weight function and then applying an appropriate Gaussian quadrature rule. The required orthogonal polynomials (i.e., their recursion coefficients) can then be generated by the algorithms of Subsection 5.2.

As an example, consider

$$I_c(\varepsilon) = \int_0^\infty \frac{\cos t}{t + \varepsilon} e^{-t} dt, \quad I_s(\varepsilon) = \int_0^\infty \frac{\sin t}{t + \varepsilon} e^{-t} dt, \tag{5.18}$$

where $\varepsilon > 0$ is small. It is easily seen that I_c and I_s are the real and imaginary parts, respectively, of

$$I(\varepsilon) = e^{(1-i)\varepsilon} E_1((1 - i)\varepsilon), \tag{5.19}$$

where E_1 is the exponential integral. For small ε, the exact answer (5.19) can easily be calculated by series expansion (cf. Gautschi [1964, Eq. 5.1.11]).

We computed (5.18) by Gaussian quadrature relative to the measure $d\hat{\sigma}(t) = d\sigma(t)/(t + \varepsilon)$, where $d\sigma(t) = e^{-t}dt$ is the Laguerre measure, generating the required orthogonal polynomials by the algorithm (5.10) where $x = -\varepsilon$. The quantity $H(-\varepsilon)$ required in (5.10) is given by $H(-\varepsilon) = e^{\varepsilon}E_1(\varepsilon)$ and can be computed by series expansion in the same way as (5.19). Some results are shown in Table 5.1. The entries in the table represent the larger of the two relative errors committed when the n-point Gaussian quadrature rule is applied to $I_c(\varepsilon)$ and $I_s(\varepsilon)$ in (5.18). As can be seen, convergence is quite rapid and essentially independent of ε, as one expects. Algorithm (5.10) produced the recursion coefficients for $d\hat{\sigma}$ quite accurately for the last three values of ε, but with some moderate loss of accuracy when $\varepsilon = .5$.

TABLE 5.1 *Relative errors in n-point Gaussian quadrature applied to the integrals (5.18)*

n	$\varepsilon = .5$	$\varepsilon = .1$	$\varepsilon = .02$	$\varepsilon = .004$
4	1.41(−3)	7.48(−4)	5.40(−4)	4.57(−4)
8	3.17(−6)	2.10(−6)	1.77(−6)	1.67(−6)
12	5.31(−9)	3.68(−9)	3.19(−9)	3.06(−9)
16	6.77(−12)	4.97(−12)	4.37(−12)	4.31(−12)

APPENDIX. A Software Package for Orthogonal Polynomials

```
subroutine recur(n,ipoly,al,be,a,b)
dimension a(n), b(n)
```

This subroutine generates the coefficients $a(k)$, $b(k)$, $k = 0,1,\ldots, n-1$, in the recurrence relation

$$p(k+1)(x) = (x - a(k)) \times p(k)(x) - b(k) \times p(k-1)(x),$$
$$k = 0,1,\ldots, n-1,$$
$$p(-1)(x) = 0, \quad p(0)(x) = 1,$$

for some classical (monic) orthogonal polynomials, and sets $b(0)$ equal to the total mass of the weight distribution. The results are stored in the arrays a,b, which hold, respectively, the coefficients $a(k-1)$, $b(k-1)$, $k = 1,2,\ldots, n$.

input: n -- the number of recursion coefficients desired; type integer
ipoly- integer identifying the polynomial as follows:
1=Legendre polynomial on [−1,1]
2=Legendre polynomial on [0,1]
3=Chebyshev polynomial of the first kind
4=Chebyshev polynomial of the second kind
5=Jacobi polynomial with parameters $al = .5$, $be = -.5$
6=Jacobi polynomial with parameters al,be
7=generalized Laguerre polynomials with parameter al
8=Hermite polynomial

al,be- parameters for Jacobi and generalized Laguerre polynomials

output: a,b - - arrays containing, respectively, the recursion coefficients
$a(k-1)$, $b(k-1)$, $k = 1,2, \ldots, n$.

The subroutine calls for a function subroutine, named *gamma*, which is to evaluate the gamma function for positive arguments. *gamma* is used only in the cases *ipoly*=6 and *ipoly*=7.

---

subroutine *gauss(n,alpha,beta,zero,weight,e,ierr)*
dimension *alpha(n)*, *beta(n)*, *zero(n)*, *weight(n)*, *e(n)*

Given the recursion coefficients $alpha(k)$, $beta(k)$, $k = 0,1, \ldots, n-1$, of the appropriate orthogonal polynomials, this subroutine generates the nodes and weights of the n-point Gaussian quadrature formula and stores them in $zero(k)$, $weight(k)$, $k = 1,2, \ldots, n$. The nodes are obtained as eigenvalues of the tridiagonal symmetric Jacobi matrix, the weights in terms of the first components of the corresponding eigenvectors. The required eigenvalues and first components of the eigenvectors are calculated using a translation and adaptation of the Algol procedure *imtql2*, Numer. Math. 12 (1968), 377-383, by Martin and Wilkinson, as modified by Dubrulle, Numer. Math. 15 (1970), 450. See also Handbook for Autom. Comput., vol. 2 – Linear Algebra, pp. 241-248, and the "Eispack" routine *imtql2*.

input: n - - the number of quadrature points; type integer
alpha,beta -- arrays to be filled with the values of $alpha(k-1)$, $beta(k-1)$, $k = 1,2, \ldots, n$

output: *zero* - - array containing the Gaussian nodes
weight - - array containing the Gaussian weights
ierr - - an error flag indicating the node for which convergence is not achieved within 30 iterations.

The array e is needed for working space.

---

subroutine *cheb(n,nd,a,b,fm,alpha,beta,s,s0,s1,s2)*
dimension *a(nd)*, *b(nd)*, *fm(nd)*, *alpha(n)*, *beta(n)*, *s(n)*, *s0(nd)*,
**s1(nd)*, *s2(nd)*

Given a set of polynomials $p(0), p(1), \ldots, p(2 \times n - 1)$ satisfying

$$p(k+1)(x) = (x - a(k)) \times p(k)(x) - b(k) \times p(k-1)(x),$$
$$p(-1)(x) = 0, \quad p(0)(x) = 1,$$

and associated modified moments

$$fm(k) = \text{integral of } p(k)(x) \times dsigma(x),$$
$$k = 0,1, \ldots, 2 \times n - 1,$$

this subroutine uses the modified Chebyshev algorithm to generate the recursion coefficients $alpha(k)$, $beta(k)$, $k = 0,1,\ldots, n-1$, for the polynomials $pi(k)$ orthogonal with respect to the integration measure $dsigma(x)$, i.e.,

$$pi(k+1)(x) = (x-alpha(k)) \times pi(k)(x)-beta(k) \times pi(k-1)(x), \quad k = 0,1,\ldots, n-1,$$
$$pi(-1)(x) = 0, \quad pi(0)(x) = 1.$$

input:
- n — — the number of recursion coefficients desired; type integer
- nd — — twice the number n; type integer
- a,b — — arrays to be filled with the values of $a(k-1)$, $b(k-1)$, $k = 1,2,\ldots,2 \times n - 1$
- fm — — array to be filled with the values of the modified moments $fm(k-1)$, $k = 1,2,\ldots,2 \times n$

output:
- $alpha, beta$ — — arrays containing, respectively, the recursion coefficients $alpha(k-1)$, $beta(k-1)$, $k = 1,2,\ldots,n$, where $beta(0)$ is the total mass
- s — — array containing the reciprocal squares of the leading coefficients of the orthonormal polynomials of degrees $0,1,2,\ldots,n-1$.

The arrays $s0$, $s1$, $s2$ are needed for working space.

```
subroutine stielt(n,ncap,x,w,p0,p1,p2,alpha,beta,ierr)
dimension x(ncap), w(ncap), p0(ncap), p1(ncap), p2(ncap), alpha(n),
*beta(n)
```

This routine applies Stieltjes' procedure to generate the recursion coefficients $alpha(k)$, $beta(k)$, $k = 0,1,\ldots, n-1$, for the discrete (monic) orthogonal polynomials associated with the inner product

$$(f,g) = \text{sum over } k \text{ from 1 to } ncap \text{ of } w(k) \times f(x(k)) \times g(x(k)).$$

The integer n must be between 1 and $ncap$, inclusive; otherwise, there is an error exit with $ierr=1$.

input:
- n — — the number of recursion coefficients desired; type integer
- $ncap$ — — the number of support points in the inner product; type integer
- x — — array to be filled with the abscissae of the inner product
- w — — array to be filled with the weights of the inner product

output:
- $alpha, beta$ — — arrays containing, respectively, the recursion coefficients $alpha(k-1)$, $beta(k-1)$, $k = 1,2,,\ldots,n$, where $beta(0)$ is the total mass
- $ierr$ — — an error flag, equal to 0 on normal return.

Stieltjes' procedure is implemented here with a view toward making it time-efficient, but it is costly in extra memory space (the working arrays $p0, p1, p2$).

---

subroutine $mcsti(n,ncapm,ncapmm,mc,mp,xp,yp,quad,eps,alpha,$
$*beta,ierr,be,x,w,xm,wm,p0,p1,p2)$
dimension $xp(mp), yp(mp), alpha(n), beta(n), x(ncapm), w(ncapm),$
$*xm(ncapmm), wm(ncapmm), p0(ncapmm), p1(ncapmm), p2(ncapmm)$
data $idegwf/0/$

This is a multiple component discretized Stieltjes procedure. It generates to a relative accuracy of eps the recursion coefficients $alpha(k)$, $beta(k)$, $k = 0,1,\ldots, n-1$, for the polynomials orthogonal with respect to a weight distribution consisting of the sum of mc continuous components and a discrete component with mp points. The continuous part of the spectrum is made up of mc weight functions, each supported on its own interval. These intervals may or may not be disjoint. The discretization of the inner product on the ith interval is furnished by the subroutine $quad$. More specifically, $quad(n,x,w,i,ierr)$ produces the abscissae $x(k)$ and weights $w(k)$, $k = 1,2,\ldots, n$, to be used in approximating the ith inner product

$$\text{integral of } p(x)\times q(x)\times wf(x,i)\, dx$$

by the

$$\text{sum over } k \text{ from 1 to } n \text{ of } w(k)\times p(x(k))\times q(x(k)),$$
$$i = 1,2,\ldots, mc.$$

The point spectrum is given through its abscissae xp and jumps yp.

If each component interval is finite and carries as weight function a polynomial, and if the maximum degree is d, the procedure is designed to converge after one iteration, provided the variable $idegwf$ in the data declaration is set equal to d and one takes for $quad$ the general purpose routine qgp.

input:	n	- - the number of recursion coefficients desired; type integer
	$ncapm$	- - a discretization parameter indicating an upper limit of the fineness of the discretization; $ncapm=500$ will usually be satisfactory; type integer
	$ncapmm$	- - to be set equal to $mc\times ncapm + mp$; type integer
	mc	- - the number of component intervals in the continuous part of the spectrum; type integer
	mp	- - the number of points in the discrete part of the spectrum; type integer; if there is no point spectrum, set $mp=0$
	xp	- - an array of dimension mp containing the abscissae of the point spectrum
	yp	- - an array of dimension mp containing the jumps of the point spectrum
	$quad$	- - a subroutine determining the discretization of the inner product on each component interval

eps   -- the desired relative accuracy of the nonzero recursion coefficients; type real

output: *alpha,beta*   -- arrays of dimension *n*, holding as *k*th element $alpha(k-1)$, $beta(k-1)$, $k = 1,2, \ldots , n$, respectively

ierr   -- an error flag, equal to 0 on normal return, equal to *i* if there is an error condition in the discretization of the *i*th interval and equal to *ncapm* if the discretized Stieltjes procedure does not converge within the discretization resolution specified by *ncapm*.

The arrays $be,x,w,xm,wm,p\,0,p\,1,p\,2$ are used for working space. The routine calls upon the subroutine *stielt*.

```
subroutine qgp(n,x,w,i,ierr)
dimension x(n), w(n)
logical finleft,finrite
common/s/mc,finleft,finrite,endleft(1),endrite(1),xfer(500),
*wfer(500)
```

This is a general purpose discretization routine that can be used as subroutine *quad* in the (multiple component) discretized Stieltjes procedure *mcsti*. It takes no account of the special nature of the weight function involved and hence may result in slow convergence of the Stieltjes procedure. This routine, therefore, should be used only as a last resort, if no better, more natural discretization can be found.

It is assumed that there are $mc \geq 1$ disjoint component intervals. The discretization is effected by the Fejér quadrature rule, suitably transformed to the respective interval. An interval that extends to minus infinity has to be indexed by 1; one that extends to plus infinity has to be indexed by *mc*.

The additional input parameters and working space used by this routine are listed in the common declaration. Their meaning is as follows:

mc   -- the number of component intervals; type integer

finleft   -- a logical variable to be set .true. if the extreme left interval is finite, and .false. otherwise

finrite   -- a logical variable to be set .true. if the extreme right interval is finite, and .false. otherwise

endleft   -- an array of dimension *mc* containing the left endpoints of the component intervals; if the first of these extends to minus infinity, *endleft*(1) can be set to an arbitrary value

endrite   -- an array of dimension *mc* containing the right endpoints of the component intervals; if the last of these extends to plus infinity, *endrite*(*mc*) can be set to an arbitrary value

xfer,wfer   -- working arrays of dimension *ncapm* holding the Fejér nodes and weights, respectively, for the interval [−1,1].

The arrays *endleft* and *endrite* in the common declaration have to be dimensioned with the actual value of *mc*, the arrays *xfer, wfer* with the actual value of *ncapm*. The user has to supply the function routine *wf(.,i)* for the weight function on the *i*th component interval. The routine also uses auxiliary subroutines, *fejer, symmtr* and *transf.*

---

subroutine *chris(n,np1,iopt,a,b,x,y,fmom,hr,hi,alpha,beta,iflag)*
dimension *a(np1), b(np1), alpha(n), beta(n)*

This subroutine implements the Christoffel or generalized Christoffel theorem. Given the recursion coefficients $a(k)$, $b(k)$, $k = 0, 1, \ldots, n$, for the (monic) orthogonal polynomials with respect to some measure $dsigma(t)$, it generates the recursion coefficients $alpha(k)$, $beta(k)$, $k = 0, 1, \ldots, n - 1$, for the measure

$$
\begin{array}{lll}
\text{abs}(t-x)dsigma(t) & \text{if} & iopt=1 \\
[(t-x)**2+y**2]dsigma(t) & \text{if} & iopt=2 \\
(t**2+y**2)dsigma(t) & \text{if} & iopt=3 \\
\text{with } dsigma(t) \text{ symmetric} & & \\
dsigma(t)/\text{abs}(t-x) & \text{if} & iopt=4 \\
dsigma(t)/[(t-x)**2+y**2] & \text{if} & iopt=5 \\
dsigma(t)/(t**2+y**2) & \text{if} & iopt=6 \\
\text{with } dsigma(t) \text{ symmetric.} & &
\end{array}
$$

input:
- *n* — — the number of recursion coefficients required; type integer
- *np* 1 — — an integer to be set equal to $n+1$
- *iopt* — — an integer selecting the desired weight distribution
- *a,b* — — arrays of dimension *np* 1 containing the recursion coefficients $a(k-1)$, $b(k-1)$, $k = 1, 2, \ldots, n + 1$, of the polynomials orthogonal with respect to the given measure $dsigma(t)$
- *x,y* — — real parameters defining the linear and quadratic factors, or divisors, of $dsigma(t)$
- *fmom* — — the first moment of $dsigma(t)$, that is, *fmom*=integral of $dsigma(t)$
- *hr, hi* — — the real and imaginary part, respectively, of the integral $h(z)$=integral of $dsigma(t)/(z-t)$, where $z=x+iy$; the parameter *hr* is used only if *iopt*=4 or 5, the parameter *hi* only if *iopt*=4, 5, or 6

output:
- *alpha, beta* — — arrays of dimension *n* containing the desired recursion coefficients $alpha(k-1)$, $beta(k-1)$, $k = 1, 2, \ldots, n$
- *iflag* — — an error flag, equal to 0 on normal return.

It is assumed that *n* is larger than or equal to 2. Otherwise, the routine exits immediately with the error flag *iflag* set equal to 1.

References

Chebyshev, P.L. [1859]: 'Sur l'interpolation par la méthode des moindres carrés', *Mém. Acad. Impér. Sci. St. Pétersbourg* (7) **1**, no. 15, pp. 1–24. [*Oeuvres* I, pp. 473–498.]

Chihara, T.S. [1978]: *An Introduction to Orthogonal Polynomials*, Gordon and Breach, New York.

Clenshaw, C.W. [1955]: 'A note on the summation of Chebyshev series', *Math. Tables Aids Comput.* **9**, pp. 118–120.

Danloy, B. [1973]: 'Numerical construction of Gaussian quadrature formulas for $\int_0^1 (-\text{Log } x)x^\alpha f(x)dx$ and $\int_0^\infty E_m(x)f(x)dx$', *Math. Comp.* **27**, pp. 861–869.

Forsythe, G.E. [1957]: 'Generation and use of orthogonal polynomials for data-fitting with a digital computer', *J. Soc. Indust. Appl. Math.* **5**, pp. 74–88.

Fox, L. and Parker, I.B. [1968]: *Chebyshev Polynomials in Numerical Analysis*, Oxford University Press, London.

Fransén, A. [1979]: 'Accurate determination of the inverse gamma integral', *BIT* **19**, pp. 137–138.

Frontini, M., Gautschi, W. and Milovanović, G.V. [1987]: 'Moment-preserving spline approximation on finite intervals', *Numer. Math.* **50**, pp. 503–518.

Galant, D. [1969]: 'Gauss quadrature rules for the evaluation of $2\pi^{-1/2}\int_0^\infty \exp(-x^2)f(x)dx$', *Math. Comp.* **23**, Review 42, pp. 676–677. Loose microfiche suppl. E.

Galant, D. [1971]: 'An implementation of Christoffel's theorem in the theory of orthogonal polynomials', *Math. Comp.* **25**, pp. 111–113.

Gautschi, W. [1964]: 'Exponential integral and related functions', in *Handbook of Mathematical Functions* (M. Abramowitz and I.A. Stegun, eds.), pp. 227–251. NBS Applied Math. Ser. **55**, U.S. Government Printing Office, Washington, D.C.

Gautschi, W. [1967a]: 'Computational aspects of three-term recurrence relations', *SIAM Rev.* **9**, pp. 24-82.

Gautschi, W. [1967b]: 'Numerical quadrature in the presence of a singularity', *SIAM J. Numer. Anal.* **4**, pp. 357–362.

Gautschi, W. [1975]: 'Computational methods in special functions – a survey', in *Theory and Application of Special Functions* (R.A. Askey, ed.), pp. 1–98. Academic Press, New York.

Gautschi, W. [1979a]: 'On generating Gaussian quadrature rules', in *Numerische Integration* (G. Hämmerlin, ed.), pp. 147–154. Internat. Ser. Numer. Math. **45**, Birkhäuser, Basel.

Gautschi, W. [1979b]: 'On the preceding paper "A Legendre polynomial integral" by James L. Blue', *Math. Comp.* **33**, pp. 742–743.

Gautschi, W. [1981a]: 'A survey of Gauss-Christoffel quadrature formulae', in *E.B. Christoffel; The Influence of his Work in Mathematics and the Physical Sciences* (P.L. Butzer and F. Fehér, eds.), pp. 72–147. Birkhäuser, Basel.

Gautschi, W. [1981b]: 'Minimal solutions of three-term recurrence relations and orthogonal polynomials', *Math. Comp.* **36**, pp. 547–554.

Gautschi, W. [1982a]: 'On generating orthogonal polynomials', *SIAM J. Sci. Stat. Comput.* **3**, pp. 289–317.

Gautschi, W. [1982b]: 'Polynomials orthogonal with respect to the reciprocal gamma function', *BIT* **22**, pp. 387–389.

Gautschi, W. [1982c]: 'An algorithmic implementation of the generalized Christoffel theorem', in *Numerical Integration* (G. Hämmerlin, ed.), pp. 89–106. Internat. Ser. Numer. Math. **57**, Birkhäuser, Basel.

Gautschi, W. [1983]: 'How and how not to check Gaussian quadrature formulae', *BIT* **23**, pp. 209–216.

Gautschi, W. [1984a]: 'Questions of numerical condition related to polynomials', in *Studies in Numerical Analysis* (G.H. Golub, ed.), pp. 140–177. Studies in Mathematics **24**, The Mathematical Association of America.

Gautschi, W. [1984b]: 'Discrete approximations to spherically symmetric distributions', *Numer. Math.* **44**, pp. 53–60.

Gautschi, W. [to appear]: 'On the remainder term for analytic functions of Gauss-Lobatto and Gauss-Radau quadratures', *Rocky Mountain J. Math.*

Gautschi, W. and Milovanović, G.V. [1985]: 'Gaussian quadrature involving Einstein and Fermi functions with an application to summation of series', *Math. Comp.* **44**, pp. 177–190.

Gautschi, W. and Milovanović, G.V. [1986]: 'Spline approximations to spherically symmetric distributions', *Numer. Math.* **49**, pp. 111–121.

Gautschi, W. and Varga, R.S. [1983]: 'Error bounds for Gaussian quadrature of analytic functions', *SIAM J. Numer. Anal.* **20**, pp. 1170–1186.

Gautschi, W., Kovačević, M.A. and Milovanović, G.V. [1987]: 'The numerical evaluation of singular integrals with coth-kernel', *BIT* **27**, pp. 389–402.

Gautschi, W., Tychopoulos, E. and Varga, R.S. [to appear]: 'A note on the contour integral representation of the remainder term for a Gauss-Chebyshev quadrature rule', *SIAM J. Numer. Anal.*

Golub, G.H. and Welsch, J.H. [1969]: 'Calculation of Gauss quadrature rules', *Math. Comp.* **23**, pp. 221–230.

Kahaner, D., Tietjen, G. and Beckman, R. [1982]: 'Gaussian-quadrature formulas for $\int_0^\infty e^{-x^2} g(x) dx$', *J. Statist. Comput. Simul.* **15**, pp. 155–160.

Kautsky, J. and Golub, G.H. [1983]: 'On the calculation of Jacobi matrices', *Linear Algebra Appl.* **52/53**, pp. 439–455.

King, H.F. and Dupuis, M. [1976]: 'Numerical integration using Rys polynomials', *J. Computational Phys.* **21**, pp. 144–165.

Lee, S.-Y. [1980]: 'The inhomogeneous Airy functions Gi(z) and Hi(z)', *J. Chem. Phys.* **72**, pp. 332–336.

Lin, J.-C. [1988]: *Rational L^2-Approximation with Interpolation*, Ph.D. Thesis, Purdue University.

Mach, R. [1984]: 'Orthogonal polynomials with exponential weight in a finite interval and application to the optical model', *J. Mathem. Phys.* **25**, pp. 2186–2193.

Micchelli, C.A. [1988]: 'Monosplines and moment preserving spline approximation', in *Numerical Integration III* (H. Braß and G. Hämmerlin, eds.), pp. 130–139. Internat. Ser. Numer. Math. **85**, Birkhäuser, Basel.

Milovanović, G.V. and Kovačević. M.A. [1988]: 'Moment-preserving spline approximation and Turán quadratures', in *Numerical Mathematics: Singapore 1988* (R. Agarwal, Y. Chou and S. Wilson, eds.), pp. 357–365. Internat. Ser. Numer. Math. **86**, Birkhäuser, Basel.

Nevai, P.G. [1979]: 'Orthogonal polynomials', *Mem. Amer. Math. Soc.* **18**, No. 213, 185 pp.

Paškovskii, S. [1983]: *Numerical Applications of Chebyshev Polynomials and Series* (Russian), "Nauka", Moscow.

Rivlin, T.J. [1974]: *The Chebyshev Polynomials*, Wiley, New York.

Russon, A.E. and Blair, J.M. [1969]: *Rational Function Minimax Approximations for the Bessel Functions $K_0(x)$ and $K_1(x)$*, Rep. AECL-3461, Atomic Energy of Canada Limited, Chalk River, Ontario.

Sack, R.A. and Donovan, A.F. [1971/72]: 'An algorithm for Gaussian quadrature given modified moments', *Numer. Math.* **18**, pp. 465–478.

Shizgal, B. [1981]: 'A Gaussian quadrature procedure for use in the solution of the Boltzmann equation and related problems', *J. Computational Phys.* **41**, pp. 309–328.

Smith, B.T., Boyle, J.M., Dongarra, J.J., Garbow, B.S., Ykebe, Y., Klema, V.C. and Moler, C.B. [1976]: *Matrix Eigensystem Routines – EISPACK Guide*, Lecture Notes Comp. Sci. **6**, 2nd ed., Springer, Berlin.

Stieltjes, T.J. [1884]: 'Quelques recherches sur la théorie des quadratures dites mécaniques', *Ann. Sci. Éc. Norm. Paris*, Sér. 3, **1**, pp. 409–426. [*Oeuvres* I, pp. 377–396.]

Stroud, A.H. and Secrest, D. [1966]: *Gaussian Quadrature Formulas*, Prentice-Hall, Englewood Cliffs, N.J.

Szegö, G. [1975]: *Orthogonal Polynomials*, AMS Colloquium Publications **23**, 4th ed., American Mathematical Society, Providence, R.I.

Uvarov, V.B. [1959]: 'Relation between polynomials orthogonal with different weights' (Russian), *Dokl. Akad. Nauk SSSR* **126**, pp. 33–36.

Uvarov, V.B. [1969]: 'The connection between systems of polynomials that are orthogonal with respect to different distribution functions' (Russian), *Ž. Vyčisl. Mat. i Mat. Fiz.* **9**, pp. 1253–1262. [English translation in *U.S.S.R. Computational Math. and Phys.* **9**, 1969, No.6, pp. 25–36.]

Wheeler, J.C. [1974]: 'Modified moments and Gaussian quadrature', *Rocky Mountain J. Math.* **4**, pp. 287–296.

Wilkinson, J.H. and Reinsch, C. [1971]: *Linear Algebra, Handbook for Automatic Computation*, Vol. II, Springer, New York.

Wimp, J. [1984]: *Computation with Recurrence Relations*, Pitman, Boston.

Wong, R. [1982]: 'Quadrature formulas for oscillatory integral transforms', *Numer. Math.* **39**, pp. 351–360.

THE RECURSION METHOD AND THE SCHROEDINGER EQUATION

Roger Haydock
Department of Physics &
Materials Science Institute
University of Oregon
Eugene, OR 97403, U.S.A.

Abstract.   The use of generalized orthogonal polynomials is described for the calculation of quantum mechanical properties of physical systems. Quantum mechanics and its mathematical representation are reviewed. Expressions for various physical quantities are related to the orthogonal polynomials obtained from the action of an observable on particular states. Polynomial sets for which weight distributions are known may be used as exact models form which the solutions of other models may be approximated by perturbations. The finite precision, orthogonal polynomial can be constructed numerically even for infinite dimensional models.

1. Quantum Mechanics

The most fundamental theory of the motion of matter and energy is quantum mechanics [1] which was developed to describe such phenomena as the spectrum of light emitted by hot objects, the spectrum of light emitted and absorbed by atoms, the diffractions patterns produced when electrons scatter off the surfaces of crystals, and many others. Each of these phenomena violates the laws of classical Newtonian mechanics: by unequal distribution of the thermal energy among the electromagnetic modes of the blackbody, by the existence of specific energies or quanta which an atom will absorb or emit, and by the wavelike interference of scattered electrons.

Quantum mechanics has been tested on a wide variety of systems, and in all cases it agrees with what is observed to the accuracy of current experimental technique. In other words, if the quantum mechanics is violated by any physical phenomena, that phenomena lies beyond the capabilities of current experiment. Despite its unequaled level of verification, the application of quantum mechanics on the human scale leads to the existence of states which are superpositions of mutually exclusive events. This sort of consequence does not conflict with experiment, but it does clash with the human view of reality which allows for only one of a conflicting pair of events. However, the resolution of such philosophical problems is not the purpose of this work.

Quantum mechanics has a simple logical structure which allows it to describe an enormous variety of phenomena, from the discrete which gives quantum mechanics its name, to the continuous of the macroscopic world. This structure is embodied in two assumptions which are stated below in a form similar to that of Dirac.

P. Nevai (ed.), Orthogonal Polynomials, 217–228.
© 1990 by Kluwer Academic Publishers.

(i) The first assumption is that physical states can be linearly superposed. It is conventional to treat quantum mechanical states as if they were complex waves so that if $\mathbf{x}$ and $\mathbf{y}$ represent two different states, then a third state $\mathbf{z}$ can be represented by

$$\mathbf{z} = \mathbf{x}\sin\theta + \mathbf{y}\cos\theta e^{i\phi} \tag{1.1}$$

where θ and ϕ are respectively the mixing and phase angles between $\mathbf{x}$ and $\mathbf{y}$ in $\mathbf{z}$. The proportion of $\mathbf{x}$ in $\mathbf{z}$ is $|\sin\theta|^2$ and the proportion of $\mathbf{y}$ in $\mathbf{z}$ is $|\cos\theta e^{i\phi}|^2$.

Consistent with the physical convention, it is mathematically convenient to represent physical states by vectors with complex scalars, although two vectors represent different physical states only if they are linearly independent. Depending on the system described, the dimensionality of this vector space need not be finite, or even countable. In order to obtain mixing and phase angles between the different physical states represented by a pair of complex vectors $\mathbf{y}$ and $\mathbf{z}$, it is necessary to introduce an inner product which is conventionally Hermitian,

$$\mathbf{y}^*\mathbf{z} = (\mathbf{z}^*\mathbf{y})^*, \tag{1.2}$$

and here is the same as Dirac's except for the notation in which * takes a vector to its dual, and a scalar to its complex conjugate. The mixing and phase angles, θ and ϕ, are given by

$$\cos\theta e^{i\phi} = (\mathbf{y}^*\mathbf{z})(\mathbf{z}^*\mathbf{z})^{-1/2}(\mathbf{y}^*\mathbf{y})^{-1/2}. \tag{1.3}$$

(ii) The second assumption is that physical quantities or observables correspond to linear transformations of the states. If P is an observable with stationary state $\mathbf{p}$, then it satisfies

$$P\mathbf{p} = \mathrm{p}\mathbf{p} \tag{1.4}$$

for stationary value p. Just as the states may be represented as vectors, the observables may be represented as linear transformations of vectors, and the above equation defines stationary vectors.

The vectors which represent physical states are restricted to those whose expectation values are real and definite for all observables. The expectation value of observable Q on vector $\mathbf{x}$ is

$$q_x = \mathbf{x}^*Q\mathbf{x}/\mathbf{x}^*\mathbf{x}. \tag{1.5}$$

Thus the zero vector 0 is unphysical because its expectation values are indefinite, and there are other vectors which are unphysical because they have infinite expectation values for some observables.

Without loss of generality and in order to ensure that the expectation values are real, the observables may be taken to the Hermitian with respect to the inner product

$$\mathbf{y}^*P\mathbf{x} = (\mathbf{x}^*P\mathbf{y})^* \tag{1.6}$$

for all physical vectors $\mathbf{x}$ and $\mathbf{y}$. A physical stationary state of P is then one whose expectation value for P is definite and coincides with its stationary value, and whose other expectation values are definite.

Quantum mechanics is used to relate observables to one another as, for example, when a beam of electrons is scattered off a target. If the momentum of the electrons are known and the forces between the target and the electrons are also known, then quantum mechanics is used to determine the proportion of electrons scattered at various angles by the target. Another example is the electronic structure of a solid where the position and scattering properties of each atom is known, and quantum mechanics is used to find the momentum and energy of the electronic states of the whole system. Fundamentally, both quantum mechanics and its applications are very simple, the complications arise in the description of specific observables such as energy, momentum, etc. for various systems, and then in the calculation of stationary states and values for these observables.

2. Representations

The assumptions of quantum mechanics are embodied in the representation of physical states and observables by vector spaces and linear transformations with the proviso that the vectors and transformations satisfy additional requirements to be physical. The first step in applying quantum mechanics is therefore to construct these representations.

The simplest of these is Heisenberg's matrix mechanics in which the states of the system are represented as N-tuples of numbers in the form of column vectors with an inner product for which * converts a column vector to its complex conjugate row vector, and the observables are represented as NxN matrices. For example, an electron has total spin 1/2 and thus has two spin states, up and down, whose coefficients are represented by the two elements of column vector. The observables of the system are the components of spin in various spacial directions, which are represented as 2x2 matrices. If the spin component along the Cartesian z-axis S_z is taken to be

$$S_z = \begin{pmatrix} 1 & 0 \\ 0 & -1 \end{pmatrix}, \tag{2.1}$$

components in other directions are

$$S_y = \begin{pmatrix} 0 & 1 \\ 1 & 0 \end{pmatrix} \tag{2.2}$$

and

$$S_x = \begin{pmatrix} 0 & i \\ -i & 0 \end{pmatrix}. \tag{2.3}$$

Because spin transforms like a vector under rotations, the stationary values of these observables are the same, but their stationary vectors are very different from one another illustrating the point that stationary states of one observable can be mixed states of another observable. Although this representation is mathematically simple, it is abstract in that none of the vectors or matrices is expressed in terms of any familiar physical quantities such as position or time.

A more sophisticated quantum mechanical representation which does relate the vectors and transformations to familiar physical quantities is Schroedinger's wave mechanics in which the states are represented as complex functions of position r and time t. For example, the plane wave

$$\Psi(r,t) = e^{i(k \cdot r - \omega t)} \tag{2.4}$$

represents a particle with momentum proportional to k and energy proportional to ω. The inner product of two wave functions is the integral over all space and time of the product of one function and the complex conjugate of the other. The observables are represented by differential operators; for example, momentum in wave mechanics is represented by

$$P = -i\hbar \nabla \tag{2.5}$$

where ∇ is the spacial gradient operator, and $\hbar$ is Planck's constant dividied by 2π. $\psi(r, t)$ is a stationary state of momentum, but not of the position observable R, because the wave is spread over all space.

The central problem in quantum mechanics is to find the stationary states and values of a given observable. Because physical systems can have large numbers of degrees of freedom which lead to large numbers of stationary states and values, this problem has to be well organized in order to make calculations possible.

For large physical systems only a small proportion of the stationary states of an observable contribute components to any one physical state, most are orthogonal to it. For example, there is negligible probability that a macroscopic diamond will spontaneously transform into graphite without the application of heat. This is because the overlap (inner product) of the diamond and graphite states is the product of the overlap between states for each electron and nucleus. Even though individual electrons and nuclei have substantial overlap from diamond to graphite, this overlap cannot be greater than unity, and when an infinite number of these are multiplied together, the result is zero – this tendency for many particle states to have zero inner product is sometimes known as the overlap catastrophe. In general the mean square overlap between two randomly chosen states decreases as the inverse of the number of degrees of freedom. Thus, calculations are greatly simplified by limiting the stationary states and values to those having non-zero overlap with a particular initial state.

3.   Electronic Structure of Solids

The electronic structure of solids is an ideal area for investigating the quantum mechanics of many particles. The relationship between the small mass of the electron and the low temperatures at which solids form, makes the behavior of electrons in solids strongly quantum mechanical. Unlike the physics of subatomic particles, electrons in solids are readily accessible to experimental investigation.

Most properties of solids are determined by their electronic structure. The geometrical structure of the solid depends on the nature of the electronic binding between atoms, and the mechanical properties depend on the way these bonds deform under stress. The magnetic properties mainly arise from the electrons' spin and orbital angular momentum. Finally, the electrical properties such as metallic conductivity or even superconductivity result from the way electronic states transport charge.

The electronic structure of solids can be understood in terms of the sharing of electrons between the atoms which make up the solid. The electrons of an isolated atom occupy individual orbitals, the same ones which give rise to the discrete atomic spectra. As the atoms come together to form a solid, electrons from one atom move to another atom to produce the bonding forces. The electronic structure of the bonded solid is a superposition of all the different arrangements or configurations of the electrons as they move from one orbital to another. This can be viewed as a kind of dance in which the electrostatic repulsion keeps the electrons apart form one another.

A simpler way of looking at this motion is in terms of quasi-electrons consisting of the bare electron and the hole in the distribution of other electrons, which moves with it. This hole which surrounds each electron contains exactly one positive charge so that the combined object is electrically neutral and thus interacts only weakly with other quasi-electrons. The motion of these renormalized electrons is approximately independent of one another, provided the exclusion principle – no two quasi-electrons may occupy the same state – is obeyed. This is called the independent electron model, and provides an excellent description of electrons in many solids.

Instead of having to consider the motion of all electrons at the same time, it is only necessary to calculate the motion of single quasi-electronics is the independent electron approximation. This may be done in terms of atomic orbitals and the coupling between them. The states of this model consist of the linear combinations of atomic or atomic-like orbitals, and the energy observable, Hamiltonian, may be represented as a matrix of couplings between orbitals on nearby atoms. The stationary states of energy are the linear combinations of orbitals which are stationary when multiplied by the Hamiltonian matrix.

In metals the electronic states are spread over macroscopic regions, and so the Hamiltonian matrix must be treated as infinite. For crystalline materials, the translation symmetries of the structure can be used to reduce the problem, but for structures with surfaces or other defects, the infinite nature of Hamiltonian must be faced directly.

It is not practical to calculate the stationary states of an infinite Hamiltonian describing extended electrons, however, the physically interesting aspects of these states are captured in a quantity called the projected density of states

$$n_0(E) = (-1/\pi)Im(\phi, [EI - H]^{-1}\phi) \qquad (3.1)$$

where H is the Hamiltonian matrix, I is the identity matrix, ϕ is an atomic orbital, and E is the energy. This quantity is the density of states in energy with each state weighted by its projection on the atomic orbital ϕ.

It can easily be seen that the power moments of the projected density of states are simply matrix elements of the powers of H,

$$\int E^n n_0(E)dE = (\phi, H^n\phi). \qquad (3.2)$$

This matrix elements of the nth power of the Hamiltonian can be calculated by summing the products of sequences of n matrix elements of H which couple a succession of orbitals starting and finishing at ϕ. Each sequence can be thought of as a path along which the electron hops. Each path contributes the product of its matrix elements, and the total moment is the sum of contributions from all paths. This appealing interpretation of the moments of the projected density of states, due to Fredel and Cyrot-Lackmann, is called the method of moments [2]. Numerically moments are difficult because details of the projected density of states depend on the higher moments with increasing sensitivity. Beyond about four of five moments, calculations become difficult because of the precision required.

Given a set of power moments, the classical moment problem [3] is solved by constructing a three term recurrence relation for the polynomials orthogonal with respect to integrals over the unknown distribution. The recurrence may be solved to express the distribution as the imaginary part of a continued fraction.

In the recursion method [4], the three term recurrence is constructed directly from the Hamiltonian by orthogonalization of successive powers of H on f to produce linear combinations of the atomic orbitals in which the Hamiltonian is tridiagonal. This construction is the same as that suggested by Lanczos [5] for the iterative tridiagonalization of matrices, and although the basis produced is numerically unstable, the stationary states, stationary energies, and other physical quantities are stable.

The remainder of this paper describes the recursion method in the context of general quantum mechanical problems rather than the electronic structure of solids. Special emphasis is placed on the role of orthogonal polynomials [6] in the method.

4. Stationary Subspaces

The basic problem of quantum mechanics may now be limited to that of finding for some observable S the stationary values and states which overlap a given initial state u. That these stationary states span the smallest subspace which is stationary with respect to S and contains u, can be seen from the argument that any stationary state outside this subspace is orthogonal to u. This subspace has the additional important property that it is non-degenerate; for each stationary value of S there is at most one state in the subspace. If there were two, then any linear combination would also be stationary, and in particular the one orthogonal to u could be discarded.

The states of real physical systems are stable in the sense that for all observables, the magnitudes of components of their stationary states decrease rapidly as their stationary values go to infinity. This property of physical states corresponds precisely to the mathematical condition on initial states necessary for their stationary subspaces to be representable by orthogonal polynomials. Mathematically, stability of an initial state u with respect to an observable S is the condition that its weight distribution have an infinite set of moments; that

$$\mu_n = \mathbf{u}^* S^n \mathbf{u} \tag{4.1}$$

be finite, but not necessarily bounded.

A consequence of the stability condition is that the stationary subspace of S generated by u is spanned by the states $\{\mathbf{u}, S\mathbf{u}, S^2\mathbf{u}, \ldots, S^n\mathbf{u}, \ldots\}$ which may be linearly dependent. The stability condition ensures that none of these states is infinite relative to the others, and the subspace is stationary because the product of S on any of them lies within the subspace.

5. Orthogonal Polynomials, Three Term Recurrences, and Continued Fractions

Within the stationary subspace generated from u, the stationary states of S may be expressed in terms of polynomials defined by the determinant,

$$P_n(s) = \frac{1}{[D_n D_{n-1}]^{1/2}} \det \begin{vmatrix} \mu_0 & \mu_1 & \cdots & \mu_n \\ \mu_1 & \mu_2 & \cdots & \mu_{n+1} \\ \vdots & \vdots & & \vdots \\ \mu_{n-1} & \mu_n & \cdots & \mu_{2n-1} \\ 1 & s & \cdots & s^n \end{vmatrix}, \tag{5.1}$$

where D_{n-1} is the determinant of the first n rows and columns of moments in the above matrix. The polynomials are orthonormal with respect to states generated from u when S

is substituted for s,

$$[D_n(S)\mathbf{u}]^* [D_m(S)\mathbf{u}] = \delta_{n,m}. \tag{5.2}$$

This can be seen by expanding the above determinants in powers of s, substituting S, and evaluating the inner product in (5.2). The inner products are moments of S on $\mathbf{u}$, and the expressions can then be rewritten as determinants which are zero if any two rows of moments are the same. The above orthogonality may also be expressed in terms of the distribution $d\mu$ of stationary values s of S in $\mathbf{u}$,

$$\int P_n(s)P_m(s)d\mu = \delta_{n,m} \tag{5.3}$$

where $d\mu$ has the same moments

$$\mu_n = \int s^n d\mu$$

as $\mathbf{u}$ in (4.1). An explicit form for $d\mu$ will be given below.

The polynomials may also be expressed as solutions of a symmetric three term recurrence relation which leads to representation of S as a symmetric tridiagonal matrix. Multiplication of a polynomials state

$$\mathbf{u}_n = P_n(S)\mathbf{u} \tag{5.4}$$

by S can be expressed as a linear combination of the polynomial states up to one degree higher. Since S is Hermitian with respect to the inner product, this linear combination can also only go to one degree lower, so the polynomial states satisfy the symmetric recurrence

$$S\mathbf{u}_n = b_{n+1}\mathbf{u}_{n+1} + a_n\mathbf{u}_n + b_n\mathbf{u}_{n-1} \tag{5.5}$$

where $\{b_n\}$ are real and non-negative, $\{a_n\}$ are real, $\mathbf{u}_{-1}$ is zero, and $\mathbf{u}_0$ is $\mathbf{u}$. This recurrence is equivalent to representation of S as a tridiagonal matrix J with diagonal elements $\{a_n\}$, elements $\{b_n\}$ on the two main sub-diagonals. In the above recurrence, the basis of this matrix has been taken to be the polynomial states $\{\mathbf{u}_n\}$.

The above three term recurrence defines a set of orthogonal polynomials in a very general way. For an argument X for which multiplication is defined, and for the initial conditions that $P_{-1}(X)$ is zero and $P_0(X)$ is a non-zero object which can be repeatedly multiplied by X, the recurrence

$$P_{n+1}(X) = [(X - a_n)P_n(X) - b_n P_{n-1}(X)]/b_{n+1}, \tag{5.6}$$

produces the set of orthonormal polynomials $\{P_n(X)\}$. This recurrence can be solved directly, and

$$P_{n+1}(X) = det|XI - J|/(b_0 b_1 \ldots b_{n+1}), \tag{5.7}$$

where I is the identity matrix, and J is the tridiagonal matrix of recurrence coefficients.

The weight distribution $d\mu$ can be expressed directly in terms of the recurrence coefficients by noting that

$$\mu_n = \frac{1}{2\pi i} \int s^n \mathbf{u}^*(s - S)^{-1}\mathbf{u}\,ds \tag{5.8}$$

where the integral is taken around a contour which encloses all the singularities of the integrand. Thus the distribution $d\mu$ is the imaginary part of $\mathbf{u}^*(s - S)^{-1}\mathbf{u}$, which can be expressed in terms of determinants and cofactors of the matrix $sI - J$ to give the continued fraction

$$\mathbf{u}^*(s - S)^{-1}\mathbf{u} = \cfrac{1}{s - a_0 - \cfrac{b_1^2}{s - a_1 - \cfrac{b_2^2}{s - a_2 - \cdots}}}. \tag{5.9}$$

The three term recurrence J has two linearly independent solutions for each value of the argument X. The polynomials defined above are often called the regular polynomial solutions, and there is a set of irregular polynomial solutions $\{Q_n(X)\}$ which satisfy the initial conditions that Q_0 is zero, and Q_1 is unity. These can be used to express the above continued fraction as the limit of a ratio of the two polynomials

$$\mathbf{u}^*(s - S)^{-1}\mathbf{u} = \lim_{n \to \infty} Q_n(s)/[b_0 P_n(s)] \tag{5.10}$$

where P_0 is unity.

As a consequence of the properties of the polynomials, the states

$$\mathbf{s} = \sum P_n(s)\mathbf{u}_n \tag{5.11}$$

satisfy the stationary condition

$$S\mathbf{s} = \mathbf{s}s \tag{5.12}$$

as can be seen by direct substitution of (5.11) in (5.12), and application of (5.5) and its analogue for the polynomials in s. Although these states are defined for all values of s, the physical stationary states are only those for which $\mathbf{s}$ has real, finite expectation values for all observables. In most cases the requirement of finite expectation values for $\mathbf{s}$ implies that the orthogonal polynomials $\{P_n(s)\}$ are bounded. This in turn implies that $d\mu$ is positive at s.

Because of its conservation, the most important observables for any system is the energy H which is

$$H = -\mathbf{P}^2/(2m) + V(\mathbf{r}) \tag{5.13}$$

in wave mechanics where $\mathbf{P}$ is the momentum observable of a particle of mass m moving in a potential $V(\mathbf{r})$. The equation for the stationary states Ψ and their stationary energies E is called the time independent Schroedinger equation

$$H\Psi = E\Psi. \tag{5.14}$$

Supposing that $\mathbf{u}$ is stable, possessing an infinite set of energy moments, then the unnormalized stationary states of H which overlap $\mathbf{u}$ are

$$\Psi E = \sum P_n(E)\mathbf{u}_n. \tag{5.15}$$

As an illustration, the polynomials can be used to express the time development of a state which is $\mathbf{u}$ at time $t = 0$. In wave mechanics this is

$$\mathbf{u}(t) = e^{-iHt/\hbar}\mathbf{u} \tag{5.16}$$

where $\hbar$ is again Planck's constant over 2π. In terms of the energy polynomials

$$\mathbf{u}(t) = \sum \Pi_n(t)\mathbf{u}_n \tag{5.17}$$

where $\{\Pi_n(t)\}$ are the Fourier transforms of the polynomials with respect to the measure

$$\Pi_n(t) = \int P_n(E)e^{-iEt/\hbar}d\mu. \tag{5.18}$$

In a similar way, any function of an observable applied to $\mathbf{u}$ can be expressed as a sum over polynomials with coefficients which are integrals of the function with respect to the measure over which they are orthogonal. For example, the resolvent of an observable is useful for calculating the distribution of stationary values. In the case of energy, this resolvent is

$$(\lambda - H)^{-1}\mathbf{u} = \sum R_n(\lambda)\mathbf{u}_n \tag{5.19}$$

where

$$R_n(\lambda) = \int P_n(E)(\lambda - E)^{-1}d\mu, \tag{5.20}$$

provided λ lies outside the support of the measure μ. In particular, $R_0(\lambda)$ is just the continued fraction whose weight distribution is $d\mu$.

6. Polynomial Models

Reversing the point of view of previous sections, every set of orthogonal polynomials $\{P_n(x)\}$, whose measure $\mu(x)$ and recurrence relation J is known, make up an exact quantum mechanical model for a stationary subspace of some observable. In the polynomial basis the stationary state with stationary value x is $\{P_0(x), P_1(x), \dots\}$, and if $d\mu/dx$ is positive then the components of the state are bounded.

An example of such a model is the set of Laguerre polynomials which are orthogonal with respect to a weight distribution,

$$d\mu(x)/dx = x^{1/2}e^{-x}, \qquad x > 0. \tag{6.1}$$

This is also the form of weight distribution of energy for a Gaussian initial state

$$\mathbf{u}(\mathbf{r}) = \exp(-\mathbf{r} \cdot \mathbf{r}), \tag{6.2}$$

and the isotropic oscillator Hamiltonian

$$H = -\nabla^2 + \frac{1}{2}\kappa\mathbf{r} \cdot \mathbf{r}. \tag{6.3}$$

There are many other examples of polynomial models for which a physical interpretation is not so simple.

7. Perturbation Theory

In physics there only approximate problems because the physical models are only approximate. However, the models with exact solutions such as those provided by orthogonal polynomial sets, can be used as starting points for approximations of which one of the most useful is perturbation theory.

Perturbation theory within the recursion method begins with a Hamiltonian T for which the exact set of polynomials generated from u is known, and to which a potential V is added to give a total Hamiltonian

$$H = T + \lambda V \tag{7.1}$$

for a coupling constant λ. Provided H possesses a full set moments with respect to u in some domain of λ, the smallest stationary subspace containing u varies with λ and is spanned by the powers of H on u.

An unnormalized or non-symmetric form of the orthogonal polynomial in (4.1) is

$$\Delta_n(E, \lambda) = [D_{n-1} D_{n-2}]^{1/2} P_n(E) \tag{7.2}$$

which is a determinant of moments and thus is a polynomial of degree n in E and of degree $n^2 - n$ in λ. Because they are polynomials, each $\Delta_n(E, \lambda)$ converges as a power series in λ over the entire domain of the λ-plane for which the moments are finite. It will be shown in the next section that physical quantities can be expressed as ratios of these unnormalized polynomials in the limit as n goes to infinity. The convergence of the power series in λ for these quantities then depends on the location of zeros of $\Delta_n(E, \lambda)$ for complex λ, and thus on the relative strength of T and V in the subspace generated by u.

This perturbation method has been applied to the problem of electron localization in random potentials [7]. The Hamiltonian is that of (5.13) with a potential $V(\mathbf{r})$ which is taken from a distribution of random potentials. In the limit of weak disorder (the potential varies over a range of energies much narrower than that of kinetic energy), the recurrence relation is expanded in powers of the random potential. This produces a distribution of recurrences some of whose stationary states are current carrying and some not. When the strength of the potential is increased so that it has the extremal accumulation point, the stationary states change character and are strongly localized with probability one.

8. Expectation Values

Next to the determination of stationary states and stationary values for observables, the most important quantum mechanical calculation is that of expectation values of one observable on the stationary states of another [8]. This can be done by application of the above perturbation theory where the expectation values of V are to be calculated for the stationary states of T.

In the stationary subspace generated by H applied to u

$$\det|E - H| = \det|EI - J| = \lim_{n \to \infty} \Delta_n(E)/D_{n-1}. \tag{8.1}$$

The logarithmic derivative of this with respect to λ gives

$$\frac{d \log(\det|E - H|)}{d\lambda}\bigg|_{\lambda=0} = tr[V(E - T)^{-1}] \tag{8.2}$$

where the trace is over the stationary subspace. Thus the measure ν of the expectation values V has a distribution $d\nu/dE$ given by the imaginary part of the above trace

$$\frac{d\nu}{dE} = \frac{-1}{\pi} Im \left\{ \lim_{n \to \infty} \left[\frac{\Delta'_n(E)}{\Delta_n(E)} - \frac{D'_{n-1}}{D_{n-1}} \right] \right\} \bigg|_{\lambda=0} \tag{8.3}$$

where the the 'primed' quantities are differentiated with respect to λ.

The weight distribution over which the polynomials are orthogonal is a special case of the above relationship when V is taken to project on $\mathbf{u}$. The expectation value of V is then the magnitude squared of its projection on $\mathbf{u}$, and the distribution of expectation values is the projected density of states.

9. Numerical Methods

In contrast to the exploration of the qualitative behavior of simple physical models, the quantitative comparison of physical theory with experiment involves many numerical parameters. This usually makes analytic methods of calculation impractical, and hence numerical implementations of the recursion method have been developed.

Numerical application of the recursion method is not simply a matter of programming the equations for orthonormalizing the polynomials obtained by applying powers of an observable to some initial state. Careful thought must be given to three important limitations of the computer: the rounding error in floating point arithmetic, the different kinds of storage available, and the time taken by each operation. The effect of these limitations depends on the choice of representation for the states and observables of the system.

The best strategy for numerical work is to generate a symmetric three term recurrence for the observable, rather than evaluating the moments and determinants which are so useful for the analytic approach. This is because direct evaluation of an $n \times n$ determinant requires n-factorial operations, and is numerically unstable because the moments diverge. The symmetric three term recurrence is a direct representation of the polynomials and does not diverge.

When quantum system has a finite number of degrees of freedom, the numerical methods are straightforward, however, in most interesting system, the observables are infinite dimensional. In the latter case, it is crucial to represent the observable as a matrix in terms of stable or sparse basis states, because then the inner product of some row or column of the Hamiltonian with a normalized vector is not only finite, but convergent so that the result can be calculated to a given accuracy in a finite number of operations.

A stable representation is one in which the observable can be expressed as an infinite matrix whose rows and columns contain only finite numbers of numerically significant elements, a sparse matrix. Since the states accessible form a gives stable initial state $\mathbf{u}$ are spanned by the powers of the observable H on that initial state, a countable basis is always sufficient.

Construction of the symmetric recurrence uses the polynomial states

$$\mathbf{u}_n = P_n(H)\mathbf{u} \tag{9.1}$$

by evaluating

$$\mathbf{u}_1 = (H - a_0)\mathbf{u}_0/b_1 \tag{9.2}$$

where a_0 is chosen to orthogonalize $\mathbf{u}_1$ to $\mathbf{u}_0$, and b_1 is chosen to normalize $\mathbf{u}_1$. Because H and $\mathbf{u}_0$ have sparse representation, then so does $\mathbf{u}_1$, and because of rounding error, the components on two sides of the above equation only agree to within rounding error relative to the larger of the two magnitudes.

In the general step of the recursion

$$\mathbf{u}_{n+1} = [(H - a_n)\mathbf{u}_n - b_n\mathbf{u}_{n-1}]/b_{n+1} \qquad (9.3)$$

where a_n is chosen to make $\mathbf{u}_{n+1}$ orthogonal to $H\mathbf{u}_n - b_n\mathbf{u}_{n-1}$, and b_{n+1} normalizes $\mathbf{u}_{n+1}$. Again, $\mathbf{u}_{n+1}$ is sparse because H, $\mathbf{u}_n$, and $\mathbf{u}_{n-1}$ are sparse, and the components on the two sides of the equation agree to within rounding error.

Rounding error determines the significance of matrix elements and vector components as well as the accuracy of the three term recurrence. In most computer, the limitation on the storage is more severe than the rounding error in the arithmetics. As a result, the accuracy of the calculation is determined by the least significant components which can be stored for the polynomial vectors. For example, if there are N storage locations for the components of a normalized vector, then the maximum error produced by storing the N largest components is $(N + 1)^{-1/2}$. This also becomes the accuracy of the recurrence relation as long as the rounding error in arithmetic operations is smaller.

The effect of precision on the kind of recursion used in this method has been the subject of much study [9]. Errors do accumulate in the $\{\mathbf{u}_n\}$ so that for n greater than some value, these states and hence the recursion parameters are determined entirely by error. However, the surprising property of method [10] is that expectation values retain the accuracy of each recurrence, and the stationary states retain the root of this accuracy, despite the instability of the polynomial states. This is because expectation values and the stationary states themselves really depend on the relationships between states in the recurrence, which are preserved in the approximate calculation, and not on the particular states in the recurrence.

Acknowledgement

This work was supported by the National Science Foundation Condensed Matter Theory Grant DMR 8712346. The author is grateful for the hospitality of the Cavendish Laboratory and Pembroke College, Cambridge.

References

1. P. A. M. Dirac, "The Principles of Quantum Mechanics," Oxford University Press, 1947.
2. F. Cyrot-Lackmann, J. Phys. (Paris), Suppl. **C1**, 67(1970).
3. J. A. Shohat and J.D. Tamarkin, "The Problem of Moments," Math. Surv. I, rev. ed., Am. Math. Soc., Providence Rhode Island, 1950.
4. R. Haydock, Solid State Physics **35**, Academic Press, New York, 215(1980).
5. C. Lanczos, J. Res. Nat. Bur. Stand. **45**, 255(1950).
6. T. S. Chihara, "An introduction to Orthogonal Polynomials," Gordon and Breach, New York, 1978.
7. R. Haydock, Phi. Mag. **53**, 545(1986).
8. W. M. C. Foulkes and R. Haydock, J. Phys. C. **19**, 6573(1986).
9. B. N. Parlett, "The Symmetric Eigenvalue Problem," Prentice Hall, Englewood Cliffs, N. J., 1980.
10. R. Haydock, Comp. Phys. Comm. **53**, 133(1989).

Birth and Death Processes and Orthogonal Polynomials

Mourad E. H. Ismail[*]
Department of Mathematics
University of South Florida
Tampa, Florida, 33620.

Jean Letessier
LPTHE[**], Université Paris 7
Tour 24-5e étage, 2 Place Jussieu,
75251 Paris Cedex 05, France.

David R. Masson[†]
Department of Mathematics,
University of Toronto, Toronto
Ontario, Canada, M5S 1A1.

Galliano Valent
LPTHE[**], Université Paris 7
Tour 24-5e étage, 2 Place Jussieu,
75251 Paris Cedex 05,France.

Abstract. We discuss connections between birth and death processes and orthogonal polynomials. We survey certain spectral properties of orthogonal polynomials and spectra of birth and death processes. Birth and death processes with linear birth and death rates are studied in some detail. We also mention some results concerning birth and death processes whose transition rates are rational functions.

1. Introduction. The purpose of this paper is to introduce mathematicians interested in orthogonal polynomials to certain applications of the subject. Our presentation uses minimal knowledge of areas outside the theory of orthogonal polynomials.

Birth and death processes (BDP's) are special stationary Markov processes whose state space is the nonnegative integers. It is useful to think of the states of a BDP as populations. The transition probabilities $\{p_{mn}(t)\}$ are defined by

(1.1) $p_{mn}(t) :=$ Probability that the system goes from state m to state n in
              time t.

The transition matrix P(t) is the matrix whose elements are $p_{mn}(t)$,

(1.2) $P(t) = (\, p_{mn}(t)\,)$, m, n = 0, 1, 2, ...

---

* Research partially supported by a grant from NSF of the U.S.A.
** Laboratoire de Physique Théorique et Hautes Energies, a laboratory affiliated with CNRS, France.
† Research partially supported by a grant from NSERC of Canada.

P. Nevai (ed.), Orthogonal Polynomials, 229–255.
© *1990 by Kluwer Academic Publishers.*

Since the process is stationary $p_{mn}(t)$ does not depend on how the system reached state m but depends only on m, n, and the time t lapsed in moving from state m to state n. This is equivalent to

(1.3) $P(s+t) = P(s)P(t)$.

For a BDP we also demand that as $t \to 0^+$ the transition probabilities obey

(1.4) $p_{mn}(t) = \begin{cases} \lambda_m t + o(t) & n = m + 1, \\ \mu_m t + o(t), & n = m - 1, \\ 1 - \lambda_m t - \mu_m t + o(t), & n = m, \end{cases}$

and $p_{mn}(t) = o(t)$ if $|m - n| > 1$. The coefficients λ_m and μ_m are birth and death rates at state m. They are assumed to satisfy

(1.5) $\lambda_m > 0$, $\mu_{m+1} > 0$, for $m \geq 0$ and $\mu_0 \geq 0$.

The above are all the assumptions describing a BDP. Assumptions (1.4) lead to a systems of differential recurrence relations called the Chapman-Kolmogorov (or C-K) equations. The forward C-K equations are

(1.6) $\dfrac{d}{dt} P_{mn}(t) = \lambda_{n-1} P_{m,n-1}(t) + \mu_{n+1} P_{m,n+1}(t) - (\lambda_n + \mu_n) P_{m,n}(t)$,

and the backward equations are

(1.7) $\dfrac{d}{dt} P_{mn}(t) = \lambda_m P_{m+1,n}(t) + \mu_m P_{m+1,n}(t) - (\lambda_m + \mu_m) P_{m,n}(t)$.

A derivation of the C-K equations will be given in Section 2. In Section 2 we will also discuss the Karlin-McGregor solution of the C-K equations.

Let A be the tridiagonal (or Jacobi) matrix

(1.8) $A = (a_{m,n}: m \geq 0, n \geq 0)$, $a_{n,n} = -\lambda_n - \mu_n$, $a_{n,n+1} = \lambda_n$, $a_{n+1,n} = \mu_{n+1}$.

We now list the properties of BDP's for future reference

(1.9)
  I $P'(t) = P(t)A$,   II $P'(t) = AP(t)$,
  III $P(0) = I = (\delta_{nm})$,   IV $P_{mn}(t) \geq 0$,
  V $\Sigma_{n \geq 0} P_{mn}(t) \leq 1$, $m \geq 0$, $t \geq 0$,   VI $P(s+t) = P(s)P(t)$.

Note that I and II are the C-K equations and that A is the infinitesimal generator of the process.

In Section 3 we treat the general BDP with linear transition rates λ_n and μ_n when μ_0 may not follow the same pattern of the rest of the μ_n's. In Section 4 we survey three applications of birth and death processes. They are the Schlögl model of Chemistry, the Moran and Robertson models of Genetics, and a model for the time evolution of two valued interacting systems in Statistical Mechanics. In Section 5 we include some qualitative properties concerning the spectra of birth and death processes. In Section 6 we discuss very briefly the bilateral BDP's, these are BDP's whose state space is all the integers, [33]. We use the theory of linear operators on Hilbert spaces to construct integral representations for the transition probabilities. Some of the material in Section 6 seems to be new.

There are interesting BDP's with $\lambda_0 = 0$ but the analysis in this becomes more delicate and will not be disscused in this paper.

This paper is only a brief introduction to the subject. Due to time and space limitations we made no attempt to write a complete treatment of the subject, nor did we intend our bibliography to be complete. For example the very interesting papers of E. Van Doorn are neither discussed in the text nor are cited in the bibliography. We hope to write a more comprehensive survey in the near future.

2. The Chapman-Kolmogorov Equations.

We start this section by deriving the C-K equations.

Proofs of (1.6) and (1.7). We compute $p_{mn}(t+\delta t)$ in two different ways then let $\delta t \to 0^+$. The system can go from state m to state n in time increments t and δt or in total time $t+\delta t$. From (1.3) we find $P(t+\delta t) = P(t) P(\delta t)$. Thus we have

$$p_{mn}(t+\delta t) = p_{m,n-1}(t) (\lambda_{n-1}\delta t) + p_{m,n+1}(t) (\mu_{n+1}\delta t) + p_{mn}(t) (1 - (\lambda_n + \mu_n)\delta t) + o(\delta t).$$

We now subtract $p_{mn}(t)$ from both sides of the above equality then divide by δt. The result establishes (1.6). We similarly prove (1.7) by applying the semigroup property $P(t+\delta t) = P(\delta t) P(t)$.

The rest of the proof is similar and the interested reader can easily fill in the details.

We next show how to guess the general solution of the C-K equations and its connection with the Stieltjes moment problem. The idea is to apply the method of separation of variables. We set

(2.1) $\qquad p_{mn}(t) = f(t) Q_m F_n$

in (1.6). Since Q_m cannot vanish identically then

(2.2) $\qquad f'(t)/f(t) = [\lambda_{n-1}F_{n-1} + \mu_{n+1}F_{n+1} - (\lambda_n + \mu_n) F_n]/F_n = -x,$

say. Therefore as a function of t

(2.3) $\qquad f(t) = e^{-xt},$

up to a multiplicative constant. The F_n's satisfy

(2.4) $\qquad F_{-1}(x) := 0, -xF_n(x) = \mu_{n+1}F_{n+1}(x) + \lambda_{n-1}F_{n-1}(x) - (\lambda_n + \mu_n) F_n(x), n \geq 0.$

We wrote $F_n(x)$ instead of F_n to exhibit the dependence of F_n on the constant x. It is clear that $F_0(x)$ is arbitrary. Thus up to a multiplicative constant, $F_n(x)$ is generated by (2.4) and the initial conditions

(2.5) $\qquad F_{-1}(x) = 0, F_0(x) = 1$

in a unique way. Next substitute the solution (2.1) in the backward C-K equations (1.7) and use (2.2) to see that that the Q_n's $(= \{Q_n(x)\})$ are generated by the initial conditions

(2.6) $Q_{-1}(x) = 0, Q_0(x) = 1$,

and the recurrence relation

(2.7) $-xQ_n(x) = \lambda_n Q_{n+1}(x) + \mu_n Q_{n-1}(x) - (\lambda_n + \mu_n) Q_n(x)$, $n \geq 0$,

up to a multiplicative function independent of n. It follows from (2.5)-(2.7) that the polynomials $\{(\lambda_0 \lambda_1 ... \lambda_{n-1}) Q_n(x)/(\mu_1 \mu_2 ... \mu_n)\}$ satisfy the same initial conditions and recurrence relation that generate $\{F_n(x)\}$. Thus

(2.8) $F_n(x) = \pi_n Q_n(x)$,

with

(2.9) $\pi_0 := 1, \pi_n := (\lambda_0 \lambda_1 ... \lambda_{n-1})/(\mu_1 \mu_2 ... \mu_n)$, $n > 0$.

Thus we have shown that the separation of variables gives

(2.10) $e^{-xt} F_m(x) F_n(x)/\pi_m$

for the right hand side of (2.1). The separation constants, which depend only on x, may represent $p_{mn}(t)$ as a sum or an integral of the terms in (2.10) multiplied by separation constants. To incorporate the cases of continuous and discrete x's we write

(2.11) $p_{mn}(t) = (1/\pi_m) \int_0^\infty e^{-xt} F_m(x) F_n(x) \, d\mu(x)$,

for some measure $d\mu$. The x's in (2.10) are nonnegative because $p_{mn}(t)$ must remain bounded as $t \to \infty$.

As $t \to 0^+$ we know that $p_{mn}(t) \to \delta_{mn}$, by (1.4). Therefore

(2.12) $\pi_m \delta_{mn} = \int_0^\infty F_m(x) F_n(x) \, d\mu(x)$,

that is, the $F_n(x)$'s are orthogonal with respect to $d\mu$.

To see that $d\mu$ is a positive measure we truncate the problem and consider a subprocess on the state space $\{0, 1, ..., N - 1\}$, then let $N \to \infty$. Let A_N and $P_N(t)$ be the principal NxN minor of A and P(t), respectively. The solution of the truncated problem is

(2.13) $P_N(t) = \exp(tA_N) = \sum_{j=0}^\infty t^j A_N^j / j!$.

From (2.4) and (2.5) one can prove that the zeros of $F_N(x)$ are real and simple, [8]. The zeros of $F_N(x)$ are also positive as we shall see below. Let

(2.14) $F_N(\xi_j) = 0, j = 1, 2, ..., N$.

From (2.4) we find that $\mathcal{F}_j := (F_0(\xi_j), F_1(\xi_j), ..., F_{N-1}(\xi_j))$ is an eigenvector of $(A_N)^T$, the transpose of A_N, with eigenvalue $- \xi_j$, for $j = 1, ..., N$. Let F be the matrix formed by the row vectors $\mathcal{F}_1, \mathcal{F}_2, ..., \mathcal{F}_N$ and D be the diagonal matrix $(- \xi_m \delta_{mn})$, $m, n = 1, ..., N$. Thus $A_N = F^{-1} D F$ and

(2.15) $P_N(t) = F^{-1} \exp(tD) F$.

The Christoffel-Darboux formula

$$\sum_{0}^{N-1} [F_n(x)]^2 / \pi_n = [F_N(x) F'_{N-1}(x) - F'_N(x) F_{N-1}(x)] / \pi_N$$

shows that F^{-1} is the matrix whose columns are

$$- \pi_N (F_0(\xi_j)/\pi_0, ..., F_{N-1}(\xi_j)/\pi_{N-1}) / F'_N(\xi_j) F_{N-1}(\xi_j), j = 1, 2, ..., N.$$

Now a simple calculation using (2.15) and the above relationships establishes the integral representation

(2.16) $P_{mn}(t) = (1/\pi_m) \displaystyle\int_{-\infty}^{\infty} e^{-xt} F_m(x) F_n(x) d\mu_N(x), m, n = 0, 1, ..., N-1,$

where $\mu_N(x)$ is a step function with jumps $-\pi_N / [F'_N(\xi_j) F_{N-1}(\xi_j)]$ at $x = \xi_j$, $1 \leq j \leq N$.
It only remains to show that the zeros of $F_N(x)$ are positive. Let
   $r_n(x) = F_n(x)/\sqrt{\pi_n}, n = 0, 1, ... N-1$.
It is straight forward to see that

$$-x r_n(x) = \sqrt{\lambda_n \mu_{n+1}} \, r_{n+1}(x) - (\lambda_n + \mu_n) r_n(x) + \sqrt{\lambda_{n-1} \mu_n} \, r_{n-1}(x),$$

which shows that $\{- \xi_j : j = 1, ..., N\}$ must be the eigenvalues of the symmetric tridiagonal matrix R_N whose m,n entry is

   $- (\lambda_n + \mu_n) \delta_{m,n} + (\lambda_m \mu_{m+1})^{1/2} \delta_{m+1,n} + (\lambda_m \mu_{m+1})^{1/2} \delta_{m,n+1}$, $m, n = 0, ..., N-1$.
It now suffices to show that R_N is negative definite. The associated quadratic form is

$$- \sum_{n=0}^{N-1} (\lambda_n + \mu_n) x_n^2 + 2 \sum_{n=0}^{N-2} \sqrt{\lambda_n \mu_{n+1}} \, x_n x_{n+1}$$

which is

$$- \mu_0 x_0^2 - \lambda_{N-1} x_{N-1}^2 - \sum_{n=0}^{N-2} [\sqrt{\lambda_n} \, x_n - \sqrt{\mu_{n+1}} \, x_{n+1}]^2.$$

Thus the quadratic form is negative definite and (2.16) now takes the following form

(2.16 -a) $p_{mn}(t) = (1/\pi_m) \displaystyle\int_0^\infty e^{-xt} F_m(x) F_n(x)\, d\mu_N(x)$,   m, n = 0, 1, ..., N -1,

A subsequence of $\{\mu_N(x)\}$ converges to a nondecreasing function $\mu(x)$ with infinitely many points of increase because the variations of $\{\mu_N(x)\}$ are uniformly bounded. Furthermore all the moments of $d\mu$ are finite. It is then plausible that as $N \to \infty$, (2.16) will tend to (2.11).

A measure $d\mu$ satisfying (2.11) is called a spectral measure of the given BDP. It is then clear that every measure the $F_n(x)$'s are orthogonal with respect to is a spectral measure and vice versa.

A fundamental problem in this subject is to ascertain whether the C-K equations have a unique solution. It can be shown that, in general I - III do not have unique solution. However, I - VI may have a unique solution. When the Stieltjes moment problem associated with the polynomials $\{F_n(x)\}$ has a unique solution the p_{mn}'s are uniquely determined from (2.11). In the cases when $d\mu$ is unique it becomes important to be able to find $d\mu$ explicitly from the knowledge of the transition rates λ_n and μ_n. We shall address this latter problem in Section 3 for linear processes and in Section 5 for general processes and general orthogonal polynomials. If however, the p_{mn}'s are not unique one would like to characterize "honest" solutions to the C-K equations, [19], [34], that is solutions $\{p_{mn}(t)\}$ satisfying

(2.17) $\Sigma_{n\geq 0}\, p_{mn}(t) = 1$,   for all $t \geq 0$, and all $m \geq 0$.

Let

$$a_1 := 1,\quad a_{2n+1} := \prod_{j=1}^{n} [\lambda_{j-1}/\mu_j],\quad a_{2n} = 1/[\mu_n\, a_{2n+1}],$$

and define numbers S, B, C in the extended real number system by

$$S := \sum_{n=1}^{\infty} a_n\ ,\quad B := \sum_{n=1}^{\infty} a_{2n} \sum_{k=0}^{n-1} a_{2k+1}\ ,\quad C := \sum_{n=1}^{\infty} a_{2n+1} \sum_{k=1}^{n} a_{2k}.$$

It can be shown that the p_{mn}'s satisfying I - VI are unique if and only if S = ∞. On the other hand a solution of the C-K equation is honest if B = ∞, [19], [34]. If B < ∞ but C = ∞ then all solutions of the C-K equations are not honest, [34]. Finally when S < ∞ all solutions, with possibly one exception, are not honest.

We conclude this section by mentioning a slight modification of the BDP's known as processes with "killing" or "absorption", [21], [22]. Such a process has an additional option, namely to be absorbed at state n with rate $\gamma_n \geq 0$. Thus (1.4) is replaced by

$$(2.18) \quad p_{mn}(t) = \begin{cases} \lambda_m t + o(t) & n = m + 1, \\ \mu_m t + o(t), & n = m - 1, \\ 1 - (\lambda_m + \mu_m + \gamma_m)t + o(t), & n = m, \end{cases}$$

and $p_{mn}(t) = o(t)$ if $|m-n| > 1$ as $t \to 0^+$. The representation (2.11) remains valid with $F_n(x)$'s generated by

$$(2.19) \quad F_0(x) = 1, F_1(x) = (\lambda_0 + \mu_0 + \gamma_0 - x)/\mu_1,$$

$$-xF_n(x) = \mu_{n+1}F_{n+1}(x) + \lambda_{n-1}F_{n-1}(x) - (\lambda_n + \mu_n + \gamma_n)F_n(x), n > 0.$$

3. Linear Processes.
The simplest cases of BDP's are the linear cases

$$(3.1) \quad \lambda_n = a(n + c + \alpha + 1), \mu_{n+1} = n + c + 1, n \geq 0, \mu_0 = 0 \text{ or } c.$$

Surprisingly the most general linear cases were not completely analyzed until recently [5], [13]. When $c = 0$ the F_n's of (2.4) and (2.5) are Laguerre polynomials when $a = 1$, [8, p. 44] or Meixner polynomials when $a \neq 1$, [8, p. 177]. When $a = 1$, $c > 0$, the F_n's are associated Laguerre polynomials. The case $\mu_0 = c$ is in [5], but [13] treats the two cases $\mu_0 = c$ and $\mu_0 = 0$. The general cases (3.1) when $a \neq 1$, $c > 0$ were treated in [13]. In 1941 Hahn [12] studied the associated Laguerre polynomials when $\mu_0 = c$ and found an explicit representation for them in terms of Whittaker functions. He also showed that they satisfy a fourth order differential equation with polynomial coefficients and gave them explicitly. He did not attempt to find their weight function. The associated Hermite polynomials are in [5], [13] and [26].

One may think that a linear BDP with linear killing rates may lead to new orthogonal polynomials. This turned out not to be true. Chihara [8, §VI.3], using a different notation and motivation, proved that we always get either the Laguerre or Meixner polynomials or the corresponding associated polynomials.

When μ_n is a given function of n (e. g. a polynomial or a rational function) we always have two natural choices for μ_0. The first choice is to let μ_0 follow the same pattern of the rest of the μ_n's. The second choice is to define μ_0 as zero. These two choices may coincide in some cases but they are usually different. This point of view seems to have appeared first in [13]. Two families of associated continuous dual Hahn polynomials are in [14]. Later it was used in [15] and [18] to study two families of Associated Wilson polynomials and associated Askey-Wilson polynomials, respectively. Masson [27] gave an alternate derivation of the measure of orthogonality of the associated Wilson polynomials. A family of associated Jacobi polynomials was introduced by Wimp [45] and a second family with $\mu_0 = 0$ was introduced in [17]. Chihara [6] studied the effect of changing initial conditions on the spectrum of a family of orthogonal polynomials. Redefining μ_0 as 0 is a special change of the initial conditions but, in the context of BDP's, it seems to be a very natural alternate choice.

We now show how to find the spectral measure of the general linear process (3.1) when a = 1. Consider the generating function

$$(3.2) \qquad F(x,w) := \sum_{n=0}^{\infty} w^n F_n(x).$$

The recurrence relation (2.4) in the present case ((3.1)) leads to the following differential equation for the generating function

$$(3.3) \qquad w(1-w)^2 \frac{\partial F}{\partial w} + [(1-w)\{c - (c + \alpha +1)w)\} + xw]F = c(1-w)^{\eta},$$

where

(3.4) $\qquad \eta = 0$ if $\mu_0 = c$ $\qquad$ and $\qquad \eta = 1$ if $\mu_0 = 0$.

The solution of the differential equation (3.3) which is bounded at w = 0 is given by

$$(3.5) \qquad F(x,w) = cw^{-c}(1-w)^{-\alpha-1} \int_0^w u^{c-1}(1-u)^{\eta+\alpha-1} \exp\left(\frac{xu}{1-u} - \frac{xw}{1-w}\right) du.$$

Due to the orthogonality of the $F_n(x)$'s, see (2.12), the boundary condition to be satisfied is

$$(3.6) \qquad \int_0^{\infty} F(x,w) \, d\mu(x) = 1.$$

If we change variables in (3.5) according to
(3.7) $\qquad z := w/(1 - w), \quad v = u/(1 - u),$
then the integral representation (3.5) will become

$$(3.8) \qquad F(x, z/(1 + z)) = cz^{-c}(1+z)^{c+\alpha+1} \int_0^z v^{c-1}(1+v)^{-\alpha-c-\eta} e^{x(v-z)} dv.$$

We then integrate both sides of (3.8) with respect to $d\mu(x)$ and take the boundary condition (3.6) into account. The result is

$$(3.9) \qquad z^c(1+z)^{-c-\alpha-1} = c \int_0^{\infty} \left\{ \int_0^z v^{c-1}(1+v)^{-\alpha-c-\eta} e^{-x(z-v)} dv \right\} d\mu(x).$$

Observe that the inner integral on the right hand side is a convolution of two functions. The integral transform associated with this convolution is the Laplace transform, so we apply the Laplace transform to both sides of (3.9). The result is

$$(3.10) \quad \int_0^\infty \frac{d\mu(x; \eta)}{x + p} = \psi(c + 1, 1 - \alpha \; ; p) / \psi(c, 1 - \alpha - \eta \; ; p), \eta = 0, 1,$$

where $\psi(a, c; x)$ is the Tricomi ψ function

$$(3.11) \quad \psi(a, c; x) = \frac{1}{\Gamma(a)} \int_0^\infty e^{-xt} t^{a-1} (1 + t)^{c-a-1} dt, \; \mathrm{Re}\, a > 0, \; -\frac{\pi}{2} < \arg x < \frac{\pi}{2},$$

[11, vol 1, § 6.5] and its analytic continuations in the complex plane.

One can combine the Perron-Stieltjes inversion formula for the Stieltjes transform, namely

$$(3.12) \quad F(p) = \int_0^\infty \frac{d\mu(x)}{x + p} \text{ if and only if } \mu(x) - \mu(y) = \lim_{\varepsilon \to 0^+} \int_y^x \frac{F(-t - i\varepsilon) - F(-t + i\varepsilon)}{2\pi i} dt$$

with technical results from special function theory to prove that $d\mu(x; \eta)$ is absolutely continuous when $c \geq 0$ and $\alpha > -1$. Furthermore the derivative of $d\mu$ is given by

$$(3.13) \quad \frac{d}{dx} \mu(x; \eta) = \frac{x^\alpha e^{-x}}{\Gamma(c + 1)\Gamma(c + \alpha + 1)} |\psi(c, 1 - \alpha - \eta; xe^{-i\pi})|^{-2}.$$

For details the interested reader may consult [5] and [13]. The polynomials orthogonal with respect to $d\mu(x; \eta)$ are studied in some details in [5] when , $\eta = 0$ and in [13] when $\eta = 0$ or $\eta = 1$.

When $a \neq 1$ in (3.1) one obtains the associated Meixner polynomials. They are orthogonal with respect to a purely discrete measure supported at the zeros of a product of a certain hypergeometric function times a quotient of gamma functions, see [13].

Letessier and Valent used the approach outlined in this section in [23] and [24] to treat certain quadratic asymptotically symmetric BDP's.

4. Some Applications of Birth and Death Processes.

Birth and death processes have applications to a variety of fields including nuclear physics, spin gas, chemical reactions, population dynamics, genetic models, to name a few. We shall illustrate some of these applications by considering the following three examples involving stochastic processes.

1. The Schlögl model in chemistry which is concerned with a problem in bifurcation theory.

2. The Moran model in genetics which describes the fluctuation of gene frequency under the influence of mutation and selection.

3. The time evolution of a system of two valued interaction elements.

4.1. **The Schlögl model**. This example [38] was analyzed by G. Nicolis and J. W. Turner [31] and illustrates the importance of fluctuations and long range correlations in problems involving cooperative behavior associated with bifurcation phenomena. This problem deals with evolution equations having a small number of observables and has many applications in Chemistry, Physics, and Biology.

Consider a chemical reaction between three compounds, A, B, and X. The number of molecules of A and B are N_A and N_B respectively, and kept constants by external controls. The number of molecules of X is n and changes with time. One molecule of A and two molecules of X produce three molecules of X at rate k_1 while 3 molecules of X produce 1 molecule of A and two molecules of B at rate k_2. Furthermore 1 molecule of X produces 1 molecule of B and 1 molecule of B produces 1 molecule of X at rates k_3 and k_4 respectively. Diagrammatically this is

$$(4.1) \quad A + 2X \; \underset{k_2}{\overset{k_1}{\rightleftarrows}} \; 3X \quad \text{and} \quad X \; \underset{k_4}{\overset{k_3}{\rightleftarrows}} \; B.$$

The differential equation describing (4.1) is

$$(4.2) \quad \frac{dn}{dt} = k_1 N_A n(n-1) - k_2 n(n-1)(n-2) - k_3 n + k_4 N_B.$$

Introducing the new variables and parameters

$$(4.3) \quad n = a(1+x), \quad N_A k_1/k_2 = 3a, \quad k_3/k_2 = (3+\delta)a^2, \quad N_B k_4/k_2 = (1+\delta')a^3,$$

we obtain, keeping only higher order terms

$$(4.4) \quad \frac{1}{k_2 a^2} \frac{dx}{dt} = -x^3 - \delta x - \delta + \delta'.$$

Equation (4.4) admits three steady state solutions when

$$(4.5) \quad \delta < 0, \quad \delta - 2(-\delta/3)^{3/2} < \delta' < \delta + 2(-\delta/3)^{3/2},$$

and only one steady state solution for other values of the parameters δ and δ'. For $\delta = \delta' = 0$ there is a bifurcation point corresponding to the triple root $x = 0$. Assuming that the cusp is located along $\delta = \delta'$, $x = 0$ is the only real solution for $\delta > 0$ but for $\delta < 0$ in addition to $x = 0$, which is now an unstable solution, we have the stable solutions $x_{\pm} = \pm\sqrt{-\delta}$, see the figure on the next page. Using the terminology of phase transition, $\sqrt{(-\delta)}$ is then called the order parameter.

The next step is to investigate the fluctuations in the neighborhood of the bifurcation point. If $p_{mn}(t)$ is the probability of having n molecules of X at time t if we start with m molecules at time 0, then we have a BDP with

(4.6) $\lambda_n = N_A k_1 n(n-1) + k_4 N_B$ and $\mu_n = k_2 n(n-1)(n-2) + k_3 n.$

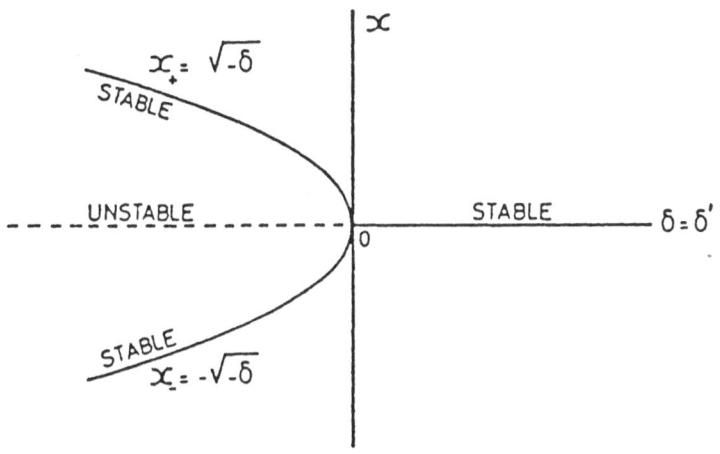

FIGURE   Bifurcation in the Schlögl model.

Since we are interested in the steady state solution we solve (1.6) in the case (4.6) using the generating function

(4.7) $G(w) := \sum_{n=0}^{\infty} w^n p_{mn}(\infty).$

The differential difference equation (1.6) is transformed to the differential equation

(4.8) $w^2[G'''(w) - 3aG''(w)] + (3 + \delta) a^2 G'(w) - (1 + \delta') a^3 G(w) = 0,$

where we have used the notation (4.3). We apply the Mellin transform to (4.8) and obtain the integral representation

(4.9) $G(w) = K \int_C \phi(u) \exp\{ a[f(u) + u \ln w]\} \, du,$

where $\phi(u)$ and $f(u)$ are known functions and K is a constant. In (4.9) C is the contour of the inverse Mellin transform. As $a \to \infty$ it is easy to see that

$$(4.10) \quad \phi(u) = \frac{h^2}{\lambda^2 u^{1/2}} \frac{u^2 + \lambda^2}{u^2 + h^2} + O(a^{-1}),$$

$$(4.11) \quad f(u) = u \left[1 - \ln (u/3)\right] + 2\text{Re} \left[(u - ih) \text{Ln} (u - ih) - (u - i\lambda) \text{Ln} (u - i\lambda)\right],$$

with

$$(4.12) \quad h^2 := \frac{1}{3} (1 + \delta') - \frac{1}{4} a^{-2}; \quad \lambda^2 := 3 + \delta - \frac{1}{4} a^{-2}.$$

Since we are interested in the macroscopic state we must let $a \to \infty$ (the thermodynamic limit). Applying the method of steepest descent to (4.9) enables us to calculate the variance of the steady state distribution

$$(4.13) \quad \sigma^2 = G''(1) + G'(1) - [G'(1)]^2.$$

We obtain

$$\lim_{a \to \infty} \sigma^2/a = 1 + 4/\delta, \text{ for } \delta > 0,$$

$$(4.14) \quad \lim_{a \to \infty} \sigma^2/a = 1 - 2/\delta - 3/\sqrt{-\delta}, \text{ for } \delta < 0,$$

$$\lim_{a \to \infty} \sigma^2/a^{3/2} = 4\Gamma(3/4)/\Gamma(1/4), \text{ for } \delta = 0.$$

This shows that the central limit theorem is applicable only if $\delta > 0$ or $\delta < 0$ when the variance is proportional to the number of molecules. The central limit theorem does not apply in the case $\delta = 0$ when the variance at the bifurcation point behaves like $a^{3/2}$. In the latter case the effect of fluctuations begins to be felt. One reason for this phenomenon to occur is that the appearance of long range correlations violates the hypothesis of independent subsystems (in Statistical Mechanics).

4.2. **The Moran Model.** Many authors have proposed various stochastic models to study frequency fluctuation in genes under the influence of mutation and selection. We will discuss some results of Karlin and McGregor [20] on the Moran model of genetics [30].

Moran studied a population of fixed total number N of gametes of two types, type a_1 and type a_2. The state of the population is characterized by the number n of a_1 gametes present at time t. We assume that the probability of a change in a state is $\lambda \Delta t$ with only one mating occurring during the interval Δt.

We assume that if an a_1 gamete (a_2 gamete) fertilizes an a_1 gamete or an

a_2 gamete then the result is an a_1 gamete (a_2 gamete). Immediately following a fertilization the progeny may mutate and γ_i, $i = 1, 2$, is the probability that an a_i gamete mutate into the other type of gamete.

The size of the population of the a_1 gamete will increase if an a_1 gamete fertilizes an a_2 gamete without mutation or if an a_2 gamete fertilizes an a_2 gamete with mutation. The corresponding birth rate is

$$(4.15) \qquad \lambda_n = \lambda \left(1 - \frac{n}{N}\right)\left[\frac{n}{N}(1 - \gamma_1) + \left(1 - \frac{n}{N}\right)\gamma_2\right].$$

Similarly the population of the a_1 gamete will decrease if an a_1 gamete fertilizes an a_1 gamete with mutation or if an a_2 gamete fertilizes an a_1 gamete without mutation. The corresponding death rate is

$$(4.16) \qquad \mu_n = \lambda \frac{n}{N}\left[\frac{n}{N}\gamma_1 + (1 - \gamma_2)\left(1 - \frac{n}{N}\right)\right].$$

This problem deals with a finite population N and this is reflected in the vanishing of μ_0 and λ_N. In the present case we have a finite set of orthogonal polynomials $\{F_n(x) : n = 0, ..., N - 1\}$. The corresponding $Q_n(x)$'s are the dual Hahn polynomials. Karlin and McGregor [20] record the transition matrix for different values of γ_1 and γ_2 and discuss probabilistic consequences of the knowledge of the spectrum of the process. They also discussed the nature of the limiting transition density as $N \to \infty$.

Observe that the birth and death rates in (4.15) and (4.16) are quadratic polynomials in n with the same leading coefficients. Such a process is called a quadratic asymptotically symmetric process. The general solution of a quadratic asymptotically symmetric BDP is given in [14], and in [23] and [24] in a special case.

Next we briefly discuss another application of BDP to genetics. In [21] and [22] Karlin and Tavaré obtained interesting result on a genetic model of Robertson [35] which involves killing and immigration. They considered a population of N individuals with three possible genotypes AA, Aa and aa. The initial population is composed only of genotypes AA and Aa.

Karlin and Tavaré addressed the interesting question of determining the time it takes to produce the first homozygote aa. They analyzed the case of linear rates

$$\lambda_n = bn, \qquad \mu_n = an, \qquad \gamma_n = cn,$$

(in (2.13)) with $a > 0, b > 0, c > 0$.

4.3. Time Evolution of Two-valued Interacting Elements. In [36] B. Roehner considered a system of N interacting particles such that:

(i) The state variable of each particle takes the values $\sigma_i = \pm 1$.

(ii) The interaction among particles preserves the symmetry of the distribution function $p(\sigma_1, \sigma_2, ..., \sigma_N ; t)$ in the σ_i's.

(iii) The stochastic evolution is defined by the transition probabilities of each particle and depends on the states of all the remaining particles.

The probability of a transition rate from a given state $\sigma(\sigma_1, ..., \sigma_i, ..., \sigma_N)$ to a different state between time t and t + Δt is

$$(4.17) \qquad \Delta t \sum_{i=1}^{N} w_i (\sigma),$$

where w_i is the transition from the state σ to the state $\sigma_i(\sigma_1, ..., -\sigma_i, ..., \sigma_N)$.

An interaction is called a one element interaction if

$$(4.18) \qquad w_i(\sigma) = v(\sigma_i),$$

a global binary interaction if

$$(4.19) \qquad w_i(\sigma) = \sum_{\substack{j=1 \\ j \neq i}}^{N} v(\sigma_i, \sigma_j),$$

a global third order (ternary) interaction if

$$(4.20) \qquad w_i(\sigma) = \sum_{\substack{j,k = 1 \\ j \neq k, j, k \neq i}}^{N} v(\sigma_i, \sigma_j, \sigma_k).$$

If the summation indices in (4.19) or (4,20) are restricted to a certain index set of neighbors of i then the interaction will be called a local interaction. We shall focus only on global interactions. Global interactions occur in fluids (or gases) due to the mobility of their particles, in neuron interactions of the brain, and also in economics.

Let $p(\sigma, t)$ be the probability that the system be at state σ at time t. The C-K equations for this discrete Markov process are

$$(4.21) \qquad \frac{d}{dt} p(\sigma, t) = - p(\sigma, t) \sum_{i=1}^{N} w_i(\sigma) + \sum_{i=1}^{N} w_i(\sigma_i) p(\sigma_i, t), \quad p(\sigma, 0) = p_0(\sigma).$$

Taking into account the symmetry of $p(\sigma, t)$ in the σ_i's one can write (4.21) in the form

$$(4.22) \qquad \frac{d}{dt} P_n(t) = \mu_{n+1} P_{n+1}(t) - (\lambda_n + \mu_n) P_n(t) + \lambda_{n-1} P_{n-1}(t)$$

where $P_n(t)$ denotes the probability that n of the σ_i's in the sequence σ are equal to one at time t and

(4.23) $\qquad \lambda_n = \dfrac{N-n}{n+1} \displaystyle\sum_{i=1}^{n+1} w_i(\sigma), \qquad \mu_n = \dfrac{n}{N-n+1} \displaystyle\sum_{i=n}^{N} w_i(\sigma).$

For independent particles, with or without external interaction $w_i(\sigma)$ is a constant

(4.24) $\qquad w_i(\sigma) = v(\sigma_i) = w,$

that is the transition probability of a particle depends only on it's own state. The corresponding BDP is linear and finite with rates

(4.25) $\qquad \lambda_n = w(N-n), \qquad \mu_n = w\, n.$

There are reflecting barriers at the states $n = 0$ and $n = N$.

In the case of binary interactions the $v(\sigma_i, \sigma_j)$'s of (4.19) are given by

(4.26) $\qquad v(\sigma_i, \sigma_j) = a + b\,\sigma_i + c\,\sigma_j + d\,\sigma_i\,\sigma_j$

where a, b, c, d are constants chosen to preserve the positiveness of $v(\sigma_i, \sigma_j)$. An easy calculation shows that the birth and death rates are now quadratic and given by

(4.27) $\qquad \lambda_n = (N-n)(\alpha n + \beta), \qquad \mu_n = n(\gamma n + \delta)$

with

$\qquad \alpha = 2(c-d), \qquad \gamma = 2(c+d), \qquad \beta = (N-1)(a-b-c+d),$

$\qquad \delta = (N-1)(a+b) - (N+1)(c+d).$

Similarly a third order interaction will lead to a BDP with cubic rates and reflecting states at $n = 0$, $n = N$.

B. Roehner [36] indicated how to modify the assumptions in order to obtain a countable state space. He also discussed a generalization to systems with state variables taking more than two values. For details consult [36].

5. Spectral Results.

As we saw in Section 2 and Section 3 it is of interest to develop methods for finding qualitative or quantitative information about the spectral measure $d\mu$ of a BDP. Chain sequences provide a very useful tool to find bounds on the spectrum of a BDP (the support of $d\mu$). Recall that a sequence $\{a_n : n \geq 1\}$ is a (positive) chain sequence if there is a parameter sequence $\{g_n : n \geq 0\}$ such that

(5.1) $\qquad a_n = g_n(1 - g_{n-1})$ with $0 \leq g_0 < 1$ and $0 < g_n < 1$, $n > 0$.

Let ξ and η be the limits of the smallest and largest zeros of a set of orthogonal polynomials $\{p_n(x)\}$. We allow $\xi = -\infty$ and $\eta = +\infty$. The interval $[\xi, \eta]$ is called the true interval of orthogonality of $\{p_n(x)\}$. A theorem of Wall and Wetzel [42] can be used to characterize the true interval of orthogonality of polynomials in terms of chain sequences. Chihara [7] used an alternate approach. He defines

(5.2) $\alpha_n(x) := a_n /[(x - c_n)(x - c_{n-1})]$, $n > 0$.

We state his result as Theorem 5.3.

Theorem 5.3. Let $\{p_n(x)\}$ be a family of monic orthogonal polynomials generated by

(5.4) $p_0(x) = 1$, $p_1(x) = x - c_0$, $p_{n+1}(x) = (x - c_n) p_n(x) - a_n p_{n-1}(x)$, $n > 0$.

Then an interval $[a, b] \supseteq [\xi, \eta]$ if and only if

(i) $a < c_n < b$, $n \geq 0$,   (ii) $\{\alpha_n(a): n > 0\}$ and $\{\alpha_n(b): n > 0\}$ are chain sequences.

Theorem 5.5 (comparison theorem, [8]). Let $\{ d_n \}$ be a chain sequence and assume $0 \leq b_n \leq d_n$, $n > 0$. Then $\{ b_n \}$ is also a chain sequence.

Theorems 5.3 and 5.5 are very useful in determining bounds for the spectra of BDP's. In view of Theorem 5.3 every known family of polynomials orthogonal on a subset of a half line generates at least one chain sequence. The totality of such known chain sequences can be used as comparison sequences to give bounds on the spectrum of orthogonal polynomials defined by recurrence relations.

The linear birth and death processes, discussed in Section 3, are the simplest processes with polynomial coefficients. The next cases to consider are BDP's with polynomial or rational rates. Pollaczek [32] considered a class of orthogonal polynomials satisfying three term recurrence relations with rational coefficients. Pollaczek [32] gave the absolutely continuous component $\mu'(x)$ of the measure $d\mu$ with respect to which his polynomials are orthogonal. The polynomials considered by Pollaczek [32] are orthogonal on a finite interval and μ' is supported on $[-1, 1]$. They include several cases of BDP orthogonal polynomials with the variable x shifted.

Maki [25] studied another class of birth and death processes with rational coefficients. Following Maki [25] we let

(5.6) $A_p(x)$, $B_q(x)$, $C_r(x)$, $D_s(x)$ be polynomials of precise degrees p, q, r, s,

(5.7) $A_p(x)/B_q(x)$ and $C_r(x+1)/D_s(x+1) > 0$, $x = 0, 1, 2, ...,$ and $C_r(0)/D_s(0) \geq 0$.

Theorem 5.8 ([25]). Consider a BDP with transition $\lambda_n = A_p(n)/B_q(n)$ and $\mu_n = C_r(n)/D_s(n)$, $n \geq 0$. If $p \leq q$ and $r \leq s$ and $p + r < q + s$ then the Stieltjes moment problem associated with this BDP has a unique solution $d\mu$ and the transition probabilities in I-VI of (1.9) are uniquely determined. Furthermore the measure $d\mu$ is a purely discrete and (2.11) holds for m, $n \geq 0$.

Theorem 5.9 ([25]). Let $\lambda_n = A_p(n)/B_q(n)$ and $\mu_n = C_r(n)/D_s(n)$, $n \geq 0$. Then in each of the following cases the transition probabilities in I-VI of (1.9) are unique.

1. $p > q$, $r \leq s$,   2. $p \leq q$, $r > s$,
3. $p > q$, $r > s$, $p - q \neq r - s$ 4. $p > q$, $r > s$, $p - q = r - s$ but $ad \neq bc$,

where a, b, c, and d are the leading coefficients of $A_p(x)$, $B_q(x)$, $C_r(x)$, $D_s(x)$, respectively.

Motivated by works of Roehner and Valent [37], and others, Chihara [10] studied spectral properties of BDP's with $\lambda_n \sim an^b$ and $\mu_n \sim an^b$. In particular, Chihara gave sufficient conditions for the presence of discrete spectrum if $b > 2$.

The aforementioned results of Chihara and Maki concern general BDP's. One can obtain more specific results for polynomials generated by specific recursions. The choices

$$\lambda_n = \frac{(1 - abq^{n+\alpha})(1 - acq^{n+\alpha})(1 - adq^{n+\alpha})(1 - abcdq^{n+\alpha-1})}{a\,(1 - abcdq^{2n+2\alpha-1}(1 - abcdq^{2n+2\alpha-2})}, n \geq 0,$$

$$\mu_n = \frac{a(1 - bcq^{n+\alpha-1})(1 - bdq^{n+\alpha-1})(1 - cdq^{n+\alpha-1})(1 - q^{n+\alpha})}{(1 - abcdq^{2n+2\alpha-1})(1 - abcdq^{2n+2\alpha-2})}, n > 0,$$

with $\mu_0 = 0$ or μ_0 follows the above pattern, lead to two families of associated Askey-Wilson polynomials [18]. As $q \to 1$ they tend to two families of associated Wilson polynomials which seem to have been first studied in this generality in [15]. The Askey-Wilson polynomials have been introduced in [4] and the Wilson polynomials are in [3], [43], [44]. The Askey-Wilson polynomials are the special case $\alpha = 0$.

We now address the question of finding the spectral measure $d\mu$ from the knowledge of the three term recurrence relation

(5.10) $x\, p_n(x) = \lambda_n\, p_{n+1}(x) + c_n\, p_n(x) + \mu_n\, p_{n-1}(x), n > 0,$

where

(5.11) $p_0(x) := 1,\ p_1(x) := (x - c_0)/\lambda_0,$

and $\{c_n\}$ are real. We assume that the positivity condition

(5.12) $\lambda_n\, \mu_{n+1} > 0, n \geq 0,$

holds to guarantee the existence of a positive spectral measure. We will also assume that the spectral measure is unique, i.e., the corresponding moment problem is determinate. The excellent book [39] of Shohat and Tamarkin contains several useful sufficient conditions for the determinacy of the moment problem. It is also assumed that the measure $d\mu$ is normalized to have a total mass of unity.

We can identify three general methods of constructing spectral measures of given polynomials. In some cases one may combine more than one method to achieve the goal. The method are: the method of moments, use of special functions, and difference equations techniques.

5.1. Method of Moments.
In this method one tries to compute the moments $\mu_n := \int_{\mathcal{R}} x^n\, d\mu(x)$ from the three term recurrence relation. This can be done for example if one can find the constant term $c_{n,0}$ in the expansion

(5.13) $x^n = \sum_{k=0}^{n} c_{n,k} \, p_k(x).$

Another way is to generate the first few polynomials, then use $\mu_0 = \int_{\mathcal{R}} d\mu(x) = 1$ and $\int_{\mathcal{R}} p_n(x) \, d\mu(x) = 0$, for $n > 0$, to generate the first few μ_n's. If a pattern emerges then one may proceed to establish an explicit expression for the moments. Once we find the moments we can then try to construct a function $F(z)$ whose asymptotic expansion is $\sum_{n \geq 0} \mu_n z^{-n-1}$ as $z \to \infty$. In the cases under consideration

(5.14) $\displaystyle \int_{-\infty}^{\infty} \frac{d\mu(t)}{z-t} =: F(z) \sim \sum_{n=0}^{\infty} \mu_n z^{-n-1}.$

Having constructed the Stieltjes transform of $d\mu$ one can use the familiar Perron-Stieltjes inversion formula, see (3.12),

(5.15) $\displaystyle F(z) = \int_{-\infty}^{\infty} \frac{d\mu(t)}{z-t}$ if and only if $\mu(x) - \mu(y) = \lim_{\varepsilon \to 0^+} \int_{y}^{x} \frac{F(t-i\varepsilon) - F(t+i\varepsilon)}{2\pi i} \, dt.$

If the interval of orthogonality is bounded the series in (5.14) converges in the exterior of any circle $|z| = r$ containing the interval of orthogonality in its interior, so we can replace $\sim$ in (5.14) by $=$, provided that $|z| > r$.

5.2. Special Function Methods. These methods depend on guessing what the measure will be then use the well-developed theory of explicit sums, integrals, and transformation theory of special functions to prove the guess already conjectured. This approach requires great insight and technical skills as can be seen from the masterful works [1], [4], [43] and [44]. In many cases it turned out that the natural quantities to compute are not the moments μ_n but the "generalized moments" $\nu_n := \int_{\mathcal{R}} r_n(x) d\mu(x)$, where $r_n(x)$ is a suitably chosen sequence of polynomials such $r_n(x)$ is of precise degree n. For example in the case of the Askey-Wilson polynomials

(5.16) $d\mu(\cos\theta) = \dfrac{(e^{2i\theta})_\infty (e^{-2i\theta})_\infty}{\displaystyle\prod_{j=1}^{4}[(a_j e^{i\theta})_\infty (a_j e^{-i\theta})_\infty]} \, \dfrac{dx}{\sqrt{1-x^2}}, \quad -1 < x < 1,$

where q is a fixed number and

(5.17) $\quad (a)_\infty = \displaystyle\prod_{n=0}^{\infty} [\, 1 - aq^n\,],\ -1 < q < 1.$

In this case

(5.18) $\quad I(a_1, a_2, a_3, a_4) := \displaystyle\int_{-1}^{1} d\mu(x) = \dfrac{2\pi\,(a_1\,a_2\,a_3\,a_4)_\infty}{(q)_\infty \displaystyle\prod_{1\le j<k\le4} (a_j\,a_k)_\infty},$

so we need to renormalize $d\mu$ to make it a probability measure. The integral in (5.18) is called the Askey-Wilson integral. A very natural choice for $r_n(x)$ is

(5.19) $\quad r_n(x) = \displaystyle\prod_{j=0}^{n-1} (\, 1 - ae^{i\theta}q^j)(1 - ae^{-i\theta}q^j),\ n > 0,\ r_0(x) = 1,\ a = a_1,\ \dots,\ \text{or } a_4,$

since $r_n(x)$ can be easily integrated with respect to $d\mu$. When we choose $a = a_1$ in $r_n(x)$ we get

(5.20) $\quad \displaystyle\int_{-1}^{1} r_n(x)\, d\mu(x) = I(a_1 q^n, a_2, a_3, a_4).$

This suggests expressing the orthogonal polynomials in terms of $\{\, r_n(x)\}$ then proving the orthogonality. This analysis leads to explicit formulas for the orthogonal polynomials and identifies the three term recurrence relation satisfied by the Askey-Wilson polynomials. The details are in [4].

5.3. Difference Equations and Generating Functions. The objective is to compute the function $F(z)$ in (5.14). Define $q_n(x)$, a second polynomial solution of (5.10) by the initial conditions

(5.21) $\quad q_0(x) = 0,\quad q_1(x) = 1/\lambda_0.$

When the moment problem is determined one can show that

(5.22) $\quad \displaystyle\lim_{n\to\infty} q_n(z)/p_n(z) = \int_{-\infty}^{\infty} \dfrac{d\mu(t)}{z - t},$

uniformly on compact subsets of the complex plane cut along the support of $d\mu$, [39]. This suggests as a possible approach to determine the asymptotic behavior of both $p_n(x)$ and $q_n(x)$ as $n \to \infty$, for complex x. In [2] it was pointed out that the asymptotic method of Darboux [41] can be applied to generating functions of $\{p_n(x)\}$ and $\{q_n(x)\}$ to determine their asymptotic behavior. This approach was also used in [2] to compute spectral measures of certain interesting families of orthogonal polynomials including the Al-Salam-Chihara polynomials, a family of random walk polynomials and their q-analogue. The same approach was also used in [14] to find the spectral measure of all quadratic asymptotically symmetric BDP's.

A contiguous relation for a hypergeometric function $F(a, b; c; z)$ is a linear recursion relation involving the functions $F(a_1 + \alpha_1, a_2 + \alpha_2; b_1 + \beta_1; z)$ where α_j or β_j take the values $0, \pm 1$. The same definition applies to hypergeometric functions of higher order. In the case of a basic hypergeometric function $a_j + \alpha_j$ and $b_j + \beta_j$ are replaced by $a_j q^{\alpha_j}$ and $b_j q^{\beta_j}$, respectively. In [43] Wilson gave a simple derivation of the three term contiguous relations for $_3F_2$ and $_4F_3$ functions. He also indicated that the same approach applies to basic hypergeometric functions. Gauss and Heine used contiguous relations to derive a continued fraction expansion for quotients of hypergeometric and basic hypergeometric functions, respectively. The work [16] started a systematic study of orthogonal polynomials generated by contiguous relations of basic hypergeometric functions. The same approach was used in [15] and [18]. More work in this direction is in progress but it seems premature to report on it at this time.

6. Bilateral Birth and Death Processes.

The theory of bilateral BDP's has been treated in detail by Pruitt [33]. Bilateral BDP's are a natural extension of the unilateral BDP's of previous sections with the usual state space of nonnegative integers replaced by the set of all integers. Certain bilateral cases related to the associated classical hypergeometric polynomials can presumably be made explicit. However, we know of no examples that have actually been worked out.

Here we indicate the connection between bilateral BDP's and the theory of symmetric linear operators and derive both the bilateral formula that replaces the unilateral formula (2.11) and Pruitt's distinguished solution to its Laplace transform. Recall that the basic problem of unilateral BDP's was to obtain a transition matrix $P(t)$ having properties I - VI of Section 1. In the bilateral case, apart from the fact that the matrix indices now take all integer values, the problem is unchanged.

We start by considering the generating matrix

(6.1) $A = (a_{m,n}: m, n = 0, \pm 1, \pm 2, ...), a_{n,n} = -\lambda_n - \mu_n, a_{n,n+1} = \lambda_n, a_{n+1,n} = \mu_{n+1},$

with

$\lambda_n, \mu_n > 0, n = 0, \pm 1, \pm 2, ... ,$

as a linear operator in a special weighted l^2 space [19]

$$l^2(\mathbf{Z}, \pi_n) := \{ \mathbf{u} = (u_n) \mid u_n \in C, \| \mathbf{u} \|^2 = \sum_{-\infty}^{\infty} |u_n|^2 \pi_n < \infty \},$$

(6.2)
$$\pi_0 = 1, \pi_n = \prod_{j=1}^{n} \frac{\lambda_{j-1}}{\mu_j} , \pi_{-n} = \prod_{j=1}^{n} \frac{\mu_{1-j}}{\lambda_{-j}}, \quad n > 0.$$

Our convention for the inner product in $l^2(\mathbf{Z}, \pi_n)$ will be

(6.3) $$(\mathbf{u}, \mathbf{v}) = \sum_{-\infty}^{\infty} \pi_n \bar{u}_n v_n,$$

where the bar denotes complex conjugation. If one takes A to have domain D(A) given by the set of all vectors (sequences) with finitely many non-zero components, it is easy to see that A is symmetric and negative, that is, D(A) is dense in $l^2(\mathbf{Z}, \pi_n)$, $(\mathbf{u}, A\mathbf{v}) = (A\mathbf{u}, \mathbf{v})$ for all $\mathbf{u}, \mathbf{v} \in D(A)$ and $(\mathbf{u}, A\mathbf{u}) < 0$ holds if $\mathbf{u} \neq 0$. In this connection note that if $\{e_n \mid -\infty < n < \infty \}$ is the standard $l^2(\mathbf{Z})$ basis, then $\{ e_n' = e_n/\sqrt{\pi_n} \mid -\infty < n < \infty \}$ is a basis for $l^2(\mathbf{Z}, \pi_n)$. Thus $(e_n, Ae_{n+1}) = \pi_n \lambda_n$ and $(e_{n+1}, Ae_n) = \pi_{n+1} \mu_{n+1}$ implies that $(e_n', Ae_{n+1}') = (e_{n+1}', Ae_n') = \sqrt{\lambda_n \mu_{n+1}}$. Now any symmetric operator has a minimal closed symmetric extension given by its double adjoint. Thus there is no loss of generality in assuming that A is closed, symmetric, and negative, see Stone [40].

Throughout the remainder of this section we make the simplifying assumption that A is self-adjoint. This is equivalent to assuming the determinacy of the Hamburger moment problems associated with both a left upper and a right lower portion of A. Because A is negative ($(\mathbf{u}, A\mathbf{u}) < 0$, $\mathbf{u} \neq 0$) a necessary and sufficient condition for such determinacy is [9], [19]

$$\sum_{n=0}^{\infty} \pi_n \left[1 + \mu_0 \sum_{k=0}^{n-1} (\lambda_k \pi_k)^{-1} \right]^2 = \infty \text{ and } \sum_{n=0}^{\infty} \pi_{-n} \left[1 + \lambda_0 \sum_{k=0}^{n-1} (\mu_k \pi_k)^{-1} \right]^2 = \infty.$$

If one or both of these series converged then A would be non-self-adjoint with infinitely many self-adjoint extensions.

With the assumption that A is self-adjoint one now has, for $t \geq 0$, a unique bounded transition matrix operator
$$P(t) = \exp(tA)$$
where the exponential is defined through the functional calculus for self-adjoint operators, [40]. It follows that

(6.4) $p_{mn}(t) = (e_m, e^{tA} e_n) / \pi_m = \dfrac{1}{\pi_m} \displaystyle\int_{-\infty}^{0} e^{xt} d\,(e_m, E_A(x)\, e_n),$

where $E_A(x)$ is the unique family of projection operators associated with the diagonalization of A.

From (6.1) one has

(6.5) $Ae_n = \lambda_{n-1}\, e_{n-1} + \mu_{n+1}\, e_{n+1} - (\lambda_n + \mu_n)\, e_n$

and it follows that

(6.6) $e_n = F_n^{(0)}(-A)\, e_0 + F_n^{(1)}(-A)\, e_1,$

where $\{F_n^{(0)}(x)\}$ and $\{F_n^{(1)}(x)\}$ are polynomials satisfying the recurrences (2.4) for both $n \geq 0$ and $n < 0$, but they are generated from different initial conditions, that is

(6.7) $-x\, F_n^{(j)}(x) = \mu_{n+1}\, F_{n+1}^{(j)}(x) - (\lambda_n + \mu_n)\, F_n^{(j)}(x) + \lambda_{n-1}\, F_{n-1}^{(j)}(x),\ n \in \mathbf{Z},\ j = 0, 1,$

with the initial conditions

$F_n^{(j)}(x) = \delta_{n,j},\ j, n = 0, 1.$

Substituting (6.6) into (6.4) and using the functional calculus for self-adjoint operators yields the integral representation

$$\pi_m p_{mn}(t) = \int_{-\infty}^{0} e^{xt}\, \{F_m^{(0)}(-x)F_n^{(0)}(-x)\, d\mu_{00}(x) + F_m^{(1)}(-x)\, F_n^{(1)}(-x)\, d\mu_{11}(x)$$

(6.8)

$$+ F_m^{(0)}(-x)\, F_n^{(1)}(-x)\, d\mu_{01}(x) + F_m^{(1)}(x)\, F_n^{(0)}(-x)\, d\mu_{10}(x)\}$$

where $d\mu_{jk}(x) = d(\, e_j, E_A(x)\, e_k),\ (-\infty, 0] \supseteq \mathrm{supp}\{\, d\mu_{jk}(x)\, \}$ and $d\mu_{jj}(x) \geq 0.$

The integral representation (6.8) should be compared with the unilateral formula (2.12). We consider (6.8) to be an important result which is missing from the Pruitt paper [33].

Pruitt's approach to obtaining $p_{mn}(t)$ is to look at the Laplace transform of P(t), [33]. With the present assumptions this yields

(6.9) $R(s) := \displaystyle\int_{0}^{\infty} e^{-ts}\, e^{tA} dt = (sI - A)^{-1},\ \mathrm{Re}\ s > 0,$

which is the resolvent of A and can be analytically continued to $s \notin \sigma(A)$, where $\sigma(A)$ is the spectrum of A and is contained in $(-\infty, 0].$

From (6.9) and (6.4), the Laplace transform of $p_{mn}(t)$ is given by

(6.10) $r_{mn}(s) = R_{mn}(s)/\pi_m$, $R_{mn}(s) = (e_m, (sI - A)^{-1} e_n)$.

Now $R_{mn}(s)$ satisfies a simple recurrence relation in both subscripts. That is, from (6.5) we get

(6.11a) $s R_{mn}(s) - \mu_{n+1} R_{m\,n+1}(s) + (\lambda_n + \mu_n) R_{mn}(s) - \lambda_{n-1} R_{m\,n-1}(s)$

$$= (e_m, (sI - A)^{-1} (sI - A)e_n) = \delta_{mn} \pi_m$$

and

(6.11b) $\overline{R_{mn}(s)} = R_{mn}(\bar{s})$.

As a solution to (6.11a) and (6.11b) one now easily obtains

$$Y_m^{(L)}(s) Y_n^{(R)}(s) / [\mu_1 W(s)], \quad m \leq n$$

(6.12) $R_{mn}(s) =$

$$Y_m^{(R)}(s) Y_n^{(L)}(s) / [\mu_1 W(s)], \quad n \leq m$$

where $Y_n^{(L)}(s)$ and $Y_n^{(R)}(s)$ are left ($n \to -\infty$) and right ($n \to \infty$) subdominant or minimal solutions, respectively, to the difference equation

(6.13) $sY_n(s) = \mu_{n+1} Y_{n+1}(s) - (\lambda_n + \mu_n) Y_n(s) + \lambda_{n-1} Y_{n-1}(s)$

and

(6.14) $W(s) = Y_0^{(R)}(s) Y_1^{(L)}(s) - Y_0^{(L)}(s) Y_1^{(R)}(s)$.

The subdominance property follows since $(e_m', (sI - A)^{-1} e_n')$ is a Fourier coefficient with regard to each subscript and hence an $l^2(Z)$ with respect to either subscript. This implies that $\{ Y_n^{(R)}/\sqrt{\pi_n} \mid n \geq 0 \}$ and $\{ Y_{-n}^{(L)}/\sqrt{\pi_{-n}} \mid n \geq 0 \}$ belong to $l^2(Z^+)$ which uniquely determines $Y_n^{(L)}$ and $Y_n^{(R)}$ up to a multiplicative constant, since for a determined Hamburger moment problem there is at most one normalized l^2 solution, [40]. Note that (6.12) is equivalent to Theorem 3.3 in Pruitt [33, p. 518] and gives what he refers to as a distinguished solution.

By comparing the Laplace transform of (6.8) with (6.10) where R_{mn} is given by (6.12), one can determine the measures $d\mu_{jk}(x)$, j, k = 0, 1 explicitly for certain cases related to the associated classical orthogonal polynomials. We shall illustrate this fact by an example.

Example: Let $\lambda_n = \lambda > 0$, $\mu_n = \mu > 0$ then

$$Y_n^{(R)}(s) = (2\mu)^{-n} \left(s + \lambda + \mu - \sqrt{(s + \lambda + \mu)^2 - 4\lambda\mu}\ \right)^n,$$

$$Y_n^{(L)}(s) = (2\lambda)^{n} \left(s + \lambda + \mu - \sqrt{(s + \lambda + \mu)^2 - 4\lambda\mu}\ \right)^{-n}.$$

where the square root is chosen such that

$$\left| s + \lambda + \mu - \sqrt{(s + \lambda + \mu)^2 - 4\lambda\mu} \right| < \left| s + \lambda + \mu + \sqrt{(s + \lambda + \mu)^2 - 4\lambda\mu} \right|,$$

when $s \notin [\,-\lambda - \mu - 2\sqrt{\lambda\mu}\,,\, -\lambda - \mu + 2\sqrt{\lambda\mu}\,]$.

We now apply (6.12) and obtain

$$R_{00}(s) = 1 / \sqrt{(s + \lambda + \mu)^2 - 4\lambda\mu}\ , \qquad R_{11}(s) = R_{00}(s) / \mu^2,$$

$$R_{01}(s) = R_{10}(s) = \frac{s + \lambda + \mu - \sqrt{(s + \lambda + \mu)^2 - 4\lambda\mu}}{2\mu\sqrt{(s + \lambda + \mu)^2 - 4\lambda\mu}},$$

and hence from the Perron-Stieltjes inversion formula we find

$$d\mu_{00}(x) = \mu^2\, d\mu_{11}(x) = \frac{dx}{\pi\sqrt{4\lambda\mu - (x + \lambda + \mu)^2}}\ ,$$

$$d\mu_{10}(x) = d\mu_{01}(x) = \frac{(x + \lambda + \mu)dx}{2\pi\mu\sqrt{4\lambda\mu - (x + \lambda + \mu)^2}}\ ,$$

and

$$R_{ij}(s) = \int_{-(\sqrt{\lambda} + \sqrt{\mu})^2}^{-(\sqrt{\lambda} - \sqrt{\mu})^2} \frac{d\mu_{ij}(x)}{s - x}.$$

This shows that the measures in (6.8) are associated with Chebyshev polynomials of the first kind while the polynomials in (6.8) are, after renormalization, given by Chebyshev polynomials of the second kind.

The above methods used to obtain R_{mn} can also be applied to other types of bilateral Jacobi matrices and also in unilateral or finite cases with suitably modified boundary conditions on $Y_n^{(L)}$ and $Y_n^{(R)}$. This will also yield representations of R_{mn} in terms of continued fractions. In the bilateral cases one can obtain explicit results for Jacobi matrices associated with the principal and supplementary series representation of the Lie algebra su(1,1) in connection with associated Meixner, Meixner-Pollaczek, and Laguerre polynomials, [28].

Acknowledgements. The authors thank Paul Nevai for his kind invitation to write this paper for the NATO ASI on orthogonal polynomials and their applications. We thank Waleed Al-Salam for bringing refence [12] to our attention, and thank Ted Chihara for his helpful comments. M. Ismail also thanks Dr Jairo Charris and the faculty members of the department of Mathematics of the National University of Colombia in Bogota for inviting him to lecture at the National University where a preliminary version of this paper was presented as part of a course on orthogonal polynomials.

Bibliography

1. G. Andrews and R. Askey, 'Classical orthogonal polynomials', in "*Polynômes Orthogonaux et Applications* ", Lecture Notes in Mathematics, volume **1171**, Springer-Verlag, New York,1985, 36-62.
2. R. Askey and M. E.H. Ismail, 'Recurrence relations, continued fractions and orthogonal polynomials', *Memoirs Amer. Math. Soc.* Number **300**, 1984.
3. R. Askey and J. A. Wilson, A set of orthogonal polynomials that generalize the Racah coefficients or 6-j symbols, *SIAM J. Math. Anal.* **10** (1979), 1008-1016.
4. R. Askey and J. A. Wilson, 'A set of orthogonal polynomials that generalize Jacobi polynomials', *Memoirs Amer. Math. Soc.* Number **319**, 1985.
5. R. Askey and J. Wimp, 'Associated Laguerre and Hermite polynomials', *Proc. Royal Soc. Edinburgh* **96A** (1984), 15-37.
6. T. Chihara, 'On co-recursive orthogonal polynomials', *Proc. Amer. Math. Soc.* **8** (1957), 899-905.
7. T. S. Chihara, 'Chain sequences and orthogonal polynomials', *Trans. Amer. Math. Soc.* **104** (1962), 1-16.
8. T. S. Chihara, *An Introduction to Orthogonal Polynomials*, Gordon and Breach, New York, 1978.
9. T. S. Chihara, 'On the spectra of certain birth and death processes', *SIAM J. Appl. Math.* **47** (1987) 662-669.
10. T. S. Chihara, 'Indeterminate symmetric moment problems', *J. Math. Anal. Appl.* **85** (1982), 331-364.
11. A. Erdelyi, W. Magnus, F. Oberhettinger, and F. G. Tricomi, *Higher Transcendental Functions*, 3 volumes, McGraw-Hill, New York, 1953.
12. W. Hahn, 'Über orthogonalpolynome mit drei parametern,' *Deutche. Math.* **5** (1940-41), 373-378.
13. M. E. H. Ismail, J. Letessier, and G. Valent, 'Linear birth and death models and associated Laguerre polynomials', *J. Approximation Theory* **56** (1988), 337-348.
14. M. E. H. Ismail, J. Letessier, and G. Valent, 'Quadratic birth and death processes and associated continuous dual Hahn polynomials', *SIAM J. Math. Anal.* **20** (1989), 727-737.
15. M. E. H. Ismail, J. Letessier, G. Valent, and J. Wimp, 'Two families of

associated Wilson polynomials', to appear.

16. M. E. H. Ismail and C. Libis, 'Contiguous relations, basic hypergeometric functions, and orthogonal polynomials I', *J. Math. Anal. Appl.* **141** (1989), to appear.

17. M. E. H. Ismail and D. R. Masson, 'Two families of orthogonal polynomials related to Jacobi polynomials', Rocky Mountain J. Math. (1990), to appear.

18. M. E. H. Ismail and M. Rahman, 'Associated Askey-Wilson polynomials', to appear.

19. S. Karlin and J. McGregor, 'The differential equations of birth and death processes, and the Stieltjes moment problem', *Transactions Amer. Math Soc.* **85** (1957), 489-546.

20. S. Karlin and J. McGregor, 'On a genetics model of Moran', *Proc. Cambridge Philos. Soc.* **58** (1958), 299-311.

21. S. Karlin and S. Tavaré, 'Linear birth and death processes with killing', *J. Appl. Prob.* **19** (1982), 477-487.

22. S. Karlin and S. Tavaré, 'A diffusion process with killing: the time to formation of recurrenent deleterious mutant genes', *Stochastic Processes and their Applications* **13** (1982), 249-261.

23. J. Letessier and G. Valent, 'The generating function method for quadratic asymptotically symmetric birth and death processes', *SIAM J. Appl. Math.* **44** (1983), 773-783.

24. J. Letessier and G. Valent, 'Dual birth and death processes and orthogonal polynomials', *SIAM J. Appl. Math.* **46** (1986), 393-405.

25. D. Maki, 'On birth-death processes with rational growth rates', *SIAM J. Math. Anal.* **7** (1976), 29-36.

26. D. R. Masson, 'The rotating harmonic oscillator eigenvalue problem. I. Continued fractions and analytic continuation', *J. Math. Physics* **24** (1983), 2074-2088.

27. D. R. Masson, 'Associated Wilson polynomials', to appear.

28. D. R. Masson and J. Repka, 'Spectral theory of Jacobi matrices in $l^2(\mathbf{Z})$', to appear.

29. P. A. P. Moran, 'The effect of selection in a haploid genetic population', *Proc. Cambridge Philos. Soc.* **54** (1958), 463-474.

30. P. A. P. Moran, 'A general theory of the distribution of gene frequencies', *Proc. Roy. Soc. London* **149 B** (1958), 102-112.

31. G. Nicolis and J. W. Turner, 'Effect of fluctuations on bifurcation phenomena', *Annals New York Acad. Sc.* **316** (1979), 251-262.

32. F. Pollaczek, 'Sur une généralisation des polynômes de Jacobi', *Mémorial des Sciences Mathématiques* **131** (1956).

33. W. Pruitt, 'Bilateral birth and death processes', *Transactions Amer. Math Soc.* **107** (1962) 508-525.

34. G. E.H. Reuter, 'Denumerable Markov processes and associated semigroupes on I', *Acta Math.* **97** (1957), 1-46.

35. A. Robertson, 'The time to detection of recessive visible genes in small populations', *Genet. Res. (Cambridge)* **31** (1978), 255-264.

36. B. Roehner, 'Time evolution of a system of two-valued interacting elements: a microscopic interpretation of birth and death equations', *Int. J. Systens Sci.* **14** (1983), 1191-1216.

37. B. Roehner and G. Valent, 'Solving the birth and death processes with qudratic asymptotically symmetric transition rates', *SIAM J. Appl. Math.* **42** (1982), 1020-1046.

38. F. Schlögl, 'On thermodynamics near a steady state', *Z. Physics* **248** (1971), 446-458.

39. J. A. Shohat and J. D. Tamarkin, *The Problem of Moments*, revised edition, Mathematical Surveys, volume 1, American Mathematical Society, Providence, Rhode Island, 1950.

40. M. H. Stone, *Linear Transformations in Hilbert Spaces and Their Applications to Analysis*, Colloquium Publications, volume 15, American Mathematical Society, Providence, Rhode Island, 1932.

41. G. Szegö, *Orthogonal Polynomials*, Colloquium Publications, fourth edition, volume 23, American Mathematical Society, Providence, Rhode Island, 1975.

42. H. S. Wall and M. Wetzel, 'Quadratic forms and convergence regions for continued fractions', *Duke Math. J.* **11** (1944), 373-397.

43. J. Wilson, *Hypergeometric series, recurrence relations and some orthogonal functions*, Doctoral Dissertation, University of Wisconsin, Madison, 1978.

44. J. Wilson, 'Some hypergeometric orthogonal polynomials', *SIAM J. Math. Anal.* **11** (1980), 690-701.

45. J. Wimp, 'Explicit formulas for the associated Jacobi polynomials and some applications', *Canadian J. Math.* **39** (1987), 983-1000.

ORTHOGONAL POLYNOMIALS IN CONNECTION WITH QUANTUM GROUPS

TOM H. KOORNWINDER
Centre for Mathematics and Computer Science
P.O. Box 4079
1009 AB Amsterdam
The Netherlands

ABSTRACT. This is a survey of interpretations of q-hypergeometric orthogonal polynomials on quantum groups. The first half of the paper gives general background on Hopf algebras and quantum groups. The emphasis in the rest of the paper is on the SU(2) quantum group. An interpretation of little q-Jacobi polynomials as matrix elements of its irreducible representations is presented. In the last two sections new results by the author on interpretations of Askey-Wilson polynomials are discussed.

1. Introduction

Quantum groups, recently introduced by Drinfeld [11], Woronowicz [49] and Jimbo [16], are fascinating objects, where many different structures meet and with applications in numerous areas. In the present paper we want to emphasize one application which was overlooked by the founding fathers of the theory: the interpretation of q-hypergeometric orthogonal polynomials, quite analogous to the interpretation of ordinary hypergeometric functions and polynomials on special Lie groups, cf. Vilenkin [47]. So the tremendous amount of new results on q-special functions during the last 15 years, cf. Askey and Wilson [6], Gasper and Rahman [14], Rahman [37] is finally matched by a satisfactory "group" theoretic setting. Before the introduction of quantum groups we only knew about interpretations of q-Hahn and q-Krawtchouk polynomials on Chevalley groups (cf. Stanton [40], [41]) and of q-ultraspherical polynomials with $q = 0$ on homogeneous trees and SL_2 over the p-adics (cf. Cartier [9]).

Until now, the best studied quantum group is $SU_q(2)$, the quantum analogue of the group $SU(2)$. Vaksman and Soibelman [45] were the first to observe that the matrix elements of the irreducible unitary representations of this quantum group can be expressed in terms of little q-Jacobi polynomials. The same observation was independently made by Masuda e.a. [27], [28] and by the author [21]. Various other classes of q-hypergeometric orthogonal polynomials have been interpreted in connection with $SU_q(2)$, see [32], [33], [17], [18] and [22]. In this last reference an addition formula for little q-Legendre polynomials is obtained which would have been hard to discover without the interpretation on the quantum group.

Very recently, significant interpretations of special functions have been found on other quantum groups. On the one hand, there are results on more general "compact" quantum

257

P. Nevai (ed.), Orthogonal Polynomials, 257–292.

groups like $SU_q(n)$, cf. Noumi e.a. [34]. On the other hand, work is starting now on quantum analogues of non-compact Lie groups like the group of plane motions (cf. Vaksman and Korogodsky [44]) and $SU(1,1)$ (cf. Masuda e.a. [29]).

During the preparation of this manuscript the author obtained an interpretation of continuous q-Legendre polynomials as "spherical" matrix elements of irreducible representations of $SU_q(2)$. For ordinary $SU(2)$ the definition of *spherical* depends on the choice of the subgroup, for instance $S(U(1) \times U(1))$ or $SO(2)$, but, still, all one-parameter subgroups are conjugate and thus yield the same spherical functions. However, in the quantum group case the "subgroups" are no longer conjugate and give rise to different types of "spherical functions", for instance little q-Legendre and continuous q-Legendre polynomials. Next, by a further generalization of the notion of *spherical* element and with some inspiration from Noumi's and Mimachi's recent preprint [32], a similar interpretation could be obtained for a two-parameter family of Askey-Wilson polynomials. As a spin-off there followed the observation that one can pass from Askey-Wilson polynomials to little or big q-Jacobi polynomials by taking suitable limits.

The connection which has thus been made between quantum groups and the "master family" of Askey-Wilson polynomials looks very promising for future research, for instance, for giving a quantum group theoretic proof of the Rahman-Verma [38] addition formula for continuous q-ultraspherical polynomials and for interpreting on quantum groups Macdonald's ([24], [25]) q-orthogonal polynomials associated with root systems.

The contents of this paper are as follows. Section 2 is a tutorial presenting the basics of Hopf algebras. In section 3 we introduce compact matrix quantum groups, in particular $SU_q(2)$. Section 4 deals with the general representation theory, due to Woronowicz, of such quantum groups. In §5 this is applied to $SU_q(2)$, and little q-Jacobi polynomials are obtained as matrix elements of the irreducible corepresentations of the corresponding Hopf algebra. Section 6 briefly reviews some further interpretations of special functions on quantum groups. Finally, sections 7 and 8 deal with the new interpretations by the author of Askey-Wilson polynomials on $SU_q(2)$, for which full proofs will be published elsewhere. For the case of continuous q-Legendre polynomials an idea of the proof is given in §7 in fairly complete detail. In §8 the main results for the larger two-parameter family of Askey-Wilson polynomials are just stated.

While reading these expository notes, it may be helpful to consult the tutorials by Rahman [37] and Stanton [41] in these Proceedings.

NOTATION. $\mathbf{Z}_+$ denotes the set of nonnegative integers.

2. Hopf Algebras

Although the term *quantum group* sounds quite attractive, the name is in fact somewhat misleading: first, because a quantum group is generally not a group, and, second, because the relationship with quantum mechanics is, in my opinion, not as clear and unambiguous as one might wish. If one still wants to get an impression of what is meant by a quantum group, one way would be to study a number of generally accepted examples of quantum groups. Another way, following Drinfeld [11], would be to define a *quantum group* as the spectrum of a (not necessarily commutative) Hopf algebra. So let us turn to Hopf algebras.

2.1. ALGEBRA OF FUNCTIONS ON A GROUP

The definition of Hopf algebras may overwhelm the uninitiated reader at first confrontation. Therefore, we will start with a detailed discussion of the guiding example: an algebra of functions on a group.

Let G be a group. Thus there is a multiplication $(x,y) \mapsto xy : G \times G \to G$, a unit element $e \in G$ and a mapping of taking the inverse $x \mapsto x^{-1} : G \to G$, which together satisfy the well-known group axioms. Let $\mathcal{A} := \mathrm{Fun}(G)$ be some complex associative algebra with unit consisting of complex-valued functions on the group G, where the multiplication $(a,b) \mapsto ab : \mathcal{A} \times \mathcal{A} \to \mathcal{A}$ and the unit $I \in \mathcal{A}$ are defined pointwise:

$$(ab)(x) := a(x)\,b(x), \quad I(x) := 1 \qquad \text{for all } x \in G.$$

Note that this is a *commutative* algebra. To start with, one may think about $\mathrm{Fun}(G)$ as the algebra of all complex-valued functions on G.

The group operations induce certain operations on the algebra $\mathcal{A}$:

(i) A *comultiplication* $\Phi : \mathrm{Fun}(G) \to \mathrm{Fun}(G \times G)$ defined by

$$(\Phi(a))(x,y) := a(xy), \qquad x,y \in G.$$

(ii) A *counit* $e : \mathrm{Fun}(G) \to \mathbf{C}$ defined by

$$e(a) := a(e).$$

(iii) An *antipode* $\kappa : \mathrm{Fun}(G) \to \mathrm{Fun}(G)$ defined by

$$(\kappa(a))(x) := a(x^{-1}), \qquad x \in G.$$

Note that the mappings Φ, e and κ are *unital algebra homomorphisms*.

We will rewrite the comultiplication by using *tensor products*. There is a linear embedding of $\mathrm{Fun}(G) \otimes \mathrm{Fun}(G)$ in $\mathrm{Fun}(G \times G)$ such that

$$(a \otimes b)(x,y) := a(x)\,b(y), \qquad x,y \in G,$$

for all $a,b \in \mathrm{Fun}(G)$. Hence also

$$\left(\sum_i a_i \otimes b_i \right)(x,y) = \sum_i a_i(x)\,b_i(y) \qquad \text{(finite sum)}.$$

(To start with, the tensor products under consideration are *algebraic*, so they involve only finite sums.) In particular, if G is an *algebraic group* over $\mathbf{C}$ and if we take for $\mathcal{A}$ the algebra $\mathrm{Pol}(G)$ of all polynomial functions on G, then the above linear embedding becomes a linear isomorphism: $\mathrm{Pol}(G) \otimes \mathrm{Pol}(G) \simeq \mathrm{Pol}(G \times G)$. By way of example let G be the algebraic group

$$SL(2,\mathbf{C}) := \left\{ \begin{pmatrix} x & y \\ u & v \end{pmatrix} \mid xv - yu = 1 \right\}.$$

Let α be the polynomial function

$$\alpha: \begin{pmatrix} x & y \\ u & v \end{pmatrix} \longmapsto x: G \longrightarrow \mathbf{C}$$

and let similarly β, γ, δ be the polynomial functions which map $\begin{pmatrix} x & y \\ u & v \end{pmatrix}$ to y, u, v, respectively. Then $\mathrm{Pol}(G)$ consists of all functions on G which can be written as polynomials in $\alpha, \beta, \gamma, \delta$.

As another example let G be a compact group and let $\mathcal{A} := C(G)$, the algebra of all continuous functions on G. Then $C(G \times G)$ can be identified with the topological tensor product of $C(G)$ with $C(G)$, i.e., embed the algebraic tensor product of $C(G)$ with $C(G)$ in $C(G \times G)$ as before and then take the completion with respect to the supremum norm on $G \times G$.

Thus, in many cases, we can view the comultiplication Φ as a unital algebra homomorphism $\Phi: \mathcal{A} \rightarrow \mathcal{A} \otimes \mathcal{A}$. Here $\mathcal{A} \otimes \mathcal{A}$ is an algebra with multiplication such that $(a \otimes b)(c \otimes d) = ac \otimes bd$, and with unit $I \otimes I$.

We can also extend the multiplication in $\mathcal{A}$, which is initially given as the bilinear mapping $(a, b) \mapsto ab: \mathcal{A} \times \mathcal{A} \rightarrow \mathcal{A}$, to the linear mapping

$$m: \sum_i a_i \otimes b_i \longmapsto \sum_i a_i b_i: \mathcal{A} \otimes \mathcal{A} \longrightarrow \mathcal{A}.$$

Note that, if $F \in \mathcal{A} \otimes \mathcal{A}$, then

$$(m(F))(x) = F(x, x).$$

We will next rewrite the group axioms for G in terms of axioms for Φ, e and κ:
(i) The associativity $(xy)z = x(yz)$ yields $a((xy)z) = a(x(yz))$ for functions a on G. Thus we have the *coassociativity* axiom

$$(\Phi \otimes \mathrm{id}) \circ \Phi = (\mathrm{id} \otimes \Phi) \circ \Phi.$$

(ii) $ex = x = xe$ for $x \in G$. Hence $a(ex) = a(x) = a(xe)$. This yields the *counit* axiom

$$(e \otimes \mathrm{id}) \circ \Phi = \mathrm{id} = (\mathrm{id} \otimes e) \circ \Phi.$$

(iii) If G is a *commutative* group then $xy = yx$ for all $x, y \in G$. Hence $a(xy) = a(yx)$. Define the linear mapping $\sigma: \mathcal{A} \otimes \mathcal{A} \rightarrow \mathcal{A} \otimes \mathcal{A}$ such that $\sigma(a \otimes b) = b \otimes a$ (the *flip automorphism*). Then we have, for commutative G, the *cocommutativity* property

$$\sigma \circ \Phi = \Phi.$$

(iv) $x^{-1}x = e = xx^{-1}$ for $x \in G$. Hence $a(x^{-1}x) = a(e) = a(xx^{-1})$. We will rewrite $a(x^{-1}x)$ by using Φ, κ and m. Observe that

$$(\Phi(a))(x, y) = a(xy), \quad \text{hence} \quad (((\kappa \otimes \mathrm{id}) \circ \Phi)(a))(x, y) = a(x^{-1}y).$$

Thus

$$((m \circ (\kappa \otimes \mathrm{id}) \circ \Phi)(a))(x) = (((\kappa \otimes \mathrm{id}) \circ \Phi)(a))(x, x) = a(x^{-1}x).$$

We can rewrite $a(xx^{-1})$ in a similar way, while $a(e) = e(a)I(x)$. Thus we have the *antipode* axiom

$$(m \circ (\kappa \otimes \mathrm{id}) \circ \Phi)(a) = e(a)I = (m \circ (\mathrm{id} \otimes \kappa) \circ \Phi)(a).$$

The group G can often be recovered from $\mathrm{Fun}(G)$. As a first example consider an algebraic group G and let $\mathcal{A} := \mathrm{Pol}(G)$, the algebra of polynomial functions on G. Then $\chi: \mathcal{A} \to \mathbf{C}$ is a unital algebra homomorphism if and only if there exists $x \in G$ such that $\chi(a) = a(x)$, and this establishes a one-to-one correspondence $\chi \leftrightarrow x$ between characters χ of $\mathcal{A}$ and elements x of G. If $\chi_1(a) = a(x_1)$ and $\chi_2(a) = a(x_2)$ then

$$((\chi_1 \otimes \chi_2) \circ \Phi)(a) = a(x_1 x_2),$$

so we can also recover the group multiplication on G from the comultiplication on $\mathcal{A}$.

As a second example let G be a compact group and let $\mathcal{A} := C(G)$, the algebra of continuous functions on G. Then $C(G)$ is a commutative C*-algebra with unit, where the *-operation is given by pointwise complex conjugation: $a^*(x) = \overline{a(x)}$. The Gelfand theory for commutative C*-algebras tells us that $\chi: \mathcal{A} \to \mathbf{C}$ is a unital *-algebra homomorphism if and only if there exists $x \in G$ such that $\chi(a) = a(x)$.

Since $\Phi: \mathcal{A} \to \mathcal{A} \otimes \mathcal{A}$ is an algebra homomorphism, it is already determined by its action on a set of generators for the algebra $\mathcal{A}$. For instance, if $G := SL(2, \mathbf{C})$ and $\mathcal{A} := \mathrm{Pol}(G)$ then the functions $\alpha, \beta, \gamma, \delta$ which send $\begin{pmatrix} x & y \\ u & v \end{pmatrix}$ to x, y, u, v, respectively, form a set of generators of $\mathcal{A}$. Let us compute, for instance, $\Phi(\alpha)$.

$$(\Phi(\alpha)) \left(\begin{pmatrix} x_1 & y_1 \\ u_1 & v_1 \end{pmatrix}, \begin{pmatrix} x_2 & y_2 \\ u_2 & v_2 \end{pmatrix} \right) = \alpha \left(\begin{pmatrix} x_1 & y_1 \\ u_1 & v_1 \end{pmatrix} \begin{pmatrix} x_2 & y_2 \\ u_2 & v_2 \end{pmatrix} \right)$$

$$= \alpha \left(\begin{pmatrix} x_1 x_2 + y_1 u_2 & * \\ * & * \end{pmatrix} \right) = x_1 x_2 + y_1 u_2$$

$$= \alpha \left(\begin{pmatrix} x_1 & y_1 \\ u_1 & v_1 \end{pmatrix} \right) \alpha \left(\begin{pmatrix} x_2 & y_2 \\ u_2 & v_2 \end{pmatrix} \right) + \beta \left(\begin{pmatrix} x_1 & y_1 \\ u_1 & v_1 \end{pmatrix} \right) \gamma \left(\begin{pmatrix} x_2 & y_2 \\ u_2 & v_2 \end{pmatrix} \right)$$

$$= (\alpha \otimes \alpha + \beta \otimes \gamma) \left(\begin{pmatrix} x_1 & y_1 \\ u_1 & v_1 \end{pmatrix}, \begin{pmatrix} x_2 & y_2 \\ u_2 & v_2 \end{pmatrix} \right).$$

Similarly we compute the action of Φ on β, γ, δ. Thus we obtain:

$$\begin{aligned} \Phi(\alpha) &= \alpha \otimes \alpha + \beta \otimes \gamma, & \Phi(\beta) &= \alpha \otimes \beta + \beta \otimes \delta, \\ \Phi(\gamma) &= \gamma \otimes \alpha + \delta \otimes \gamma, & \Phi(\delta) &= \gamma \otimes \beta + \delta \otimes \delta, \end{aligned} \tag{2.1}$$

and this determines $\Phi(a)$ for all $a \in \mathcal{A}$.

The above derivation can be done in a more conceptual way if we use the notion of *matrix representation*, i.e. a homomorphism

$$\pi: x \longmapsto \begin{pmatrix} \pi_{11}(x) & \cdots & \pi_{1n}(x) \\ \vdots & & \vdots \\ \pi_{n1}(x) & \cdots & \pi_{nn}(x) \end{pmatrix}$$

of the group G into the group of invertible complex $n \times n$ matrices. Then the matrix elements π_{ij} are functions on G. Under further requirements on π these functions may be continuous, C^∞, analytic, polynomial, etc. Let us consider how Φ, e and κ act on the π_{ij}:

(i) We have $\pi(xy) = \pi(x)\pi(y)$, hence $\pi_{ij}(xy) = \sum_k \pi_{ik}(x) \otimes \pi_{kj}(y)$. Thus

$$\Phi(\pi_{ij}) = \sum_k \pi_{ik} \otimes \pi_{kj}. \tag{2.2}$$

(ii) We have $\pi(e) = I$, hence $\pi_{ij}(e) = \delta_{ij}$. Thus

$$e(\pi_{ij}) = \delta_{ij}.$$

(iii) We have $\pi(x^{-1})\pi(x) = \pi(e) = \pi(x)\pi(x^{-1})$, hence

$$\sum_k \pi_{ik}(x^{-1})\pi_{kj}(x) = \pi_{ij}(e) = \sum_k \pi_{ik}(x)\pi_{kj}(x^{-1}).$$

Thus

$$\sum_k \kappa(\pi_{ik})\,\pi_{kj} = \delta_{ij}I = \sum_k \pi_{ik}\,\kappa(\pi_{kj}).$$

In the example of $SL(2,\mathbf{C})$, with $g := \begin{pmatrix} x & y \\ u & v \end{pmatrix}$, we have a two-dimensional matrix representation

$$g \longmapsto \begin{pmatrix} \alpha(g) & \beta(g) \\ \gamma(g) & \delta(g) \end{pmatrix},$$

which explains (2.1) in view of (2.2).

2.2. DEFINITION OF HOPF ALGEBRA

The reader may now be sufficiently prepared to digest the general definition of a Hopf algebra. Let $\mathcal{A}$ be a complex linear space. Then $\mathcal{A}$ is a *Hopf algebra* if the following four properties are satisfied:

(i) $\mathcal{A}$ is an associative *algebra* with unit I, where the multiplication $(x,y) \mapsto xy$ is linearly extended to $m: \mathcal{A} \otimes \mathcal{A} \to \mathcal{A}$.

(ii) $\mathcal{A}$ is a coassociative *coalgebra* with comultiplication $\Phi: \mathcal{A} \to \mathcal{A} \otimes \mathcal{A}$ and counit $e: \mathcal{A} \to \mathbf{C}$ (linear mappings) satisfying

$$(\Phi \otimes \mathrm{id}) \circ \Phi = (\mathrm{id} \otimes \Phi) \circ \Phi \quad \text{and} \quad (e \otimes \mathrm{id}) \circ \Phi = \mathrm{id} = (\mathrm{id} \otimes e) \circ \Phi.$$

(iii) $\mathcal{A}$ is a *bialgebra*, i.e., $\mathcal{A}$ satisfies both (i) and (ii), and the mappings $\Phi: \mathcal{A} \to \mathcal{A} \otimes \mathcal{A}$ and $e: \mathcal{A} \to \mathbf{C}$ are unital algebra homomorphisms.

(iv) There is an *antipode* $\kappa: \mathcal{A} \to \mathcal{A}$ (linear mapping) satisfying

$$(m \circ (\kappa \otimes \mathrm{id}) \circ \Phi)(a) = e(a)I = (m \circ (\mathrm{id} \otimes \kappa) \circ \Phi)(a) \qquad \text{for all } a \in \mathcal{A}.$$

In general, $\kappa \circ \kappa \neq \mathrm{id}$ (κ is not involutive) and $\kappa(ab) \neq \kappa(a)\kappa(b)$ (κ is not multiplicative). However, it can be shown that

$$\kappa(ab) = \kappa(b)\kappa(a), \quad \kappa(I) = I, \quad \sigma \circ (\kappa \otimes \kappa) \circ \Phi = \Phi \circ \kappa, \quad e \circ \kappa = e,$$

so κ is an anti-multiplicative unital algebra homomorphism and an anti-comultiplicative counital coalgebra homomorphism.

2.3. COREPRESENTATIONS OF HOPF ALGEBRAS

Let $\mathcal{A}$ be a Hopf algebra. A *matrix corepresentation* of $\mathcal{A}$ is a matrix

$$u = \begin{pmatrix} u_{11} & \cdots & u_{1n} \\ \vdots & & \vdots \\ u_{n1} & \cdots & u_{nn} \end{pmatrix} \quad \text{with } u_{ij} \in \mathcal{A}$$

such that

$$\Phi(u_{ij}) = \sum_{k=1}^{n} u_{ik} \otimes u_{kj} \quad \text{and} \quad e(u_{ij}) = \delta_{ij}. \tag{2.3}$$

If $\mathcal{A} = \mathrm{Fun}(G)$ for some group G then this definition is equivalent to the definition of matrix representation of G. It is possible to give a more abstract definition of corepresentation without use of matrices, for which we refer to the literature.

It follows from the antipode axiom that

$$\sum_{k=1}^{n} \kappa(u_{ik}) u_{kj} = \delta_{ij} I = \sum_{k=1}^{n} u_{ik} \kappa(u_{kj}). \tag{2.4}$$

Two corepresentations u and v of $\mathcal{A}$ are called *equivalent* if u and v are matrices of the same size $n \times n$ and if there is an invertible complex $n \times n$ matrix s such that

$$su = vs \qquad \text{(matrix products)}.$$

A corepresentation u of $\mathcal{A}$ is called *irreducible* if u is not equivalent to a corepresentation v of the form

$$v = \begin{pmatrix} * & * \\ 0 & * \end{pmatrix},$$

i.e., if not for some m, $1 \le m \le n-1$, we have $v_{ij} = 0$ for all (i,j) such that $m+1 \le i \le n$, $1 \le j \le m$.

2.4. NOTES

There are two textbooks on Hopf algebras: Sweedler [42] and Abe [1]. A concise introduction to Hopf algebras is given in Hazewinkel [15, §37.1]. An informal account of some basic facts and examples can be found in Bergman [8]. Manin [26] discusses Hopf algebras in connection with quantum groups. We have introduced notation for Hopf algebra operations as in Woronowicz [49]. In fact, the notation Δ for comultiplication, ε for counit and S for antipode is more common, cf. [1], [11]. This last notation will be used in §7 for dual Hopf algebras, in particular for (quantized) universal enveloping algebras.

3. Quantum Groups

In this section we will introduce our main examples of quantum groups: the quantum analogues of the groups $SL(2, \mathbf{C})$ and $SU(2)$. We will also give the definition of Hopf ∗-algebras and of compact matrix quantum groups.

3.1. GENERATORS AND RELATIONS

Many special Hopf algebras are introduced by means of generators and relations. For instance, let $G := SL(2, \mathbf{C})$ and $\mathcal{A} := \text{Pol}(G)$, the algebra of polynomial functions on the algebraic group G. Then $\mathcal{A}$ as unital algebra is isomorphic to the commutative unital algebra with generators $\alpha, \beta, \gamma, \delta$ and with relation $\alpha\delta - \beta\gamma = I$. We can equivalently describe this algebra as the free commutative unital algebra generated by $\alpha, \beta, \gamma, \delta$ divided out by the (necessarily two-sided) ideal generated by $\alpha\delta - \beta\gamma - I$.

We might also describe $\mathcal{A}$ as the unital algebra with a priori non-commuting generators $\alpha, \beta, \gamma, \delta$ and relations

$$\begin{aligned}
&\alpha\beta - \beta\alpha = 0, \quad \alpha\gamma - \gamma\alpha = 0, \quad \beta\delta - \delta\beta = 0, \quad \gamma\delta - \delta\gamma = 0, \\
&\beta\gamma - \gamma\beta = 0, \quad \alpha\delta - \delta\alpha = 0, \quad \alpha\delta - \beta\gamma - I = 0.
\end{aligned} \tag{3.1}$$

So $\mathcal{A}$ is the free non-commutative unital algebra with generators $\alpha, \beta, \gamma, \delta$ divided out by the two-sided ideal generated by the left hand sides in (3.1).

Note also that the generators are precisely the matrix elements of the corepresentation $\begin{pmatrix} \alpha & \beta \\ \gamma & \delta \end{pmatrix}$. This determines the Hopf algebra structure of $\mathcal{A}$ completely.

In general, if a Hopf algebra is presented as algebra by means of generators and relations then it is sufficient to specify Φ, e and κ by their action on the generators, as these operations are multiplicative or anti-multiplicative with respect to the multiplication. If it is already given that the generators are the matrix elements of a corepresentation then the action of Φ and e on the generators follows by (2.3), while κ acting on the generators, if it exists, is uniquely determined by (2.4).

3.2. THE QUANTUM SL(2,C) GROUP

This celebrated quantum group is presented by a deformation of the relations (3.1). We will use a deformation parameter q, which we will fix at some nonzero complex value. For $q = 1$ the Hopf algebra $\text{Pol}(SL(2, \mathbf{C}))$ will be recovered.

Let $\mathcal{A}_q$ be the unital algebra with non-commuting generators $\alpha, \beta, \gamma, \delta$ and with relations

$$\alpha\beta = q\beta\alpha, \quad \alpha\gamma = q\gamma\alpha, \quad \beta\delta = q\delta\beta, \quad \gamma\delta = q\delta\gamma, \quad \beta\gamma = \gamma\beta,$$
$$\alpha\delta - q\beta\gamma = \delta\alpha - q^{-1}\beta\gamma = I. \tag{3.2}$$

We define a comultiplication Φ and a counit e on $\mathcal{A}_q$ by the requirement that $\begin{pmatrix} \alpha & \beta \\ \gamma & \delta \end{pmatrix}$ is a corepresentation of $\mathcal{A}_q$. This yields, by (2.3), again (2.1) for Φ acting on the generators, and

$$e(\alpha) = e(\delta) = 1, \quad e(\beta) = e(\gamma) = 0.$$

Also, if we put

$$\kappa(\alpha) = \delta, \quad \kappa(\delta) = \alpha, \quad \kappa(\beta) = -q^{-1}\beta, \quad \kappa(\gamma) = -q\gamma, \tag{3.3}$$

then (2.4) is satisfied in view of the relations (3.2). The only thing left for the proof that $\mathcal{A}_q$ is a Hopf algebra, is to show that the relations (3.2) are preserved by Φ and e extended as homomorphisms and by κ extended as anti-homomorphism. See for instance Woronowicz [50] for a proof that this is indeed the case.

We say that the Hopf algebra $\mathcal{A}_q$ is associated with the quantum group $SL_q(2, \mathbf{C})$. Observe that $\mathcal{A}_q$ is a deformation of the commutative Hopf algebra $\mathcal{A}_1 = \mathrm{Pol}(SL(2, \mathbf{C}))$ into a non-commutative Hopf algebra. As such, $\mathcal{A}_q$ is a kind of quantization of $\mathcal{A}_1$, which motivates the name quantum group. Observe also that $SL(2, \mathbf{C})$ is rigid within the category of complex Lie groups, but has a nontrivial deformation within the wider category of quantum groups.

3.3. HOPF *-ALGEBRAS

We will now pass from complex to real by introducing a *-operation. Recall that a *-algebra is a complex associative algebra with anti-linear mapping $a \mapsto a^*$ which is involutive and anti-multiplicative, i.e. $(a^*)^* = a$ and $(ab)^* = b^* a^*$. If the algebra possesses a unit I (which we will always assume) then it is also required that $I^* = I$.

A Hopf *-algebra is a Hopf algebra $\mathcal{A}$ with a mapping $a \mapsto a^*: \mathcal{A} \to \mathcal{A}$ such that $\mathcal{A}$ as an algebra becomes a *-algebra, the mappings $\Phi: \mathcal{A} \to \mathcal{A} \otimes \mathcal{A}$ and $e: \mathcal{A} \to \mathbf{C}$ are *-homomorphisms, and κ satisfies

$$\kappa \circ * \circ \kappa \circ * = \mathrm{id}.$$

The two mappings κ and $*$ will not necessarily commute.

A matrix corepresentation u of $\mathcal{A}$ is called unitary if

$$u_{ij}^* = \kappa(u_{ji}), \tag{3.4}$$

which can, in view of (2.4), be equivalently written as

$$\sum_k u_{ki}^* u_{kj} = \delta_{ij} I = \sum_k u_{ik} u_{jk}^*. \tag{3.5}$$

If $\mathcal{A}$ is presented by generators and relations, then the $*$-operation is already characterized by its action on the generators. If the generators are moreover the matrix elements of a unitary corepresentation of $\mathcal{A}$ then (3.4) shows how $*$ acts on them.

There is a one-to-one correspondence between the choice of a real form of a complex algebraic group G and the choice of a Hopf $*$-operation on $\mathrm{Pol}(G)$. For instance, if $\mathcal{A} = \mathrm{Pol}(SL(2, \mathbf{C}))$ then the group

$$SU(2) := \left\{ \begin{pmatrix} x & -\overline{u} \\ u & \overline{x} \end{pmatrix} \mid |x|^2 + |u|^2 = 1 \right\}$$

is a compact real form of $SL(2, \mathbf{C})$. Now define $a \mapsto a^*$ on $\mathcal{A}$ by first restricting the polynomial a to $SU(2)$, then taking pointwise complex conjugates and finally extending the resulting function to a holomorphic polynomial on $SL(2, \mathbf{C})$. Thus $\alpha^* = \delta$ and $\beta^* = -\gamma$. On the other hand, the real form $SU(2)$ can be recovered from our knowledge of the $*$-operation: $\chi: \mathcal{A} \to \mathbf{C}$ is a unital $*$-homomorphism if and only if there is $x \in SU(2)$ such that $\chi(a) = a(x)$ for all $a \in \mathcal{A}$.

3.4. THE QUANTUM SU(2) GROUP

From now on consider the Hopf algebra $\mathcal{A}_q$ with $0 < q < 1$. Often, the results will remain valid for the classical case $q = 1$, sometimes the case $q = 1$ has to be interpreted by taking a suitable limit. Our restriction is mainly for convenience—the assumption that q is real and nonzero would also be possible.

Let $a \mapsto a^*: \mathcal{A}_q \to \mathcal{A}_q$ make $\mathcal{A}_q$ into a Hopf $*$-algebra such that $\begin{pmatrix} \alpha & \beta \\ \gamma & \delta \end{pmatrix}$ is a unitary corepresentation. This yields

$$\alpha^* = \delta, \quad \delta^* = \alpha, \quad \beta^* = -q\gamma, \quad \gamma^* = -q^{-1}\beta.$$

It can be verified that the continuation of $*$ to $\mathcal{A}_q$ as an anti-linear anti-multiplicative mapping is well-defined in view of the relations (3.2). Thus $\mathcal{A}_q$ becomes a Hopf $*$-algebra, which we say to be *associated with the quantum group* $SU_q(2)$.

3.5. COMPACT MATRIX QUANTUM GROUPS

Let us bring a little analysis into this algebraic story. Given a Hopf $*$-algebra $\mathcal{A}$, generated by the matrix elements of a unitary corepresentation u of $\mathcal{A}$, we want to make $\mathcal{A}$ into a normed $*$-algebra, by analogy to $\mathrm{Pol}(SU(2))$ with respect to the sup norm.

For a Hilbert space $\mathcal{H}$ let $\mathcal{L}(\mathcal{H})$ be the algebra of all bounded linear operators on $\mathcal{H}$. It is a C*-algebra, as we have $\|T T^*\| = \|T\|^2$. By a $*$-representation π of $\mathcal{A}$ on $\mathcal{H}$ we mean a homomorphism $\pi: \mathcal{A} \to \mathcal{L}(\mathcal{H})$ of unital $*$-algebras. Define

$$\|a\| := \sup_{\pi} \|\pi(a)\|, \qquad \pi \text{ running over all } *\text{-representations of } \mathcal{A}.$$

Since u is a unitary corepresentation, we have $\|\pi(u_{ij})\| \le 1$ for all $*$-representations π and for all indices i, j. Hence, for all $a \in \mathcal{A}$, $\|a\|$ will be finite, so $\|.\|$ defines a seminorm on $\mathcal{A}$ and

$$\|a^* a\| = \|a\|^2, \qquad a \in \mathcal{A}. \tag{3.6}$$

If, moreover, $a = 0$ whenever $\|a\| = 0$, or if, equivalently, $\mathcal{A}$ has a faithful *-representation, then $\| \cdot \|$ is a norm on $\mathcal{A}$ satisfying (3.6). Let A be the norm completion of $\mathcal{A}$ as a normed linear space. Then the *-algebra operations will also extend to A and A will become a C*-algebra. Moreover, with a suitable definition of C*-tensor product $A \otimes A$, Φ also extends to a C*-homomorphism of A to $A \otimes A$.

We will say that $\mathcal{A}$ is the Hopf *-algebra and A is the Hopf C*-algebra associated with a *compact matrix quantum group*. Here the term *matrix quantum group* is used because $\mathcal{A}$ is generated by the matrix elements of a corepresentation and the quantum group is called *compact* because the *-algebra can be made into a normed *-algebra.

By way of example, let G be a compact Lie group. Equivalently, G is isomorphic to a closed subgroup of $SU(n)$, so we have a faithful unitary matrix representation of G. Let $\mathcal{A}$ be the Hopf *-algebra of polynomials in the matrix elements of this representation. Then the above construction makes $\mathcal{A}$ into a normed *-algebra with norm given by the supremum norm on G. The C*-algebra completion A of $\mathcal{A}$ can be identified with the (commutative) C*-algebra $C(G)$ of continuous functions on G. As we observed in §2.1, the group G can be recovered from A by considering the *-homomorphisms of A to $\mathbb{C}$. In fact, each commutative Hopf C*-algebra fits into this example, cf. [49, Theorem 1.5].

With the above norm and C*-algebra construction we have made contact with Woronowicz' general theory [49] of compact matrix quantum groups. In §4 we will state the main results of this theory. However, for special quantum groups as $SU_q(2)$, it is possible to derive such results without reference to the general theory, by only using the Hopf *-algebra structure, cf. Masuda e.a. [28], Vaksman and Soibelman [45].

3.6. IRREDUCIBLE *-REPRESENTATIONS OF $\mathcal{A}_q$

For a Hopf *-algebra $\mathcal{A}$ associated with a compact matrix quantum group there are two interesting representation theories: the irreducible unitary corepresentations of $\mathcal{A}$ on finite dimensional Hilbert spaces and the irreducible *-representations of $\mathcal{A}$ on possibly infinite-dimensional Hilbert spaces. If $\mathcal{A} = \mathrm{Pol}(G)$ with G a compact Lie group, then the irreducible *-representations of $\mathcal{A}$ are one-dimensional and correspond to point evaluations on the elements of G. By analogy, for non-commutative $\mathcal{A}$, the irreducible *-representations of $\mathcal{A}$ may be considered as the elements of the underlying quantum group.

Let us consider the classification of the irreducible *-representations of $\mathcal{A}_q$, cf. [45]. There is a family of one-dimensional representations χ^θ ($0 \le \theta < 2\pi$) and a family of infinite-dimensional representations τ^θ ($0 \le \theta < 2\pi$). The first family is given by

$$\chi^\theta(\alpha) = e^{i\theta}, \quad \chi^\theta(\alpha^*) = e^{-i\theta}, \quad \chi^\theta(\gamma) = \chi^\theta(\gamma^*) = 0.$$

The second family is defined on a Hilbert space $\mathcal{H}$ with orthonormal basis $v_0, v_1, \ldots$:

$$\tau^\theta(\gamma)\, v_n = e^{i\theta}\, \mu^n\, v_n, \quad \tau^\theta(\gamma^*)\, v_n = e^{-i\theta}\, \mu^n\, v_n,$$

$$\tau^\theta(\alpha)\, v_n = \begin{cases} (1 - \mu^{2n})^{1/2}\, v_{n-1} & \text{if } n > 0, \\ 0 & \text{if } n=0, \end{cases}$$

$$\tau^\theta(\alpha^*)\, v_n = (1 - \mu^{2n+2})^{1/2}\, v_{n+1}.$$

Furthermore, it can be shown that the representation τ defined by the direct integral

$$\tau := \int_{0 \le \theta < 2\pi}^{\oplus} \tau^\theta \, d\theta$$

is faithful and that $\|a\| = \|\tau(a)\|$ for all $a \in \mathcal{A}$. Finally, each *-representation of $\mathcal{A}_q$ is a direct integral of irreducible *-representations (so is a type I representation).

3.7. QUANTUM SUBGROUPS

Let $K := S(U(1) \times U(1))$ the subgroup of $SU(2)$ of diagonal matrices $\begin{pmatrix} e^{i\theta} & 0 \\ 0 & e^{-i\theta} \end{pmatrix}$. This subgroup is isomorphic to $U(1)$. The algebra $B := \mathrm{Pol}(K)$ has generators α, δ (functions evaluating left upper respectively right lower matrix element) with relations $\alpha\delta = \delta\alpha = I$ and *-operation given by $\alpha^* = \delta$. There is a comultiplication $\Psi: B \to B \otimes B$ given by $\Psi(\alpha) := \alpha \otimes \alpha$, $\Psi(\delta) := \delta \otimes \delta$. Now one way of expressing that K is a closed subgroup of $SU(2)$ is that there is a *surjective* *-homomorphism F of $\mathrm{Pol}(SU(2))$ onto $\mathrm{Pol}(K)$ (namely the restriction of functions on $SU(2)$ to K), which is moreover intertwining between the comultiplications on the two algebras.

In an analogous way we can consider K as a *quantum subgroup* of the quantum group $SU_q(2)$. The mapping $F: \mathcal{A}_q \to B$, for which $F(\alpha) := \alpha$, $F(\delta) := \delta$, $F(\beta) := F(\gamma) := 0$, defines a surjective *-homomorphism and the following diagram is commutative:

$$
\begin{array}{ccc}
\mathcal{A}_q & \xrightarrow{\;\Phi\;} & \mathcal{A}_q \otimes \mathcal{A}_q \\
\downarrow{\scriptstyle F} & & \downarrow{\scriptstyle F \otimes F} \\
B & \xrightarrow{\;\Psi\;} & B \otimes B
\end{array}
$$

Note that the one-dimensional *-representation χ^θ applied to $a \in \mathcal{A}_q$ is precisely the point evaluation of $F(a)$ at $\mathrm{diag}(e^{i\theta}, e^{-i\theta})$.

3.8. NOTES

Several motivations have been given for the particular way of defining the $SL(2, \mathbf{C})$ and $SU(2)$ quantum groups, cf. Woronowicz [50, Appendix A1] and Manin [26]. A strong reason for this particular choice is also that the Hopf algebra $\mathcal{A}_q$ is contained in the dual Hopf algebra of Jimbo's [16] quantized universal Hopf algebra for root system A_1, cf. §7.3. Thus, in principle, quantum analogues can be constructed of all complex or compact semisimple Lie groups.

See for instance Nijenhuis and Richardson [30] for rigidity of semisimple Lie algebras.

In [27], [28] and [32] the authors write y, v, u, x instead of our $\alpha, \beta, \gamma, \delta$, respectively.

Our use of the term *Hopf C*-algebras* (cf. §3.5) is informal. Different definitions of Hopf C*-algebras have been given in literature, cf. for instance Vallin [46].

Woronowicz [49] called compact matrix quantum groups originally *compact matrix pseudogroups*. He does not use Hopf algebras, but he starts with a pair (A, u) of a C*-algebra A with comultiplication and a corepresentation u, such that the matrix elements of u generate a dense *-subalgebra $\mathcal{A}$ of A and an antipode exists on $\mathcal{A}$. Our construction of the Hopf

C*-algebra A from the Hopf *-algebra $\mathcal{A}$ in §3.5 yields less general A than considered in [49]. For instance, in our approach the counit on $\mathcal{A}$ always extends to A, while this is not necessarily the case in [49].

Our definition of Hopf *-algebra in §3.3 was highly suggested by [49]. The same definition occurs in [44] and a quite similar definition in [26].

In [50] an explicit realization is given of the faithful representation τ of §3.6. The operators $\tau^\theta(\gamma\gamma^*)$, $\tau^\theta(\alpha)$ and $\tau^\theta(\alpha^*)$ of §3.6 form a q-analogue of the *Schrödinger representation of the Heisenberg algebra* and were studied already by Arik and Coon [3] and Feinsilver [13], long before quantum groups were introduced. These authors also considered a q-Fock representation. In this connection it is possible to give interpretations of q-analogues of Hermite polynomials.

4. Representation Theory of Compact Matrix Quantum Groups

Let $\mathcal{A}$ and A be a Hopf *-algebra, respectively Hopf C*-algebra associated with a compact matrix quantum group. In this section we summarize the powerful results of Woronowicz [49] on the representation theory of compact matrix quantum groups, i.e. the corepresentation theory of $\mathcal{A}$ and A.

4.1. HAAR FUNCTIONAL

The key to harmonic analysis, both on compact groups and on compact matrix quantum groups, is the Haar functional. Recall that, on a compact group G, we have a unique measure dx, the *Haar measure*, and corresponding functional h on $C(G)$, the *Haar functional*

$$h(a) := \int_G a(x)\, dx, \qquad a \in C(G),$$

with the following properties:

(i) $\int_G dx = 1$.

(ii) $\int_G a(x)\, dx \geq 0$ if $a \in C(G)$ is nonnegative.

(iii) $\int_G a(xy)\, dx = \int_G a(x)\, dx = \int_G a(yx)\, dx$ if $a \in C(G)$, $y \in G$.

THEOREM 4.1 (Woronowicz [49]). Let A be a Hopf C*-algebra. Then there is a unique linear functional h on A such that:

(i) $h(I) = 1$.

(ii) $h(a^* a) \geq 0$ for all $a \in A$.

(iii) $(h \otimes \mathrm{id})(\Phi(a)) = h(a)I = (\mathrm{id} \otimes h)(\Phi(a))$ for all $a \in A$.

Moreover, if $a \in A$ and $h(a^* a) = 0$ then $a = 0$.

Let us give h explicitly on $\mathcal{A}_q$. For this we use a certain basis of $\mathcal{A}_q$:

PROPOSITION 4.2 (Woronowicz [50]). The elements $\alpha^k \gamma^m (\gamma^*)^n$ $(k, m, n \geq 0)$ and $(\alpha^*)^k \gamma^m (\gamma^*)^n$ $(k > 0,\ m, n \geq 0)$ form a basis of $\mathcal{A}_q$.

Now we have (cf. Woronowicz [49, Appendix A.1]):

$$h(\alpha^k \gamma^m (\gamma^*)^n) = 0 = h((\alpha^*)^k \gamma^m (\gamma^*)^n) \qquad \text{if } k > 0 \text{ or } m \neq n,$$

$$h((\gamma \gamma^*)^n) = \frac{1 - q^2}{1 - q^{2n+2}}.$$

Hence, for a polynomial p we have

$$h(p(\gamma^* \gamma)) = (1 - q^2) \sum_{k=0}^{\infty} p(q^{2k}) q^{2k} = \int_0^1 p(x) \, d_{q^2} x, \tag{4.1}$$

where we used the notation for q-integrals (cf. [37]). However, quite surprisingly, it is also shown in [50, Appendix A.1] that

$$h(p(\tfrac{1}{2}(\alpha + \alpha^*))) = \frac{2}{\pi} \int_{-1}^1 p(x) \sqrt{1 - x^2} \, dx, \tag{4.2}$$

an ordinary integral.

The polynomials in $\gamma\gamma^*$ respectively $(\alpha + \alpha^*)/2$ are the $U(1)$-biinvariant respectively cocentral elements of $\mathcal{A}_q$. Here a is called $U(1)$-biinvariant if

$$(\chi^\theta \otimes \mathrm{id})(\Phi(a)) = a = (\mathrm{id} \otimes \chi^\theta)(\Phi(a)) \qquad \text{for all } \theta,$$

and a is called cocentral if $\sigma \circ \Phi(a) = \Phi(a)$.

4.2. SCHUR TYPE ORTHOGONALITY RELATIONS

Let G be a compact Lie group. Choose for each equivalence class of irreducible unitary representations of G a matrix representation $(t^\sigma_{mn})_{m,n=1,\ldots,d_\sigma}$ as representative. Then the well-known *Schur orthogonality relations* state that

$$\int_G t^\sigma_{mk}(x) \, \overline{t^\tau_{nl}(x)} \, dx = d_\sigma^{-1} \delta_{\sigma\tau} \delta_{mn} \delta_{kl}. \tag{4.3}$$

If, moreover, G is a closed subgroup of some $U(n)$ and if $\mathcal{A}$ is the $*$-algebra generated by the matrix elements of the natural representation of G then all t^σ_{mn} are contained in $\mathcal{A}$ and they form a basis of $\mathcal{A}$.

Woronowicz [49] obtained the quantum group analogue of (4.3). In order to formulate this, let $\mathcal{A}$ be a Hopf $*$-algebra associated with a compact matrix quantum group and choose for each equivalence class of irreducible unitary corepresentations of $\mathcal{A}$ a representative $(t^\sigma_{mn})_{m,n=1,\ldots,d_\sigma}$.

THEOREM 4.3. There exists a unique unital multiplicative linear functional f on $\mathcal{A}$ such that

$$h(t^\sigma_{mk} (t^\tau_{nl})^*) = \frac{\delta_{\sigma\tau} \delta_{mn} f(t^\sigma_{lk})}{f(\sum_{l=1}^{d_\sigma} t^\sigma_{ll})},$$

$$h((t^\sigma_{km})^* t^\tau_{ln}) = \frac{\delta_{\sigma\tau} \delta_{mn} \overline{f((t^\sigma_{lk})^*)}}{f(\sum_{l=1}^{d_\sigma} t^\sigma_{ll})}.$$

Furthermore, the t^σ_{mn} form a basis of $\mathcal{A}$. Any finite-dimensional matrix corepresentation of $\mathcal{A}$ is equivalent to a direct sum of irreducible corepresentations of $\mathcal{A}$.

The occurrence of f in this quantum Schur theorem is a new phenomenon, by which matrix elements belonging to the same representation are no longer orthogonal in a straightforward way. This phenomenon is related to the fact that the Haar functional is not necessarily central, i.e. $h(ab) \neq h(ba)$ in general.

For A_q the unital multiplicative linear functional f was computed in [49, Appendix A.1]. The result is:

$$f(\alpha) = q^{-1}, \quad f(\delta) = q, \quad f(\beta) = f(\gamma) = 0. \tag{4.4}$$

5. Little q-Jacobi Polynomials Interpreted on $SU_q(2)$

Given an explicit compact matrix quantum group like $SU_q(2)$ one may try to realize the following program:
(i) Classify the irreducible unitary corepresentations t^σ.
(ii) Choose suitable bases of the corepresentation spaces, by which one obtains matrix corepresentations (t^σ_{mn}).
(iii) Compute the t^σ_{mn} as polynomials in the generators or other suitable primitives.
(iv) Recognize these polynomials as special functions and rewrite the Schur type orthogonality relations as orthogonality relations for these special functions.

We will discuss this program here for the case $SU_q(2)$. Let us first recall the classical case of $SU(2)$, cf. for instance Vilenkin [47, Ch. 3] or Stanton's tutorial [41].

5.1. IRREDUCIBLE UNITARY REPRESENTATIONS OF SU(2)

Let $l \in \frac{1}{2}\mathbf{Z}_+$. Consider the $(2l+1)$-dimensional space of homogeneous polynomials of degree $2l$ in two complex variables, with inner product such that the polynomials

$$e^l_n(\xi, \eta) := \binom{2l}{l-n}^{1/2} \xi^{l-n} \eta^{l+n}, \quad n = -l, -l+1, \ldots, l,$$

form an orthonormal basis. Define a representation t^l of $SU(2)$ on this vector space by

$$\left(t^l \begin{pmatrix} x & y \\ u & v \end{pmatrix} f \right)(\xi, \eta) := f(x\xi + u\eta, y\xi + v\eta).$$

If t^l has matrix elements t^l_{mn} with respect to the basis vectors e^l_n then it follows that

$$\binom{2l}{l-n}^{1/2} (x\xi + u\eta)^{l-n} (y\xi + v\eta)^{l+n}$$
$$= \sum_{m=-l}^{l} t^l_{mn}\left(\begin{pmatrix} x & y \\ u & v \end{pmatrix} \right) \binom{2l}{l-m}^{1/2} \xi^{l-m} \eta^{l+m}.$$

The basis is such that the matrix $(t^l_{mn}(g))$ becomes diagonal when g is in the diagonal subgroup $U(1)$:

$$t^l_{mn}\left(\begin{pmatrix} e^{i\theta} & 0 \\ 0 & e^{-i\theta} \end{pmatrix} \right) = e^{-2in\theta} \delta_{mn}.$$

We know that the representations $t^l = (t^l_{mn})$ are unitary and irreducible and that each irreducible unitary representation of $SU(2)$ is equivalent to a t^l. Furthermore, the t^l_{mn} are expressible in terms of Jacobi polynomials, and the Schur orthogonality relations

$$\int_G t^l_{mn}(g) \overline{t^{l'}_{mn}(g)}\, dg = \frac{\delta_{l,l'}}{2l+1}$$

are equivalent to orthogonality relations for Jacobi polynomials.

5.2. IRREDUCIBLE UNITARY COREPRESENTATIONS OF $\mathcal{A}_q$

These can be classified by various methods, cf. Woronowicz [50], Vaksman and Soibelman [45], Masuda e.a. [27], [28], and the author [21]. Here we follow the approach of [21].

Fix $0 < q < 1$. Let

$$\begin{bmatrix} n \\ k \end{bmatrix}_q := \frac{(q;q)_n}{(q;q)_k\,(q;q)_{n-k}}.$$

We will use the following lemma, which follows easily by complete induction:

LEMMA 5.1. Let $xy = qyx$. Then

$$(x+y)^n = \sum_{k=0}^{n} \begin{bmatrix} n \\ k \end{bmatrix}_q y^{n-k}\, x^k = \sum_{k=0}^{n} \begin{bmatrix} n \\ k \end{bmatrix}_{q^{-1}} x^k\, y^{n-k}. \tag{5.1}$$

Let $l, n \in \frac{1}{2}\mathbf{Z}_+$. By definition of Φ acting on $\mathcal{A}_q$ we have

$$\Phi\left(\begin{bmatrix} 2l \\ l-n \end{bmatrix}_{q^{-2}}^{1/2} \alpha^{l-n} \gamma^{l+n} \right)$$
$$= \begin{bmatrix} 2l \\ l-n \end{bmatrix}_{q^{-2}}^{1/2} (\alpha \otimes \alpha + \beta \otimes \gamma)^{l-n} (\gamma \otimes \alpha + \delta \otimes \gamma)^{l+n}. \tag{5.2}$$

Now expand the right hand side of (5.2) by use of (5.1) and (3.2), such that we get monomials $\alpha^{l-m} \gamma^{l+m}$ in the tensor factors on the right. Then

$$\begin{bmatrix} 2l \\ l-n \end{bmatrix}_{q^{-2}}^{1/2} (\alpha \otimes \alpha + \beta \otimes \gamma)^{l-n} (\gamma \otimes \alpha + \delta \otimes \gamma)^{l+n}$$
$$= \sum_{m=-l}^{l} t^l_{nm} \otimes \begin{bmatrix} 2l \\ l-m \end{bmatrix}_{q^{-2}}^{1/2} \alpha^{l-m} \gamma^{l+m}, \tag{5.3}$$

where the sum runs over $m = -l, -l+1, \ldots, l$ and the t^l_{nm} are certain elements of $\mathcal{A}_q$. Application of the coassociativity and counit axioms to the left hand side of (5.2) and the right hand side of (5.3) shows that $t^l = (t^l_{nm})$ is a matrix corepresentation of $\mathcal{A}_q$. One also sees that (t^l_{nm}) becomes diagonal with respect to the quantum subgroup $U(1)$:

$$\chi^\theta(t^l_{nm}) = e^{-2in\theta}\, \delta_{nm}.$$

Formula (5.3) can be considered as a generating function for the matrix elements t_{nm}^l. From (5.3) another generating function can be derived which sums both over m and n.

The coefficients $\left[\begin{smallmatrix} 2l \\ l-n \end{smallmatrix}\right]_{q^{-2}}^{1/2}$ in (5.2), (5.3) were inserted, because it will turn out that they make the corepresentations unitary. They also make the matrix elements more symmetric, as we will state now. Let $a(\alpha,\beta,\gamma,\delta)$ be some algebraic expression in the generators $\alpha,\beta,\gamma,\delta$ of $\mathcal{A}_q$, which yields an element of $\mathcal{A}_q$. Let $\tilde{a}(\alpha,\beta,\gamma,\delta)$ be the expression obtained by reversing the order of the factors in all terms of $a(\alpha,\beta,\gamma,\delta)$. Now it follows from the relations (3.2) that the mapping

$$a(\alpha,\beta,\gamma,\delta) \longmapsto a(\alpha,\gamma,\beta,\delta)$$

is an algebra isomorphism of $\mathcal{A}_q$, while the mapping

$$a(\alpha,\beta,\gamma,\delta) \longmapsto \tilde{a}(\delta,\beta,\gamma,\alpha)$$

is an algebra anti-isomorphism of $\mathcal{A}_q$. Write $t_{nm}^l(\alpha,\beta,\gamma,\delta)$ in order to emphasize that t_{nm}^l is some algebraic expression in $\alpha,\beta,\gamma,\delta$. It turns out, just by inspecting (5.3) and the double sum generating function, that the t_{nm}^l satisfy the following symmetries:

$$t_{nm}^l(\alpha,\beta,\gamma,\delta) = t_{mn}^l(\alpha,\gamma,\beta,\delta) \tag{5.4}$$

$$= (t_{-n,-m}^l)\tilde{\ }(\delta,\gamma,\beta,\alpha) \tag{5.5}$$

$$= (t_{-m,-n}^l)\tilde{\ }(\delta,\beta,\gamma,\alpha). \tag{5.6}$$

THEOREM 5.2. The corepresentations t^l are unitary and irreducible. Moreover, each irreducible unitary matrix corepresentation of $\mathcal{A}_q$ is equivalent to some t^l.

Here the unitariness follows from (5.4), the irreducibility from reduction to the quantum subgroup $U(1)$ and the nonvanishing of the t_{nl}^l, the completeness by observing from a somewhat more explicit expression for the t_{nm}^l that these elements form a basis of $\mathcal{A}_q$.

5.3. LITTLE q-JACOBI POLYNOMIALS

Recall (cf. [2]) that *little q-Jacobi polynomials* are defined by

$$p_n(x;a,b;q) := {}_2\phi_1\left(\begin{matrix} q^{-n}, q^{n+1}ab \\ aq \end{matrix}; q, qx\right) \tag{5.7}$$

and that they satisfy orthogonality relations

$$\frac{(q^{\alpha+1};q)_\infty (q^{\beta+1};q)_\infty}{(1-q)(q;q)_\infty (q^{\alpha+\beta+2};q)_\infty} \int_0^1 p_n(x) p_m(x)\, x^\alpha\, \frac{(qx;q)_\infty}{(q^{\beta+1}x;q)_\infty}\, d_q x$$

$$= \frac{q^{n(\alpha+1)}(1-q^{\alpha+\beta+1})(q^{\beta+1};q)_n (q;q)_n}{(1-q^{2n+\alpha+\beta+1})(q^{\alpha+1};q)_n (q^{\alpha+\beta+1};q)_n}\, \delta_{nm}, \qquad \alpha,\beta > -1. \tag{5.8}$$

We call the special polynomials $p_n(x;1,1;q)$ *little q-Legendre polynomials*. Now we obtain from (5.3), by straightforward but somewhat tedious computations:

THEOREM 5.3.   We have

$$t^l_{nm} = c^l_{nm} (\alpha^*)^{n+m} \, p_{l-n}(\gamma\gamma^*; q^{2(n-m)}, q^{2(n+m)}; q^2) \, \gamma^{n-m},$$ (5.9)

where $n \geq m \geq -n$ and

$$c^l_{nm} := \begin{bmatrix} l-m \\ n-m \end{bmatrix}_{q^2}^{1/2} \begin{bmatrix} l+n \\ n-m \end{bmatrix}_{q^2}^{1/2} q^{-(n-m)(l-n)}.$$ (5.10)

This theorem was successively but independently proved by Vaksman and Soibelman [45], Masuda e.a. [27], [28], and the author [21]. Expressions for t^l_{nm} in case of the other three possibilities for n and m follow from (5.9) and the symmetries (5.4)–(5.6). In particular, for $l \in \mathbf{Z}_+$ we have

$$t^l_{00} = p_l(\gamma\gamma^*; 1, 1; q^2),$$ (5.11)

a little q-Legendre polynomial of argument $\gamma\gamma^*$.

The Schur type orthogonality (cf. Theorem 4.3 and formula (4.4)) yields

$$h((t^k_{pr})^* \, t^l_{nm}) = \delta_{kl} \, \delta_{pn} \, \delta_{rm} \, \frac{q^{2(l-n)} \, (1-q^2)}{1 - q^{2(2l+1)}}.$$ (5.12)

By substitution of (5.9) and (4.1) this is seen to be equivalent to the orthogonality relations (5.8) for little q-Jacobi polynomials.

It is also possible to identify the matrix elements as special orthogonal polynomials, when we use (5.12) and have some a priori information about the algebraic structure of the matrix elements. For instance, if we already know that, for $l \in \mathbf{Z}_+$, t^l_{00} is a polynomial p_l of degree l in $\gamma\gamma^*$ then we obtain from (5.12), for $k \neq l$:

$$0 = h((t^k_{00})^* \, t^l_{00}) = \int_0^1 p_l(x) \, p_k(x) \, d_{q^2}x,$$

from which we conclude that $p_l(x) = \text{const.} \, p_l(x; 1, 1; q^2)$.

5.4. NOTES

Lemma 5.1 is a folk lemma which was often reproved in the literature. In a different formulation it seems to be present already in the works of Netto and MacMahon. It can also be found in Schützenberger [39], Cigler [10] and Feinsilver [12].

In [45], [27] and [28] the second order q-difference eigenvalue equation for the little q-Jacobi polynomials was obtained from the interpretation on $SU_q(2)$ by use of the Casimir element in the dual Hopf algebra to $\mathcal{A}_q$.

6. Summary of Further Interpretations on Quantum Groups

In this section we briefly indicate some other classes of q-hypergeometric orthogonal polynomials and functions which have an interpretation on quantum groups.

6.1. q-MEIXNER-KRAWTCHOUK POLYNOMIALS

For fixed $g \in SU(2)$ the representation matrix $(t^l_{mn}(g))$ dicussed in §5.1 is unitary. It is possible to express these matrix elements in terms of Krawtchouk polynomials, such that the orthogonality relations between the rows or columns of the unitary matrix are equivalent to the orthogonality relations for the Krawtchouk polynomials, cf. [31, §12.7] and [20, §2]. In a similar way (cf. Koornwinder [21]) we can rewrite the orthogonality relations (3.5) for the unitary matrix corepresentation (t^l_{nm}) of $\mathcal{A}_q$ as orthogonality relations for the q-Meixner-Krawtchouk polynomials

$$K_n(q^{-x}; b, N; q) := {}_2\phi_1(q^{-n}, q^{-x}; q^{-N}; q, bq^{n+1}), \qquad n, x = 0, 1, \ldots, N,$$

(an ad hoc notation). Before [21] these polynomials had not been recognized in literature as a separate family of orthogonal polynomials. They are q-analogues of Krawtchouk polynomials which can be obtained from the q-Meixner polynomials by specializing a parameter such that the support of the orthogonality measure becomes finite. Note that the orthogonality relations (3.5) are identities in $\mathcal{A}_q$. These can be made into scalar identities by first rewriting them as operator identities by means of the representations τ^θ of §3.6, and next taking matrix elements of these operators with respect to the basis vectors v_n of the representation space.

6.2. q-HAHN and q-RACAH POLYNOMIALS

Clebsch-Gordan coefficients obtained by decomposing the tensor product of two irreducible representations of $SU(2)$ as a direct sum of irreducible representations, can be expressed in terms of Hahn polynomials (cf. for instance [19]). Racah coefficients, which give the transformation between two canonical ways of decomposing a threefold tensor product of irreducible representations of $SU(2)$ as a direct sum of irreducible representations, can be expressed in terms of Racah polynomials (cf. [48]). There are analogous results for $SU_q(2)$.

Define the *tensor product* of two matrix corepresentations (u_{ij}) and (v_{ij}) of a Hopf algebra $\mathcal{A}$ as the matrix corepresentation $(w_{ik,jl})$, where $w_{ik,jl} := u_{ij}v_{kl}$. Then, for the tensor product of matrix corepresentations (t^l_{nm}) of $\mathcal{A}_q$ we have the direct sum decomposition

$$t^{l_1} \otimes t^{l_2} = \bigoplus_{l = l_1 + l_2, l_1 + l_2 - 1, \ldots, |l_1 - l_2|} t^l. \tag{6.1}$$

Clebsch-Gordan coefficients for this decomposition were considered by Kirillov and Reshetikhin [17], Vaksman [43] and Koelink and Koornwinder [18]. Here we sketch the approach of [18]. Consider the linear subspace of $\mathcal{A}_q$ with basis vectors

$$\alpha^{l_1 - n_1} \gamma^{l_1 + n_1} \beta^{l_2 - n_2} \delta^{l_2 + n_2}, \qquad n_i = -l_i, -l_i + 1, \ldots, l_i, \quad i = 1, 2.$$

There is a natural realization of the corepresentation $t^{l_1} \otimes t^{l_2}$ on this space. Another basis of this space is given by the matrix elements $t^l_{n, l_2 - l_1}$, where $l = l_1 + l_2, l_1 + l_2 - 1, \ldots, |l_1 - l_2|$, $n = -l, -l+1, \ldots, l$, and the direct summands in (6.1) have a natural realization on vectors in this second basis. The matrix elements of the transformation matrix from the first to the second basis are called Clebsch-Gordan coefficients. It was shown in [18] that they can be expressed in terms of q-Hahn polynomials.

Racah coefficients associated with decompositions of threefold tensor products of corepresentations t^l of $\mathcal{A}_q$ were studied by Kirillov and Reshetikhin [17]. They expressed them in terms of q-Racah polynomials.

6.3. ADDITION FORMULA AND WALL POLYNOMIALS

Let $l \in \mathbf{Z}_+$, $g, h \in SU(2)$, t^l as in §5.1. Then

$$t^l_{00}(gh) = \sum_{k=-l}^{l} t^l_{0k}(g)\, t^l_{k0}(h).$$

By expressing g and h in a suitable way in terms of coordinates on the group, we get the addition formula for Legendre polynomials from this identity, cf. for instance [47, Ch. 3].

Something analogous for $SU_q(2)$ was done in Koornwinder [22]. For the corepresentation t^l of $\mathcal{A}_q$ we have

$$\Phi(t^l_{00}) = \sum_{k=-l}^{l} t^l_{0k} \otimes t^l_{k0}.$$

If we substitute (5.9), (5.10) and, in particular (5.11), then we get an identity in $\mathcal{A}_q \otimes \mathcal{A}_q$ which can be considered as an expansion of the little q-Legendre polynomial of degree l and of argument $\Phi(\gamma\gamma^*)$, quite analogous to the addition formula for Legendre polynomials. The passage to a scalar identity is along similar lines as in §6.1. However, now we have to take matrix elements of operators on $\mathcal{H} \otimes \mathcal{H}$, where $\mathcal{H}$ is the representation space of the representation τ^θ of §3.6. These matrix elements are taken on one side with respect to the standard basis, but on the other side with respect to a basis defined by means of *Wall polynomials*

$$p_n(x; a, 0; q) := {}_2\phi_1(q^{-n}, 0; aq; q, qx),$$

(specialization of little q-Jacobi polynomials (5.7)). The resulting addition formula for little q-Jacobi polynomials, cf. [22, Theorem 4.1], expands a left hand side

$$p_l(q^x; 1, 1; q) p_y(q^z; q^x, 0; q),$$

considered as a function of q^z, in terms of Wall polynomials $p_{y+k}(q^z; q^x, 0; q)$. A typical term in the expansion equals

$$\text{const.}\, p_{l-k}(q^{x+y}; q^k, q^k; q)\, p_{l-k}(q^y; q^k, q^k; q)\, p_{y+k}(q^z; q^x, 0; q).$$

Subsequently, Rahman [36] has given an analytic proof of this addition formula, while van Assche and Koornwinder [7] have shown that the formula tends to the addition formula for Legendre polynomials as q tends to 1.

In yet unpublished work the author has given a conceptual interpretation of the occurrence of Wall polynomials in the addition formula. It turns out that Wall polynomials occur as Clebsch-Gordan coefficients in the direct integral decomposition for $\tau^{\theta_1} \otimes \tau^{\theta_2}$.

6.4. SPHERICAL FUNCTIONS FOR $SU_q(n+1)/SU_q(n)$

Recently, Noumi, Yamada and Mimachi [34] announced an interpretation of little q-Jacobi polynomials $p_m(z; q^{2(n-1)}, q^{2(l-m)}; q^2)$ as matrix elements of irreducible corepresentations of the quantum group $SU_q(n+1)$ which are biinvariant with respect to the quantum subgroup $SU_q(n)$. This result strengthens the expectation that, parallel to the theory of spherical functions on compact symmetric (or other homogeneous) spaces, a similar theory can be developed for quantum groups.

6.5. QUANTUM 2-SPHERES AND BIG q-JACOBI POLYNOMIALS

Let $\mathcal{A}$ be a Hopf algebra and $\mathcal{B}$ be an associative algebra. By a *coaction* of $\mathcal{A}$ on $\mathcal{B}$ we mean a unital algebra homomorphism $\Psi\colon \mathcal{A} \to \mathcal{A} \otimes \mathcal{B}$ such that (i) $(\Phi \otimes \mathrm{id}) \circ \Psi = (\mathrm{id} \otimes \Psi) \circ \Psi$ and (ii) $(e \otimes \mathrm{id}) \circ \Psi) = \mathrm{id}$. The guiding example is the case that G is a group acting on a space X, $\mathcal{A} = \mathrm{Fun}(G)$, $\mathcal{B} = \mathrm{Fun}(X)$ and $(\Psi(b))(g,x) := b(g.x)$. If $\mathcal{A}$ is a Hopf $*$-algebra and $\mathcal{B}$ a $*$-algebra we define a $*$-coaction as a coaction Ψ which is also a $*$-homomorphism. One can think about $\mathcal{A}$ as the dual of a quantum group G and about $\mathcal{B}$ as the dual of a *quantum space* X. Then one has a *quantum action* of G on X. A coaction of $\mathcal{A}$ on $\mathcal{B}$ defines a (usually infinite-dimensional) corepresentation of $\mathcal{A}$ on $\mathcal{B}$. One can try to decompose this into irreducible subspaces.

For the case $SU_q(2)$ Podleś [35] has defined quantum actions on so called quantum spheres. Next, Noumi and Mimachi [32] have given explicit orthogonal bases for the irreducible subspaces of the algebras corresponding to these quantum spheres. These bases are chosen such that the action of the quantum subgroup $U(1)$ is diagonalized. They obtain big q-Jacobi polynomials $P_n^{(\alpha,\beta)}(x;c,d;q)$ and also, for certain quantum spheres, q-Hahn polynomials $Q_n(x;q^\alpha,q^\beta,N;q)$, in both cases with $\alpha = \beta$ and q replaced by q^2. Next, in [33], Noumi and Mimachi also find a realization for the nonsymmetric case $\alpha \neq \beta$, this time on quantum 3-spheres.

6.6. QUANTUM GROUP OF PLANE MOTIONS AND q-BESSEL FUNCTIONS

Vaksman and Korogodsky [44] studied the quantum analogue of the group of Euclidean motions of the plane. This work is extremely interesting, since it is the first example of harmonic analysis on a quantum analogue of a noncompact Lie group. Just as the irreducible unitary representations of the group of plane motions have matrix elements expressible in terms of Bessel functions (cf. for instance [47]), so the matrix elements in the quantum case are expressible in terms of q-Bessel functions given as $_1\phi_1$ q-hypergeometric series, i.e., different from the more common notion (cf. [14]) of q-Bessel functions as $_0\phi_1$ functions.

6.7. QUANTUM SU(1,1) GROUP

In [29] Masuda e.a. studied the quantum group $SU_q(1,1)$, the quantum analogue of the noncompact semisimple Lie group $SU(1,1)$. The authors obtained series of infinite-dimensional unitary representations of this quantum group. They computed matrix elements of these representations in terms of $_2\phi_1$ q-hypergeometric functions.

6.8. q-EXPONENTIAL FUNCTION

If $xy = qyx$ then $e_q(x+y) = e_q(y)e_q(x)$, where $e_q(x)$ is the q-exponential function $\sum_{k=0}^{\infty} x^k/(q;q)_k$. This result, due to Schützenberger [39], follows easily from Lemma 5.1. In yet unpublished work the author has shown that the functions $x \mapsto e_q(cx)$ occur as one-dimensional representations of a quantum group version of the additive group $\mathbf{R}$ of real numbers. For the construction of this quantum group one needs a slight generalization of the definition of Hopf algebra $\mathcal{A}$. The usual definition of multiplication on $\mathcal{A} \otimes \mathcal{A}$ can be viewed as the mapping $(m \otimes m) \circ (\mathrm{id} \otimes \sigma \otimes \mathrm{id})$ from $\mathcal{A} \otimes \mathcal{A} \otimes \mathcal{A} \otimes \mathcal{A}$ to $\mathcal{A} \otimes \mathcal{A}$. Here σ is the flip automorphism. In our generalization we change the definition of this flip.

Let $\mathcal{A}$ be the free unital algebra generated by one indeterminate α. Fix $q \neq 0$. Define

$$\sigma(\alpha^k \otimes \alpha^l) := q^{2kl} \alpha^l \otimes \alpha^k.$$

Then multiplication on $\mathcal{A} \otimes \mathcal{A}$ satisfies

$$(\alpha^k \otimes \alpha^l)(\alpha^m \otimes \alpha^n) = q^{2lm} \alpha^{k+m} \otimes^{l+n} .$$

We can extend the comultiplication, defined on the generator by

$$\Phi(\alpha) := \alpha \otimes I + I \otimes \alpha,$$

to an algebra homomorphism $\Phi: \mathcal{A} \to \mathcal{A} \otimes \mathcal{A}$ by putting

$$\Phi(\alpha^n) := \sum_{k=0}^{n} \begin{bmatrix} n \\ k \end{bmatrix}_{q^2} \alpha^{n-k} \otimes \alpha^k.$$

Now $\Phi(a) = a \otimes a$ if $a = e_{q^2}(c\alpha)$.

7. Continuous q-Legendre Polynomials

In the sections 5 and 6 we listed an impressive collection of q-hypergeometric orthogonal polynomials admitting an interpretation on the quantum group $SU_q(2)$. What we were still badly missing there, are the *Askey-Wilson polynomials* [6] themselves, i.e., polynomials

$$p_n(\cos\theta; a, b, c, d \mid q) := a^{-n} (ab; q)_n (ac; q)_n (ad; q)_n \times$$
$$\times {}_4\phi_3 \left(\begin{matrix} q^{-n}, q^{n-1} abcd, ae^{i\theta}, ae^{-i\theta} \\ ab, ac, ad \end{matrix} ; q, q \right), \qquad (7.1)$$

which are, for $q, a, b, c, d \in (-1, 1)$, orthogonal polynomials on $(-1, 1)$ with respect to a continuous weight function. There is one case where Askey-Wilson polynomials already occurred in connection with $SU_q(2)$: we have the *character formula*

$$\mathrm{tr}(t^l) := \sum_{n=-l}^{l} t^l_{nn} = U_{2l} \left(\frac{\alpha + \alpha^*}{2} \right),$$

cf. [49, Appendix A1]. Note that the right hand side is independent of q. Here the U_n are the *Chebyshev polynomials of the second kind*

$$U_n(\cos\theta) := \frac{\sin(n+1)\theta}{\sin\theta}.$$

These are usually considered as special Jacobi polynomials, but, as pointed out in [6, p.17], they can also be written as special Askey-Wilson polynomials:

$$U_n(\cos\theta) = \frac{p_n(\cos\theta; q, -q, q^{1/2}, -q^{1/2} \mid q)}{(q^{n+2}; q)_n}.$$

Note that the left hand side is independent of q, although the right hand side would suggest the contrary.

It will turn out that certain Askey-Wilson polynomials can be interpreted as spherical matrix elements of corepresentations t^l, where the notion *spherical* has to be specified. I obtained this interpretation first for the *continuous q-Legendre polynomials*, for which we use here the ad hoc notation

$$P_n(\cos\theta \mid q) := \frac{p_n(\cos\theta; q^{1/2}, -q^{1/2}, q^{1/2}, -q^{1/2} \mid q)}{p_n((q^{1/2} + q^{-1/2})/2; q^{1/2}, -q^{1/2}, q^{1/2}, -q^{1/2} \mid q)}$$

$$= {}_4\phi_3 \left(\begin{matrix} q^{-n}, q^{n+1}, q^{1/2}e^{i\theta}, q^{1/2}e^{-i\theta} \\ q, -q, -q \end{matrix} ; q, q \right).$$

By [6, (4.20) and (4.2)] these polynomials are equal, up to a constant factor, to special continuous q-ultraspherical polynomials $C_n(\cos\theta; q \mid q^2)$. They are also Macdonald's ([24], [25]) orthogonal polynomials P_λ with parameters q, t and associated with root system A_1 if $t = q^{\frac{1}{2}}$.

I obtained the key for the interpretation on $SU_q(2)$ from the formula giving the explicit expansion of $P_n(\cos\theta \mid q)$ as a finite Fourier series:

$$P_n(\cos \mid q) = q^{n/2} \sum_{k=0}^{n} \frac{(q; q^2)_k (q; q^2)_{n-k}}{(q^2; q^2)_k (q^2; q^2)_{n-k}} e^{i(n-2k)\theta}, \qquad (7.2)$$

cf. [4, (3.1)]. The $q = 1$ analogue and limit case of this formula is the following finite Fourier series for *Legendre polynomials*:

$$P_n(\cos\theta) = \sum_{k=0}^{n} \frac{(1/2)_k (1/2)_{n-k}}{k! (n-k)!} e^{i(n-2k)\theta}. \qquad (7.3)$$

It turned out that the group theoretic interpretation of (7.3) could be imitated in order to obtain a quantum group theoretic interpretation of (7.2).

7.1. THE FOURIER SERIES FOR LEGENDRE POLYNOMIALS INTERPRETED

Put $a_\theta := \begin{pmatrix} e^{i\theta} & 0 \\ 0 & e^{-i\theta} \end{pmatrix}$. Recall that, in §5.1, we considered the representation (t^l_{mn}) of $SU(2)$ with respect to an orthonormal basis e^l_n ($n = -l, -l+1, \ldots, l$) of eigenvectors for the $t^l(a_\theta)$:

$$t^l(a_\theta) e^l_n = e^{-2in\theta} e^l_n.$$

In particular, for $l \in \mathbf{Z}_+$:

$$t^l(a_\theta) e^l_0 = e^l_0,$$

and e^l_0 is, up to a constant factor, the unique $U(1)$-fixed vector in the representation space of t^l. Now consider the subgroup $K := SO(2)$ of $SU(2)$. As $U(1)$ and $SO(2)$ are conjugate subgroups of $SU(2)$, there must be a K-fixed unit vector e_K^l in the representation space of t^l ($l \in \mathbf{Z}_+$). Expand this vector in terms of the original basis:

$$e_K^l = \sum_{n=-l}^{l} c_n^l e_n^l, \tag{7.4}$$

where the c_n^l are yet unknown.

Put $g := \begin{pmatrix} x & -\bar{u} \\ u & \bar{x} \end{pmatrix}$. The function $g \mapsto P_l(x\bar{x}) = t_{00}^l(g)$ is $U(1)$-biinvariant. By conjugacy the function

$$g \mapsto P_l(\tfrac{1}{2}(x^2 + \bar{x}^2 + u^2 + \bar{u}^2)) = (t^l(g)e_K^l, e_K^l) = \sum_{n,m=-l}^{l} c_n^l \, \overline{c_m^l} \, t_{mn}^l(g)$$

is $SO(2)$-biinvariant. For $g := a_\theta$ this yields:

$$P_l(\cos 2\theta) = \sum_{n=-l}^{l} |c_n^l|^2 \, e^{-2in\theta}. \tag{7.5}$$

So the expansion coefficients in (7.3) follow from the c_n^l as defined by (7.4). We will find explicit expressions for the c_n^l by passing to the corresponding Lie algebra representation.

7.2. AN INFINITESIMAL APPROACH

The Lie algebra $\mathbf{g} := sl(2, \mathbf{C})$ of the Lie group $SL(2, \mathbf{C})$ consists of all complex 2×2 matrices of trace 0. It is the complexification of the Lie algebra of $SU(2)$. A basis for $\mathbf{g}$ is given by

$$H := \begin{pmatrix} 1 & 0 \\ 0 & -1 \end{pmatrix}, \quad B := \begin{pmatrix} 0 & 1 \\ 0 & 0 \end{pmatrix}, \quad C := \begin{pmatrix} 0 & 0 \\ 1 & 0 \end{pmatrix},$$

with commutator relations

$$[H, B] = 2B, \quad [H, C] = -2C, \quad [B, C] = H. \tag{7.6}$$

The representation t^l of $\mathbf{g}$ corresponding to the representation t^l of $SU(2)$ is defined by

$$t^l(X) := \frac{d}{dt}\bigg|_{t=0} t^l(\exp tX)$$

for X in the Lie algebra of $SU(2)$ and extended to $\mathbf{g}$ by complexification. The explicit expression for t^l acting on the basis of $\mathbf{g}$ is:

$$t^l(H) e_n^l = -2n \, e_n^l,$$

$$t^l(B) e_n^l = \begin{cases} \sqrt{(l-n+1)(l+n)} \, e_{n-1}^l & \text{if } n = -l+1, -l+2, \ldots, l, \\ 0 & \text{if } n = -l, \end{cases} \tag{7.7}$$

$$t^l(C)\,e_n^l = \begin{cases} \sqrt{(l-n)(l+n+1)}\,e_{n+1}^l & \text{if } n = -l, -l+1, \ldots, l-1, \\ 0 & \text{if } n = l. \end{cases} \tag{7.8}$$

The $SO(2)$-invariance of the vector e_K (cf. (7.4)) can be infinitesimally characterized as

$$t^l(B-C)\,e_K^l = 0. \tag{7.9}$$

When we substitute (7.4), (7.7) and (7.8) in this formula then we obtain a two-term recurrence relation for the c_n^l. Up to a constant factor this can be solved by

$$c_n^l = \begin{cases} \left(\dfrac{(1/2)_{(l-n)/2}\,(1/2)_{(l+n)/2}}{((l-n)/2)!\,((l+n)/2)!} \right)^{\frac{1}{2}}, & l-n \text{ even}, \\ 0, & l-n \text{ odd}, \end{cases} \tag{7.10}$$

where $l \in \mathbf{Z}_+$. In view of (7.5) this yields (7.3) up to a constant factor.

7.3. QUANTIZED UNIVERSAL ENVELOPING ALGEBRA

There is no quantum subgroup of $SU_q(2)$ analogous to the subgroup $SO(2)$ of $SU(2)$. However, there is a quantum analogue of the infinitesimal generator $B - C$ of $SO(2)$. For this we need Jimbo's [16] quantization of the universal enveloping algebra of the Lie algebra $sl(2, \mathbf{C})$.

Let $\mathcal{U}_q$ be a Hopf algebra with unit 1, generated as algebra by elements A, B, C, D with relations

$$AD = DA = 1, \quad AB = qBA, \quad AC = q^{-1}CA,$$
$$BC - CB = \frac{A^2 - D^2}{q - q^{-1}}. \tag{7.11}$$

We can recover (7.6) from (7.11) by substituting in (7.11)

$$A := e^{\frac{1}{2}(q-1)H}, \quad D := e^{-\frac{1}{2}(q-1)H},$$

and by letting q tend to 1. We denote the comultiplication by $\Delta : \mathcal{U}_q \to \mathcal{U}_q \otimes \mathcal{U}_q$. Its action on the generators is given by

$$\Delta(A) = A \otimes A, \quad \Delta(D) = D \otimes D,$$
$$\Delta(B) = A \otimes B + B \otimes D,$$
$$\Delta(C) = A \otimes C + C \otimes D.$$

The counit, denoted by $\varepsilon : \mathcal{U}_q \to \mathbf{C}$, is given by

$$\varepsilon(A) = \varepsilon(D) = 1, \quad \varepsilon(B) = \varepsilon(C) = 0.$$

The antipode $S : \mathcal{U}_q \to \mathcal{U}_q$ is such that

$$S(A) = D, \quad S(D) = A, \quad S(B) = -q^{-1}B, \quad S(C) = -qC.$$

It can be easily shown that Δ, ε and S have well-defined extensions to $\mathcal{U}_q$ as Hopf algebra operations.

We can think about $\mathcal{U}_q$ as a dual Hopf algebra to $\mathcal{A}_q$, i.e., $\mathcal{U}_q$ is embedded in the linear dual of $\mathcal{A}_q$ such that the following rules have to be satisfied for $X, Y \in \mathcal{U}_q$, $a, b \in \mathcal{A}_q$:

$$
\begin{aligned}
(XY)(a) &= (X \otimes Y)(\Phi(a)), \quad (\Delta(X))(a \otimes b) = X(ab), \\
\varepsilon(X) &= X(I), \quad e(a) = a(1), \quad (S(X))(a) = X(\kappa(a)).
\end{aligned}
\tag{7.12}
$$

In view of the first two rules, it is sufficient to specify $X(a)$ if X is a generator A, B, C or D and a is a generator α, β, γ or δ. This we declare to yield 0 except in the following cases:

$$
\begin{aligned}
A(\alpha) &= q^{\frac{1}{2}}, \quad A(\delta) = q^{-\frac{1}{2}}, \quad D(\alpha) = q^{-\frac{1}{2}}, \quad D(\delta) = q^{\frac{1}{2}}, \\
B(\beta) &= 1, \quad C(\gamma) = 1.
\end{aligned}
\tag{7.13}
$$

It can be shown (cf. Vaksman and Soibelman [45]) that (7.12) and (7.13) yield a well-defined nondegenerate bilinear pairing $\langle X, a \rangle := X(a)$ between $\mathcal{U}_q$ and $\mathcal{A}_q$.

The following observation will be important:

$$
\Delta(X) = A \otimes X + X \otimes D \quad \text{if } X = B, C \text{ or } A - D.
\tag{7.14}
$$

7.4. CONTINUOUS q-LEGENDRE POLYNOMIALS INTERPRETED

A corepresentation t^l of $\mathcal{A}_q$ yields a representation of $\mathcal{U}_q$ by the rule

$$
t^l_{nm}(X) := X(t^l_{nm}).
$$

Indeed, we find that

$$
t^l_{nm}(XY) = \sum_{k=-l}^{l} t^l_{nk}(X) \, t^l_{km}(Y).
$$

Let the vectors e^l_n, $n = -l, -l+1, \ldots, l$, form the standard basis of the representation space of t^l, so

$$
t^l(X) e^l_m = \sum_{n=-l}^{l} t^l_{nm}(X) e^l_n.
$$

Then the action of t^l for the generators of $\mathcal{U}_q$ becomes

$$
\begin{aligned}
t^l(A) e^l_n &= q^{-n} e^l_n, \quad t^l(D) e^l_n = q^n e^l_n, \\
t^l(B) e^l_n &= \frac{(q^{-l+n-1} - q^{l-n+1})^{\frac{1}{2}} (q^{-l-n} - q^{l+n})^{\frac{1}{2}}}{q^{-1} - q} e^l_{n-1}, \\
t^l(C) e^l_n &= \frac{(q^{-l+n} - q^{l-n})^{\frac{1}{2}} (q^{-l-n-1} - q^{l+n+1})^{\frac{1}{2}}}{q^{-1} - q} e^l_{n+1},
\end{aligned}
\tag{7.15}
$$

where, as in (7.8) and (7.11) we suppose e^l_{-l-1} and e^l_{l+1} to be zero.

In the case of $SU(2)$ we found the $SO(2)$-invariant vector (7.4) by solving (7.9), which yielded coefficients c^l_n given by (7.10). We imitate this in the quantum case and look for a solution vector $v := \sum_{n=-l}^{l} b^{l,\lambda}_n e^l_n$ of

$$t^l(q^\lambda B - q^{-\lambda}C)v = 0,$$

where the exponent λ is yet to be specified. In view of (7.15) this yields a two-term recurrence relation for the coefficients $b^{l,\lambda}_n$, which has only the zero solution if $l \in \frac{1}{2} + \mathbf{Z}_+$ and which gives in the case $l \in \mathbf{Z}_+$:

$$v = \text{const.} \sum_{\substack{n=-l \\ l-n \text{ even}}}^{l} q^{-\lambda n} c^l_n e^l_n \tag{7.16}$$

where, for $l \in \mathbf{Z}_+$ and $n = -l, -l+2, \ldots, l$,

$$c^l_n := \left(\frac{(q^2;q^4)_{(l-n)/2}\, (q^2;q^4)_{(l+n)/2}}{(q^4;q^4)_{(l-n)/2}\, (q^4;q^4)_{(l+n)/2}} \right)^{\frac{1}{2}}. \tag{7.17}$$

Now compare with (7.2). We recognize the expansion coefficients in (7.2) as squares of coefficients c^l_n in (7.17):

$$P_l(\cos\theta \mid q^2) = q^n \sum_{\substack{n=-l \\ l-n \text{ even}}}^{l} (c^l_n)^2\, e^{in\theta}. \tag{7.18}$$

This is the crucial observation opening the road to quantum group interpretations of Askey-Wilson polynonomials.

A function $a \in \text{Pol}(SU(2))$ is right invariant under $SO(2)$ iff

$$\left. \frac{d}{dt} \right|_{t=0} a(x \exp(t(B - C))) = 0, \qquad x \in SU(2), \tag{7.19}$$

and left invariant under $SO(2)$ iff

$$\left. \frac{d}{dt} \right|_{t=0} a(\exp(t(B - C))\, x) = 0, \qquad x \in SU(2). \tag{7.20}$$

We propose as quantum analogues of the conditions (7.19) and (7.20):

$$(\text{id} \otimes (q^\lambda B - q^{-\lambda}C))(\Phi(a)) = 0, \tag{7.21}$$

respectively

$$((q^\lambda B - q^{-\lambda}C) \otimes \text{id})(\Phi(a)) = 0. \tag{7.22}$$

Here λ is a constant which has yet to be specified. We might say that (7.21) respectively (7.22) express the right respectively left invariance of $a \in \mathcal{A}_q$ under the virtual quantum subgroup $SO(2)$.

If $X \in \mathcal{U}_q$ then $\Delta(X)$ will be a finite sum of elements of the form $Y \otimes Z$, where $Y, Z \in \mathcal{U}_q$. We express this formally as

$$\Delta(X) = \sum_{(X)} X_{(1)} \otimes X_{(2)}. \tag{7.23}$$

Then it can easily be shown that:

LEMMA 7.1. If $\Delta(X)$ is given by (7.23) and $a, b \in \mathcal{A}_q$ then

$$(\mathrm{id} \otimes X)(\Phi(ab)) = \sum_{(X)} (\mathrm{id} \otimes X_{(1)})(\Phi(a))\,(\mathrm{id} \otimes X_{(2)})(\Phi(b)),$$

$$(X \otimes \mathrm{id})(\Phi(ab)) = \sum_{(X)} (X_{(1)} \otimes \mathrm{id})(\Phi(a))\,(X_{(2)} \otimes \mathrm{id})(\Phi(b)).$$

Now it follows from (7.14) that:

PROPOSITION 7.2. The elements a satisfying (7.21) respectively (7.22) form a unital subalgebra of $\mathcal{A}_q$.

It is not difficult to show from (7.16) that the elements $a \in \mathrm{Span}\{t^l_{nm}\}$ which satisfy both (7.21) and (7.22), are just the null element if $l \in \frac{1}{2} + \mathbf{Z}_+$, and form for $l \in \mathbf{Z}_+$ a one-dimensional subspace of elements

$$a = \mathrm{const.} \sum_{\substack{n,m=-l \\ l-n, l-m \text{ even}}}^{l} q^{(n-m)\lambda}\, c^l_n\, c^l_m\, t^l_{nm}, \tag{7.24}$$

where c^l_n is given by (7.17). Let us compute (7.24) explicitly for $l = 1$. From Theorem 5.3 we get

$$t^1_{1,1} = (\alpha^*)^2, \quad t^1_{2,2} = \alpha^2, \quad t^1_{1,-1} = \gamma^2, \quad t^1_{-1,1} = q^2\,(\gamma^*)^2.$$

Thus the element $a \in \mathrm{Span}\{t^1_{nm}\}$ satisfying both (7.21) and (7.22) equals, up to a constant factor,

$$a = \alpha^2 + (\alpha^*)^2 + q^{2\lambda}\gamma^2 + q^{2-2\lambda}(\gamma^*)^2.$$

Now we would like to have this element a self-adjoint, i.e., $a = a^*$. This forces us to take $\mathrm{Re}\,\lambda = \frac{1}{2}$. From the point of view of interpretations of special functions, the choice $\lambda := \frac{1}{2}$ will be sufficient for our purposes. For the moment we call an element $a \in \mathcal{A}_q$ spherical if a satisfies (7.21) and (7.22) with $\lambda = \frac{1}{2}$. Put

$$\rho := \frac{1}{2}(\alpha^2 + (\alpha^*)^2 + q\gamma^2 + q(\gamma^*)^2) = \rho^*.$$

Then ρ spans the spherical elements in $\mathrm{Span}\{t^1_{nm}\}$. All $a := p(\rho)$, with p a polynomial, are also spherical. Actually we can prove:

THEOREM 7.3. $a \in \mathcal{A}_q$ is spherical if and only if a is of the form $p(\rho)$, with p a polynomial.

In particular, we can apply this theorem to the spherical elements given by (7.24). Then

$$\sum_{\substack{n,m=-l \\ l-n,l-m \text{ even}}}^{l} q^{(n-m)/2} \, c_n^l \, c_m^l \, t_{nm}^l = \sum_k b_k \rho^k \tag{7.25}$$

for certain coefficients b_k. Now apply the characters χ^θ to both sides of (7.25). This yields

$$\sum_{\substack{n=-l \\ l-n \text{ even}}}^{l} (c_n^l)^2 \, e^{-2in\theta} = \sum_k b_k (\cos 2\theta)^k.$$

In view of (7.18) we have

$$q^{-n} \, P_l(\cos 2\theta \mid q^2) = \sum_k b_k \, (\cos 2\theta)^k.$$

Hence

$$q^{-n} \, P_l(\rho \mid q^2) = \sum_k b_k \, \rho^k. \tag{7.26}$$

So we obtain:

THEOREM 7.4.   Let $l \in \mathbf{Z}_+$. Then, up to a constant factor, the spherical element in $\mathrm{Span}\{t_{nm}^l\}$ is given by $P_l(\rho \mid q^2)$, a continuous q-Legendre polynomial of degree l in ρ.

We compare next the known orthogonality relations

$$\frac{(q^4, q^4; q^4)_\infty}{2\pi \, (q^2, q^2; q^4)_\infty} \int_0^\pi P_k(\cos \theta \mid q^2) \, P_l(\cos \theta \mid q^2) \left| \frac{(e^{2i\theta}; q^4)_\infty}{(q^2 e^{2i\theta}; q^4)_\infty} \right|^2 d\theta$$
$$= \frac{1 - q^2}{1 - q^{2(2l+1)}} \, \delta_{kl} \tag{7.27}$$

of the continuous q-Legendre polynomials with the Schur type orthogonality relations. From (7.25), (7.26) we obtain

$$P_l(\rho \mid q^2) = q^l \sum_{\substack{n,m=-l \\ l-n,l-m \text{ even}}}^{l} q^{(n-m)/2} \, c_n^l \, c_m^l \, t_{nm}^l. \tag{7.28}$$

Now substitute (7.28) twice into (7.27) and apply (5.12). This yields

$$h\big(P_k(\rho \mid q^2) \, P_l(\rho \mid q^2)\big) = \delta_{kl} \, \frac{(1-q^2)q^{2l}}{1 - q^{2(2l+1)}} \left(\sum_{\substack{n=-l \\ l-n \text{ even}}}^{l} q^{-n}(c_n^l)^2 \right)^2$$
$$= \delta_{kl} \, \frac{(1-q^2)q^{2l}}{1 - q^{2(2l+1)}} \left(P_l((q + q^{-1})/2 \mid q^2) \right)^2 \tag{7.29}$$
$$= \frac{1 - q^2}{1 - q^{2(2l+1)}}.$$

By comparing (7.27) with (7.29) we conclude:

THEOREM 7.5. Let p be a polynomial. Then

$$h(p(\rho)) = \frac{(q^4, q^4; q^4)_\infty}{2\pi (q^2, q^2; q^4)_\infty} \int_0^\pi p(\cos\theta) \left| \frac{(e^{2i\theta}; q^4)_\infty}{(q^2 e^{2i\theta}; q^4)_\infty} \right|^2 d\theta. \tag{7.30}$$

So, beside polynomials in $\gamma^*\gamma$ and $(\alpha + \alpha^*)/2$ (cf. (4.1), (4.2)), the Haar functional can now also be evaluated when acting on polynomials in ρ. Note that our derivation of (7.30) is very indirect, without computation of moments as for (4.1).

8. Askey-Wilson Polynomials

The theorems in §7 are not the end of the story. In §6.5 we mentioned the interpretation by Noumi and Mimachi [32] of big q-Jacobi polynomials

$$P_n^{(\alpha,\beta)}(x; c, d; q) := {}_3\phi_2 \left(\begin{matrix} q^{-n}, q^{n+\alpha+\beta+1}, q^{\alpha+1}x/c \\ q^{\alpha+1}, -q^{\alpha+1}d/c \end{matrix}; q, q \right)$$

with $\alpha = \beta$ on quantum 2-spheres. In particular, they found an interpretation of big q-Legendre polynomials $P_n^{(0,0)}(x; c, d; q)$ as zonal spherical elements on quantum 2-spheres. It was tempting to relate their results to the approach of §7. A first link could be made by proving the following

PROPOSITION 8.1. Let $l \in \mathbf{Z}_+$ and $a \in \mathrm{Span}\{t_{nm}^l\}$. Then a satisfies both

$$(\mathrm{id} \otimes (q^{\frac{1}{2}} B - q^{-\frac{1}{2}} C))(\Phi(a)) = 0 \quad (\text{“quantum right } SO(2)\text{-invariance”}) \tag{8.1}$$

and

$$(\chi^\theta \otimes \mathrm{id})(\Phi(a)) = a \qquad (\text{left } U(1)\text{-invariance}) \tag{8.2}$$

if and only if

$$a = \mathrm{const.}\, P_l^{(0,0)}(iq^{-1}(\alpha\gamma^* - \gamma\alpha^*); 1, 1; q^2).$$

The left $U(1)$-invariance (8.2) can also be expressed by

$$((A - D) \otimes \mathrm{id})(\Phi(a)) = 0. \tag{8.3}$$

Recall that (7.14) is also satisfied for $X := A - D$. This suggests that we may generalize Proposition 8.1 such that we get interpretations of big q-Legendre polynomials $P_n^{(0,0)}(x; c, d; q)$ with $c \neq d$ by keeping (8.2) and replacing (8.1) by

$$(\mathrm{id} \otimes (q^{\frac{1}{2}} B - q^{-\frac{1}{2}} C + \mathrm{const.}\, (A - D)))(\Phi(a)) = 0.$$

This was the starting point for some very recent results by the author (yet unpublished), which we will summarize now.

8.1. (σ,τ)-SPHERICAL ELEMENTS

Let $\sigma,\tau \in \mathbf{R}$. We will call an element $a \in \mathcal{A}_q$ (σ,τ)-spherical if

$$\left(\mathrm{id} \otimes \left(iq^{\frac{1}{2}}B - iq^{-\frac{1}{2}}C - \frac{q^{-\sigma} - q^{\sigma}}{q^{-1} - q}(A - D)\right)\right)(\Phi(a)) = 0, \tag{8.4}$$

$$\left(\left(iq^{\frac{1}{2}}B - iq^{-\frac{1}{2}}C - \frac{q^{-\tau} - q^{\tau}}{q^{-1} - q}(A - D)\right) \otimes \mathrm{id}\right)(\Phi(a)) = 0. \tag{8.5}$$

If $\tau = \pm\infty$ then we replace (8.5) by (8.3), and similarly for $\sigma = \pm\infty$. Put

$$\rho_{\sigma\tau} := \frac{1}{2}\Big(\alpha^2 + (\alpha^*)^2 + q(\gamma^2 + (\gamma^*)^2) + iq(q^{-\sigma} - q^{\sigma})(\alpha^*\gamma - \gamma^*\alpha) +$$
$$+ iq(q^{-\tau} - q^{\tau})(\alpha^*\gamma^* - \gamma\alpha) - q(q^{-\sigma} - q^{\sigma})(q^{-\tau} - q^{\tau})\gamma^*\gamma\Big) = \rho_{\sigma\tau}^*.$$

Put, for $l = 0,1,\dots$ and $n = -l,-l+1,\dots,l$:

$$c_n^{l,\sigma} := \frac{i^n\, q^{-(l+\sigma)n}\, q^{n^2/2}}{(q^2;q^2)_{l+n}^{1/2}\,(q^2;q^2)_{l-n}^{1/2}} \times$$
$$\times\ {}_3\phi_2\left(\genfrac{}{}{0pt}{}{q^{-2l+2n}, q^{-2l}, -q^{-2l-2\sigma}}{q^{-4l}, 0}; q^2, q^2\right) = c_{-n}^{l,\sigma}.$$

THEOREM 8.2. Let $a \in \mathcal{A}$. Then a is (σ,τ)-spherical if and only if it is a polynomial in $\rho_{\sigma\tau}$.

THEOREM 8.3. Let $l \in \mathbf{Z}_+$. The space of (σ,τ)-spherical elements in $\mathrm{Span}\{t_{nm}^l\}$ is one-dimensional and spanned by an element which we can represent in the following two ways:

$$\sum_{n,m=-l}^{l} q^{(n-m)/2}\, c_m^{l,\sigma}\, \overline{c_n^{l,\tau}}\, t_{nm}^l$$

$$= \frac{c_l^{l,\sigma}\, \overline{c_l^{l,\tau}}}{q^{2l+2};q^2)_l}\, p_l(\rho_{\sigma\tau}; -q^{\sigma+\tau+1}, -q^{-\sigma-\tau+1}, q^{\sigma-\tau+1}, q^{-\sigma+\tau+1} \mid q^2). \tag{8.6}$$

Here the p_l at the right hand side of (8.6) is an Askey-Wilson polynomial (7.1). So we have given a quantum group interpretation of a two-parameter family of Askey-Wilson polynomials.

What about the limit cases as σ and/or τ tend to ∞? We should get little or big q-Jacobi polynomials, but these latter polynomials have discrete orthogonality measures, while the Askey-Wilson polynomials have absolutely continuous orthogonality measure, at least as the parameters stay within $(-1,1)$. However, some parameters of the Askey-Wilson polynomials in (8.6) tend to ∞ as σ or τ tend to ∞. Then discrete mass points are added, cf. [6, Theorems 2.4, 2.5]. If we make, at the same time, a scale transformation, such that the continuous spectrum shrinks, then we will arrive in the limit at infinitely many mass points and no continuous spectrum left. In fact, it follows immediately from (7.1), with the normalization

$$\tilde{p}_n(x; a, b, c, d \mid q) := \frac{p_n(x; a, b, c, d \mid q)}{p_n((a + a^{-1}/2); a, b, c, d \mid q)},$$

that we have the two following limits:

$$\tilde{p}_n \left(\frac{-x}{2q^{(a+b-1)/2}}; -q^{(1+a+b)/2}, -q^{(1-a-b)/2}, q^{(1+a-b)/2}, q^{(1-a+b)/2} \mid q \right)$$

$$\xrightarrow{a\to\infty} {}_3\phi_2 \left(\begin{array}{c} q^{-n}, q^{n+1}, qx \\ q, -q^{b+1} \end{array}; q, q \right) = \text{big } q\text{-Jacobi polynomial} \tag{8.7}$$

and

$$\tilde{p}_n \left(\frac{-x}{2q^{a-1/2}}; -q^{1/2+a}, -q^{1/2-a}, q^{1/2}, q^{1/2} \mid q \right)$$

$$\xrightarrow{a\to\infty} {}_3\phi_2 \left(\begin{array}{c} q^{-n}, q^{n+1}, qx \\ q, 0 \end{array}; q, q \right) = \text{little } q\text{-Jacobi polynomial.} \tag{8.8}$$

R. Askey told me that he has known such limit transitions already for several years, but never published them.

We also get new expressions for the Haar functional. Let $dm(x) = dm_{a,b,c,d;q}(x)$ be the normalized orthogonality measure for the Askey-Wilson polynomials:

$$\int_{-1}^{1} (p_m p_n)(x; a, b, c, d \mid q) \, dm(x) = \delta_{mn} \frac{(1 - q^{-1}abcd)(q, ab, ac, ad, bc, bd, cd; q)_n}{(1 - q^{2n-1}abcd)(q^{-1}abcd; q)_n}.$$

THEOREM 8.4. Let p be a polynomial. Let h be the Haar functional on $\mathcal{A}$. Then

$$h(p(\rho_{\sigma\tau})) = \int_0^1 p(x) \, dm_{a,b,c,d;q^2}(x),$$

where $a = -q^{\sigma+\tau+1}$, $b = -q^{-\sigma-\tau+1}$, $c = q^{\sigma-\tau+1}$, $d = q^{-\sigma+\tau+1}$.

By putting $\sigma = \tau = 0$ in the Theorems 8.2–8.4, we get back the results of §7.4. If we put $\sigma = \tau$ and let $\sigma \to \infty$ then we approach, by (8.8), the little q-Legendre case. If we fix σ and let $\tau \to \infty$ then, by (8.7), we approach the big q-Jacobi case. It should be possible to relate this last case to the results in [32].

8.2. DUAL q-KRAWTCHOUK POLYNOMIALS INTERPRETED

It is possible to give an explicit matrix for the transition in the representation space of t^l from the basis of eigenvectors e_n^l for $t^l(A - D)$ to a basis of eigenvectors for $t^l(H_\sigma)$, where

$$H_\sigma := D \left(iq^{1/2}B - iq^{-1/2}C - \frac{q^{-\sigma} - q^\sigma}{q^{-1} - q}(A - D) \right).$$

(Observe that (8.4) can be written as $(\mathrm{id} \otimes H_\sigma)(\Phi(a)) = 0$ and that H_σ is self-adjoint if we make $\mathcal{U}_q$ into a $*$-algebra such that $A^* = A$, $D^* = D$, $B^* = C$. Then t^l is a $*$-representation of $\mathcal{U}_q$.)

We define *dual q-Krawtchouk polynomials* by

$$R_n(q^{-x} - q^{x-N-c}; q^c, N \mid q) := {}_3\phi_2(q^{-n}, q^{-x}, -q^{x-N-c}; 0, q^{-N}; q, q).$$

These are special q-Racah polynomials and satisfy the orthogonality relations

$$\frac{1}{(-q^c; q)_N} \sum_{x=0}^{N} (R_n R_m)(q^{-x} - q^{x-N-c}; q^c, N \mid q) \times$$

$$\times \frac{(1 + q^{2x-N-c})(-q^{-N-c}, q^{-N}; q)_x}{(1 + q^{-N-c})(q, -q^{-c-1}; q)_x (-q^{x-2N-c})^x} = \delta_{nm} \frac{(q; q)_n}{(q^{-N}; q)_n} (-q^{-N-c})^n,$$

where $n, m = 0, \ldots, N$. See Askey and Wilson [5] and Stanton [40].

THEOREM 8.5. $t^l(H_\sigma)$ has simple spectrum consisting of eigenvalues

$$x_j := \frac{q^{2j-\sigma} - q^{\sigma-2j} + q^\sigma - q^{-\sigma}}{q^{-1} - q}, \qquad j = -l, -l+1, \ldots, l.$$

An eigenvector corresponding to eigenvalue x_j is given by

$$\sum_{n=0}^{2l} i^{-n} q^{n\sigma} q^{n(n+1)/2} (q^2; q^2)_n^{-1/2} (q^{4l}; q^{-2})_n^{1/2} \times$$

$$\times R_n(q^{-2l-2j} - q^{2j-2l-2\sigma}; q^{2\sigma}, 2l \mid q^2) e_{n-l}^l.$$

Noumi and Mimachi told me that, in a follow-up to [32], they have also obtained such an interpretation of q-Krawtchouk polynomials.

References

[1] E. Abe, *Hopf Algebras*, Cambridge University Press, 1980.

[2] G. E. Andrews an R. Askey, 'Enumeration of partitions: the role of Eulerian series and q-orthogonal polynomials', pp. 3–26 in *Higher Combinatorics* (M. Aigner, ed.), Reidel, 1977.

[3] M. Arik and D. D. Coon, 'Hilbert spaces of analytic functions and generalized coherent states', *J. Math. Phys.* 17 (1976), 524–527.

[4] R. Askey and M. E. H. Ismail, 'A generalization of ultraspherical polynomials', pp. 55–78 in *Studies in Pure Mathematics* (P. Erdös, ed.), Birkhäuser, 1983.

[5] R. Askey and J. Wilson, 'A set of orthogonal polynomials that generalize the Racah coefficients or 6-j symbols', *SIAM J. Math. Anal.* 10 (1979), 1008–1016.

[6] R. Askey and J. Wilson, *Some basic hypergeometric orthogonal polynomials that generalize Jacobi polynomials*, Memoirs Amer. Math. Soc. 54 (1985) No. 319.

[7] W. Van Assche and T. H. Koornwinder, *Asymptotic behaviour for Wall polynomials and the addition formula for little q-Legendre polynomials*, preprint, 1989.

[8] G. M. Bergman, 'Everybody knows what a Hopf algebra is', *Contemp. Math.* **43** (1985), 25–48.

[9] P. Cartier, *Harmonic analysis on trees*, Proc. Sympos. Pure Math. **26** (1973), 419-424.

[10] J. Cigler, 'Operatormethoden für q-Identitäten', *Monatsh. Math.* **88** (1979), 87–105.

[11] V. G. Drinfeld, 'Quantum groups', pp. 798–820 in *Proceedings of the International Congress of Mathematicians, Berkeley, 1986*, American Mathematical Society, 1987.

[12] Ph. Feinsilver, 'Commutators, anti-commutators and Eulerian calculus', *Rocky Mountain J. Math.* **12** (1982), 171–183.

[13] Ph. Feinsilver, 'Discrete analogues of the Heisenberg-Weyl algebra', *Monatsh. Math.* **104** (1987), 89–108.

[14] G. Gasper and M. Rahman, *Basic hypergeometric series*, Cambridge University Press, 1989.

[15] M. Hazewinkel, *Formal groups and applications*, Academic Press, 1978.

[16] M. Jimbo, 'A q-difference analogue of $U(\mathbf{g})$ and the Yang-Baxter equation', *Lett. Math. Phys.* **10** (1985), 63–69.

[17] A. N. Kirillov and N. Yu. Reshetikhin, *Representations of the algebra $U_q(sl(2))$, q-orthogonal polynomials and invariants of links*, LOMI Preprints E-9-88, Leningrad, 1988.

[18] H. T. Koelink and T. H. Koornwinder, 'The Clebsch-Gordan coefficients for the quantum group $S_\mu U(2)$ and q-Hahn polynomials', *Nederl. Akad. Wetensch. Proc. Ser. A*, to appear.

[19] T. H. Koornwinder, 'Clebsch-Gordan coefficients for $SU(2)$ and Hahn polynomials', *Nieuw Archief Wisk. (3)* **29** (1981), 140–155.

[20] T. H. Koornwinder, 'Krawtchouk polynomials, a unification of two different group theoretic interpretations', *SIAM J. Math. Anal.* **13** (1982), 1011–1023.

[21] T. H. Koornwinder, 'Representations of the twisted $SU(2)$ quantum group and some q-hypergeometric orthogonal polynomials', *Nederl. Akad. Wetensch. Proc. Ser. A* **92** (1989), 97–117.

[22] T. H. Koornwinder, *The addition formula for little q-Legendre polynomials and the $SU(2)$ quantum group*, CWI Rep. AM-R8906, preprint, 1989.

[23] T. H. Koornwinder, 'Continuous q-Legendre polynomials are spherical matrix elements of irreducible representations of the quantum $SU(2)$ group', *CWI Quarterly*, to appear.

[24] I. G. Macdonald, *Orthogonal polynomials associated with root systems*, preprint, 1988.

[25] I. G. Macdonald, 'Orthogonal polynomials associated with root systems', *These Proceedings*.

[26] Yu. I. Manin, *Quantum groups and non-commutative geometry*, Centre de Recherches Mathématiques, Montréal, 1988.

[27] T. Masuda, K. Mimachi, Y. Nakagami, M. Noumi and K. Ueno, 'Representations of quantum groups and a q-analogue of orthogonal polynomials', *C. R. Acad. Sci. Paris, Sér. I Math.* **307** (1988), 559–564.

[28] T. Masuda, K. Mimachi, Y. Nakagami, M. Noumi and K. Ueno, 'Representations of the quantum group $SU_q(2)$ and the little q-Jacobi polynomials', *J. Functional Anal.*, to appear.

[29] T. Masuda, K. Mimachi, Y. Nakagami, M. Noumi, Y. Saburi and K. Ueno, *Unitary representations of the quantum group $SU_q(1,1)$, I, II*, preprint, 1989.

[30] A. Nijenhuis and R. W. Richardson, 'Deformations of Lie algebra structures', *J. Math. Mech.* **17** (1967), 89–105.

[31] A. F. Nikiforov and V. B. Ugarov, *Special Functions of Mathematical Physics*, Birkhäuser, 1988.

[32] M. Noumi and K. Mimachi, *Quantum 2-spheres and big q-Jacobi polynomials*, preprint, 1989.

[33] M. Noumi and K. Mimachi, *Big q-Jacobi polynomials, q-Hahn polynomials and a family of quantum 3-spheres*, preprint, 1989.

[34] M. Noumi, H. Yamada and K. Mimachi, *Zonal spherical functions on the quantum homogeneous space $SU_q(n+1)/SU_q(n)$*, Proc. Japan Acad. **65** (1989), 169–171.

[35] P. Podleś, 'Quantum spheres', *Lett. Math. Phys.* **14** (1987), 193–202.

[36] M. Rahman, *A simple proof of Koornwinder's addition formula for the little q-Legendre polynomials*, preprint, 1988.

[37] M. Rahman, 'Some extensions of the beta integral and the hypergeometric function', *These Proceedings*.

[38] M. Rahman and A. Verma, 'Product and addition formula for the continuous q-ultraspherical polynomials', *SIAM J. Math. Anal.* **17** (1986), 1461–1474.

[39] M. P. Schützenberger, 'Une interprétation de certaines solutions de l'équation fonctionnelle: $F(x+y) = F(x)F(y)$', *C. R. Acad. Sci. Paris* **236** (1953), 352–353.

[40] D. Stanton, 'Orthogonal polynomials and Chevalley groups' pp. 87–128 in *Special Functions: Group Theoretic Aspects and Applications* (R. A. Askey, T. H. Koornwinder and W. Schempp, eds.), Reidel, 1984.

[41] D. Stanton, 'An introduction to group representations and orthogonal polynomials', *These Proceedings*.

[42] M. E. Sweedler, *Hopf Algebras*, Benjamin, 1969.

[43] L. L. Vaksman, *q-Analogues of Clebsch-Gordan coefficients in the algebra of functions on the quantum group $SU(2)$*, 1989.

[44] L. L. Vaksman and L. I. Korogodsky, 'Algebra of bounded functions on the quantum group of plane motions and q-analogues of Bessel functions', *Dokl. Akad. Nauk SSSR* **304** (1989), 1036–1040 (in Russian).

[45] L. L. Vaksman and Ya. S. Soibelman, 'Algebra of functions on the quantum group $SU(2)$', *Functional Anal. Appl.* **22** (1988), 170–181.

[46] J.-M. Vallin, 'C*-algèbres de Hopf et C*-algèbres de Kac', *Proc. London Math. Soc.* **(3) 50** (1985), 131–174.

[47] N. Ya. Vilenkin, *Special Functions and the Theory of Group Representations*, Amer. Math. Soc. Transl. of Math. Monographs, Vol. 22, 1968.

[48] J. A. Wilson, 'Some hypergeometric orthogonal polynomials', *SIAM J. Math. Anal.* **11** (1980), 690–701.

[49] S. L. Woronowicz, 'Compact matrix pseudogroups', *Comm. Math. Phys.* **111** (1987), 613–665.

[50] S. L. Woronowicz, 'Twisted $SU(2)$ group. An example of a non-commutative differential calculus', *Publ. Res. Inst. Math. Sci.* **23** (1987), 117–181.

THE APPROXIMATE APPROACH TO ORTHOGONAL POLYNOMIALS FOR WEIGHTS ON (−∞,∞)

D. S. Lubinsky
Department of Mathematics
Witwatersrand University
P.O. Wits 2050
Republic of South Africa

ABSTRACT. Orthogonal polynomials play an important role in the quantitative and qualitative theorems of polynomial approximation. Conversely the theory of orthogonal polynomials draws heavily on one–sided and uniform approximations. In this paper, we review some of these historical and modern connections, within the context of orthogonal polynomials for weights on (−∞,∞).

1. Introduction

In this paper, we review some historical and modern connections between polynomial approximation and the now very active area of orthogonal polynomials for weights on (−∞,∞). Section 2 contains the main notation. Section 3 discusses approximate consequences of the moment problem and Section 4 reviews quantitative approximation. Section 5 outlines recent results on pointwise asymptotics on (−∞,∞) of orthonormal polynomials, and one–sided approximation. The presentation emphasizes results and a few ideas, omitting proofs.

2. Notation

Throughout, $\alpha : \mathbb{R} \to \mathbb{R}$ denotes a right continuous, monotone increasing function with infinitely many points of increase for which all moments

$$(2.1) \qquad \rho_n := \int_{-\infty}^{\infty} x^n \, d\alpha(x), \quad n = 0,1,2,\ldots,$$

are finite. We call $d\alpha$ a *mass distribution.* The sequence of *orthonormal polynomials* for $d\alpha$ are denoted by

$$(2.2) \qquad p_n(d\alpha,x) = \gamma_n(d\alpha)x^n + \ldots, \quad \gamma_n(d\alpha) > 0,$$

$n = 0,1,2,\ldots,$ satisfying

P. Nevai (ed.), *Orthogonal Polynomials,* 293–310.
© 1990 by Kluwer Academic Publishers.

$$(2.3) \qquad \int_{-\infty}^{\infty} p_n p_m \, d\alpha = \delta_{mn}, \quad m, n = 0,1,2,\dots .$$

If $d\alpha$ is absolutely continuous, say $d\alpha(x) = W^2(x)\,dx$, we write $p_n(W^2,x)$, $\gamma_n(W^2)$ and so on.

The set of real polynomials is denoted by **P**, and the real polynomials of degree at most n by $\mathbf{P}_n$. The nth *Christoffel function* for $d\alpha$ is

$$(2.4) \qquad \lambda_n(d\alpha,x) := \inf_{P \in \mathbf{P}_{n-1}} \int_{-\infty}^{\infty} P^2(t)\,d\alpha(t) / P^2(x),$$

$n = 1,2,3,\dots$. Throughout, C, C_1, C_2 ... denote positive constants independent of n and x, which are not necessarily the same in different occurrences. Given real sequences $\{s_n\}_{n=1}^{\infty}$ and $\{t_n\}_{n=1}^{\infty}$, we write $s_n \sim t_n$, if there exist C_1 and C_2 such that

$$C_1 \leq s_n/t_n \leq C_2, \quad n \text{ large enough.}$$

We shall need a suitable class of *Freud weights*, for which a broad class of theorems hold. Individual theorems, quoted only for FR_Λ, may have been proved in greater generality in the original sources.

Definition 2.1. Let $W := e^{-Q}$, where $Q : \mathbb{R} \to \mathbb{R}$ is even, continuous and Q' is continuous in $(0,\infty)$. Further, assume that $Q(0) = 0$, that $Q' > 0$ in $(0,\infty)$, and that for some Λ, $C > 0$,

$$(2.5) \qquad \Lambda - 1 \leq xQ''(x)/Q'(x) \leq C, \quad x \in (0,\infty).$$

Then we write $W \in FR_\Lambda$. If in addition, Q'' exists for large x, and for some $C_1 > 0$,

$$(2.6) \qquad x^2 |Q''(x)| / Q'(x) \leq C_1,$$

then we write $W \in FR_\Lambda (3)$.

Note that if $W = e^{-Q} \in FR_\Lambda$, then $xQ'(x)$ is increasing in $(0,\infty)$ with limits 0 and ∞ at 0 and ∞ respectively. Furthermore, from (2.5) follows

$$C_3 x^\Lambda \leq Q(x) \leq C_4 x^C, \quad x \text{ large enough,}$$

so Q is of smooth polynomial growth. The simplest exponential weights

$$(2.7) \qquad W_\Lambda(x) := \exp(-|x|^\Lambda), \quad x \in \mathbb{R}, \quad \Lambda > 0,$$

belong to FR_Λ and FR_Λ (3).

If, for example, $W = e^{-Q} \in FR_\Lambda$, its nth *Mhaskar–Rahmanov–Saff number* is the root a_n of the equation

$$(2.8) \qquad n = \frac{2}{\pi} \int_0^1 a_n t Q'(a_n t)(1-t^2)^{-1/2} dt.$$

Note that $a_n \uparrow$ as $n \uparrow$ and grows roughly like $Q^{[-1]}(n)$, where $Q^{[-1]} : [0,\infty) \to [0,\infty)$ is the inverse of Q. In particular, if $W = W_\Lambda$, $a_n = Cn^{1/\Lambda}$, $n \geq 1$.

3. The Moment Problem

Given a sequence of real numbers or *moments* $\{\rho_n\}_0^\infty$, is there a mass distribution $d\alpha$ satisfying (2.1)? Furthermore, if $d\alpha$ *exists*, is it *unique*?

This *moment problem* occupied mathematicians of the stature of Chebyshev, Stieltjes, M. Riesz and R. Nevanlinna from the 1880's through the 1920's. Its theory is probably the most elegant of any chapter in orthogonal polynomials, and it has had ramifications in functional analysis and complex function theory. Perhaps its most important historical association is the development of the Riemann–Stieltjes integral: In investigating certain continued fractions, Stieltjes found it essential to introduce the concept of integration with respect to $d\alpha$, and thereafter to consider the moment problem in the full generality of (2.1).

Hamburger's theorem [9, p. 60] will be familiar to most readers:

Theorem 3.1 (Hamburger, 1921–2). Let $\{\rho_n\}_0^\infty \subset \mathbb{R}$. The following are equivalent:

(a) There exists a mass distribution $d\alpha$ solving (2.1).

(b) $\det(\rho_{j+k})_{j,k=0}^n > 0$, $n = 0,1,2,\ldots$.

The standard proofs involve positive linear functionals, Gauss quadrature, and Helly's Selection Theorem. A somewhat deeper question is uniqueness. Let us say that mass distributions $d\alpha$ and $d\beta$ are equivalent (or identical) if $\alpha - \beta$ is constant in $\mathbb{R}$. Then we can say that the moment problem (2.1) has a *unique solution* or is *determinate* if any two mass distributions solving (2.1) are equivalent. Otherwise, we say that the moment problem is *indeterminate*.

Amongst the several criteria for determinacy, we recall [9, p. 63].

Theorem 3.2 (Hamburger, 1921–22). Let $d\alpha$ be a mass distribution solving the moment problem (2.1). The following are equivalent:
(a) The moment problem (2.1) is determinate.
(b) There exists a real number x such that

(3.1)
$$\lim_{n \to \infty} \lambda_n(d\alpha, x) = 0.$$

(c) The limit (3.1) holds except at points x of discontinuity of α.

Since the definition (2.4) of $\lambda_n(d\alpha, x)$ can be rewritten as

(3.2)
$$\lambda_n(d\alpha, x) = \inf_{P \in P_{n-2}} \int_{-\infty}^{\infty} \left(1 - (t-x)P(t)\right)^2 d\alpha(t),$$

we see that (3.1) asserts the possibility of approximating 1 by polynomials vanishing at x. Another form for $\lambda_n(d\alpha, x)$ is

(3.3)
$$\lambda_n(d\alpha, x) = 1 / \sum_{j=0}^{n-1} p_j^2(d\alpha, x),$$

so (3.1) can also be reformulated as

(3.4)
$$\sum_{j=0}^{\infty} p_j^2(d\alpha, x) = \infty.$$

Among the important ingredients of the proof of Theorem 3.2 are Posse–Markov–Stieltjes inequalities for Gauss quadrature sums [9].
    A less implicit *sufficient* condition for determinacy is due to T. Carleman [29, p. 19]:

(3.5)
$$\sum_{j=1}^{\infty} \rho_{2j}^{-1/(2j)} = \infty \Rightarrow \text{determinacy}.$$

It is powerful enough to yield determinacy of the moment problem for $d\alpha(x) = \exp(-|x|^{\Lambda}) dx$, $\Lambda \geq 1$. When $0 < \Lambda < 1$, the corresponding moment problem is indeterminate.
    Determinacy is closely related to the possibility of one–sided polynomial approximation [9, p. 73]:

Theorem 3.3 (M. Riesz, 1922). Let $d\alpha$ be the unique solution of its moment problem. Let $f: \mathbb{R} \to \mathbb{R}$ be Riemann–Stieltjes integrable with respect to $d\alpha$ in each fixed finite interval, and of at most polynomial growth at infinity. Let $\epsilon > 0$. Then there exist $P_1, P_2 \in \mathbf{P}$ such that

$$P_1 \le f \le P_2 \quad \text{in} \quad \mathbb{R},$$

and

$$\int_{-\infty}^{\infty} (P_2 - P_1)\, d\alpha < \epsilon.$$

The restriction of polynomial growth can be weakened when more is assumed about $d\alpha$. For $d\alpha$ without jumps, determinacy and density of polynomials are equivalent [9, p. 77]:

Theorem 3.4 (M. Riesz, 1922). Let $d\alpha$ be a mass distribution with α continuous in $\mathbb{R}$. The following are equivalent:
(a) $d\alpha$ is the unique solution of its moment problem.
(b) For each $\epsilon > 0$ and each $f: \mathbb{R} \to \mathbb{R}$ satisfying

$$\int f^2\, d\alpha < \infty,$$

there exists $P \in \mathbf{P}$ satisfying

$$\int (f - P)^2\, d\alpha < \epsilon.$$

The proofs of Theorems 3.3 and 3.4 involve clever, but elementary, constructions of "good jump functions" and rational approximation. Somewhat deeper are the connections to Nevanlinna matrices of entire functions and related topics [2, 29]. There are many applications of Theorem 3.4, to convergence of partial sums of orthonormal expansions, to convergence of Gauss quadratures, and to the Erdös–Turán theorem on Lagrange interpolation at the zeros of p_n [9].

It is natural to ask for an L_∞ analogue of M. Riesz' Theorem 3.4. Very often in polynomial approximation problems, the results are the same in any L_p space: the particular p does not matter. This is partly true here. Nevertheless, the L_∞ weighted approximation problem (often called *Bernstein's approximation problem*) was fully resolved only in the 1950's, some thirty years after the L_2-problem. Probably Akhiezer and Pollard are the main protagonists of the following [1, p. 104, p. 118]:

Theorem 3.5. Let $W: \mathbb{R} \to (0, \infty)$ be continuous, with

$$\lim_{|x| \to \infty} x^n W(x) = 0, \quad n = 0,1,2,\dots .$$

The following are equivalent:
(a) For each continuous $f: \mathbb{R} \to \mathbb{R}$ satisfying

$$\lim_{|x| \to \infty} (fW)(x) = 0,$$

and for each $\epsilon > 0$, there exists $P \in \mathbf{P}$ such that

$$\|(f - P)W\|_{L_\infty(\mathbb{R})} < \epsilon.$$

(b) $\sup \left\{ \int_{-\infty}^{\infty} \frac{\ln|P(x)|}{1+x^2} \, dx : \|PW\|_{L_\infty(\mathbb{R})} \leq 1, \quad P \in \mathbf{P} \right\} = \infty.$

(c) There exist $\{P_n\}_{n=1}^{\infty} \subset \mathbf{P}$ with

$$\lim_{n \to \infty} (P_n W)(x) = 1 \text{ in } \mathbb{R}, \quad \|P_n W\|_{L_\infty(\mathbb{R})} \leq C, \quad n \geq 1,$$

and

(3.6)
$$\int_{-\infty}^{\infty} \frac{|\ln W(x)|}{1 + x^2} \, dx = \infty.$$

In the special case where W is even and $\ln(1/W(e^x))$ is convex in $(0,\infty)$, (3.6) is alone a necessary and sufficient condition for density of $\mathbf{P}$. In particular, if $W(x) = W_\Lambda(x) = \exp(-|x|^\Lambda)$, the polynomials are dense iff $\Lambda \geq 1$.

4. Quantitative Approximation

In the early 1950's, with the Bernstein approximation problem finally settled, Dzrbasyan and Tavadyan began to investigate rates of weighted approximation. But in the late 1950's, Dzrbasyan turned to other topics, and only in the late 1960's, did Freud renew energetic activity in the area. Some of Dzrbasyan's important results were established only in fixed finite intervals. Freud's fundamental contribution, together with P. Nevai, was to introduce *infinite–finite range inequalities* that allowed treatment of the whole real line.

This type of inequality showed that the interesting features of a weighted polynomial PW, $P \in \mathbf{P}_n$, occur on an interval that depends on n, but not on the particular P. Freud and others obtained the correct rate of growth of the interval, but not the (often crucial) sharp constant.

A fundamental recent advance, due to Mhaskar and Saff [19, 20], is the sharp L_∞ infinite–finite range inequality, involving the Mhaskar–Rahmanov–Saff number a_n, defined by (2.8). The proofs involve potential theory. For weights on $\mathbb{R}$, one form of this is [16, p. 49]:

Theorem 4.1. Let $W:= e^{-Q} \in FR_\Lambda$, some $\Lambda > 0$.

(a) For $n \geq 1$ and $P \in \mathbf{P}_n$,

$$(4.1) \qquad \|PW\|_{L_\infty(\mathbb{R})} = \|PW\|_{L_\infty[-a_n, a_n]}.$$

Moreover, if $P \not\equiv 0$,

$$|PW|(x) < \|PW\|_{L_\infty[-a_n, a_n]}, \quad |x| > a_n.$$

(b) This is sharp in the following sense: Given $L > 0$, there exists $s > 0$ such that if

$$\chi_n := a_n \left(1 - s\left[\frac{\ln n}{n}\right]^{2/3}\right), \quad n \geq 1,$$

then for all n large enough and some $P \in \mathbf{P}_n$,

$$\|PW\|_{L_\infty(\mathbb{R})} \geq n^L \|PW\|_{L_\infty[-\chi_n, \chi_n]}.$$

The L_p norm of PW lives on an interval slightly larger than $[-a_n, a_n]$, if $p \neq \infty$ [16, pp. 50–51],[21]:

Theorem 4.2. Let $0 < p < \infty$. Let $W:= e^{-Q} \in FR_\Lambda$, some $\Lambda > 0$. Given $L > 0$, there exists $s > 0$ such that if

$$(4.2) \qquad c_n := a_n \left(1 + s\left[\frac{\ln n}{n}\right]^{2/3}\right), \quad n \geq 1,$$

then for n large enough and all $P \in \mathbf{P}_n$,

$$\|PW\|_{L_p(\mathbb{R})} \leq \left(1 + n^{-L}\right) \|PW\|_{L_p[-c_n, c_n]}.$$

Theorem 4.2 is sharp in a sense similar to Theorem 4.1. It is now known that a_n plays an important descriptive role in all the finer aspects of orthonormal polynomials and weighted approximation, for example, *Markov–Bernstein inequalities* [12, Thms 1.1, 1.3]:

Theorem 4.3. Let $W \in FR_\Lambda$, some $\Lambda > 0$.

(a) *Markov Inequality:* There exists C such that for $n \geq 1$ and $P \in \mathbf{P}_n$,

$$(4.3) \qquad \|P'W\|_{L_\infty(\mathbb{R})} \leq \left\{ \int_1^{Cn} ds/Q^{[-1]}(s) \right\} \|PW\|_{L_\infty(\mathbb{R})}.$$

(b) *Bernstein Inequality:* Let $0 < \eta < 1$. There exists C such that for $n \geq 1$, $P \in \mathbf{P}_n$ and $|x| \geq \eta a_n$,

$$(4.4) \qquad |(PW)'(x)| \leq C\|PW\|_{L_\infty(\mathbb{R})} \frac{n}{a_n} \max\left\{ n^{-2/3}, 1 - \frac{|x|}{a_n} \right\}^{1/2}.$$

For $\Lambda > 1$, (4.4) is valid for all $x \in \mathbb{R}$. In the important special case $W := W_\Lambda$,

$$\int_1^{Cn} ds/Q^{[-1]}(s) \sim \begin{cases} 1 & , \quad \Lambda < 1, \\ \log(n+1), & \Lambda = 1, \\ n^{1 - 1/\Lambda} & , \quad \Lambda > 1, \end{cases}$$

and (4.3) was obtained somewhat earlier by Freud ($\Lambda \geq 2$), Levin and Lubinsky ($1 < \Lambda < 2$), and Nevai and Totik ($0 < \Lambda \leq 1$), together with L_p analogues.

Freud's methods involved Christoffel functions and old Fourier series methods of T. Carleman [10]. The other authors used various types of entire functions imitating W, while the present form of Theorem 4.3 involves majorisation techniques, and Cauchy's integral formula for derivatives.

Remarkably enough, Dzrbasyan had (4.3) in the 1950's, but for a fixed finite subinterval of $\mathbb{R}$, and his proofs did not extend. The missing ingredient was the (crude) infinite–finite range inequalities that were sufficient for Freud's direct and converse (or Jackson–Bernstein) theorems.

Traditionally, the latter employed moduli of continuity. In recent years, K–functionals (introduced originally for interpolation between spaces) have been helpful in equivalent, and often more elegant, formulations. If $r \geq 1$, $1 \leq p \leq \infty$ and $fW \in L_p(\mathbb{R})$, the associated K–functional is

$$K_r(f,t):= \inf_{g^{(r)}W\in L_p(\mathbb{R})} \left\{ \|(f-g)W\|_{L_p(\mathbb{R})} + t\|g^{(r)}W\|_{L_p(\mathbb{R})} \right\},$$

$t > 0$. The error in polynomial approximation from $\mathbf{P}_n$ is

$$E_n(f):= \inf_{P\in \mathbf{P}_n} \|(f-P)W\|_{L_p(\mathbb{R})}.$$

The necessary estimates are now available for the following form of a theorem of Freud and Mhaskar (originally proved under more stringent conditions) [6, p. 185],[11]:

Theorem 4.4. Let $W \in FR_\Lambda$, some $\Lambda > 1$. Let $r \geq 1$, $1 \leq p \leq \infty$, and let $fW \in L_p(\mathbb{R})$. If $p = \infty$, we assume also that f is continuous. There exists C independent of n and f such that for $n \geq 1$,

$$E_n(f) \leq CK_r(f,(a_n/n)^r),$$

and

$$K_r(f,(a_n/n)^r) \leq C(a_n/n)^r \sum_{k=1}^{n} k^{r-1} a_k^{-r} E_k(f).$$

Ditzian and Totik [6, p. 185] established the following corollary involving the (perhaps more familiar) symmetric difference operator,

$$\Delta_h^r f(x):= \sum_{k=0}^{r} \binom{r}{k}(-1)^k f(x + rh/2 - kh).$$

Theorem 4.5. Assume the hypotheses of Theorem 4.4. Given $h > 0$, let h^* denote the positive root of the equation

$$Q'(h^*) = 1/h.$$

Then for $0 < \alpha < r$, the following are equivalent:

(a) $E_n(f) = O((a_n/n)^\alpha)$, $n \to \infty$.

(b) $\|W\Delta_h^r f\|_{L_p[-h^*,h^*]} = O(h^\alpha)$, $h \to 0+$.

Note that since $\Lambda > 1$, one can show that

$$\lim_{n \to \infty} a_n/n = 0.$$

The delicate case $\Lambda = 1$, including $W_1(x) = \exp(-|x|)$, is still not completely resolved [7].

When $f^{(r)} W \in L_p(\mathbb{R})$, Theorem 4.4 asserts that

$$E_n(f) = O(a_n/n)^r, \quad n \to \infty.$$

Conversely, if for some $\epsilon > 0$,

$$E_n(f) = O(a_n/n)^{r+\epsilon}, \quad n \to \infty,$$

Theorem 4.5 implies that $f^{(r)}$ exists a.e. and $f^{(r)} W \in L_p(\mathbb{R})$.

For further equivalences involving K-functionals, moduli of continuity and Jackson–Bernstein theorems, see [6, Ch. 11]. For an extensive and entertaining survey of Christoffel functions, Nikolskii and Markov–Bernstein inequalities, and the circle of ideas of this section, see [24]. Some more recent developments appear in [25] and [15].

5. Asymptotics on $(-\infty,\infty)$ of Orthogonal Polynomials

The orthonormal polynomials for the Chebyshev weight $w(x) := (1-x^2)^{-1/2}$ on $[-1,1]$ are given by the identity

$$p_n(w, \cos \theta) = (2/\pi)^{1/2} \cos n\theta, \quad \theta \in [0,\pi], \quad n \geq 1.$$

In fact, many weights w on $[-1,1]$ admit the asymptotic

(5.1) $$p_n(w, \cos \theta) w(\cos \theta)^{1/2} |\sin \theta|^{1/2}$$

$$= (2/\pi)^{1/2} \cos(n\theta + \Gamma(\theta)) + o(1), \quad \theta \in (0,\pi), \quad n \to \infty,$$

where the phase change $\Gamma(\theta)$ depends on w.

It is a remarkable feature of weights $W^2 = e^{-2Q}$ on $\mathbb{R}$, that many admit what Paul Nevai calls a *Plancherel–Rotach* asymptotic:

$$(5.2) \qquad a_n^{1/2} p_n(W^2, a_n \cos\theta) W(a_n \cos\theta) |\sin\theta|^{1/2}$$

$$= (2/\pi)^{1/2} \cos(n\theta + \Gamma_n(\theta)) + o(1), \quad \theta \in (0,\pi), \quad n \to \infty.$$

Here $\Gamma_n(\theta)$ depends on n and W, and may be of the same size as $n\theta$. Of course, (5.2) can be viewed as a generalisation of (5.1), with suitably chosen a_n.

Establishing asymptotics such as (5.1) or (5.2) has always been viewed as one of the finer problems in orthogonal polynomials, because of its inherent technical difficulties. Apart from intrinsic interest, it implies bounds on the orthonormal polynomials that have applications in convergence of orthonormal expansions, Lagrange interpolation, quadrature and other processes. From a methodological point of view, obtaining the bounds is usually as difficult as proving the deeper asymptotics. So one might simply aim for the latter.

Szegő (in 1922) and Bernstein (in 1930) were the first to establish (5.1) for quite general w. The central tools in their analysis were uniform or one–sided polynomial approximations of a function related to w; and the following explicit formula for orthonormal polynomials for special weights. We shall need in the sequel one more piece of notation: If $f \geq 0$ and $\log f \in L_1[-\pi,\pi]$, we introduce for $\theta \in (-\pi,\pi)$,

$$(5.3) \qquad \Gamma(f;\theta) := \frac{1}{4\pi} \int_{-\pi}^{\pi} \{\log f(t) - \log f(\theta)\} \cot\left(\frac{\theta-t}{2}\right) dt,$$

whenever defined.

Theorem 5.1 (Bernstein, Szegő, Akhiezer). Let $\nu > 0$ and $S \in P_\nu$, positive in $[-1,1]$, except possibly for simple zeros at ± 1. Let

$$(5.4) \qquad w(x) := (1-x^2)^{1/2}/S(x), \quad x \in (-1,1),$$

and

$$f(\theta) := w(\cos\theta)|\sin\theta|, \quad \theta \in [-\pi,\pi].$$

Then for $n \geq \nu/2$, $x = \cos\theta$, $\theta \in [-\pi,\pi]$,

$$(5.5) \qquad p_n(w,\cos\theta) w(\cos\theta)^{1/2} |\sin\theta|^{1/2}$$

$$= (2/\pi)^{1/2} \cos(n\theta + \Gamma(f;\theta)).$$

Further, for $n > \nu/2$,

(5.6) $\qquad \pi \lambda_n^{-1}(w, \cos\theta) w(\cos\theta) |\sin\theta|$

$$= n - \frac{1}{2} + \{2\sin\theta\}^{-1} \sin((2n-1)\theta + 2\Gamma(f,\theta)) + \Gamma'(f,\theta).$$

We note that (5.6) is incorrectly stated in [9, p. 251]. See [15, p. 118].

For weights on $(-\infty, \infty)$, sharp infinite–finite range inequalities have recently permitted application of Szegő's device of one–sided approximation (by weights of the form (5.4)) to weights on $\mathbb{R}$. In particular, they permitted partial resolution of [24, p. 141]:

P.Nevai's Conjecture. Let $W := W_\Lambda$, some $\Lambda > 1$. Let $0 < \epsilon < \frac{\pi}{2}$. Then (5.2) holds uniformly for $\theta \in [\epsilon, \pi-\epsilon]$, with $o(1)$ replaced by $O(n^{-1})$.

For the Hermite weight W_2^2, Nevai's conjecture is a theorem that goes back to Plancherel and Rotach in 1929. P. Nevai and his students W. Bauldry and R. Sheen proved his conjecture for $\Lambda = 4, 6$, and an analogue for $W(x) := \exp(-x^4 + P(x))$, $P \in \mathbf{P}_3$ [3, 23, 28]. Their methods involve a second order differential equation and careful analysis of the coefficients in the three term recurrence relation.

The drawback of this procedure is the increasing complexity of the differential equation as Λ (restricted to positive even integers) increases. Stan Bonan, a student of P. Nevai, circumvented this by introducing an "approximate" d.e. This in turn was an important ingredient of [4].

Theorem 5.2 (Bonan and Clark). Let Λ be a positive even integer. Then as $n \to \infty$,

$$\|p_n(W_\Lambda^2, x) W_\Lambda(x)\|_{L_\infty(\mathbb{R})} \sim n^{1/6 - 1/(2\Lambda)}.$$

The future of the differential equation approach may well lie in H.N. Mhaskar's recent [18]:

Theorem 5.3. Let $W := e^{-Q} \in FR_\Lambda$, some $\Lambda \geq 2$. For $x, t \in \mathbb{R}$ and $n \geq 1$, define

$$q(x,t) := \frac{Q'(x) - Q'(t)}{x - t};$$

$$\eta_n := \gamma_{n-1}/\gamma_n;$$

$$A_n(x) := 2\eta_n \int_{-\infty}^{\infty} q(x,t) p_n^2(t) W^2(t) dt;$$

$$B_n(x) := 2\eta_n \int_{-\infty}^{\infty} q(x,t) p_{n-1}(t) p_n(t) W^2(t) dt,$$

where $\gamma_n := \gamma_n(W^2)$ and $p_n(x) := p_n(W^2, x)$. Then

$$p_n' = A_n p_{n-1} - B_n p_n,$$

and

$$p_n'' + M_n p_n' + N_n p_n = 0,$$

where

$$M_n := -2Q' - A_n'/A_n,$$

and

$$N_n := A_{n-1} A_n \eta_n / \eta_{n-1} - A_{n-1} B_n x / \eta_n + B_{n-1} B_n + B_n' - A_n' B_n / A_n.$$

At present, the most generally successful method for establishing (5.2) is based on Szegő's old ideas. These require one–sided approximations on the interval on which the L_2 norm of PW, $P \in P_n$, really lives. A simplified form would be the following, which bears more than a passing resemblance to old polynomial approximation problems of Ed Saff. Ed Saff's contributions now live everywhere in the theory of weights on $(-\infty,\infty)$, in contrast to his weighted polynomials that live only on $[-a_n, a_n]$.

Problem 5.4. Let $W \in FR_\Lambda$, some $\Lambda > 0$. Let $s \geq 0$ and $\{c_n\}_{n=1}^{\infty}$ be defined by (4.2). Find $P \in P_{2n}$, n large enough, such that

$$\tau_n(u) := P_n(u) W^2(c_n u)$$

satisfies

(5.7) $$\lim_{n \to \infty} \int_{-1}^{1} |\log \tau_n(u)| (1-u^2)^{-1/2} du = 0,$$

and either

(5.8) $$\tau_n(u) \geq 1, \quad u \in [-1,1],$$

or

(5.9) $$0 \leq \tau_n(u) \leq 1, \quad u \in [-1,1].$$

Note that (5.7) ensures that $\tau_n(u) \to 1$ a.e. in $[-1,1]$ as $n \to \infty$. Of course, much more than this is required, and at present the proofs of the requisite estimates are extremely technical. The main part of the approximation comes (as in [16]) from Lagrange interpolation to $1/W$ at zeros of L_∞ extremal polynomials.

To show that the error in interpolation $\to 0$ as $n \to \infty$, one uses Hermite's contour integral error formula for Lagrange interpolation. Since the latter requires analyticity, one replaces $1/W$ by an entire function introduced by the author [15, p. 53]:

Theorem 5.5. Let $W \in FR_\Lambda$ (3), some $\Lambda > 0$. For $n \geq 1$, define *Freud's number* q_n to be the positive root of the equation

$$n = q_n Q'(q_n),$$

and define

(5.10) $$G_{Q/2}(x) := 1 + \sum_{n=1}^{\infty} \left(\frac{x}{q_{2n}}\right)^{2n} e^{Q(q_{2n})} n^{-1/2}.$$

Let

$$T(x) := 1 + xQ''(x)/Q'(x), \quad x \in (0,\infty).$$

Then as $x \to \infty$,

(5.11) $$G_{Q/2}(x) = W^{-1}(x)\{\pi \, T(x)\}^{1/2}\{1 + 0(Q(x)^{-1/2})\}.$$

We note that (2.5) ensures $T(x) \sim 1$ in $\mathbb{R}$. A 1968 result of Clunie and Kövari [5] constructs, for very general W, an even entire function G with non–negative Maclaurin series coefficients satisfying

$$GW \sim 1 \text{ in } \mathbb{R}.$$

(The author is grateful to Prof. D. Shea for this reference). However, the sharper (5.11) is needed in proving [15, p. 200]:

Theorem 5.6. Let $W \in FR_\Lambda$ (3), some $\Lambda > 0$. Let $s \geq 0$ and $\{c_n\}_{n=1}^\infty$ be defined by (4.2), and $G_{Q/2}$ be defined by (5.10). There exist $P_n \in P_{2n}$, n large enough, such that

$$\tau_n(u) := P_n(u) / G_{Q/2}^2(c_n u)$$

satisfies (5.7) and either (5.8) or (5.9).

For $\Lambda > 3$, Theorem 5.6 implies a positive solution to Problem 5.4. Armed with these one–sided approximations, one can first establish asymptotics for $p_n(G_{Q/2}^{-2}, x)$ and then pass to the asymptotics for $p_n(W^2, x)$ via the *relativized Szegő theory* introduced by P. Nevai in [22] and developed by Máté, Nevai, Totik and Rahmanov [17, 27]. This yields [15, p. 187]:

Theorem 5.7. Let $W \in FR_\Lambda$ (3), some $\Lambda > 3$. Let $s > 0$ be fixed, but large enough, and let $\{c_n\}_{n=1}^\infty$ be defined by (4.2), and for $n \geq 1$, define

$$f_n(\theta) := W^2(c_n \cos \theta) |\sin \theta|, \quad \theta \in [-\pi, \pi].$$

Then there exists $\delta > 0$ such that for n large enough, and uniformly for $\theta \in [n^{-\delta}, \pi - n^{-\delta}]$,

$$c_n^{1/2} p_n(W^2, c_n \cos \theta) W(c_n \cos \theta) |\sin \theta|^{1/2}$$

$$= (2/\pi)^{1/2} \cos(n\theta + \Gamma(f_n;\theta)) + 0(n^{-\delta}).$$

For Christoffel functions, the corresponding asymptotic is [15, p. 183–184]:

Theorem 5.8. Let $W \in FR_\Lambda$ (3), some $\Lambda > 1$. For $n \geq 1$, define

$$f_n(\theta) := W^2(a_n \cos \theta) |\sin \theta|, \quad \theta \in [-\pi, \pi].$$

Then, given $\eta > 0$, there exists $\delta > 0$ such that for n large enough, and uniformly for $\theta \in [n^{-\delta}, \pi - n^{-\delta}]$,

$$\pi \lambda_n^{-1}(W^2, a_n \cos \theta) W^2(a_n \cos \theta) |\sin \theta| \, a_n/n$$

$$= 1 + n^{-1} \Gamma'(f_n;\theta) + 0(n^{\eta - \min\{1/2, 1 - 1/\Lambda\}}).$$

Other and sharper forms of the above asymptotics, for weights $W^2 = e^{-2Q}$, where Q is of polynomial growth (Freud weights) or of faster than polynomial growth (Erdös weights) appear in [15]. For $W^2 = W_\Lambda^2$, $\Lambda > 0$, E.A. Rahmanov announced in [13] a form of (5.2), based also on polynomial approximation. It will be of interest to see the generality of Rahmanov's methods when the proofs are published.

It is to be emphasised that the above is the tip of the iceberg. For asymptotics of recurrence coefficients, leading coefficients, and $p_n(W^2, z)$ for complex z, in various senses, see [13, 14, 15, 16, 20, 21, 26, 31], and the other papers in this proceedings.

References

1. N.I. Akhiezer, *On the Weighted Approximation of Continuous Functions by Polynomials on the Entire Real Axis*, Amer. Math. Soc. Transl., 22(1962), 95–137.

2. N.I. Akhiezer, *The Classical Moment Problem*, Oliver and Boyd, Edinburgh, 1965.

3. W.C. Bauldry, *Orthogonal Polynomials Associated with Exponential Weights*, Ph.D dissertation, Ohio State University, Columbus, 1985.

4. S.S. Bonan and D.S. Clark, *Estimates of the Hermite and Freud Polynomials*, to appear in J. Approx. Th.

5. J. Clunie and T. Kövari, *On Integral Functions having Prescribed Asymptotic Growth* II, Canadian J. Math, 20(1968), 7–20.

6. Z. Ditzian and V. Totik, *Moduli of Smoothness*, Springer Series in Computational Mathematics, Vol. 9, Springer, New York, 1987.

7. Z. Ditzian, D.S. Lubinsky, P. Nevai and V. Totik, *Polynomial Approximation with Exponential Weights*, Acta Math. Hung., 50(1987), 165–175.

8. M.M. Dzrbasyan, *Some Questions of the Theory of Weighted Polynomial Approximations in a Complex Domain*, (in Russian) Mat. Sb., 36(78)(1955), 353–440.

9. G. Freud, *Orthogonal Polynomials*, Akademiai Kiado/Pergamon Press, Budapest/Oxford, 1971.

10. G. Freud, *On Markov–Bernstein Type Inequalities and their Applications*, J. Approx. Th., 19(1977), 22–37.

11. G. Freud and H.N. Mhaskar, *K–functionals and moduli of continuity in weighted polynomial approximation*, Ark. Mat., 21(1983), 145–161.

12. A.L. Levin and D.S. Lubinsky, *Refined L_∞ Markov–Bernstein Inequalities for Freud Weights*, to appear in SIAM J. Math. Anal.

13. G. Lopez and E.A. Rahmanov, *Rational Approximations, Orthogonal Polynomials and Equilibrium Distributions*, Springer Lecture Notes, Vol. 1329, pp. 125–157, Springer, Berlin, 1988.

14. D.S. Lubinsky, *A Survey of General Orthogonal Polynomials for Weights on Finite and Infinite Intervals*, Acta Applicandae Mathematicae, 10(1987), 237–296.

15. D.S. Lubinsky, *Strong Asymptotics for Extremal Errors and Polynomials Associated with Erdös–Type Weights*, Pitman Research Notes, Vol. 202, Longmans, Harlow, 1989.

16. D.S. Lubinsky and E.B. Saff, *Strong Asymptotics for Extremal Polynomials Associated with Weights on* ℝ, Springer Lecture Notes, Vol. 1305, Springer, Berlin, 1988.

17. A. Máté, P. Nevai and V. Totik, *Extensions of Szegő's Theory of Orthogonal Polynomials* III, Constructive Approximation, 3(1987), 73–96.

18. H.N. Mhaskar, *Bounds for Certain Freud–Type Orthogonal Polynomials*, to appear in J. Approx. Th.

19. H.N. Mhaskar and E.B. Saff, *Extremal Problems for Polynomials with Exponential Weights*, Trans. Amer. Math. Soc., 285(1984), 203–234.

20. H.N. Mhaskar and E.B. Saff, *Where does the Sup–norm of a Weighted Polynomial Live?*, Constructive Approximation, 1(1985), 71–91.

21. H.N. Mhaskar and E.B. Saff, *Where does the L_p–norm of a Weighted Polynomial Live?*, Trans. Amer. Math. Soc., 303(1987), 109–124.

22. P. Nevai, *Orthogonal Polynomials*, Mem. Amer. Math. Soc., Vol. 213, Providence, 1979.

23. P. Nevai, *Asymptotics for Orthogonal Polynomials Associated with* $\exp(-x^4)$, SIAM J. Math. Anal., 15(1984), 1177–1187.

24. P. Nevai, *Geza Freud, Orthogonal Polynomials and Christoffel Functions, A Case Study*, J. Approx. Th., 48(1986), 3–167.

25. P. Nevai and V. Totik, *Sharp Nikolskii Inequalities with Exponential Weights*, Anal. Math., 13(1987), 261–267.

26. E.A. Rahmanov, *On Asymptotic Properties of Polynomials Orthogonal on the Real Axis*, Math. USSR. Sbornik, 47(1984), 155–193.

27. E.A. Rahmanov, *On Asymptotic Properties of Orthogonal Polynomials on the Circle with Weights not Satisfying Szegő's Condition*, Math. USSR. Sbornik, 58(1987), 149–167.

28. R.C. Sheen, *Plancherel–Rotach Type Asymptotics for Orthogonal Polynomials Associated with* $\exp(-x^6/6)$, J. Approx. Th., 50(1987), 232–293.

29. J.A. Shohat and J.D. Tamarkin, *The Problem of Moments*, Math. Surveys. No.1., Amer. Math. Soc., Providence, R.I., 1943, 1970.

30. G. Szegő, *Orthogonal Polynomials*, Amer. Math. Soc. Colloq. Publs., Vol. 23, Amer. Math. Soc., Providence, 1939, 4th edn., 1975.

31. W. Van Assche, *Norm Behavior and Zero Distribution for Orthogonal Polynomials with Non–Symmetric Weights*, Constructive Approximation, 5(1989), 329–345.

ORTHOGONAL POLYNOMIALS ASSOCIATED WITH ROOT SYSTEMS

I. G. Macdonald
School of Mathematical Sciences
Queen Mary College, University of London
London E1 4NS, England

Introduction. The orthogonal polynomials that are the subject of these lectures are Laurent polynomials in several variables. They depend rationally on two parameters q and t, and there is a family of them attached to each root system R. For particular values of the parameters q and t, these polynomials reduce to objects familiar in representation theory:

(i) when $q = t$, they are independent of q and are the Weyl characters for the root system R.

(ii) when $q = 0$ they are (up to a scalar factor) the polynomials that give the values of zonal spherical functions on a semisimple p-adic Lie group G relative to a maximal compact subgroup K, such that the restricted root system of (G, K) is the dual root system $R^{\check{}}$.

(iii) when q and t both tend to 1, in such a way that $(1-t)/(1-q)$ tends to a definite limit k, then (for certain values of k) our polynomials give the values of zonal spherical functions on a real (compact or noncompact) symmetric space G/K arising from finite-dimensional spherical representations of G, that is to say representations having a non zero K-fixed vector. Here the root system R is the restricted root system of G/K, and the parameter k is half the root multiplicity (assumed to be the same for all restricted roots).

Thus these two-parameter families of orthogonal polynomials constitute a sort of bridge between harmonic analysis on real symmetric spaces and on their p-adic analogs.

All this is in fact a simplified description. The general picture is more elaborate and involves parameters q_α and t_α for each root $\alpha \in R$, such that $q_\alpha = q_\beta$ and $t_\alpha = t_\beta$ when α and β are roots of the same length. The necessary modifications will be given in the concluding remarks.

Root Systems

Let V be a real vector space of finite dimension, endowed with a positive-definite symmetric bilinear from $< u, v >$. For each nonzero $\alpha \in V$ let w_α denote the orthogonal reflection in the hyperplane through the origin perpendicular to α, so that

$$w_\alpha(v) = v - < v, \alpha^{\check{}} > \alpha \qquad (v \in V)$$

P. Nevai (ed.), Orthogonal Polynomials, 311–318.
© 1990 by Kluwer Academic Publishers.

where $\alpha^{\check{}} = 2\alpha/\langle\alpha,\alpha\rangle$.

A *root system* R in V is a finite non empty set of nonzero vectors (called *roots*) that span V and are such that for each pair $\alpha,\beta \in R$ we have $\langle\alpha^{\check{}},\beta\rangle \in \mathbf{Z}$ and $w_\alpha(\beta) \in R$. Thus each reflection $w_\alpha(\alpha \in R)$ permutes R, and the group of orthogonal transformations of V generated by the w_α is a finite group W called the Weyl group of R.

The vectors $\alpha^{\check{}}$ for $\alpha \in R$ form a root system $R^{\check{}}$, the *dual* of R.

If $\alpha \in R$, then $-\alpha \in R$ (because $-\alpha = w_\alpha(\alpha)$). R is said to be *reduced* if for each $\alpha \in R$ the only scalar multiples of α that belong to R are $\pm\alpha$. Furthermore, R is said to be *irreducible* if it is not possible to partition R into two non-empty subsets R_1, R_2 such that each root in R_1 is orthogonal to each root in R_2 (which would imply that R_1 and R_2 are themselves root systems). Until further notice, we shall assume that R is both reduced and irreducible.

For those to whom these notions are unfamiliar, some examples to bear in mind are the following. Let $\varepsilon_1,\cdots,\varepsilon_n$ be the standard basis of $\mathbf{R}^n(n \geq 2)$, with the usual scalar product. Then the vectors

$$(A_{n-1}) \qquad\qquad\qquad \varepsilon_i - \varepsilon_j$$

where $i \neq j$, form a root system (and V is the hyperplane perpendicular to $\varepsilon_1 + \cdots + \varepsilon_n$). The Weyl group is the symmetric group S_n of all permutations of the ε_i.

Moreover, each of the sets of vectors

$$(B_n) \qquad\qquad \pm\varepsilon_i(1 \leq i \leq n),\ \pm\varepsilon_i \pm \varepsilon_j\ (1 \leq i < j \leq n)$$

$$(C_n) \qquad\qquad \pm 2\varepsilon_i(1 \leq i \leq n),\ \pm\varepsilon_i \pm \varepsilon_j\ (1 \leq i < j \leq n)$$

$$(D_n) \qquad\qquad\qquad \pm\varepsilon_i \pm \varepsilon_j\ (1 \leq i < j \leq n)$$

is a root system. For B_n and C_n the Weyl group is the group of all signed permutations of the ε_i, of order $2^n n!$ For D_n, it is a subgroup of index 2 of this group. The root systems B_n and C_n are duals of each other, and A_{n-1}, D_n are self dual.

In fact, the root systems $A_n(n \geq 1), B_n(n \geq 2), C_n(n \geq 3)$ and $D_n(n \geq 4)$ almost exhaust the catalog of reduced irreducible root systems up to isomorphism. Apart from these, there are just five others, the "exceptional" root systems E_6, E_7, E_8, F_4 and G_2.

Now let R be any (reduced, irreducible) root system and let $v \in V$ be such that $\langle\alpha,v\rangle \neq 0$ for all $\alpha \in R$. Then the set R^+ of roots $\alpha \in R$ such that $\langle\alpha,v\rangle > 0$ is called a *system of positive roots* in R. Of course R^+ depends on the choice of the vector v, but one can show that the possible sets R^+ are permuted transitively by the Weyl group, so that there is really no loss of generality in fixing R^+ once and for all.

The abelian group Q generated by R, whose elements are the integral linear combinations of the roots, is a lattice in V (i.e., a free abelian group of rank equal to the dimension of V) called the *root lattice*. We denote by Q^+ the subsemigroup of Q consisting of all sums $\sum_{\alpha \in R^+} m_\alpha\alpha$, where the coefficients m_α are integers ≥ 0.

Next, the set P of all $\lambda \in V$ such that $\langle \lambda, \alpha \check{} \rangle \in \mathbf{Z}$ for all $\alpha \in R$ is another lattice in V, called the *weight lattice*. We denote by P^{++} the set of *dominant weights* $\lambda \in P$ such that $\langle \lambda, \alpha \check{} \rangle \in \mathbf{N}$ for all $\alpha \in R^+$. We have $P \supset Q$ (but in general $P^{++} \not\supset Q^+$), and P/Q is a finite group. Both P and Q are stable under the action of the Weyl group W.

If $\lambda, \mu \in P$ we shall write $\lambda \geq \mu$ to mean that $\lambda - \mu \in Q^+$. This is a partial order on the lattice P.

We next introduce the (real) group algebra $A = \mathbf{R}P$ of the lattice P. Since the group operation in P is addition, we use an exponential notation in A, and we denote by e^λ the element of A corresponding to $\lambda \in P$. These "formal exponentials" e^λ from an $\mathbf{R}$-basis of A, such that $e^\lambda e^\mu = e^{\lambda+\mu}$. In particular, $e^0 = 1$ is the identity element of A.

The Weyl group W acts on P and therefore also on A: $w(e^\lambda) = e^{w\lambda}$ for $\lambda \in P$ and $w \in W$. We denote by A^W the subalgebra of W-invariant elements of A.

Two Bases of A^W

Each W-orbit in P meets P^{++} exactly once, and hence the *orbit sums*

$$m_\lambda = \sum_{\mu \in W\lambda} e^\mu \quad (\lambda \in P^{++})$$

form an $\mathbf{R}$-basis of A^W.

Another $\mathbf{R}$-basis of A^W is obtained as follows: Let

$$\rho = \frac{1}{2} \sum_{\alpha \in R^+} \alpha$$

and define

$$\delta = \prod_{\alpha \in R^+} (e^{\alpha/2} - e^{-\alpha/2}).$$

In fact, $\rho \in P^{++}$ and $\delta \in A$. Moreover, δ is skew-symmetric for W, that is to say $w\delta = \varepsilon(w)\delta$ for all $w \in W$, where $\varepsilon(w) = \det(w) = \pm 1$. For each $\lambda \in P^{++}$, the sum

$$\sum_{w \in W} \varepsilon(w) e^{w(\lambda+\rho)}$$

is likewise skew-symmetric, and is divisible by δ in A; the quotient

$$\chi_\lambda = \delta^{-1} \sum_{w \in W} \varepsilon(w) e^{w(\lambda+\rho)}$$

is an element of A^W called the *Weyl character* corresponding to λ, and is of the form

$$\chi_\lambda = m_\lambda + \text{lower terms}$$

where by "lower terms" is meant a linear combination of the orbit-sums m_μ such that $\mu \in P^{++}$ and $\mu < \lambda$ (for the partial order defined above). The $\chi_\lambda (\lambda \in P^{++})$ form another $\mathbf{R}$-basis of A^W.

Scalar Product

Let $q \in (0,1)$ and write

$$(x;q)_\infty = \prod_{i=0}^{\infty}(1 - xq^i),$$

$$(x;q)_k = \prod_{i=0}^{k-1}(1 - xq^i)$$

for any integer $k \geq 0$. We now define, for any $t \geq 0$,

$$\Delta = \Delta(q,t) = \prod_{\alpha \in R}(e^\alpha;q)_\infty/(te^\alpha;q)_\infty.$$

Suppose first that $t = q^k$, where k is an integer ≥ 0. Then

$$\Delta = \prod_{\alpha \in R}(e^\alpha;q)_k \in A^W.$$

We shall use Δ to define a (symmetric, positive-definite) scalar product on A, as follows. If $f \in A$, say

$$f = \sum_{\lambda \in P} f_\lambda e^\lambda$$

let

$$\bar{f} = \sum_{\lambda \in P} f_\lambda e^{-\lambda}$$

and let $[f]_1$ denote the constant term f_0 of f. We then define the scalar product of f and g to be

$$\langle f,g \rangle = |W|^{-1}[f\bar{g}\Delta]_1.$$

(If t is not a power of q, a different definition is needed. Let T be the torus $V/Q^{\check{}}$, where $Q^{\check{}}$ is the root lattice of $R^{\check{}}$; then the character group of T may be identified with the weight lattice P, and we may regard each $e^\lambda, \lambda \in P$, as a character of T by the rule

$$e^\lambda(\dot{x}) = \exp 2\pi i \langle \lambda, x \rangle$$

where $\dot{x} \in T$ is the image of $x \in V$, and exp is the exponential function. Then $f\bar{g}\Delta$ is a continuous function on the torus T, and we define

$$\langle f,g \rangle = |W|^{-1} \int_T f\bar{g}\Delta,$$

the integration being with respect to normalized Haar measure on T.)

The Existence Theorem

Theorem. *There is a unique* **R**-*basis* $(P_\lambda)_{\lambda \in P^{++}}$ *of* A^W *such that*

(i) $P_\lambda = m_\lambda + \sum\limits_{\substack{\mu < \lambda \\ \mu \in P^{++}}} u_{\lambda\mu} m_\mu$

where the coefficients $u_{\lambda\mu}$ *are rational functions of* q *and* t;

(ii) $\langle P_\lambda, P_\mu \rangle = 0$ *if* $\lambda \neq \mu$.

It is easy to see that the P_λ, if they exist, are unique. Their existence is not so obvious. If the partial order $\lambda \geq \mu$ on P^{++} were a *total* ordering, existence would follow directly from the Gram-Schmidt orthogonalization process. But it isn't a total ordering (unless $R = A_1$) and we should therefore have to extend it to a total ordering before applying the Gram-Schmidt mechanism. Thus the content of the theorem is that however we extend the partial order $\geq$ to a total order, we always obtain the same basis.

The theorem is a consequence of the

Proposition. *There is a linear operator* $E : A^W \to A^W$ *with the following three properties:*

(a) $\langle Ef, g \rangle = \langle f, Eg \rangle$ *for all* $f, g \in A^W$;

(b) *For all* $\lambda \in P^{++}$, Em_λ *is of the form*

$$Em_\lambda = \sum_{\substack{\mu \leq \lambda \\ \mu \in P^{++}}} c_{\lambda\mu} m_\mu;$$

(c) *If* $\lambda, \mu \in P^{++}$ *and* $\lambda \neq \mu$, *then* $c_{\lambda\lambda} \neq c_{\mu\mu}$.

Property (a) says that E is self-adjoint; (b) says that the matrix of E is triangular, relative to the basis (m_λ) of A^W; and (c) says that the eigenvalues of E are all distinct.

Given an operator E with these three properties, for each $\lambda \in P^{++}$ let P_λ be the eigenfunction of E with eigenvalue $c_{\lambda\lambda}$, normalized so that the coefficient of m_λ in P_λ is equal to 1. Then the P_λ satisfy condition (i) of the theorem, and since

$$c_{\lambda\lambda}\langle P_\lambda, P_\mu \rangle = \langle EP_\lambda, P_\mu \rangle = \langle P_\lambda, EP_\mu \rangle = c_{\mu\mu}\langle P_\lambda, P_\mu \rangle$$

it follows that $\langle P_\lambda, P_\mu \rangle = 0$ if $\lambda \neq \mu$, so that (ii) is satisfied.

Thus it remains to construct a suitable operator E for each root system R, having the three properties (a), (b), (c) above. The simplest case is that in which the weight lattice P is bigger than the root lattice Q. In that case there is a nonzero vector $\pi \in V$ such that $\langle \pi, \alpha \rangle = 0$ or 1 for all $\alpha \in R^+$ (in fact, there are $f - 1$ such vectors, where f is the index of Q in P). For each such π we define

$$E_\pi f = \sum_{w \in W} w \left(\Delta_+^{-1} T_\pi (\Delta^+ f) \right)$$

where $T_\pi : A \to A$ is defined by $T_\pi(e^\lambda) = q^{<\lambda, \pi>} e^\lambda$, and $\Delta_+ = \prod\limits_{\alpha \in R^+} (e^\alpha; q)_\infty / (te^\alpha; q)_\infty$.

The desired operator E is then a linear combination of these operators E_π.

This construction works in all cases except E_8, F_4 and G_2, these being the only reduced irreducible root systems for which $P = Q$. In these cases another construction is needed, for the details of which I refer to [4].

Special Cases

(1) When $t = 1$, we have $\Delta = 1$ and P_λ is the orbit-sum m_λ.

(2) When $t = q$, we have $\Delta = \prod_{\alpha \in R} (1 - e^\alpha) = \delta\bar{\delta}$, where as before δ is the denominator of Weyl's character formula. Thus we have

$$< f, g >= |W|^{-1}[f\delta \cdot \overline{g\delta}]_1$$

from which it follows that P_λ is the Weyl character χ_λ.

(3) When R is of type A_1, the P_λ are the q-ultraspherical polynomials of Askey and Ismail [1], suitably normalized.

(4) When $q \to 0$, t being arbitrary, we have

$$\Delta = \prod_{\alpha \in R} (1 - e^\alpha)/(1 - te^\alpha).$$

In this case there is an explicit formula for P_λ, namely

$$P_\lambda = W_\lambda(t)^{-1} \sum_{w \in W} w \left(e^\lambda \prod_{\alpha \in R^+} \frac{1 - te^{-\alpha}}{1 - e^{-\alpha}} \right)$$

where $W_\lambda(t)$ is the polynomial

$$\sum_{\substack{w \in W \\ w\lambda = \lambda}} t^{\ell(w)}$$

and $\ell(w)$ is the number of positive roots $\alpha \in R^+$ such that $w\alpha \notin R^+$. These polynomials arise in nature as the values of zonal spherical functions on a p-adic Lie group, when t^{-1} is a prime power.

(5) Let $t = q^k$, fix k (which need not be an integer) and let $q \to 1$ (so that $t \to 1$ also). Then in the limit we have

$$\Delta = \prod_{\alpha \in R} (1 - e^\alpha)^k.$$

In this limiting case the polynomials P_λ were defined (by quite different methods) by Heckman and Opdam [2]. They called the P_λ "Jacobi polynomials" and established the limiting case of the conjecture stated below for the value of $\langle P_\lambda, P_\lambda \rangle$.

For particular values of the parameter k, the polynomials P_λ again occur in nature as zonal spherical functions, but this time on a real semisimple Lie group G, relative to a maximal compact subgroup K. For example, if $G = SL_n(\mathbf{R})$, $K = SO(n)$, and $k = \frac{1}{2}$, the P_λ are essentially the zonal polynomials familiar to statisticians.

A Conjecture

If one has a family of orthogonal polynomials P_λ, the first thing one needs to know is the value of the scalar product $< P_\lambda, P_\lambda >$. Here I can only offer a conjecture. Recall that the q-gamma function $\Gamma_q(x)$ is defined (for $x \in \mathbf{R}, x \notin -\mathbf{N}$) by

$$\Gamma_q(x) = \frac{(q;q)_\infty}{(q^x;q)_\infty}(1-q)^{1-x}.$$

It is convenient also to define

$$\Gamma_q^*(x) = 1/\Gamma_q(1-x).$$

Let $t = q^k$ and define, for $\lambda \in V$,

$$c(\lambda) = \prod_{\alpha \in R^+} \frac{\Gamma_q(\langle \lambda, \alpha^\vee \rangle)}{\Gamma_q(\langle \lambda, \alpha^\vee \rangle + k)},$$

$$c^*(\lambda) = \prod_{\alpha \in R^+} \frac{\Gamma_q^*(\langle \lambda, \alpha^\vee \rangle)}{\Gamma_q^*(\langle \lambda, \alpha^\vee \rangle + k)}.$$

Conjecture. For all $\lambda \in P^{++}$,

$$\langle P_\lambda, P_\lambda \rangle = \frac{c^*(-\lambda - \rho_k)}{c(\lambda + \rho_k)}$$

where $\rho_k = \frac{1}{2}k \sum_{\alpha \in R^+} \alpha$.

If k is a non-negative integer, an equivalent statement is

$$\langle P_\lambda, P_\lambda \rangle = \prod_{\alpha \in R^+} \prod_{i=0}^{k-1} \frac{1 - q^{\langle \lambda + k\rho, \alpha^\vee \rangle + i}}{1 - q^{\langle \lambda + k\rho, \alpha^\vee \rangle - i}}.$$

Notice that when $\lambda = 0$ we have $P_\lambda = 1$, so that in this case the conjecture reduces to the constant term conjectures of [3] and [5]. We remark also that the conjecture is true in all the special cases (1) - (5) considered above, and for all values of q and t when R is of type A_{n-1}.

Concluding Remarks

All this is in fact a simplified version of the theory. In the general version we take as starting point two root systems R and S in the same vector space V, such that S (but not necessarily R) is reduced, and such that R and S have the same Weyl group W. In this situation, for each $\alpha \in R$ there exists a unique positive real number u_α such that $u_\alpha^{-1}\alpha \in S$; we have $u_\alpha = u_\beta$ if α and β are in the same W-orbit, and the mapping $\alpha \mapsto u_\alpha^{-1}\alpha$ from R to S is surjective. Let q be a real number such that $0 \le q < 1$ as before, and for each $\alpha \in R$ let $q_\alpha = q^{u_\alpha}$. Moreover, for each $\alpha \in R$ let t_α be a real number ≥ 0, such that $t_\alpha = t_\beta$ if α and β are in the same W-orbit. The weight function Δ is now

$$\Delta = \prod_{\alpha \in R} \left(t_{2\alpha}^{1/2} e^\alpha;\ q_\alpha \right)_\infty \bigg/ \left(t_\alpha t_{2\alpha}^{1/2} e^\alpha; q_\alpha \right)_\infty$$

where $t_{2\alpha} = 1$ if $2\alpha \notin R$, and the scalar product on A^W is defined by

$$\langle f, g \rangle = |W|^{-1}[f\bar{g}\Delta]_1$$

as before. With these modifications the existence theorem is still valid.

The definition of $c(\lambda)$ is now

$$c(\lambda) = \prod_{\alpha \in R^+} \frac{\Gamma_{q_\alpha}(\langle \lambda, \alpha^\vee \rangle + \frac{1}{2}k_{\alpha/2})}{\Gamma_{q_\alpha}(\langle \lambda, \alpha^\vee \rangle + \frac{1}{2}k_{\alpha/2} + k_\alpha)} ,$$

where k_α is defined by $t_\alpha = q_\alpha^{k_\alpha}$, and $k_{\alpha/2} = 0$ if $\frac{1}{2}\alpha \notin R$. With $c^*(\lambda)$ defined in the same way, by replacing each Γ by Γ^*, the conjectured value of $\langle P_\lambda, P_\lambda \rangle$ is

$$c^*(-\lambda - \rho_k)/c(\lambda + \rho_k)$$

where now $\rho_k = \frac{1}{2} \sum_{\alpha \in R^+} k_\alpha \alpha$.

In particular, when R is of type BC_1 the polynomials P_λ are, up to a scalar factor, the polynomials $P_n^{(\alpha,\beta)}(x;q)$ introduced by Rahman [6]. Thus they are a particular case of the Askey-Wilson polynomials.

References

1. R. Askey and M. E. H. Ismail, *'A generalization of ultraspherical polynomials'*, pp. 55-78 in Studies in Pure Mathematics (ed. P. Erdős), Birkhäuser, 1983.

2. G. J. Heckman and E. M. Opdam, *Root systems and hypergeometric functions I-IV*, Comp. Math. **64** (1987) 329-352, 353-373; **67** (1988) 21-50, 191-209.

3. I. G. Macdonald, *'Some conjectures for root systems'*, SIAM Journal Math. Analysis **13** (1982) 988-1007.

4. I. G. Macdonald, *Orthogonal polynomials associated with root systems*, preprint (1988).

5. W. G. Morris, *Constant term identities for finite and affine root systems: conjectures and theorems*, Thesis, Madison (1982).

6. M. Rahman, *'The linearization of the product of continuous q-Jacobi polynomials'*, Can. J. Math. **33** (1981) 961-987.

SOME EXTENSIONS OF THE BETA INTEGRAL AND THE HYPERGEOMETRIC FUNCTION

Mizan Rahman*
Department of Mathematics & Statistics
Carleton University
Ottawa, Ontario K1S 5B6, CANADA

ABSTRACT.   This is a brief survey of the theory of basic hypergeometric series with particular emphasis on some extensions of Euler's beta integral and Gauss' hypergeometric function. From Heine's q-analogue of the binomial theorem we develop the summation and transformation formulas for balanced and well-poised basic hypergeometric series. To illustrate the usefulness of these formulas some examples are drawn from orthogonal polynomials.

1. THE BETA INTEGRAL AND THE HYPERGEOMETRIC FUNCTION

Two of the most important formulas in Special Functions are Newton's binomial theorem:

$$(1.1) \qquad (1-z)^{-a} = \sum_{n=0}^{\infty} \frac{(a)_n}{n!} z^n, \quad |z| < 1,$$

where $(a)_n = \Gamma(a+n)/\Gamma(a)$ is the shifted factorial, and Euler's beta integral

$$(1.2) \qquad \int_0^1 t^{b-1}(1-t)^{c-1} dt = B(b,c) = \frac{\Gamma(b)\Gamma(c)}{\Gamma(b+c)},$$

provided Re $b > 0$, Re $c > 0$. The main objective of this article is to consider some extensions of these formulas, old and new, particularly the so-called q-extensions, and to indicate their applications to orthogonal polynomials.

For Re $c >$ Re $b > 0$ we may combine (1.1) and (1.2) in a very elementary way to obtain the following:

$$(1.3) \qquad [B(b, c-b)]^{-1} \int_0^1 t^{b-1}(1-t)^{c-b-1}(1-zt)^{-a} dt$$

$$= [B(b, c-b)]^{-1} \int_0^1 t^{b-1}(1-t)^{c-b-1} \sum_{n=0}^{\infty} \frac{(a)_n}{n!} (zt)^n dt$$

$$= \sum_{n=0}^{\infty} \frac{(a)_n}{n!} \frac{B(b+n, c-b)}{B(b, c-b)} z^n = F(a, b; c; z), \quad |z| < 1,$$

* This work was partially supported by the NSERC of Canada under grant #A6197.

319

P. Nevai (ed.), Orthogonal Polynomials, 319–344.
© *1990 by Kluwer Academic Publishers.*

where

(1.4)
$$F(a,b;c;z) = \sum_{n=0}^{\infty} \frac{(a)_n (b)_n}{n!(c)_n} z^n$$

is the ordinary or Gaussian hypergeometric function. Although the above derivation is valid only for Re $c >$ Re $b > 0$ and $|z| < 1$, the infinite series in (1.4) is absolutely convergent for z inside the unit circle and for all complex a, b, c, except $c = 0, -1, -2, \ldots$. Since, by Stirling's formula,

(1.5)
$$\frac{(a)_n (b)_n}{n!(c)_n} = O(n^{a+b-c-1}) \text{ as } n \to \infty,$$

the hypergeometric series (1.4) converges at $z = 1$ if Re $(c - a - b) > 0$. On the other hand, for Re $b > 0$ and Re $(c - a - b) > 0$ the integral in (1.3) gives

(1.6)
$$F(a,b;c;1) = \frac{B(b, c-a-b)}{B(b, c-b)} = \frac{\Gamma(c)\Gamma(c-a-b)}{\Gamma(c-a)\Gamma(c-b)}$$

which is known as Gauss' summation formula. The restriction Re $b > 0$ is not necessary for this formula to hold although its derivation by means of (1.3) is difficult to justify without this. Gauss [1813] used one of his contiguous relations for $F(a,b;c;z)$ to prove that (1.6) holds if Re $(c - a - b) > 0$. If a or b is a nonpositive integer, say, $a = -n$, $n = 0, 1, \ldots$, then the infinite series in (1.4) becomes a polynomial in z (provided $-c$ is not a positive integer less than n) and (1.6) reduces to the Chu-Vandermonde sum

(1.7)
$$F(-n, b; c; 1) = \frac{(c-b)_n}{(c)_n}$$

which is simply an alternative form of the discrete binomial formula

(1.8)
$$(a+b)_n = \sum_{k=0}^{n} \binom{n}{k} (a)_k (b)_{n-k}.$$

Formula (1.7) used to be attributed to Vandermonde [1772] alone, but it was discovered long ago by the Chinese mathematician Chu [1303]. Also, some formulas involving Gauss' hypergeometric function were given earlier by Euler [1748] including the important transformation formula:

(1.9)
$$F(a,b;c;z) = (1-z)^{c-a-b} F(c-a, c-b; c; z), \quad |z| < 1.$$

Expressible in terms of the hypergeometric function are the classical orthogonal polynomials such as the Tchebichef polynomials of the first and second kinds

(1.10)
$$T_n(x) = F\left(-n, n; \frac{1}{2}; \frac{1-x}{2}\right),$$

$$U_n(x) = (n+1)F\left(-n, n+2; \frac{3}{2}; \frac{1-x}{2}\right),$$

the Legendre polynomials

(1.11)
$$P_n(x) = F\left(-n, n+1; 1; \frac{1-x}{2}\right),$$

the ultraspherical polynomials

(1.12) $$C_n^\lambda(x) = \frac{(2\lambda)_n}{n!} F\left(-n, n+2\lambda; \lambda+\frac{1}{2}; \frac{1-x}{2}\right), \quad \lambda > -\frac{1}{2},$$

and the more general Jacobi polynomials

(1.13) $$P_n^{(\alpha,\beta)}(x) = \frac{(\alpha+1)_n}{n!} F\left(-n, n+\alpha+\beta+1; \alpha+1; \frac{1-x}{2}\right), \quad \alpha,\beta > -1;$$

see Chihara [1978], Erdélyi [1953] and Szegö [1975].

The bilinear transformation

(1.14) $$t = \frac{\alpha u + \beta}{\gamma u + \delta}, \quad \alpha\delta - \beta\gamma \neq 0$$

transforms the integral in (1.3) into

(1.15)
$$(\alpha\delta - \beta\gamma) \int_{-\beta/\alpha}^{(\delta-\beta)/(\alpha-\gamma)} (\alpha u+\beta)^{b-1}(\delta-\beta-(\alpha-\gamma)u)^{c-b-1}(\gamma u+\delta)^{a-c}(\delta-\beta z-(\alpha z-\gamma)u)^{-a} du$$

which gives Euler's transformation (1.9) if we choose $\delta = \beta = -\alpha, \gamma = \alpha z$. It also gives Pfaff-Kummer's transformation formulas

(1.16) $$F(a,b;c;z) = (1-z)^{-a} F\left(a, c-b; c; \frac{z}{z-1}\right)$$

$$= (1-z)^{-b} F\left(c-a; b; c; \frac{z}{z-1}\right)$$

by taking $\gamma = 0$, $\delta = \beta = -\alpha$. The transformation (1.14) also reduces the beta integral (1.2) to the form

(1.17) $$B(b,c) = (\alpha\delta - \beta\gamma) \int_{-\beta/\alpha}^{(\delta-\beta)/(\alpha-\gamma)} (\alpha u+\beta)^{b-1}(\delta-\beta-(\alpha-\gamma)u)^{c-1}(\gamma u+\delta)^{-b-c} du$$

from which follows another form of the integral by choosing $\delta = \gamma = \beta > 0$ and letting $\alpha \to 0^-$:

(1.18) $$B(b,c) = \int_0^\infty \frac{u^{c-1} du}{(1+u)^{b+c}}, \quad \text{Re } b > 0, \text{ Re } c > 0.$$

In this article we shall consider extensions of (1.2), (1.18) as well as the following form of the beta integral

(1.19) $$\int_{-\beta}^{\gamma} (u+\beta)^{b-1}(\gamma-u)^{c-1} du = B(b,c)(\beta+\gamma)^{b+c-1}$$

which is obtained from (1.17) by setting $\gamma = 0$, $\alpha = 1$ and then replacing δ by $\beta + \gamma$. We shall also consider extensions of the hypergeometric function $F(a,b;c;z)$ by introducing more parameters. In fact, a hypergeometric series can be defined as an infinite series of the form $\sum_{n=0}^{\infty} a_n z^n$ where a_{n+1}/a_n is a rational function of n. An explicit representation of such a series is then given by

(1.20) $${}_rF_s\begin{bmatrix} a_1, a_2, \ldots, a_r \\ b_1, b_2, \ldots, b_s \end{bmatrix}; z \end{bmatrix}$$

$$= \sum_{n=0}^{\infty} \frac{(a_1, a_2, \ldots, a_r)_n}{(b_1, b_2, \ldots, b_s)_n} \frac{z^n}{n!},$$

where $(a_1, a_2, \ldots, a_r)_n = (a_1)_n (a_2)_n \cdots (a_r)_n$, so that there are r numerator and s denominator parameters. If $r > s + 1$ the series is meaningful only if $z = 0$ or if the series terminates, i.e., when one of the numerator parameters is a nonpositive integer, assuming of course, that there are no zero factors in the denominators of the terms of the series. If $r \leq s$ the series (1.20) converges for all values of z. The most important case is when $r = s + 1$. By the ratio test it is clear that the radius of convergence of the $_{s+1}F_s$ series is 1. Gauss' series (1.4) is a $_2F_1$ series and as such is also denoted by

$$_2F_1(a, b; c; z) \text{ or } _2F_1 \begin{bmatrix} a, b \\ c \end{bmatrix} ; z \end{bmatrix}.$$

2. BASIC HYPERGEOMETRIC SERIES

As a generalization of Gauss' hypergeometric series Heine [1846, 1847, 1878] introduced the series

$$(2.1) \qquad 1 + \frac{(1-q^a)(1-q^b)}{(1-q)(1-q^c)} z + \frac{(1-q^a)(1-q^{a+1})(1-q^b)(1-q^{b+1})}{(1-q)(1-q^2)(1-q^c)(1-q^{c+1})} z^2 + \cdots,$$

where it is assumed that $c \neq 0, -1, -2, \ldots$, and $|q| < 1$, $|z| < 1$ to ensure convergence of the series when it does not terminate. Since

$$(2.2) \qquad \lim_{q \to 1} \frac{1 - q^a}{1 - q} = a,$$

the series (2.1) converges to the series on the right side of (1.4) in the limit $q \to 1$.

A basic hypergeometric series is generally defined to be a series of the type $\sum_{n=0}^{\infty} a_n z^n$ where a_{n+1}/a_n is a rational function of q^n, q being a fixed complex parameter called the base of the series, usually with modulus less than 1. An explicit representation of such a series is given by

$$(2.3) \qquad _r\phi_s \begin{bmatrix} a_1, a_2, \ldots, a_r \\ b_1, b_2, \ldots, b_s \end{bmatrix} ; q, z \end{bmatrix}$$
$$= \sum_{n=0}^{\infty} \frac{(a_1, a_2, \ldots, a_r; q)_n}{(q, b_1, \ldots, b_s; q)_n} z^n \left[(-1)^n q^{\binom{n}{2}} \right]^{1+s-r},$$

where $\binom{n}{2} = n(n-1)/2$, and

$$(2.4) \qquad (a_1, a_2, \ldots, a_r; q)_n = (a_1; q)_n (a_2; q)_n \cdots (a_r; q)_n,$$

with the q-shifted factorials defined by

$$(2.5) \qquad (a; q)_n = \begin{cases} 1, & \text{if } n = 0, \\ (1-a)(1-aq) \cdots (1-aq^{n-1}), & \text{if } n = 1, 2, \ldots . \end{cases}$$

If $0 < |q| < 1$, the $_r\phi_s$ series converges absolutely for all z if $r \leq s$, and for $|z| < 1$ if $r = s + 1$. If $|q| > 1$, the series converges absolutely if $|z| < |b_1 \cdots b_s|/|a_1 \cdots a_r|$. It diverges for $|z| \neq 0$ if $0 < |q| < 1$ and $r > s + 1$, and if $|q| > 1$ and $|z| > |b_1 \cdots b_s|/|a_1 \cdots a_r|$, unless it terminates.

By means of the transformation

(2.6)
$$(a; q^{-1})_n = (a^{-1}; q)_n(-a)^n q^{-\binom{n}{2}}$$

a basic hypergeometric series in base q^{-1} can be transformed to one in base q and so it is sufficient to consider the series in base q with $|q| < 1$. Unless otherwise stated we shall therefore assume throughout the paper that $|q| < 1$.

Some elementary properties of the q-shifted factorials that we shall need in the subsequent sections are:

(2.7)
$$(a; q)_{m+n} = (a; q)_m(aq^m; q)_n,$$

(2.8)
$$(aq^{-n}; q)_k = \frac{(a; q)_k(q/a; q)_n}{(q^{1-k}/a; q)_n} q^{-nk},$$

(2.9)
$$(aq^{1-n}; q)_n = (a^{-1}; q)_n(-a)^n q^{-\binom{n}{2}},$$

(2.10)
$$(a; q)_{2n} = (a, aq; q^2)_n,$$

(2.11)
$$(a^2; q^2)_n = (a, -a; q)_n, \quad n = 0, 1, \ldots .$$

We shall also use the notation

(2.12)
$$(a; q)_\alpha = (a; q)_\infty/(aq^\alpha; q)_\infty$$

where α is arbitrary, real or complex, and

(2.13)
$$(a; q)_\infty = \prod_{n=0}^{\infty}(1 - aq^n).$$

3. THE q-BINOMIAL THEOREM, THE q-GAMMA AND q-BETA FUNCTIONS

The most fundamental summation formula in the theory of basic hypergeometric series is the q-binomial formula:

(3.1)
$$_1\phi_0(a; -; q, z) \equiv {}_1\phi_0 \left[\begin{matrix} a \\ - \end{matrix} ; q, z \right]$$

$$= \sum_{n=0}^{\infty} \frac{(a; q)_n}{(q; q)_n} z^n$$

$$= \frac{(az; q)_\infty}{(z; q)_\infty}, \qquad |z| < 1.$$

This approaches the binomial formula (1.1) when we replace a by q^a and take the limit $q \to 1^-$. Heine's [1847] proof of (3.1) was based on the observation that both sides of (3.1) satisfy the functional equation

(3.2)
$$(1 - z)f(z) = (1 - az)f(zq),$$

which, together with $f(0) = 1$, determines a unique analytic function $f(z)$ inside the unit circle. For an earlier proof see Cauchy [1843], and for a better understanding of Heine's proof see Askey [1980].

For $0 < q < 1$, the q-gamma function defined by

$$(3.3) \qquad \Gamma_q(x) = \frac{(q;q)_\infty}{(q^x;q)_\infty}(1-q)^{1-x}, \quad \text{Re } x > 0,$$

was first introduced by Thomae [1869] and later by F.H. Jackson [1904]. Wm. Gosper gave a simple proof of the limit property

$$(3.4) \qquad \lim_{q \to 1^-} \Gamma_q(x) = \Gamma(x),$$

see Andrews [1986]. For a rigorous proof of (3.4) see Koornwinder [1989]. Askey [1978] studied various properties of $\Gamma_q(x)$ and Moak [1980a,b] considered extensions of (3.3) for $q > 1$ and for $x < 0$.

In view of (3.3) it is now natural to define the q-beta function by

$$(3.5) \qquad B_q(x, y) = \frac{\Gamma_q(x)\Gamma_q(y)}{\Gamma_q(x+y)}, \quad 0 < q < 1, \text{ Re } x, y > 0.$$

By (3.3) and (3.1) we can write

$$(3.6) \qquad B_q(x,y) = (1-q)\frac{(q, q^{x+y}; q)_\infty}{(q^x, q^y; q)_\infty}$$

$$= (1-q)\frac{(q;q)_\infty}{(q^y;q)_\infty} \sum_{n=0}^{\infty} \frac{(q^y;q)_n}{(q;q)_n}q^{nx}$$

$$= (1-q)\sum_{n=0}^{\infty} \frac{(q^{n+1};q)_\infty}{(q^{n+y};q)_\infty}q^{nx}.$$

This can be written in a form resembling the integral formula (1.2) in the following manner. Thomae [1869, 1870] and Jackson [1910a] introduced the q-integral

$$(3.7) \qquad \int_0^1 f(t)d_qt = (1-q)\sum_{n=0}^{\infty} f(q^n)q^n$$

which was generalized by Jackson to

$$(3.8) \qquad \int_a^b f(t)d_qt = \int_0^b f(t)d_qt - \int_0^a f(t)d_qt$$

with

$$(3.9) \qquad \int_0^a f(t)d_qt = a(1-q)\sum_{n=0}^{\infty} f(aq^n)q^n.$$

Jackson also gave the definition

$$(3.10) \qquad \int_0^\infty f(t)d_qt = (1-q)\sum_{n=-\infty}^{\infty} f(q^n)q^n.$$

If $f(t)$ is continuous on $[0, a]$ then it easily follows that

$$(3.11) \qquad \lim_{q \to 1^-} \int_0^a f(t)d_qt = \int_0^a f(t)dt.$$

By (2.12), (3.6) and (3.7) it follows that

$$(3.12) \qquad B_q(x,y) = \int_0^1 t^{x-1}(qt;q)_{y-1}\,d_q t,$$

Re $x > 0$, $y \neq 0,-1,-2,\ldots$, which clearly approaches the beta-integral (1.2) as $q \to 1^-$.

4. HEINE'S FORMULAS FOR THE $_2\phi_1$ SERIES: THE q-GAUSS SUM

Our principal aim in this section is to derive a q-analogue of Gauss' summation formula (1.6). We begin by giving a proof of Heine's [1847, 1878] transformation formula

$$(4.1) \qquad {}_2\phi_1(a,b;c;q,z) \equiv {}_2\phi_1 \left[\begin{array}{c} a,b \\ c \end{array} ;q,z \right]$$
$$= \frac{(b,az;q)_\infty}{(c,z;q)_\infty}\, {}_2\phi_1(c/b,z;az;q,b),$$

where $\max(|q|,|z|,|b|) < 1$, By (3.1)

$$(4.2) \qquad \frac{(cq^n;q)_\infty}{(bq^n;q)_\infty} = \sum_{m=0}^{\infty} \frac{(c/b;q)_m}{(q;q)_m}(bq^n)^m$$

and hence

$$
\begin{aligned}
{}_2\phi_1(a,b;c;q,z) &= \frac{(b;q)_\infty}{(c;q)_\infty} \sum_{n=0}^{\infty} \frac{(a;q)_n}{(q;q)_n} \frac{(cq^n;q)_\infty}{(bq^n;q)_\infty} z^n \\
&= \frac{(b;q)_\infty}{(c;q)_\infty} \sum_{m=0}^{\infty} \frac{(c/b;q)_m}{(q;q)_m} b^m \sum_{n=0}^{\infty} \frac{(a;q)_n}{(q;q)_n}(zq^m)^n \\
&= \frac{(b;q)_\infty}{(c;q)_\infty} \sum_{m=0}^{\infty} \frac{(c/b;q)_m}{(q;q)_m} b^m \frac{(azq^m;q)_\infty}{(zq^m;q)_\infty} \\
&= \frac{(b,az;q)_\infty}{(c,z;q)_\infty}\, {}_2\phi_1(c/b,z;az;q,b).
\end{aligned}
$$

By (2.12), (3.3), (3.5) and (3.7) this can be expressed in a q-integral form

$$(4.3) \qquad {}_2\phi_1(q^a,q^b;q^c;q,z) = [B_q(b,c-b)]^{-1} \int_0^1 t^{b-1} \frac{(qt;q)_{c-b-1}}{(zt;q)_a}\,d_q t$$

which is clearly a q-analogue of (1.3).

By iterating the transformation formula (4.1) we obtain two more formulas:

$$
{}_2\phi_1(a,b;c;q,z)
$$
$$(4.4) \qquad = \frac{(c/b,bz;q)_\infty}{(c,z;q)_\infty}\, {}_2\phi_1(abz/c,b;bz;q,c/b)$$

$$(4.5) \qquad = \frac{(abz/c;q)_\infty}{(z;q)_\infty}\, {}_2\phi_1(c/a,c/b;c;q,abz/c).$$

Note that (4.5) is a q-analogue of (1.9) and that by setting $abz/c = 1$ in (4.4) we obtain the q-Gauss summation formula

$$(4.6) \qquad {}_2\phi_1(a,b;c;q,c/ab) = \frac{(c/a,c/b;q)_\infty}{(c,c/ab;q)_\infty}, \quad |c/ab| < 1,$$

which is clearly a q-analogue of (1.6). This formula is usually attributed to Heine [1847], but it was given earlier by Jacobi [1846].

When $a = q^{-n}, n = 0, 1, \ldots$, (4.6) reduces to the q-Vandermonde formula

$$(4.7) \qquad {}_2\phi_1(q^{-n},b;c;q,cq^n/b) = \frac{(c/b;q)_n}{(c;q)_n}.$$

By changing the order of summation in (4.7) we also find that

$$(4.8) \qquad {}_2\phi_1(q^{-n},b;c;q,q) = \frac{(c/b;q)_n}{(c;q)_n}b^n.$$

Both (4.7) and (4.8) are q-analogues of the Chu-Vandermonde summation formula (1.7). It may be remarked that ${}_2\phi_1(a,b;c;q,q)$ which is also a q-analogue of $F(a,b;c;1)$, cannot be summed as ratios of infinite products unless a or b has the form $q^{-n}, n = 0, 1, \ldots$. What we have instead is a different kind of summation formula:

$$(4.9) \qquad {}_2\phi_1(a,b;c;q,q) + \frac{(q/c,a,b;q)_\infty}{(c/q,aq/c,bq/c;q)_\infty} {}_2\phi_1(aq/c,bq/c;q^2/c;q,q)$$

$$= \frac{(q/c,abq/c;q)_\infty}{(aq/c,bq/c;q)_\infty},$$

which we shall prove in a later section. This can be expressed as a q-integral

$$(4.10) \qquad \int_{qd/c}^{d} \frac{(qt/d,ct/d;q)_\infty}{(at/d,bt/d;q)_\infty} d_q t = \frac{d(1-q)(q,c,q/c,abq/c;q)_\infty}{(a,b,aq/c,bq/c;q)_\infty},$$

where d is an arbitrary nonzero parameter. The same formula also yields Andrews and Askey's [1981] q-beta integral

$$(4.11) \qquad \int_{-a}^{b} \frac{(-qt/a,qt/b;q)_\infty}{(-ct/a,dt/b;q)_\infty} d_q t$$

$$= \frac{b(1-q)(q,-a/b,-bq/a,cd;q)_\infty}{(c,d,-bc/a,-ad/b;q)_\infty}$$

which is a q-analogue of the beta integral (1.19).

Since, by (4.1)

$$(4.12) \qquad {}_2\phi_1(a,b;c;q,q) = \frac{(b,aq;q)_\infty}{(c,q;q)_\infty} \sum_{n=0}^{\infty} \frac{(c/b;q)_n}{(aq;q)_n}b^n, \quad |b| < 1$$

and

$${}_2\phi_1(aq/c,bq/c;q^2/c;q,q) = \frac{(aq/c,bq^2/c;q)_\infty}{(q^2/c,q;q)_\infty} \sum_{n=0}^{\infty} \frac{(q/a;q)_n}{(bq^2/c;q)_n}(aq/c)^n, \quad |aq/c| < 1,$$

$$(4.13) \qquad = \frac{(1-c/q)(aq/c,bq/c;q)_\infty}{(1-a)(q/c,q;q)_\infty} \sum_{n=1}^{\infty} \frac{(1/a;q)_n}{(bq/c;q)_n}(aq/c)^n,$$

we find by use of the identity

$$(4.14) \qquad (a;q)_{-n} = \frac{(a;q)_\infty}{(aq^{-n};q)_\infty} = \frac{(-a^{-1})^n q^{n(n+1)/2}}{(q/a;q)_n},$$

that formula (4.9) also contains Ramanujan's celebrated summation formula

$$(4.15) \qquad \sum_{n=-\infty}^{\infty} \frac{(a;q)_n}{(b;q)_n} x^n = \frac{(q,b/a,ax,q/ax;q)_\infty}{(x,q/a,b,b/ax;q)_\infty},$$

where $|b/a| < |x| < 1$. There are many proofs of (4.15), see in particular, Andrews and Askey [1978], Askey [1980], Ismail [1977] and Kadell [1987]. Replacing a, b, x by $-q^a, -q^b$ and q^x, respectively, we may express (4.15) in the form

$$(4.16) \qquad \int_0^\infty \frac{t^{x-1} d_q t}{(-tq^a;q)_{b-a}} = \frac{(-q^{1-a-x};q)_x}{(-q^a;q)_x} B_q(x, b-a-x)$$

by (3.10), which is clearly a q-analogue of (1.18).

The transformation formula

$$(4.17) \qquad {}_2\phi_1(a,b;c;q,z) = \frac{(az;q)_\infty}{(z;q)_\infty} \, {}_2\phi_2 \left[\begin{matrix} a, c/b \\ c, az \end{matrix} ; q, bz \right]$$

was given by Jackson [1910b] as a q-analogue of Pfaff- Kummer transformation formulas (1.16). It is proved by using (4.7) to get

$$(4.18) \qquad \frac{(b;q)_k}{(c;q)_k} = \sum_{j=0}^{k} \frac{(q^{-k}, c/b;q)_j}{(q,c;q)_j} (bq^k)^j$$

and then interchanging the order of summation in the series on the left side of (4.17). If $a = q^{-n}, n = 0, 1, \ldots$, then we may reverse the order of summation in (4.17) to get Sears' [1951a] transformation formula

$$(4.19) \qquad {}_2\phi_1(q^{-n},b;c;q,z) = \frac{(c/b;q)_n}{(c;q)_n}(bz/q)^n \, {}_3\phi_2 \left[\begin{matrix} q^{-n}, q^{1-n}/c, q/z \\ bq^{1-n}/c, 0 \end{matrix} ; q, q \right].$$

5. q-CONTIGUOUS RELATIONS FOR THE ${}_2\phi_1$ SERIES

Let $\phi(a,b,c)$ denote the series ${}_2\phi_1(a,b;c;q,z)$. Then, by equating the coefficients of z^n on both sides one can easily verify the following contiguous relations of Heine [1847]:

$$(5.1) \qquad \phi(a,b,cq^{-1}) - \phi(a,b,c) = cz\frac{(1-a)(1-b)}{(q-c)(1-c)} \phi(aq,bq,cq),$$

$$(5.2) \qquad \phi(aq,b,c) - \phi(a,b,c) = az\frac{(1-b)}{(1-c)} \phi(aq,bq,cq),$$

$$(5.3) \qquad \phi(aq,b,cq) - \phi(a,b,c) = az\frac{(1-b)(1-c/a)}{(1-c)(1-cq)} \phi(aq,bq,cq^2)$$

$$(5.4) \qquad \phi(aq,bq^{-1},c) - \phi(a,b,c) = az\frac{(1-b/aq)}{(1-c)} \phi(aq,b,cq).$$

From (5.4) one can deduce that

$$(5.5) \qquad z\phi(a,b,c) = A\phi(aq,bq^{-1},c) - (A+B)\phi(a,b,c) + B\phi(aq^{-1},bq,c),$$

where

$$(5.6) \qquad A = \frac{(b-c)(1-a)}{(b-a)(a-b/q)}, \quad B = \frac{(a-c)(1-b)}{(a-b)(b-a/q)},$$

and from (5.3) it follows that

$$(5.7) \qquad \phi(aq^{-1}, b, cq^{-1}) + z\frac{(1-a)(b-c)}{(1-c)(q-c)}\,\phi(aq, b, cq)$$

$$= \left(1 - \frac{a-b}{q-c}z\right)\,\phi(a, b, c).$$

Also, denoting the expressions $_2\phi_1(a, b; c; q, z)$, $_2\phi_1(aq^{\pm 1}, b; c; q, z)$, $_2\phi_1(a, bq^{\pm 1}; c; q, z)$ and $_2\phi_1(a, b; cq^{\pm 1}; q, z)$ by ϕ, $\phi(aq^{\pm 1})$, $\phi(bq^{\pm 1})$ and $\phi(cq^{\pm 1})$, respectively, one can show that

$$(5.8) \qquad b(1-a)\phi(aq) - a(1-b)\phi(bq) = (b-a)\phi,$$

$$(5.9) \qquad a(1-b/c)\phi(bq^{-1}) - b(1-a/c)\phi(aq^{-1}) = (a-b)(1-abz/cq)\phi,$$

$$(5.10) \qquad q(1-a/c)\phi(aq^{-1}) + (1-a)(1-abz/c)\phi(aq)$$

$$= [1 + q - a - aq/c + (1-b/a)za^2/c]\phi,$$

$$(5.11) \qquad (1-c)(q-c)(abz-c)\phi(cq^{-1}) + (c-a)(c-b)z\phi(cq)$$

$$= (c-1)[c(q-c) + (ca+cb-ab-abq)z]\phi.$$

It follows from (5.5) and (5.6) that Hahn's [1949] little q-Jacobi polynomials (see also Andrews and Askey [1977]) defined by

$$(5.12) \qquad p_n(x) = {}_2\phi_1(q^{-n}, abq^{n+1}; aq; q, qx)$$

satisfy the 3-term recurrence relation

$$(5.13) \qquad xp_n(x) = A_n p_{n+1}(x) - (A_n + B_n)p_n(x) + B_n p_{n-1}(x),$$

$n = 0, 1, 2, \ldots$, $p_{-1}(x) = 0$, where

$$(5.14) \qquad A_n = \frac{(1-aq^{n+1})(1-abq^{n+1})}{(1-abq^{2n+1})(1-abq^{2n+2})}(-q^n),$$

$$B_n = \frac{(1-q^n)(1-bq^n)}{(1-abq^{2n})(1-abq^{2n+1})}(-aq^n).$$

It also follows from (5.7) that Rogers' [1893, 1894, 1895] continuous q-ultraspherical polynomials (see also Askey and Ismail [1983]) defined by

$$(5.15) \qquad C_n(x; \beta|q)$$

$$= \sum_{k=0}^{n} \frac{(\beta; q)_k (\beta; q)_{n-k}}{(q; q)_k (q; q)_{n-k}} e^{i(n-2k)\theta}$$

$$= \frac{(\beta; q)_n}{(q; q)_n} e^{in\theta}\, {}_2\phi_1(q^{-n}, \beta; \beta^{-1}q^{1-n}; q, q\beta^{-1}e^{-2i\theta}),$$

$x = \cos\theta$, $0 \le \theta \le \pi$, satisfy the 3-term recurrence relation

$$(5.16) \qquad 2xC_n(x; \beta|q) = \frac{1-q^{n+1}}{1-\beta q^n}C_{n+1}(x; \beta|q) + \frac{1-\beta^2 q^{n-1}}{1-\beta q^n}C_{n-1}(x; \beta|q),$$

$n = 0, 1, \ldots$, $C_{-1}(x; \beta|q) = 0$.

6. BALANCED AND WELL-POISED SERIES: THE q-SAALSCHÜTZ FORMULA

The origin of the balanced and well-poised series can be traced to the trivial identities

$$(6.1) \qquad \frac{(ax;q)_\infty (x;q)_\infty}{(x;q)_\infty (ax;q)_\infty} = 1$$

and

$$(6.2) \qquad (x;q)_a(-x;q)_a = (x^2;q^2)_a,$$

respectively. If we assume that $|x| < 1$, $|ax| < 1$, expand the product on the left side of (6.1) by (3.1) and collect the coefficient of x^n we find that

$$(6.3) \qquad 1 \equiv \sum_{n=0}^{\infty} \frac{(a;q)_n}{(q;q)_n} x^n \, {}_2\phi_1(q^{-n}, a^{-1}; a^{-1}q^{1-n}; q, q),$$

which gives the summation formula

$$(6.4) \qquad {}_2\phi_1(q^{-n}, a^{-1}; a^{-1}q^{1-n}; q, q) = \delta_{n,0}, \quad n = 0, 1, \dots,$$

where $\delta_{m,n}$ is the Kronecker delta. Observe that the argument of the ${}_2\phi_1$ series is q, and that $q \cdot q^{-n} \cdot a^{-1} = a^{-1}q^{1-n}$. A basic hypergeometric series of the form

$$(6.5) \qquad {}_{r+1}\phi_r \left[\begin{matrix} a_1, a_2, \dots, a_{r+1} \\ b_1, \dots, b_r \end{matrix} ; q, z \right]$$

is called k-balanced if $z = q$ and

$$(6.6) \qquad b_1 b_2 \cdots b_r = q^k a_1 a_2 \cdots a_{r+1}, \quad k = 0, 1, \dots .$$

A 1-balanced series is simply called a balanced series. Formula (6.4), which also follows directly from (4.8), gives the sum of the simplest terminating balanced basic hypergeometric series. To obtain the sum of a terminating balanced ${}_3\phi_2$ series we simply expand both sides of (4.5) in powers of z and collect the coefficients of z^n to get

$$\frac{(a,b;q)_n}{(q,c;q)_n} = \frac{(ab/c;q)_n}{(q;q)_n} \sum_{k=0}^{n} \frac{(q^{-n}, c/a, c/b; q)_k}{(q, c, cq^{1-n}/ab; q)_k} q^k$$

which can be rewritten as

$$(6.7) \qquad {}_3\phi_2 \left[\begin{matrix} q^{-n}, & a, & b \\ & c, & d \end{matrix} ; q, q \right] = \frac{(c/a, c/b; q)_n}{(c, c/ab; q)_n},$$

where $cd = abq^{1-n}$. This is known as the q-Saalschütz formula since it is a q-analogue of the well-known formula

$$(6.8) \qquad {}_3F_2 \left[\begin{matrix} -n, & a, & b \\ & c, & 1+a+b-c-n \end{matrix} ; 1 \right] = \frac{(c-a)_n(c-b)_n}{(c)_n(c-a-b)_n},$$

which was originally found by Pfaff [1797] and rediscovered by Saalschütz [1890]. Note that the product of the numerator parameters on the right side of (6.7) equals that of the denominator parameters, expressing a balanced property of the ${}_3\phi_2$ sum.

Turning now to the identity (6.2), if we use (2.12) and (3.1) we obtain the following formula

(6.9)
$$\sum_{n=0}^{\infty} \frac{(q^{-a};q)_n}{(q;q)_n}(xq^a)^n \; {}_2\phi_1(q^{-n},q^{-a};q^{a+1-n};q,-q^{a+1})$$

$$= \sum_{n=0}^{\infty} \frac{(q^{-2a};q^2)_n}{(q^2;q^2)_n}(xq^a)^{2n}.$$

This immediately leads to the summation formula

(6.10)
$$\begin{aligned}
& {}_2\phi_1(q^{-n},q^{-a};q^{a+1-n};q,-q^{a+1}) \\
& = \begin{cases} 0, & \text{if } n \text{ is odd,} \\[2mm] \dfrac{(q;q)_n(q^{-2a};q^2)_{n/2}}{(q^{-a};q)_n(q^2;q^2)_{n/2}}, & \text{if } n \text{ is even.} \end{cases}
\end{aligned}$$

Observe that the product of q^{-a} and q^{a+1-n} in the ${}_2\phi_1$ series in (6.10) is q times q^{-n}, a property usually described as the well-poised property. The basic hypergeometric series (6.5) is called well-poised if

(6.11)
$$qa_1 = a_2b_1 = a_3b_2 = \ldots = a_{r+1}b_r;$$

it is called very-well-poised (a name coined by R. Askey [1975]) if, in addition, $a_2 = qa_1^{\frac{1}{2}}$, $a_3 = -qa_1^{\frac{1}{2}}$. The series (6.5) is called a nearly-poised series of the first kind if

(6.12)
$$qa_1 \neq a_2b_1 = a_3b_2 = \ldots = a_{r+1}b_r,$$

and it is called a nearly-poised series of the second kind if

(6.13)
$$qa_1 = a_2b_1 = a_3b_2 = \ldots = a_rb_{r-1} \neq a_{r+1}b_r.$$

There is a close relationship between balanced and very-well-poised series. One can get an indication of this relationship if one uses (2.9) to rewrite the product on the right side of (6.7) as

$$\frac{(c/a,bq^{1-n}/c;q)_n}{(abq^{1-n}/c,c;q)_n}a^n,$$

where the product of the parameters in one vertical column equals that in another column. Analogous to (6.4) one also has the summation formula for a very-well-poised ${}_4\phi_3$ series:

(6.14)
$$\begin{aligned}
& {}_4\phi_3\left[\begin{matrix} a,qa^{\frac{1}{2}},-qa^{\frac{1}{2}},q^{-n} \\ a^{\frac{1}{2}},-a^{\frac{1}{2}},aq^{n+1} \end{matrix} ; q,q^n\right] \\
& = \sum_{k=0}^{n} \frac{(1-aq^{2k})(a,q^{-n};q)_k}{(1-a)(q,aq^{n+1};q)_k}q^{nk} = \delta_{n,0}.
\end{aligned}$$

To prove (6.14), note that the series on the left side clearly equals 1 when $n = 0$ and that for $n = 1,2,\ldots$, we may use the identity $1 - aq^{2k} = q^k(1-aq^k) + (1-q^k)$ to split the ${}_4\phi_3$ series into two ${}_2\phi_1$ series and then use (4.1). A bibasic generalization of (6.14), namely,

(6.15)
$$\sum_{k=0}^{n} \frac{(1-ap^kq^k)(a;p)_k(q^{-n};q)}{(1-a)(q;q)_k(apq^n;p)_k}q^{nk} = \delta_{n,0}$$

was found by Carlitz [1973] and made extensive use of by Gessel and Stanton [1986] and Rahman [1989].

For the sake of abbreviation we shall use the following notation for a very-well-poised basic hypergeometric series

$$(6.16) \qquad {}_{r+1}W_r(a; a_1, \ldots, a_{r-2}; q, z)$$
$$= {}_{r+1}\phi_r \left[\begin{array}{c} a, qa^{\frac{1}{2}}, -qa^{\frac{1}{2}}, a_1, \ldots, a_{r-2} \\ a^{\frac{1}{2}}, -a^{\frac{1}{2}}, aq/a_1, \ldots, aq/a_{r-2} \end{array} ; q, z \right].$$

Closely related to the q-Saalschütz formula (6.7) is Sears' [1951b] transformation formula for a terminating and balanced ${}_4\phi_3$ series:

$$(6.17) \qquad {}_4\phi_3 \left[\begin{array}{c} q^{-n}, a, b, c, \\ d, e, f \end{array} ; q, q \right] = \frac{(e/a, f/a; q)_n}{(e, f; q)_n} a^n \, {}_4\phi_3 \left[\begin{array}{c} q^{-n}, a, d/b, d/c \\ d, aq^{1-n}/e, aq^{1-n}/f \end{array} ; q, q \right],$$

where $def = abcq^{1-n}$. To prove this important formula we observe that by (4.5)

$$(6.18) \qquad {}_2\phi_1(a, b; c; q, z) \, {}_2\phi_1(d, e; f; q, abz/c)$$
$$= \frac{(abdez/cf; q)_\infty}{(z; q)_\infty} \, {}_2\phi_1(c/a, c/b; c; q, abz/c) \, {}_2\phi_1(f/d, f/e; f; q, abdez/cf)$$

and hence, if $cf = abde$, (6.17) follows immediately by equating the coefficients of z^n on both sides.

7. A CONTIGUOUS RELATION FOR BALANCED AND TERMINATING ${}_4\phi_3$ SERIES

Let one of the parameters a, b, c, d be a nonnegative integer power of q^{-1} and let

$$(7.1) \qquad \phi(a, b) \equiv {}_4\phi_3 \left[\begin{array}{c} a, b, c, d \\ e, f, g \end{array} ; q, q \right],$$

where $efg = abcdq$. It can be shown by using (6.17) and some straightforward manipulations with the q-shifted factorials that $\phi(a, b)$ satisfies the q-contiguous relation

$$(7.2) \qquad (1 - c)(1 - d)\phi(a, b) = A[\phi(aq, bq^{-1}) - \phi(a, b)] - B[\phi(a, b) - \phi(aq^{-1}, bq)],$$

where

$$(7.3) \qquad A = (1 - a)(e - b)(f - b)(g - b)/b(a - b)(aq - b)$$
$$B = (1 - b)(e - a)(f - a)(g - a)/a(b - a)(bq - a);$$

see Askey and Wilson [1979]. It follows that the q-Racah polynomials defined in Askey and Wilson [1979] by

$$(7.4) \qquad W_n(x) \equiv W_n(x; a, b, c, N; q)$$
$$= {}_4\phi_3 \left[\begin{array}{c} q^{-n}, abq^{n+1}, q^{-x}, cq^{x-N} \\ aq, q^{-N}, bcq \end{array} ; q, q \right]$$

satisfy the 3-term recurrence relation

$$(7.5)$$
$$- (1 - q^{-x})(1 - cq^{x-N})W_n(x) = A_n[W_{n+1}(x) - W_n(x)] - B_n[W_n(x) - W_{n-1}(x)],$$
$$W_{-1}(x) \equiv 0, \text{ with}$$

$$A_n = (1 - aq^{n+1})(1 - q^{n-N})(1 - bcq^{n+1})(1 - abq^{n+1})/(1 - abq^{2n+1})(1 - abq^{2n+2}),$$
(7.6)
$$B_n = cq^{-N}(1 - bq^n)(1 - abq^{N+n+1})(1 - aq^n/c)(1 - q^n)/(1 - abq^{2n})(1 - abq^{2n+1}).$$

Similarly, (7.2) gives the recurrence relation

(7.7) $$(2x - a - a^{-1})r_n(x) = A_n'[r_{n+1}(x) - r_n(x)] - B_n'[r_n(x) - r_{n-1}(x)],$$

$r_{-1}(x) \equiv 0$, for the Askey-Wilson [1985] polynomials

(7.8) $$r_n(x) = {}_4\phi_3 \left[\begin{matrix} q^{-n}, abcdq^{n-1}, ae^{i\theta}, ae^{-i\theta} \\ ab, ac, ad \end{matrix} ; q, q \right], \quad x = \cos\theta,$$

with

(7.9)
$$A_n' = a^{-1}(1 - abq^n)(1 - acq^n)(1 - adq^n)(1 - abcdq^{n-1})/(1 - abcdq^{2n-1})(1 - abcdq^{2n}),$$
$$B_n' = a(1 - bcq^{n-1})(1 - bdq^{n-1})(1 - cdq^{n-1})(1 - q^n)/(1 - abcdq^{2n-2})(1 - abcdq^{2n-1}).$$

In particular, if we take $b = aq^{\frac{1}{2}} = -d$, $c = -a$, then (7.7) reduces to

(7.10) $$(2x - a - a^{-1})r_n(x) = \frac{a^{-1}(1 - a^4 q^n)}{(1 - a^2 q^n)}[r_{n+1}(x) - r_n(x)] - \frac{a(1 - q^n)}{(1 - a^2 q^n)}[r_n(x) - r_{n-1}(x)].$$

Setting $a = \beta^{\frac{1}{2}}$ and

(7.11) $$p_n(x) = \frac{(\beta^2; q)_n}{(q; q)_n} \beta^{-n/2} r_n(x)$$

we find that (7.10) can be written as

(7.12) $$2x p_n(x) = \frac{1 - q^{n+1}}{1 - \beta q^n} p_{n+1}(x) + \frac{1 - \beta^2 q^{n-1}}{1 - \beta q^n} p_{n-1}(x),$$

$p_{-1}(x) \equiv 0$, $p_0(x) = 1$, which is exactly the same as the recurrence relation (5.16). Since this $p_n(x)$ and the q-ultraspherical polynomials $C_n(x; \beta|q)$ satisfy the same initial conditions we may conclude that they are the same and hence it follows that

(7.13) $$C_n(x; \beta|q) = \frac{(\beta; q)_n}{(q; q)_n} e^{in\theta} {}_2\phi_1(q^{-n}, \beta; q^{1-n}/\beta; q, qe^{-2i\theta}/\beta)$$

$$= \frac{(\beta^2; q)_n}{(q; q)_n} \beta^{-n/2} {}_4\phi_3 \left[\begin{matrix} q^{-n}, \beta^2 q^n, \beta^{\frac{1}{2}} e^{i\theta}, \beta^{\frac{1}{2}} e^{-i\theta} \\ \beta q^{\frac{1}{2}}, -\beta q^{\frac{1}{2}}, -\beta \end{matrix} ; q, q \right].$$

8. SUMMATION AND TRANSFORMATION FORMULAS FOR WELL-POISED SERIES

Since by (6.7),

(8.1) $$\frac{(b, c; q)_k}{(aq/b, aq/c; q)_k} = \left(\frac{bc}{aq} \right)^k {}_3\phi_2 \left[\begin{matrix} q^{-k}, aq^k, aq/bc \\ aq/b, aq/c \end{matrix} ; q, q \right],$$

$k = 0, 1, \ldots$, we have, for an arbitrary square $\{\alpha_k\}$, the expansion formula

(8.2)
$$\sum_{k=0}^{n} \frac{(b, c, q^{-n}; q)_k}{(q, aq/b, aq/c; q)_k} \alpha_k$$

$$= \sum_{j=0}^{n} \frac{(aq/bc, q^{-n}; q)_j (a; q)_{2j}}{(q, a, aq/b, aq/c; q)_j} (-1)^j q^{-\binom{j}{2}}$$

$$\cdot \sum_{i=0}^{n-j} \frac{(aq^{2j}, q^{j-n}; q)_i}{(q, aq^j; q)_i} q^{-ij} \left(\frac{bc}{aq}\right)^{i+j} \alpha_{i+j}, \quad n = 0, 1, \ldots .$$

Choosing

(8.3)
$$\alpha_k = \frac{(a, a_1, \ldots, a_r; q)_k}{(b_1, \ldots, b_{r+1}; q)_k} z^k$$

we find that

(8.4)
$$_{r+4}\phi_{r+3} \left[\begin{array}{c} a, b, c, a_1, \ldots, a_r, q^{-n} \\ aq/b, aq/c, b_1, \ldots, b_r, b_{r+1} \end{array} ; q, z \right]$$

$$= \sum_{j=0}^{n} \frac{(aq/bc, a_1, \ldots, a_r, q^{-n}; q)_j (a; q)_{2j}}{(q, aq/b, aq/c, b_1, \ldots, b_r, b_{r+1}; q)_j} \left(-\frac{bcz}{aq}\right)^j q^{-\binom{j}{2}}$$

$$\cdot _{r+2}\phi_{r+1} \left[\begin{array}{c} aq^{2j}, a_1 q^j, \ldots, a_r q^j, q^{j-n} \\ b_1 q^j, \ldots, b_r q^j, b_{r+1} q^j \end{array} ; q, \frac{bczq^{-j-1}}{a} \right].$$

If we now set $a_1 = qa^{\frac{1}{2}} = -a_2$, $b_1 = a^{\frac{1}{2}} = -b_2$, $b_k = a_k$, $k = 3, \ldots, r$, $b_{r+1} = aq^{n+1}$, and $z = aq^{n+1}/bc$ and then use (6.14) for the very-well-poised $_4\phi_3$ series on the right side, then (8.4) gives the sum of a terminating very-well-poised $_6\phi_5$ series:

(8.5)
$$_6W_5(a; b, c, q^{-n}; q, aq^{n+1}/bc) = \frac{(aq, aq/bc; q)_n}{(aq/b, aq/c; q)_n}.$$

Let us now set $a_1 = qa^{\frac{1}{2}} = -a_2$, $b_1 = a^{\frac{1}{2}} = -b_2$, $a_3 = d$, $a_4 = e$, $b_3 = aq/d$, $b_4 = aq/e$, $b_k = a_k$, $k = 5, \ldots, r$, $b_{r+1} = aq^{n+1}$, $z = a^2 q^{n+2}/bcde$ in (8.4) and use the sum (8.5) on the right side to obtain Watson's [1929] transformation formula

(8.6)
$$_8W_7(a; b, c, d, e, q^{-n}; q, a^2 q^{n+2}/bcde)$$

$$= \frac{(aq, aq/de; q)_n}{(aq/d, aq/e; q)_n} {}_4\phi_3 \left[\begin{array}{c} q^{-n}, d, e, aq/bc \\ aq/b, aq/c, deq^{-n}/a \end{array} ; q, q \right]$$

which expresses a terminating very-well-poised $_8\phi_7$ series in terms of a terminating balanced $_4\phi_3$ series. In the special case $aq/bc = deq^{-n}/a$, that is, when the $_8\phi_7$ series is balanced, the $_4\phi_3$ series on the right side reduces to a terminating balanced $_3\phi_2$ series which can be summed by (6.7). Thus we obtain Jackson's [1921] summation formula

(8.7)
$$_8W_7(a; b, c, d, e, q^{-n}; q, q)$$

$$= \frac{(aq, aq/bc, aq/bd, aq/cd; q)_n}{(aq/b, aq/c, aq/d, aq/bcd; q)_n},$$

where $a^2 q^{n+1} = bcde$. Many of the summation formulas for basic hypergeometric series are special or limiting cases of (8.7). For example, (8.5) is the $d \to \infty$ limit case of (8.7), and

(6.7) is obtained from (8.7) by taking the limit $a \to 0$ after replacing d by aq/d. Also in the limit $n \to \infty$, we derive the following nonterminating $_6\phi_5$ summation formula

$$(8.8) \qquad {}_6W_5(a; b, c, d; q, aq/bcd)$$

$$= \frac{(aq, aq/bc, aq/bd, aq/cd; q)_\infty}{(aq/b, aq/c, aq/d, aq/bcd; q)_\infty}, \quad |aq/bcd| < 1.$$

Setting $d = a^{\frac{1}{2}}$ in (8.8) we also obtain a q-analogue of Dixon's summation formula:

$$(8.9) \qquad {}_4\phi_3\left[\begin{array}{c} a, -qa^{\frac{1}{2}}, b, c \\ -a^{\frac{1}{2}}, aq/b, aq/c \end{array}; q, qa^{\frac{1}{2}}/bc\right]$$

$$= \frac{(aq, aq/bc, qa^{\frac{1}{2}}/b, qa^{\frac{1}{2}}/c; q)_\infty}{(aq/b, aq/c, qa^{\frac{1}{2}}, qa^{\frac{1}{2}}/bc; q)_\infty}, \quad |qa^{\frac{1}{2}}/bc| < 1.$$

9. BAILEY'S TRANSFORMATION FORMULAS

Using Jackson's formula (8.7) it can be shown that

$$(9.1) \qquad \frac{(a, b, c; q)_k}{(q, aq/b, aq/c; q)_k} = \frac{(\lambda bc/a; q)_k}{(qa^2/\lambda bc; q)_k} \sum_{j=0}^{k} \frac{(1 - \lambda q^{2j})(\lambda, \lambda b/a, \lambda c/a, aq/bc; q)_j}{(1 - \lambda)(q, aq/b, aq/c, \lambda bc/a; q)_j}$$

$$\cdot \frac{(a; q)_{k+j}(a/\lambda; q)_{k-j}}{(\lambda q; q)_{k+j}(q; q)_{k-j}}(a/\lambda)^j,$$

where λ is an arbitrary parameter. Thus, for an arbitrary sequence $\{\alpha_k\}$,

$$(9.2) \qquad \sum_{k=0}^{\infty} \frac{(a, b, c; q)_k}{(q, aq/b, aq/c; q)_k}\alpha_k$$

$$= \sum_{j=0}^{\infty} \frac{(1 - \lambda q^{2j})(\lambda, \lambda b/a, \lambda c/a, aq/bc; q)_j(a; q)_{2j}}{(1 - \lambda)(q, aq/b, aq/c, qa^2/\lambda bc; q)_j(\lambda q; q)_{2j}}(a/\lambda)^j$$

$$\cdot \sum_{k=0}^{\infty} \frac{(aq^{2j}, a/\lambda, \lambda bcq^j/a; q)_k}{(q, \lambda q^{2j+1}, a^2q^{j+1}/\lambda bc; q)_k}\alpha_{j+k},$$

provided all the series terminate or are absolutely convergent.

Choosing $\lambda = qa^2/bcd$, and

$$(9.3) \qquad \alpha_k = \frac{(1 - aq^{2k})(d, e, f, g, h; q)_k}{(1 - a)(aq/d, aq/e, aq/f, aq/g, aq/h; q)_k}q^k,$$

where at least one of the parameters d, e, f, g, h is of the form q^{-n}, $n = 0, 1, \ldots$, and

$$(9.4) \qquad q^2 a^3 = bcdefgh,$$

we obtain, by using (8.7) for the inner series on the right side of (9.2), Bailey's [1929] transformation formula for a terminating, very-well-poised balanced $_{10}\phi_9$ series,

$$(9.5) \qquad {}_{10}W_9(a; b, c, d, e, f, a^3q^{n+2}/bcdef, q^{-n}; q, q)$$

$$= \frac{(aq, aq/ef, a^2q^2/bcde, a^2q^2/bcdf; q)_n}{(aq/e, aq/f, a^2q^2/bcdef, a^2q^2/bcd; q)_n}$$

$$\cdot {}_{10}W_9(qa^2/bcd; aq/bc, aq/bd, aq/cd, e, f, a^3q^{n+2}/bcdef, q^{-n}; q, q).$$

An iteration of this formula gives

(9.6) $\quad {}_{10}W_9(a;b,c,d,e,f,a^3q^{n+2}/bcdef,q^{-n};q,q)$

$$= \frac{(aq,aq/de,aq/df,aq/ef;q)_n}{(aq/d,aq/e,aq/f,aq/def;q)_n}$$

$$\cdot \, {}_{10}W_9(defq^{-n-1}/a;aq/bc,d,e,f,bdefq^{-n-1}/a^2,cdefq^{-n-1}/a^2,q^{-n};q,q).$$

If we now take the limit $n \to \infty$ in (9.5) we immediately get the transformation formula for a nonterminating very-well-poised ${}_8\phi_7$ series:

(9.7) $\quad {}_8W_7(a;b,c,d,e,f;q,\lambda q/ef)$

$$= \frac{(aq,aq/ef,\lambda q/e,\lambda q/f;q)_\infty}{(aq/e,aq/f,\lambda q/ef,\lambda q;q)_\infty} \, {}_8W_7(\lambda;\lambda b/a,\lambda c/a,\lambda d/a,e,f;q,aq/ef),$$

where $\lambda = qa^2/bcd$, provided $\max(|aq/ef|, |\lambda q/ef|) < 1$. However, one must be careful in taking the $n \to \infty$ limit on the right side of (9.6) where the terms near both ends of the series are large compared to those in the middle for large n, which prevents us from taking the term-by-term limit directly. To circumvent this difficulty, Bailey [1935] chose n to be an odd integer, say $n = 2m + 1$, and divided the series on the right side of (9.6) into two halves, each containing $m + 1$ terms and then reversed the order of the second series. This procedure followed by the limit $m \to \infty$ yields the transformation formula

(9.8) $\quad {}_8W_7(a;b,c,d,e,f;q,a^2q^2/bcdef)$

$$= \frac{(aq,aq/de,aq/df,aq/ef;q)_\infty}{(aq/d,aq/e,aq/f,aq/def;q)_\infty} \, {}_4\phi_3 \left[\begin{array}{c} d,e,f,aq/bc \\ aq/b,aq/c,def/a \end{array} ;q,q \right]$$

$$+ \frac{(aq,aq/bc,d,e,f,a^2q^2/bdef,a^2q^2/cdef;q)_\infty}{(aq/b,aq/c,aq/d,aq/e,aq/f,a^2q^2/bcdef,def/aq;q)_\infty}$$

$$\cdot \, {}_4\phi_3 \left[\begin{array}{c} aq/de,aq/df,aq/ef,a^2q^2/bcdef \\ a^2q^2/bdef,a^2q^2/cdef,aq^2/def \end{array} ;q,q \right].$$

Both ${}_4\phi_3$ series on the right side are balanced and, in general, nonterminating, and hence (9.8) is a nonterminating extension of Watson's formula (8.6).

Al-Salam and Verma [1982] pointed out that (9.8) is equivalent to the q-integral representation

(9.9) $\quad \displaystyle\int_a^b \frac{(qt/a,qt/b,ct,dt;q)_\infty}{(et,ft,gt,ht;q)_\infty} d_q t$

$$= b(1-q)\frac{(q,a/b,bq/a,cd/eh,cd/fh,cd/gh,bc,bd;q)_\infty}{(ae,af,ag,be,bf,bg,bh,bcd/h;q)_\infty}$$

$$\cdot \, {}_8W_7(bcd/hq;be,bf,bg,c/h,d/h;q,ah),$$

provided $cd = abefgh$ and $|ah| < 1$.

If we set $h = d$ in (9.9) and replace g by d then we obtain Sears' [1951c] extension of the q-Saalschütz formula (6.7):

(9.10) $\quad \displaystyle\int_a^b \frac{(qt/a,qt/b,ct;q)_\infty}{(dt,et,ft;q)_\infty} d_q t$

$$= \frac{b(1-q)(q,a/b,bq/a,c/d,c/e,c/f;q)_\infty}{(ad,ae,af,bd,be,bf;q)_\infty},$$

provided $c = abdef$. Specializing further by letting $c \to 0$, $f \to 0$ with $c/f = abde$, we obtain the summation formula (4.10).

10. A QUADRATIC TRANSFORMATION FORMULA

For $n = 0, 1, \ldots$, let us choose $\alpha_k = (q^{-n}; q)_k q^k / (b^2 c^2 q^{-n-1}/a; q)_k$ in (9.2), and use (6.7) on the right side to obtain the transformation formula

$$(10.1) \quad {}_4\phi_3 \left[\begin{matrix} a, b, c, q^{-n} \\ aq/b, aq/c, b^2 c^2 q^{-n-1}/a \end{matrix} ; q, q \right] = \frac{(a^2 q^2/b^2 c^2, a^{\frac{1}{2}} q^{\frac{3}{2}}/bc; q)_n}{(aq^2/b^2 c^2, (aq)^{\frac{3}{2}}/bc; q)_n}$$

$$\cdot {}_{10}W_9 \left(a^{\frac{3}{2}} q^{\frac{1}{2}}/bc; a^{\frac{1}{2}}, -a^{\frac{1}{2}}, -(aq)^{\frac{1}{2}}, \frac{(aq)^{\frac{1}{2}}}{b}, \frac{(aq)^{\frac{1}{2}}}{c}, \frac{a^2 q^{n+2}}{b^2 c^2}, q^{-n}; q, q \right),$$

where, it may be observed, that the ${}_4\phi_3$ series on the left is a terminating, balanced and nearly-poised series of the second kind, while the ${}_{10}\phi_9$ series on the right is terminating, balanced and very-well-poised. Using (9.5) we now transform the ${}_{10}\phi_9$ series on the right side of (10.1) to a multiple of ${}_{10}W_9(-cq^{-n-1}; a^{\frac{1}{2}}, -a^{\frac{1}{2}}, (aq)^{\frac{1}{2}}/b, -(aq)^{\frac{1}{2}}/b, -bcq^{-n-1}/a, bc^2 q^{-n-1}/a, q^{-n}; q, q)$, replace c by cq^n, take the limit $n \to \infty$, use (2.10) and (2.11) to simplify the coefficients, and apply (9.7) once to derive the following quadratic transformation formula

$$(10.2) \quad {}_2\phi_1(a, b; aq/b; q, qx/b)$$

$$= \frac{(x(aq)^{\frac{1}{2}}, -x(aq)^{\frac{1}{2}}, qxa^{\frac{1}{2}}/b, -qxa^{\frac{1}{2}}/b; q)_\infty}{(xq^{\frac{1}{2}}, -xq^{\frac{1}{2}}, -xq/b, aqx/b; q)_\infty}$$

$$\cdot {}_8W_7(ax/b; a^{\frac{1}{2}}, -a^{\frac{1}{2}}, (aq)^{\frac{1}{2}}/b, -(aq)^{\frac{1}{2}}/b, x; q, qx/b), \quad |qx/b| < 1,$$

where, in the final step, we have replaced c by aqx/b. This is a q-analogue of the transformation formula

$$(10.3) \quad F(a, b; 1 + a - b; x) = (1 - x)^{-a} F\left(\frac{1}{2}a, \frac{1}{2} + \frac{1}{2}a - b; 1 + a - b; -\frac{4x}{(1-x)^2} \right),$$

which is valid when $|x| < 1$ and $|4x/(1-x)^2| < 1$.

For further extensions of (10.2) see Gasper and Rahman [1986, 1989a,b].

11. ASKEY-WILSON'S q-BETA INTEGRAL

In a beautiful piece of work Askey and Wilson [1985] gave the following q-extension of the beta integral

$$(11.1) \quad \int_{-1}^{1} \frac{h(x; 1, -1, q^{\frac{1}{2}}, -q^{\frac{1}{2}})}{h(x; a, b, c, d)} \frac{dx}{\sqrt{1 - x^2}}$$

$$= \frac{2\pi (abcd; q)_\infty}{(q, ab, ac, ad, bc, bd, cd; q)_\infty},$$

where

$$h(x; a_1, a_2, \ldots, a_m) = h(x; a_1) h(x; a_2) \cdots h(x; a_m),$$

$$(11.2) \quad h(x; a) = \prod_{n=0}^{\infty} (1 - 2axq^n + a^2 q^{2n}) = (aq^{i\theta}, ae^{-i\theta}; q)_\infty, \quad x = \cos\theta,$$

and $\max(|a|, |b|, |c|, |d|, |q|) < 1$.

If we set $a = q^{\frac{1}{2}}$, $b = q^{\alpha + \frac{1}{2}}$, $c = -q^{\beta + \frac{1}{2}}$, $d = -q^{\frac{1}{2}}$, then it can be shown that (11.1) reduces to the beta integral

$$(11.3) \qquad \int_{-1}^{1} (1-x)^{\alpha}(1+x)^{\beta} dx = 2^{\alpha + \beta + 1} B(\alpha + 1, \beta + 1)$$

in the limit $q \to 1^{-}$. Askey and Wilson [1985] gave a contour integral proof of (11.1) as well as a direct evaluation in the special case $c = q^{\frac{1}{2}} = -d$. Simpler evaluations of (11.1) were given in Rahman [1984], Askey [1988] and Ismail and Stanton [1988]. Here we shall give a brief sketch of the proof given in Rahman [1984] and Gasper and Rahman [1989a].

Assuming, for the moment, that a, b, c, d and their pairwise quotients are not of the form q^n, $n = 0, 1, \ldots$, and observing that by (9.10)

$$(11.4) \qquad h(x;1)/h(x;a,b) = A \int_{a}^{b} \frac{(qu/a, qu/b, u; q)_{\infty}}{(u/ab; q)_{\infty} h(x; u)} d_q u,$$

$$(11.5) \qquad h(x;-1)/h(x;c,d) = B \int_{c}^{d} \frac{(qv/c, qv/d, -v; q)_{\infty}}{(-v/cd; q)_{\infty} h(x; v)} d_q v,$$

$$(11.6) \qquad h(x;-q^{\frac{1}{2}})/h(x;u,v) = C \int_{uq^{-\frac{1}{2}}}^{vq^{-\frac{1}{2}}} \frac{(tq^{\frac{3}{2}}/u, tq^{\frac{3}{2}}/v, -qt; q)_{\infty}}{(-qt/uv; q)_{\infty} h(x; tq^{\frac{1}{2}})} d_q t,$$

where

$$(11.7) \qquad A = \frac{(a^{-1}, b^{-1}; q)_{\infty}}{b(1-q)(q, a/b, bq/a, ab; q)_{\infty}}, \qquad B = \frac{(-c^{-1}, -d^{-1}; q)_{\infty}}{d(1-q)(q, c/d, dq/c, cd; q)_{\infty}},$$

$$C = \frac{q^{\frac{1}{2}}(-q^{\frac{1}{2}}u^{-1}, -q^{\frac{1}{2}}v^{-1}; q)_{\infty}}{v(1-q)(q, u/v, vq/u, uv; q)_{\infty}},$$

and that, by symmetry, the integral in (11.1) can be written as

$$(11.8) \qquad I(a,b,c,d) = \frac{1}{2} \int_{-\pi}^{\pi} \frac{h(\cos\theta; 1, -1, q^{\frac{1}{2}}, -q^{\frac{1}{2}})}{h(\cos\theta; a, b, c, d)} d\theta,$$

we find that the evaluation of $I(a,b,c,d)$ is reduced to that of the integral

$$\frac{1}{2} \int_{-\pi}^{\pi} \frac{h(\cos\theta; q^{\frac{1}{2}})}{h(\cos\theta; tq^{\frac{1}{2}})} d\theta, \qquad |tq^{\frac{1}{2}}| < 1.$$

Using the q-binomial formula (3.1) and the q-Gauss formula (4.6) this integral can be easily evaluated to get $\pi(qt, qt; q)_{\infty}/(q, qt^2; q)_{\infty}$. (11.1) then follows by repeated applications of (9.10). By analytic continuation, the temporary restrictions imposed above can be removed.

Use of (9.10) also enabled Nassrallah and Rahman [1985] to obtain the following integral representation of an $_8\phi_7$ series:

$$(11.9) \qquad {}_8W_7(g^2/q; g/a, g/b, g/c, g/d, g/f; q, abcdf/g)$$

$$= \frac{(q, ab, ac, ad, af, bc, bd, bf, cd, cf, df, g^2; q)_{\infty}}{2\pi(ag, bg, cg, dg, fg, abcdf/g; q)_{\infty}}$$

$$\cdot \int_{-1}^{1} \frac{h(x; 1, -1, q^{\frac{1}{2}}, -q^{\frac{1}{2}}, g)}{h(x; a, b, c, d, f)} \frac{dx}{\sqrt{1-x^2}},$$

when $|abcdf/g| < 1$ and $\max(|a|, |b|, |c|, |d|, |f|, |q|) < 1$. This can be viewed as a q-analogue of (1.3). In the special case $g = abcdf$, it was pointed out in Rahman [1986b] that

(11.10)
$$\int_{-1}^{1} \frac{h(x; 1, -1, q^{\frac{1}{2}}, -q^{\frac{1}{2}}, abcdf)}{h(x; a, b, c, d, f)} \frac{dx}{\sqrt{1 - x^2}}$$
$$= \frac{2\pi(abcd, abcf, abdf, acdf, bcdf; q)_\infty}{(q, ab, ac, ad, af, bc, bd, bf, cd, cf, df; q)_\infty}$$

which reduces to (11.1) when $f = 0$.

Askey and Wilson [1985] showed that the polynomials $r_n(x)$ defined in (7.8) are orthogonal with respect to the weight function in the integral (11.1).

12. LINEARIZATION AND PRODUCT FORMULAS FOR THE q-ULTRASPHERICAL POLYNOMIALS

As a final application of the summation and transformation formulas developed in this article we shall now derive Rogers' [1895] linearization formula

(12.1)
$$C_m(x; \beta|q) \, C_n(x; \beta|q)$$
$$= \sum_{k=0}^{\min(m,n)} \frac{(\beta; q)_{m-k}(\beta; q)_{n-k}(\beta; q)_k(q; q)_{m+n-2k}(\beta^2; q)_{m+n-k}}{(q; q)_{m-k}(q; q)_{n-k}(q; q)_k(\beta^2; q)_{m+n-2k}(\beta q; q)_{m+n-k}} \frac{1 - \beta q^{m+n-2k}}{1 - \beta}$$
$$\cdot C_{m+n-2k}(x; \beta|q),$$

and Rahman and Verma's [1986] product formula

(12.2)
$$C_n(\cos\theta; \beta|q) \, C_n(\cos\phi; \beta|q)$$
$$= \int_0^\pi K(\cos\theta, \cos\phi, \cos\psi|q) \frac{(\beta^2; q)_n}{(q; q)_n} \beta^{-n/2} C_n(\cos\psi; \beta|q) d\psi,$$

where

(12.3)
$$K(\cos\theta, \cos\phi, \cos\psi|q)$$
$$= \frac{(q, \beta, \beta; q)_\infty |(\beta e^{2i\theta}, \beta e^{2i\phi}, e^{2i\psi}; q)_\infty|^2}{2\pi(\beta^2; q)_\infty h(\cos\psi; \beta^{\frac{1}{2}} e^{i\theta+i\phi}, \beta^{\frac{1}{2}} e^{i\theta-i\phi}, \beta^{\frac{1}{2}} e^{-i\theta-i\phi}, \beta^{\frac{1}{2}} e^{i\phi-i\theta})},$$

with $-1 < \beta < 1$, $0 \le \theta$, $\phi \le \pi$, and $C_n(x; \beta|q)$ is Rogers' q-ultraspherical polynomials defined in (5.15) and (7.13).

To prove (12.1) let us set

(12.4)
$$f_\alpha(x) = {}_2\phi_1(\alpha, \beta; \alpha q/\beta; q, qx^2/\beta)$$

which, by (4.5), transforms to

(12.5)
$$f_\alpha(x) = \frac{(\beta x^2; q)_\infty}{(qx^2/\beta; q)_\infty} {}_2\phi_1(q/\beta, \alpha q/\beta^2; \alpha q/\beta; q, \beta x^2).$$

Assuming, for the time being that $|\beta x^2| < 1$, we then have

(12.6)
$$C_m(x; \beta|q) C_n(x; \beta|q)$$

$$= A_{m,n} \sum_{k=0}^{\infty} \frac{(q/\beta, q^{1-m}/\beta^2; q)_k}{(q, q^{1-m}/\beta; q)_k} (\beta e^{-2i\theta})^k$$

$$\cdot {}_4\phi_3 \left[\begin{matrix} q^{-k}, q^{-n}, \beta, \beta q^{m-k} \\ \beta q^{-k}, q^{1-n}/\beta, \beta^2 q^{m-k} \end{matrix} ; q, q \right],$$

where

(12.7) $$A_{m,n} = \frac{(\beta e^{-2i\theta}; q)_\infty}{(q e^{-2i\theta}/\beta; q)_\infty} \frac{(\beta; q)_m (\beta; q)_n}{(q; q)_m (q; q)_n} e^{i(m+n)\theta}.$$

The important point to observe is that the ${}_4\phi_3$ series in (12.6) is balanced and terminating and so, by (8.6),

(12.8) $${}_4\phi_3 \left[\begin{matrix} q^{-k}, q^{-n}, \beta, \beta q^{m-k} \\ \beta q^{-k}, q^{1-n}/\beta, \beta^2 q^{m-k} \end{matrix} ; q, q \right] = \frac{(q^{1-m-n}/\beta^2, q^{1-m}/\beta; q)_k}{(q^{1-m-n}/\beta, q^{1-m}/\beta^2; q)_k}$$

$$\cdot {}_8 W_7(q^{-m-n}/\beta; \beta, q^{1+k-m-n}/\beta^2, q^{-m}, q^{-n}, q^{-k}; q, q/\beta).$$

Substituting this into (12.6) and interchanging the order of summation, we obtain

(12.9)

$$C_m(x; \beta|q) \, C_n(x; \beta|q)$$

$$= A_{m,n} \sum_{k=0}^{\min(m,n)} \frac{(1 - q^{2k-m-n}/\beta)(q^{-m-n}/\beta, \beta, q^{-m}, q^{-n}; q)_k}{(1 - q^{-m-n}/\beta)(q, q^{1-m-n}/\beta^2, q^{1-n}/\beta, q^{1-m}/\beta; q)_k}$$

$$\cdot \frac{(q^{1-m-n}/\beta^2; q)_{2k}}{(q^{1-m-n}/\beta; q)_{2k}} (q e^{-2i\theta}/\beta)^k \cdot {}_2\phi_1(q/\beta, q^{1+2k-m-n}/\beta^2; q^{1+2k-m-n}/\beta; q, \beta e^{-2i\theta})$$

which gives (12.1) by using (12.5) and (12.7) and observing that both sides of (12.1) are polynomials in x. This simple proof of (12.1) was given by Gasper [1985]. For other proofs see Bressoud [1981] and Rahman [1981].

To prove (12.2) we have to proceed more cautiously. First, let us assume that $|\alpha| < 1$ and $|\alpha q/\beta^2| < 1$. Then, using (4.1) we get

(12.10) $$f_\alpha(x) = \frac{(\alpha, qx^2; q)_\infty}{(\alpha q/\beta, qx^2/\beta; q)_\infty} \, {}_2\phi_1(q/\beta, qx^2/\beta; qx^2; q, \alpha),$$

(12.11) $$f_\alpha(y) = \frac{(\alpha, qy^2; q)_\infty}{(\alpha q/\beta, qy^2/\beta; q)_\infty} \, {}_2\phi_1(q/\beta, qy^2/\beta; qy^2; q, \alpha)$$

$$= \frac{(\alpha q/\beta^2, qy^2; q)_\infty}{(\alpha q/\beta, qy^2/\beta; q)_\infty} \, {}_2\phi_1(\beta, \beta y^2; qy^2; q, \alpha q/\beta^2), \quad \text{by (4.5).}$$

Hence

(12.12) $$f_\alpha(x) f_\alpha(y) = B_\alpha \sum_{k=0}^{\infty} \frac{(q/\beta, qx^2/\beta; q)_k \alpha^k}{(q, qx^2; q)_k} \, {}_4\phi_3 \left[\begin{matrix} q^{-k}, q^{-k}/x^2, \beta, \beta y^2 \\ \beta q^{-k}/x^2, \beta q^{-k}, qy^2 \end{matrix} ; q, q \right],$$

where

(12.13) $$B_\alpha = \frac{(\alpha, \alpha q/\beta^2, qx^2, qy^2; q)_\infty}{(\alpha q/\beta, \alpha q/\beta, qx^2/\beta, qy^2/\beta; q)_\infty}.$$

Once again, the terminating $_4\phi_3$ series in (12.12) is balanced and so, by (8.6), equals

(12.14)
$$\frac{(qx^2y^2, qx^2; q)_k}{(\beta qx^2y^2, qx^2/\beta; q)_k} \, _8W_7(\beta x^2y^2; \beta, \beta x^2, \beta y^2, x^2y^2q^{k+1}, q^{-k}; q, q/\beta).$$

Substituting this into (12.12) and interchanging the order of summation we find that

(12.15)
$$f_\alpha(x)f_\alpha(y) = B_\alpha \sum_{j=0}^{\infty} \frac{(1 - \beta x^2y^2q^{2j})(\beta x^2y^2, \beta, \beta x^2, \beta y^2; q)_j(qx^2y^2; q)_{2j}}{(1 - \beta x^2y^2)(q, qx^2y^2, qy^2, qx^2; q)_j(\beta qx^2y^2; q)_{2j}}$$
$$\cdot (\alpha q/\beta^2)^j \, _2\phi_1(x^2y^2q^{2j+1}, q/\beta; \beta x^2y^2q^{2j+1}; q, \alpha).$$

Since the $_2\phi_1$ series on the right side of (12.15) is well-poised we will now use (10.2) to express it in terms of an $_8\phi_7$ series which, in turn, will be written as the sum of two nonterminating balanced $_4\phi_3$ series via (9.8). Thus

(12.16)
$$f_\alpha(x)f_\alpha(y) = B_\alpha \frac{(q\alpha xy/\beta; q)_\infty}{(\alpha/\beta xy; q)_\infty} \sum_{k=0}^{\infty} \frac{(\beta xy, xyq^{\frac{1}{2}}, -xyq^{\frac{1}{2}}, -xyq; q)_k}{(q, \alpha xyq/\beta, \beta xyq/\alpha, q\beta x^2y^2; q)_k} q^k$$
$$\cdot \, _8W_7(\beta x^2y^2; \beta, \beta x^2, \beta y^2, qxy, q^{-k}; q, xyq^{k+1}/\beta)$$
$$+ B_\alpha \frac{(\alpha q/\beta, qx^2y^2, \beta xy, \alpha xyq; q)_\infty}{(\alpha, q\beta xy, qxy, \beta xy/\alpha; q)_\infty} \sum_{k=0}^{\infty} \frac{(\alpha, \alpha q^{\frac{1}{2}}/\beta, -\alpha q^{\frac{1}{2}}/\beta, -\alpha q/\beta; q)_k}{(q, \alpha q^2/\beta, \alpha xyq, \alpha q/\beta xy; q)_k} q^k$$
$$\cdot \, _8W_7(\beta x^2y^2; \beta, \beta x^2, \beta y^2, qxy, \beta xyq^{-k}/\alpha; q, \alpha q^{k+1}/\beta^2).$$

When $x = y$, both $_8\phi_7$ series above reduce to very-well-poised $_6\phi_5$ series which can be summed by (8.8) resulting in a nonterminating q-Clausen formula

(12.17)
$$f_\alpha^2(x) = \frac{(\alpha, \alpha q/\beta^2, qx^2, qx^2, \alpha qx^2/\beta; q)_\infty}{(\alpha q/\beta, \alpha q/\beta, qx^2/\beta, \alpha/\beta x^2; q)_\infty} \, _5\phi_4 \left[\begin{array}{c} \beta x^2, qx^2/\beta, x^2q^{\frac{1}{2}}, -x^2q^{\frac{1}{2}}, -qx^2 \\ qx^2, qx^4, \alpha qx^2/\beta, \beta qx^2/\alpha \end{array} ; q, q \right]$$
$$+ \frac{(\beta x^2, \alpha qx^2/\beta; q)_\infty}{(qx^2/\beta, \beta x^2/\alpha; q)_\infty} \, _5\phi_4 \left[\begin{array}{c} \alpha, \alpha q^{\frac{1}{2}}/\beta, -\alpha q^{\frac{1}{2}}/\beta, -\alpha q/\beta, \alpha q/\beta^2 \\ \alpha q/\beta, q\alpha^2/\beta^2, \alpha qx^2/\beta, \alpha q/\beta x^2 \end{array} ; q, q \right],$$

where, by analytic continuation, the restrictions $|\alpha| < 1$, $|\alpha q/\beta^2| < 1$ are no longer necessary. For another derivation of (12.17) see Gasper and Rahman [1989a,b] where it is shown that in the limit $q \to 1^-$, it approaches the Clausen [1828] formula

(12.18)
$$F^2(a, b; a + b + 1/2; x) = \, _3F_2 \left[\begin{array}{c} 2a, 2b, a + b \\ 2a + 2b, a + b + 1/2 \end{array} ; x \right].$$

For $x \neq y$, we apply (9.8) to both the $_8\phi_7$ series in (12.16) and note that the resulting 4 double series are absolutely convergent even without the conditions $|\alpha| < 1$, $|\alpha q/\beta^2| < 1$. We then set $\alpha = q^{-n}$, $n = 0, 1, \ldots$, which results in the vanishing of the coefficients of 3 of these series, and the fourth one gives, after some manipulations, via (12.4) and (5.15), the following product formula:

(12.19)
$$C_n(\cos\theta; \beta|q) \, C_n(\cos\phi; \beta|q)$$
$$= \frac{(\beta^2, \beta^2; q)_n}{(q, q; q)_n} \beta^{-n} \sum_{k=0}^{n} \frac{(q^{-n}, \beta^2 q^n, \beta e^{-i\theta - i\phi}, e^{i\theta + i\phi}; q)_k}{(q, \beta q^{\frac{1}{2}}, -\beta q^{\frac{1}{2}}, -\beta; q)_k} q^k$$

$$\cdot \; {}_4\phi_3 \left[\begin{array}{c} q^{-k}, \beta, \beta e^{-2i\theta}, \beta e^{-2i\phi} \\ \beta^2, \beta e^{-i\theta-i\phi}, e^{-i\theta-i\phi}q^{1-k} \end{array} ; q, q \right].$$

However, by (11.1),

(12.20)
$$\frac{(\beta, \beta e^{-2i\theta}, \beta e^{-2i\phi}; q)_j}{(\beta^2; q)_j} = \int_0^\pi K(\cos\theta, \cos\phi, \cos\psi | q)(\beta^{\frac{1}{2}} e^{i\psi-i\theta-i\phi}, \beta^{\frac{1}{2}} e^{-i\psi-i\theta-i\phi}; q)_j d\psi$$

and hence (12.2) follows by use of (6.7) and (7.13). For product formulas for q-Racah and Askey-Wilson polynomials see Rahman [1982] and Gasper and Rahman [1989a].

REFERENCES

Al-Salam, W. and Verma, A. (1982). Some remarks on q-beta integral, Proc. Amer. Math. Soc. **85**, 360-362.

Andrews, G.E. (1986). q-Series: Their Development and Application in Analysis, Number Theory, Combinatorics, Physics, and Computer Algebra, CBMS Regional Conference Lecture Series #66, Amer. Math. Soc., Providence, R.I.

Andrews, G.E. and Askey, R. (1977). Enumeration of partitions: the role of Eulerian series and q-orthogonal polynomials, in Higher Combinatorics, ed. by M. Aigner, Reidel, Boston, pp. 3- 26.

Andrews, G.E. and Askey, R. (1978). A simple proof of Ramanujan's summation of the ${}_1\psi_1$, Aequationes Math. **18**, 333-337.

Andrews, G.E. and Askey, R. (1981). Another q-extension of the beta function, Proc. Amer. Math. Soc. **81**, 97-100.

Askey, R. (1975). Orthogonal Polynomials and Special Functions, Regional Conference Series in Applied Mathematics, 21, SIAM, Philadelphia.

Askey, R. (1978). The q-gamma and q-beta functions, Applicable Analysis, **8**, 125-141.

Askey, R. (1980). Ramanujan's extensions of the gamma and beta functions, Amer. Math. Monthly, **87**, 346-359.

Askey, R. (1988). Beta integrals and q-extensions, Proceedings of the Ramanujan Centennial International Conference, ed. by R. Balakrishnan et al., Published by Ramanujan Mathematical Society, Annamalainagar, India, pp. 85-102.

Askey, R. and Ismail, M.E.-H. (1983). A generalization of ultraspherical polynomials, in Studies in Pure Mathematics, ed. by P. Erdős, Birkhäuser, Boston, pp. 55-78.

Askey, R. and Wilson, J. (1979). A set of orthogonal polynomials that generalize the Racah coefficients or 6-j symbols, SIAM J. Math. Anal. **10**, 1008-1016.

Askey, R. and Wilson, J. (1985). Some basic hypergeometric orthogonal polynomials that generalize Jacobi polynomials, Mem. Amer. Math. Soc. #319.

Bailey, W.N. (1929). An identity involving Heine's basic hypergeometric series, J. London Math. Soc. **4**, 254-257.

Bailey, W.N. (1935). Generalized Hypergeometric Series, Cambridge University Press, Cambridge, reprinted by Stechert-Hafner Service Agency, New York, 1964.

Bressoud, D.M. (1981). Linearization and related formulas for q-ultraspherical polynomials, SIAM J. Math. Anal. **12**, 161-168.

Carlitz, L. (1973). Some inverse relations, Duke Math. J. **40**, 893-901.

Cauchy, A.-L. (1843). Mémoire sur les fonctions dont plusieurs valeurs sont liées entre elles par une équation linéare, et sur diverses transformations de produits composés d'un nombre indéfini de facteurs, C.R.T. XVII, p. 523, in Oeuvres de Cauchy, 1re série, T. VII, Gauthier-Villiars, Paris, 1893, pp. 42-50.

Chihara, T.S. (1978). An Introduction to Orthogonal Polynomials, Gordon and Breach, New York.

Chu, Shih-Chieh (1303). Ssu Yuan Yü Chien (Precious Mirror of the Four Elements), in Chinese; see Askey [1975, p. 59], Needham [1959, p. 138] and Takács [1973]).

Clausen, T. (1828). Ueber die Fälle, wenn die Reihe von der Form $y = 1 + \dfrac{\alpha\beta}{1\cdot\gamma}x + \dots$ ein Quadrat von der Form $z = 1 + \dfrac{\alpha'\beta'\gamma'}{1\cdot\delta'\,\epsilon'}x + \dots$ hat, J. für Math. **3**, 89-91.

Erdélyi, A., ed. (1953). Higher Transcendental Functions, Vol. II, McGraw-Hill, New York.

Euler, L. (1748). Introductio in Analysis Infinitorum, Marcum-Michaelem Bousquet, Lausanne.

Gasper, G. (1985). Rogers' linearization formula for the continuous q-ultraspherical polynomials and quadratic transformation formulas, SIAM J. Math. Anal. **16**, 1061- 1071.

Gasper, G. and Rahman, M. (1986). Positivity of the Poisson kernel for the continuous q-Jacobi polynomials and some quadratic transformation formulas for basic hypergeometric series, SIAM J. Math. Anal. **17**, 970-999.

Gasper, G. and Rahman, M. (1989a). Basic Hypergeometric Series, Cambridge University Press, to appear.

Gasper, G. and Rahman, M. (1989b). A nonterminating q-Clausen formula and some related product formulas, SIAM J. Math. Anal., to appear.

Gauss, C.F. (1813). Disquisitiones generales circa seriem infinitam ..., Comm. soc. reg. sci. Gott. rec., Vol. II; reprinted in Werke **3** (1976), pp. 123-162.

Gessel, I. and Stanton, D. (1986). Another family of q-Lagrange inversion formulas, Rocky Mountain J. Math. **16**, 373-384.

Hahn, W. (1949). Über orthogonal polynome, die Differenzengleichungen genügen, Math. Nachr. **2**, 4-34.

Heine, E. (1846). Über die Reiche ..., J. reine angew. Math. **32**, 210-212.

Heine, E. (1847). Untersuchungen über die Reiche ..., J. reine angew. Math. **34**, 285-328.

Heine, E. (1878). Handbuch der Kugelfunctionen, Theorie und Anwendungen, Vol. 1, Reimer, Berlin.

Ismail, M.E.-H. (1977). A simple proof of Ramanujan's $_1\psi_1$ sum, Proc. Amer. Math. Soc. **63**, 185-186.

Ismail, M.E.-H. and Stanton, D. (1988). On the Askey-Wilson and Rogers polynomials, Can. J. Math., **40**, 1025-1045.

Jackson, F.H. (1904). A generalization of the functions $\Gamma(n)$ and x^n, Proc. Roy. Soc. London, **74**, 64-72.

Jackson, F.H. (1910a). On q-definite integrals, Quart. J. Pure and Appl. Math. **41**, 193-203.

Jackson, F.H. (1910b). Transformations of q-series, Messenger of Math. **39**, 145-153.

Jackson, F.H. (1921). Summation of q-hypergeometric series, Messenger of Math. **50**, 101-112.

Jacobi, C.G.J. (1846). Über einige der Binomialreihe analoge Reihen, J. reine angew. Math. **32**, 197-204; reprinted in Gessamette Werke **6** (1881), 163-173, Reimer, Berlin.

Kadell, K.W.J. (1987). A probabilistic proof of Ramanujan's $_1\psi_1$ sum, SIAM J. Math. Anal. **18**, 1539-1548.

Koornwinder, T.H. (1989). Jacobi functions as limit cases of q-ultraspherical polynomials, J. Math. Anal. Appl., to appear.

Moak, D.S. (1980a). The q-gamma function for $q > 1$, Aequationes Math. **20**, 278-285.

Moak, D.S. (1980b). The q-gamma function for $x < 0$, Aequationes Math. **21**, 179-191.

Nassrallah, B. and Rahman, M. (1985). Projection formulas, a reproducing kernel and a generating function for q-Wilson polynomials, SIAM J. Math. Anal. **16**, 186-197.

Needham, J. (1959). Science and Civilization in China, Vol. 3, Mathematics and the Sciences of the Heavens and the Earth, Cambridge University Pres. Cambridge.

Pfaff, J.F. (1797). Observationes analyticae ad L. Euler Institutiones Calculi Integralis, Vol. IV, Supplem. II et IV, Historie de 1793, Nova acta acad. sci. Petropolitanae 11 (1797), pp. 38-57.

Rahman, M. (1981). The linearization of the product of continuous q-Jacobi polynomials, Can. J. Math. **33**, 255-284.

Rahman, M. (1982). Reproducing kernels and bilinear sums for q-Racah and q-Wilson polynomials, Trans. Amer. Math. Soc. **273**, 483-508.

Rahman, M. (1984). A simple evaluation of Askey and Wilson's q-beta integral, Proc. Amer. Math. Soc. **92**, 413-417.

Rahman, M. (1986). An integral representation of a $_{10}\phi_9$ and continuous biorthogonal $_{10}\phi_9$ rational functions, Can. J. Math. **38**, 605-618.

Rahman, M. (1989). Some quadratic and cubic summation formulas for basic hypergeometric series, to appear.

Rahman, M. and Verma, A. (1986). Product and addition formulas for the continuous q-ultraspherical polynomials, SIAM J. Math. Anal. **17**, 1461-1474.

Rogers, L.J. (1893). On the expansion of some infinite products, Proc. London Math. Soc. **24**, 337-352.

Rogers, L.J. (1894). Second memoir on the expansion of certain infinite products, Proc. London Math. Soc. **25**, 318-343.

Rogers, L.J. (1895). Third memoir on the expansion of certain infinite products, Proc. London Math. Soc. **26**, 15-32.

Saalschütz, L. (1890). Eine Summationformel, Zeitschr. Math. Phys. **35**, 186-188.

Sears, D.B. (1951a). On the transformaiton theory of basic hypergeometric functions, Proc. London Math. Soc. (2) **53**, 181-191.

Sears, D.B. (1951b). On the transformation theory of basic hypergeometric functions, Proc. London. Math. Soc. (2) **53**, 158-180.

Sears, D.B. (1951c). Transformations of basic hypergeometric functions of special type, Proc. London Math. Soc (2) **52**, 467-483.

Szegő, G. (1975). Orthogonal Polynomials, 4th edition, Amer. Math. Soc. Colloq. Publ. 23.

Takács, L. (1973). On an identity of Shih-Chieh Chu, Acta. Sci. Math. (Szeged) **34**, 383-391.

Thomae, J. (1869). Beiträge zur Theorie der durch die Heinesche Reiche ..., J. reine angew. math. **70**, 258-281.

Thomae, J. (1870). Les séries Heinéennes supériures, ou les séries de la forme ..., Annali di Mathematica Pura de Applicata, **4**, 105-138.

Vandermonde, A.T. (1772). Mémoire sur des irrationnales de différens ordres avec une application au cercle, Mem. Acad. Roy. Sci. Paris, 489-498.

Watson, G.N. (1924). A new proof of Rogers-Ramanujan identities, J. London Math. Soc. **4**, 4-9.

ORTHOGONAL MATRIX POLYNOMIALS

Leiba Rodman*
The College of William and Mary
Department of Mathematics
Williamsburg, Virginia 23187, U.S.A.

INTRODUCTION. An overview is given of some classical and recent results concerning zeros of orthogonal matrix polynomials on the unit circle. The basic questions are: How these zeros are located in the complex plane? Conversely, what conditions on the location of the zeros of a given matrix polynomial ensure that the polynomial is orthogonal?

For scalar orthogonal polynomials with respect to a positive weight function these are classical questions (especially the first one) and the answers are well-known [Go, A, S]. For scalar polynomials with respect to an indefinite weight function the answers were given by M. G. Krein [Kr]. Recently, there have been considerable interest in developing matrix analogues of these results, motivated primarily by a variety of applications and connections with other important problems in mathematics and engineering. Our goal here is to provide an introduction to the most basic results in this area. Because of space limitation, the applications are left out completely (the interested reader can acquaint himself with some of these in the review articles [D, Ka1, F, KVM, L]). Further development of the theory of orthogonal matrix polynomials can be found in [YK, DGK1, DGK2, Go] and very recently in [GL, AG, BG, FK].

We keep our exposition on an elementary level (thus, the weight functions are assumed to be continuous, instead of the usual setting in terms of integrals against a measure), and should be accessible to persons with solid background in undergraduate mathematics.

Notation: C is the field of complex numbers; upper bar denotes the complex conjugation. Positive semidefiniteness (resp. positive definiteness) of a square matrix X over C is indicated by $X \geq 0$ (resp. by $X > 0$). X^T is the transpose and $X^* = \overline{X}^T$ is the conjugate transpose of a matrix X. We use δ_{jk} as the Kronecker delta ($\delta_{jk} = 0$ if $j \neq k$, $\delta_{jk} = 1$ if $j = k$).

1. Zeros of Scalar Polynomials Orthogonal on the Unit Circle

The starting point of our exposition is a classical and well-known result concerning location of zeros of orthogonal polynomials with respect to the unit circle. We set up the framework for this result. Denote by C[z] the set of all polynomials in the complex variable z with complex coefficients . The set C[z] is an (infinite dimensional) vector space over C,

---

* This material is based upon research supported in part by the National Science Foundation under grant number DMS 88-02836.

345

P. Nevai (ed.), Orthogonal Polynomials, 345–362.

and the polynomials $s_n(z) = z^n$, $n = 0, 1, ...$ form a standard basis in $C[z]$. Let $f(\theta)$ be real valued continuous function of the real variable θ, $-\pi \leq \theta \leq \pi$. Assume that $f(\theta) \geq 0$ for all θ and that $f(\theta)$ is not identically zero. Given such function $f(\theta)$, we introduce the _inner product_ (also called dot product, or scalar product) in $C[z]$ by the formula

$$< p, q > = \frac{1}{2\pi} \int_{-\pi}^{\pi} p(e^{i\theta})\overline{q(e^{i\theta})}f(\theta)d\theta$$

(1.1)

where $p \in C[z]$, $q \in C[z]$. It is a genuine inner product, i.e. it is linear in the first variable, has the property that $< p, q > = \overline{< q, p >}$, and, moreover, it is _positive definite_: $< p, p > \geq 0$ for every $p \in C[z]$ with the equality $< p, p > = 0$ only when $p = 0$. The norm is defined as usual: $\|p\| = \sqrt{< p, p >}$. Using this inner product, we now apply the Gram-Schmidt orthogonalization to the standard basis $\{s_j\}_{j=0}^{\infty}$ and obtain as a result a sequence of _orthogonal polynomials_ $\{g_j\}_{j=0}^{\infty}$ with respect to f. So $< g_i, g_j > = 0$ for $i \neq j$ and g_i has degree $i (i = 0, 1, ...)$. We leave aside (for the time being) the issue of normalization of g_n, and just remark that each g_n is uniquely defined (given f) up to multiplication by a non-zero constant.

We now state and prove the result promised at the beginning of this section.

THEOREM 1.1. All the zeros of each g_n lie in the open unit disc $D = \{z \in C: |z| < 1\}$. Conversely, if g is a polynomial of degree n with all the zeros in D, then g is the nth orthogonal polynomial with respect to some continuous positive function $f(\theta)$ on $[-\pi, \pi]$ (in fact, $f(\theta)$) can be chosen so that $f(\theta)^{-1}$ is a linear combination (with constant coefficients) of the functions 1, $\cos\theta$, ... , $\cos(n\theta)$, $\sin\theta$, ... , $\sin(n\theta)$.)

Proof. Note the property of the inner product $< \cdot , \cdot >$:

$$< zp(z), zq(z) > = < p(z), q(z) > \text{ for every } p, q \in C[z].$$

(1.2)

Assume z_0 is a zero of $g_n(z)$, so $g_n(z) = (z - z_0) r(z)$ for some $r \in C[z]$ of degree $< n$. Then $g_n(z) + z_0 r(z) = zr(z)$, and since $g_n(z)$ is orthogonal to all polynomials of degree $< n$ we have

$$\|g_n(z)\|^2 + |z_0|^2 \|r(z)\|^2 = \|zr(z)\|^2 = \|r(z)\|^2,$$

where the last equality follows from (1.2). Now

$$1 - |z_0|^2 = \|r\|^{-2} \|g_n\|^2,$$

and $z_0 \in D$, as required.

Conversely, assume $g(z) = x_0 z^n + \dots + x_n$ is a polynomial of degree n (so $x_0 \neq 0$) with all zeros in D. Let

$$h(z) = \bar{x}_n z^n + \bar{x}_{n-1} z^{n-1} + \dots + \bar{x}_0,$$

i.e., $g(z) = z^n \bar{h}(z^{-1})$, where by $\bar{h}(z)$ we denote the polynomial whose coefficients are complex conjugate of the corresponding coefficients of h(z). All the zeros of h(z) are outside the closed unit disc. Now put $f(\theta) = |h(e^{i\theta})|^{-2}$, $-\pi \leq \theta \leq \pi$. For $k = 0, \dots, n-1$ we have

$$< g(z), z^k > = \frac{1}{2\pi} \int_{-\pi}^{\pi} g(e^{i\theta}) e^{-ik\theta} f(\theta) d\theta =$$

$$= \frac{1}{2\pi} \int_{-\pi}^{\pi} e^{in\theta} \overline{h}(e^{-i\theta}) e^{-ik\theta} (h(e^{i\theta}))^{-1} (\overline{h}(e^{-i\theta}))^{-1} d\theta =$$

$$= \frac{1}{2\pi} \int_{-\pi}^{\pi} e^{i(n-k)\theta} (h(e^{i\theta}))^{-1} d\theta = \frac{1}{2\pi i} \int_{\Gamma} z^{n-k-1} h(z)^{-1} dz,$$

where the latter is the contour integral over the unit circle Γ. By Cauchy's theorem,

$$\frac{1}{2\pi i} \int_{\Gamma} z^{n-k-1} h(z)^{-1} dz = 0,$$

and we are done. $\qquad\qquad\qquad\qquad\qquad\qquad\qquad\qquad\qquad\qquad\square$

The proof of Theorem 1.1 presented here is well-known in the literature (cf. Proposition 1 in [L], Theorem 11.2 in [S]).

2. Zeros of Scalar Polynomials Orthogonal on the Unit Circle (Indefinite Case)

As our next step we consider the situation when the real valued continuous function $f(\theta) \neq 0$, $-\pi \leq \theta \leq \pi$, is not necessarily nonnegative. Define the inner product in C[z] by the same formula (1.1); now this inner product is generally <u>indefinite</u> (i.e. there are polynomials p for which $< p, p > < 0$ as well as polynomials q for which $< q, q > > 0$) and <u>nondegenerate</u> (meaning that $< p, q > = 0$ for all $q \in$ C[z] happens only when $p = 0$).

It is evident that the proof of Theorem 1.1 cannot be extended to indefinite $f(\theta)$ (and actually the result itself is false in this situation). We need additional tools to study orthogonal scalar polynomials with respect to such $f(\theta)$.

Given the function $f(\theta)$ as above, let

$$c_j = \frac{1}{2\pi} \int_{-\pi}^{\pi} e^{-ij\theta} f(\theta) d\theta; \quad j = 0, \pm 1, \pm 2, \dots \qquad\qquad (2.1)$$

be the Fourier coefficients of $f(\theta)$. Then we can rewrite the formula (1.1) in the form

$$<p, q> = [p_0 \cdots p_n] \begin{bmatrix} c_0 & c_{-1} & \cdots & c_{-n} \\ c_{+1} & c_0 & \cdots & c_{-n+1} \\ \vdots & \vdots & & \vdots \\ c_{+n} & c_{n-1} & \cdots & c_0 \end{bmatrix} \begin{bmatrix} \bar{q}_0 \\ \bar{q}_1 \\ \vdots \\ \bar{q}_n \end{bmatrix} \tag{2.2}$$

where $p(z) = p_0 z^n + p_1 z^{n-1} + \ldots + p_n$; $q(z) = q_0 z^n + q_1 z^{n-1} + \ldots + q_n$ and n is the largest of the degrees of p and q. The Toeplitz matrices $T_n = [c_{j-k}]_{j,k=0}^n$ (n = 0, 1, ...) are Hermitian. (A square matrix X is called <u>Toeplitz</u> if its (i, j) entry depends only on the difference i - j.) The <u>orthogonal polynomials</u> $\{g_n(z)\}_{n=0}^\infty$ with respect to f are defined as in Section 1: $g_n(z)$ is a polynomial of degree n (with non-zero leading coefficient), $< g_n, f > = 0$ for all polynomials f of degree < n; and $< g_n, g_n > \neq 0$. In contrast with positive $f(\theta)$, here orthogonal polynomials need not exist (example: let $f(\theta) = \cos\theta$; then there is no constant g_0, i.e. polynomial of degree 0, for which $< g_0, g_0 > \neq 0$). If the nth orthogonal polynomial $g_n(z) = x_0 z^n + x_1 z^{n-1} + \ldots + x_n$ exists (we suppress here the dependence of the coefficients x_j on n) then, up to a constant multiple, it has the form

$$g_n(z) = \det \begin{bmatrix} c_0 & c_{-1} & \cdots & c_{-n+1} & c_{-n} \\ c_1 & c_0 & \cdots & c_{-n+2} & c_{-n+1} \\ \vdots & & \vdots & & \vdots \\ c_{n-1} & c_{n-2} & \cdots & c_0 & c_{-1} \\ 1 & z & \cdots & z^{n-1} & z^n \end{bmatrix}. \tag{2.3}$$

Indeed, using formula (2.2) we have for k = 0, ... , n - 1

$$< g_n(z), z^k > = [x_0 \, x_1 \, \ldots \, x_n] \, [c_{-n+k} \, c_{-n+k+1} \, \cdots \, c_k]^T$$

$$= \det \begin{bmatrix} c_0 & c_{-1} & \cdots & c_{-n+1} & c_{-n} \\ \vdots & & \vdots & & \vdots \\ c_{n-1} & c_{n-2} & \cdots & c_0 & c_{-1} \\ c_k & c_{k-1} & \cdots & c_{-n+k+1} & c_{-n+k} \end{bmatrix} = 0.$$

We can (and will) normalize g_n by requiring

$$[c_{j-k}]_{j,k=0}^n [x_0 x_1 \ldots x_n]^T = [1 \, 0 \ldots 0]^T. \tag{2.4}$$

Observe that x_0 is a non-zero real number. There is a complete description of zeros of orthogonal polynomials, as well as a characterization of such polynomials (for indefinite $f(\theta)$):

THEOREM 2.1.(a). <u>Given a real-valued not identically zero continuous function</u> $f(\theta)$, $-\pi \le \theta \le \pi$. <u>Let</u> $n \ge 1$, <u>and assume that the Hermitian Toeplitz matrices</u>

$$T_m = \left[c_{j-k}\right]_{j,k=0}^{m} \quad (m = 0, 1, \ldots, n)$$

<u>are nonsingular. Further, let</u> β_n <u>(respectively,</u> γ_n) <u>denote the number of permanences (respectively, changes) of sign in the sequence</u> $1, \det T_0, \ldots, \det T_{n-1}$. <u>If</u> $\det T_n \det T_{n-1} > 0$, <u>then the</u> n^{th} <u>orthogonal polynomial</u> g_n <u>with respect to f exists and has exactly</u> β_n <u>zeros counting with multiplicities in the open unit disc D. If</u> $\det T_n \det T_{n-1} < 0$, <u>then</u> g_n <u>exists and has exactly</u> γ_n <u>zeros counting with multiplicities in D.</u>

(b). <u>A polynomial</u> $g(z) = x_0 z^n + \ldots + x_n$ <u>with</u> $x_0 \ne 0$ <u>is the</u> n^{th} <u>orthogonal polynomial with the normalization equality (2.4) with respect to some real valued not identically zero continuous function</u> $f(\theta)$, $-\pi \le \theta \le \pi$, <u>if and only if the following conditions are satisfied:</u>

(i) x_0 <u>is real;</u>

(ii)   <u>no pair of zeros of</u> $g(z)$ <u>are symmetric relative to the unit circle (i.e. if</u> $g(z_0) = 0$, z_0

$\ne 0$ <u>then</u> $g(\bar{z}_0^{-1}) \ne 0$); <u>in particular,</u> $g(z)$ <u>has no zeros on the unit circle.</u>

This result is due to Krein [Kr]. The original proof was based on ideas and results from the theory of linear transformations in finite dimensional vector spaces with an indefinite inner product. A different proof is found in [EGL]. Still another proof of part (a) (based on Schur parametrization) is given in [DG].

Here we prove only the following lemma which is crucial for the proof of Theorem 2.1 and is also independently interesting.

LEMMA 2.2. <u>Let</u> $g(z) = x_0 z^n + \ldots + x_n$ <u>be a polynomial with real non-zero leading coefficient</u> x_0 <u>and assume that no pair of zeros of</u> $g(z)$ <u>is symmetric relative to the unit circle. Then there exists unique Hermitian Toeplitz matrix</u> $T = [c_{i-j}]_{i,j=0}^{n}$ <u>such that the equality (2.4) holds. Moreover, the matrix T is invertible.</u>

<u>Proof.</u> Introduce the polynomial $\tilde{g}(z) = \bar{x}_n z^n + \ldots + \bar{x}_0$. Without loss of generality we can assume $x_n \ne 0$ (otherwise apply Lemma 2.2 for the polynomial $z^{-1}g(z)$; then c_n is uniquely defined by $c_n x_0 + c_{n-1}x_1 + \ldots + c_1 x_{n-1} = 0$). Consider the system of $2n + 2$ equations with $2n + 1$ unknowns $c_{-n}, \ldots, c_0, \ldots, c_n$:

$$\sum_{j=0}^{n} c_{i-j}\bar{x}_{n-j} = \delta_{in}\,(i = 0,\ldots,n);\ \sum_{j=0}^{n} c_{i-j}x_j = \delta_{io}\,(i = 0,\ldots,n).$$

(2.5)

The coefficient matrix of this system is the $(2n + 2) \times (2n + 1)$ matrix M given by

$$M = \begin{bmatrix} \bar{x}_0 & \cdots & \bar{x}_{n-1} & \bar{x}_n & \cdots & 0 & 0 \\ & \ddots & \vdots & \vdots & \ddots & & \vdots \\ 0 & & \bar{x}_0 & \bar{x}_1 & \cdots & \bar{x}_n & 0 \\ 0 & \cdots & 0 & \bar{x}_0 & \cdots & \bar{x}_{n-1} & \bar{x}_n \\ x_n & \cdots & x_1 & x_0 & & 0 & 0 \\ & \ddots & \vdots & \vdots & \ddots & & \vdots \\ 0 & & x_n & x_{n-1} & & x_0 & 0 \\ 0 & \cdots & 0 & x_n & \cdots & x_1 & x_0 \end{bmatrix}$$

If the $(n + 1)$st row of M is deleted and the determinant of the resulting $(2n + 1) \times (2n + 1)$ matrix S is expanded by the cofactors of the last column, the result is $x_0 \det R(g, \tilde{g})$, where $R(g, \tilde{g})$ is the resultant (or Sylvester) matrix of g and $\tilde{g}$ (see e.g., [U] for the basic properties of the resultant matrices). As no pair of the zeros of $g(z)$ are symmetric with respect to the unit circle, the polynomials g and $\tilde{g}$ have no common zeros, and therefore $\det R(g, \tilde{g}) \neq 0$. It follows that rank $S = $ rank $M = 2n + 1$. Let Q be the $(2n + 2) \times (2n + 2)$ matrix [M q], where the column q represents the right-hand side of (2.5), i.e. has zero in all its entries except the $(n + 1)^{th}$ and the $(n + 2)^{th}$ entry which are 1. Using the expansion of $\det Q$ by the cofactors of its last column, we obtain

$$\det Q = (-1)^{n+1} x_0 \det R(g, \tilde{g}) + (-1)^{n+2} \bar{x}_0 \det R(g, \tilde{g})$$

$$= (-1)^n (\bar{x}_0 - x_0) \det R(g, \tilde{g}) = 0.$$

Thus, rank $Q = $ rank $M = 2n + 1$. We conclude that the system of equations (2.5) has unique solution $(c_{-n}, \ldots, c_0, \ldots, c_n)$. As $(\bar{c}_n, \ldots, \bar{c}_0, \ldots, \bar{c}_{-n})$ is also a solution, it follows that $c_j = \bar{c}_{-j}\,(j = 0,\ldots, n)$. We have proved that there exists unique Hermitian Toeplitz matrix T satisfying (2.4).

Next, we show that T is nonsingular. Observe that (2.4) implies

$$T[\bar{x}_n\,\bar{x}_{n-1}\,\ldots\,\bar{x}_0]^T = [0\,\ldots\,0\,1]^T.$$

(2.6)

Arguing by contradiction assume that T is singular. So there is $y = [y_0, \ldots, y_n]$, $y \neq 0$ such that $[y_0, y_1, \ldots, y_n]\,T = 0$, i.e. y is orthogonal to the columns of T. Because of (2.4) and (2.6) we have $y_0 = y_n = 0$. Let p be the largest index such that $y_p \neq 0$. Consider the vector

$w = [0, y_0, y_1, ..., y_{n-1}]$. Since T is Toeplitz and $y_n = 0$, w is orthogonal to each column of T except possibly the first column t_0 of T. But (2.4) (taking into account that $x_0 \neq 0$) shows that t_0 is a linear combination of the other columns of T and the vector $[1\ 0\ ...\ 0]^T$. It follows that $wT = 0$. Repeating this argument (with y replaced by w) several times we finally obtain the equality $[0\ 0\ ...\ y_0\ ...\ y_p]T = 0$, a contradiction with (2.6). $\quad\square$

The proof of Lemma 2.2 is based on the exposition in [EGL]. Formulas for T^{-1} are provided in [EGL] as well.

Using Theorem 1.1, we can relate the number of positive (or negative) eigenvalues of the Toeplitz matrix T to the number of zeros of g inside the unit circle.

PROPOSITION 2.3. Assume the hypotheses of Lemma 2.2, and let α be the number of zeros (counted with multiplicites) of $g(z)$ inside the unit circle. Then the number π_+ of positive eigenvalues (counted with multiplicities) of the invertible Hermitian Toeplitz matrix $T = [c_{i-j}]_{i,j=0}^n$ which satisfies (2.4) is given as follows:

(i) $\pi_+ = \alpha + 1$, if the leading coefficient x_0 of $g(z)$ is positive;

(ii) $\pi_+ = n - \alpha$, if $x_0 < 0$.

Proof. Assume first that the matrices $T_m = [c_{i-j}]_{i,j=0}^m$ (m = 0, ... , n - 1) are invertible. We have by Cramer's formulas $x_0 = (\det T_{n-1})(\det T)^{-1}$. Using Jacobi theorem (see, e.g., Section 8.6 in [LT]), in the notation of Theorem 1.1, we obtain that T has $\beta_n + 1$ positive eingenvalues if $x_0 > 0$ and T has $n - \gamma_n$ positive eigenvalues if $x_0 < 0$. Now application of Theorem 1.1(a) gives the desired result. If not all matrices $T_0, ... , T_{n-1}$ are invertible, argue as follows. It is easy to see that the unique solution T of (2.4) is a continuous function of $(x_0, ... , x_n) \in W$, where the set W is defined by $W = \{(y_0, ..., y_n)$

where $y_0 > 0$; $y_j \in C$ for $j = 1, ... , n$ and the polynomial $\sum_{j=0}^{n} y_{n-j} z^j$ has no pair of zeros symmetric relative to the unit circle}.

Write $g(z) = x_0 z^n + ... + x_n$, and let W_0 be the connected component of W that contains $(x_0, ... , x_n)$. As W_0 is a connected set, the number of positive eigenvalues of T is the same for all $(y_0, ..., y_n) \in W$. It remains to observe that there is $(y_0, ... , y_n) \in W_0$ for which all the corresponding matrices T_m (m = 0, ..., n - 1) are invertible. This fact can be proved, for example, by noting that locally the map $(y_0, ..., y_n) \rightarrow T$ is one-to-one and onto. The case when $x_0 < 0$ is dealt with analogously. $\quad\square$

3.   Matrix Polynomials

We consider now the matrix valued orthogonal polynomials. Thus, let $C_{m \times m}[z]$ be the set of polynomials in the complex variable z whose coefficients are $m \times m$ matrices with complex entries (in short, <u>matrix polynomials</u>). The set $s_n(z) = z^n I$, $n = 0, 1, ...,$ will serve as basic polynomials. Introduce a continuous on $[-\pi, \pi]$ function $F(\theta)$ whose values are m × m Hermitian matrices (continuity of $F(\theta)$ is understood as entrywise continuity). It will be assumed that det $F(\theta)$ is not identically zero. Again, one could allow more general Hermitian valued functions $F(\theta)$ such as entrywise Lebesgue integrable but we do not need this generality here. Define the matricial <u>inner product</u> on $C_{m\times m}[z]$ as follows:

$$< P(z), Q(z) > = \frac{1}{2\pi} \int_{-\pi}^{\pi} P(e^{i\theta}) F(\theta) (Q(e^{i\theta}))^* d\theta \qquad (3.1)$$

where $P(z), Q(z) \in C_{m\times m}[z]$. The integral $\int_{-\pi}^{\pi} G(\theta) d\theta$ of a continuous m × m matrix valued function $G(\theta) = [g_{ij}(\theta)]_{i,j=1}^{m}$ is understood as the m × m matrix of integrals

$$\left[\int_{-\pi}^{\pi} g_{ij}(\theta) d\theta \right]_{i,j=1}^{m}.$$

The following properties of this matricial inner product are easily observed:

(i) $< C_1 P_1 + C_2 P_2, Q > = C_1 < P_1, Q > + C_2 < P_2, Q >$, where C_1, C_2 are constant m x m matrices and $P_1, P_2, Q \in C_{m\times m}[z]$;

(ii) $< P, Q > = (< Q, P >)^*$ for $P, Q \in C_{m\times m}[z]$;

The positive definiteness property:

(iii) $< P, P >$ is positive semidefinite for every $P \in C_{m\times m}[z]$, and $< P, P > = 0$ only when P is identically zero

is valid if and only if $F(\theta)$ is positive semidefinite for all $\theta \in [-\pi, \pi]$. (In this case the blanket assumption det $F(\theta) \neq 0$ guarantees positive definiteness of $F(\theta_0)$ for some θ_0).

As a side remark note that a theory of matrix integrals against a matrix valued measure was developed in [R].

As in the scalar case, we now apply (if possible) the Gram-Schmidt orthogonalization to $\{s_n(z)\}_{n=0}^{\infty}$ with respect to the matricial inner product (3.1). The result is a sequence of orthogonal matrix polynomials $\{G_i(z)\}_{i=0}^{\infty}$ with the following properties:

(α) $G_i(z)$ is a matrix polynomial of degree i with the leading coefficient a nonsingular matrix;

(β) $< G_i, G_j > = 0$ if $i \neq j$; $< G_i, G_i >$ is nonsingular.

Clearly, $G_i(z)$ is defined only up to multiplication on the left by an invertible matrix.

In the sequel we will consider each orthogonal matrix polynomial individually, without reference to other orthogonal matrix polynomials. Accordingly, we somewhat modify the definition and say that $G_n(z)$ is an nth <u>orthogonal matrix polynomial</u> if $G_n(z)$ is a matrix polynomial of degree n with nonsingular leading coefficient, $< G_n, F > = 0$ for every matrix polynomial of degree $< n$, and the matrix $< G_n, G_n >$ is nonsingular.

Introduce the matrix Fourier coefficients:

$$C_k = \frac{1}{2\pi} \int_{-\pi}^{\pi} e^{-ik\theta} F(\theta) d\theta; \; k = 0, \pm 1, \pm 2, \ldots$$

Furthermore, write the nth orthogonal matrix polynomial $G_n(z)$ in the form

$$G_n(z) = X_0 z^n + X_1 z^{n-1} + \ldots + X_n, \tag{3.2}$$

where X_j are $m \times m$ matrices (which depend on n, of course). Then we have

$$\sum_{m=0}^{n} C_{p-m} X_{n-m} = \frac{1}{2\pi} \int_{-\pi}^{\pi} G_n(e^{i\theta}) F(\theta) \overline{(e^{-ip\theta} I)} d\theta = 0 \tag{3.3}$$

for $p = 0, \ldots, n - 1$, and

$$\sum_{m=0}^{n} C_{n-m} X_m = \left(\frac{1}{2\pi} \int_{-\pi}^{\pi} G_n(e^{i\theta}) F(\theta) (G_n(e^{i\theta}))^* d\theta\right) X_0^{*-1}$$

$$= < G_n, G_n > X_0^{*-1}. \tag{3.4}$$

We normalize $G_n(z)$ so that $< G_n, G_n > X_0^{*-1} = I$ (by replacing $G_n(z)$ with $SG_n(z)$ where $S = (< G_n, G_n > X_0^{*-1})^{-1})$. The equalities (3.3), (3.4) take now the form

$$\begin{bmatrix} C_0 & C_{-1} & \cdots & C_{-n} \\ C_1 & C_0 & \cdots & C_{-n+1} \\ \vdots & \vdots & & \vdots \\ C_n & C_{n-1} & \cdots & C_0 \end{bmatrix} \begin{bmatrix} X_0 \\ X_1 \\ \vdots \\ X_n \end{bmatrix} = \begin{bmatrix} I \\ 0 \\ \vdots \\ 0 \end{bmatrix}, \tag{3.5}$$

precisely the matrix analogue of (2.4). Thus existence of the nth orthogonal matrix polynomial is equivalent to the solvability of (3.5) with invertible matrix X_0. Observe also that the block Toeplitz matrix $[C_{i-j}]_{i,j=0}^{n}$ which appears in the left-hand side of (3.5) is Hermitian. Our attention will be focussed on the inverse problem: Given matrix polynomial

$$G_n(z) = \sum_{j=0}^{n} X_{n-j} z^j$$ with invertible X_0, when there exists a continuous Hermitian valued

function $F(\theta)$, $-\pi \leq \theta \leq \pi$ with det $F(\theta) \neq 0$ such that G_n is the nth orthogonal polynomial with respect to $F(\theta)$? In other words, given $m \times m$ matrices $X_0, X_1, ..., X_n$ with invertible X_0, when there is a Hermitian block Toeplitz matrix $[C_{i-j}]_{i,j=0}^{n}$ satisfying (3.5)?

We shall assume in the sequel that the matrix $[C_{i-j}]_{i,j=0}^{n}$ is invertible. Also observe that, by using the polar decomposition of X_0, the invertible matrix X_0 can be assumed positive definite without loss of generality.

As should be expected by analogy with the scalar case, the solution to this problem should involve zeros of certain matrix polynomials. This is indeed the case; however, we postpone this formulation to the next section. Here the main result is stated in a less technical (and also less insightful) way.

We need the notion of coprimeness for matrix polynomials. Two matrix polynomials P, Q $\in C_{m \times m}[z]$ with not identically zero determinants are called right coprime if the simultaneous equalities

$$P(z_0)x_0 = 0; \quad Q(z_0)x_0 = 0; \quad z_0 \in C; \quad x_0 \in C^m \tag{3.6}$$

are possible only when $x_0 = 0$. The definition of left coprimeness is obtained if the first two equalities in (3.6) are replaced by $x_0^T P(z_0) = x_0^T Q(z_0) = 0$. A sufficient (but not necessary) condition for left and right coprimeness of P and Q is that det $P(z)$ and det $Q(z)$ are coprime as scalar polynomials. Note also that matrix polynomials can be right coprime without being left coprime, and vice versa.

Given a sequence $\Omega = (X_0, ..., X_n)$ of $m \times m$ complex matrices with $X_0 > 0$, we introduce the associated matrix polynomial

$$A_\Omega(z) = X_0^{-\frac{1}{2}} \sum_{j=0}^{n} z^j X_{n-j}$$

and the matrix function

$$F_\Omega(z) = (A_\Omega(\bar{z}^{-1}))^* A_\Omega(z).$$

Observe that the matrix function $F_\Omega(z)$ can be represented in the form

$$F_\Omega(z) = \sum_{j=-n}^{n} z^j F_j$$

for certain $m \times m$ matrices F_j, and that $F_\Omega(z) \geq 0$ for every $|z| = 1$. Also, the leading coefficient of $A_\Omega(z)$ is invertible, and therefore det $A_\Omega(z)$ is a polynomial of degree mn (in particular, det $A_\Omega(z) \not\equiv 0$). Consequently, det $F_\Omega(z) \not\equiv 0$, and $F(z)$ is positive definite on the unit circle with the possible exception of at most mn points.

THEOREM 3.1 ([GL]) Let $\Omega = (X_0, ..., X_n)$ be a sequence of $m \times m$ matrices with $X_0 > 0$. Then there exists invertible Hermitian block Toeplitz matrix $T = [C_{i-j}]_{i,j=0}^{n}$ satisfying (3.5) if and only if $F_\Omega(z)$ admits factorization of the form

$$F_\Omega(z) = B(z)(B(\bar{z}^{-1}))^*, \qquad z \in \mathbb{C}\backslash\{0\} \tag{3.7}$$

where $B(z)$ is a matrix polynomial of degree n with positive definite leading coefficient such that $A_\Omega(z)$ and $z^n(B(\bar{z}^{-1}))^*$ are right coprime.

In the scalar case ($m = 1$) we recover Theorem 2.1(b). Indeed, in this case the polynomial $z^n(B(\bar{z}^{-1}))^*$ picks up a unimodular zero of $A_\Omega(z)$ (if any) and one zero from each pair of zeros of $A_\Omega(z)$ symmetric relative to the unit circle (if any). So $A_\Omega(z)$ and $z^n(B(\bar{z}^{-1}))^*$ can be coprime only if $A_\Omega(z)$ has no unimodular zeros and no pair of symmetric zeros. Conversely, for a scalar polynomial $A_\Omega(z)$ with these properties (3.7) holds with $B(z) = A_\Omega(z)$.

In the matrix case a sufficient condition for existence of invertible Hermitian block Toeplitz matrix satisfying (3.5) is that $\det A_\Omega(z) = 0$, $z_0 \neq 0$ implies $\det A_\Omega(\bar{z}_0^{-1}) \neq 0$. However, this condition is not necessary, as the following example (borrowed from ([GL]) shows. Let

$$X_0 = \begin{bmatrix} 2 & 0 \\ 0 & \frac{1}{2} \end{bmatrix}; X_1 = \begin{bmatrix} 1 & 0 \\ 0 & 1 \end{bmatrix}.$$

Then (3.5) is satisfied with

$$C_0 = \begin{bmatrix} \frac{2}{3} & 0 \\ 0 & -\frac{2}{3} \end{bmatrix}, C_1 = \begin{bmatrix} -\frac{1}{3} & 0 \\ 0 & \frac{4}{3} \end{bmatrix}.$$

On the other hand,

$$A_\Omega(z) = X_0^{-\frac{1}{2}}(zX_0 + X_1) = \sqrt{2}\begin{bmatrix} z + \frac{1}{2} & 0 \\ 0 & \frac{z}{2} + 1 \end{bmatrix},$$

and $\det A_\Omega(z)$ has a pair of symmetric zeros -2, -1/2.

4. Jordan Chains, Null Pairs and Factorization

In this section we discuss some of the ideas and techniques involved in the proof of Theorem 3.1, as well as restate this theorem in different terms. The reader is referred to the monographs [GLR1, GLR2, BGK, BGR, Ka2] for thorough expositions of this and related material, including many applications and extensions. Let

$$L(z) = \sum_{j=0}^{l} z^j A_j$$

be a matrix polynomial, so the coefficients A_j are $m \times m$ (complex) matrices and z is a complex variable. The point $z_0 \in C$ is called an <u>eigenvalue</u> of L(z) if det $L(z_0) = 0$. The set of all eigenvalues is called the <u>spectrum</u> of L(z) and is denoted $\sigma(L)$. The polynomial L(z) is said to be <u>regular</u> if $\sigma(L) \neq C$ (in the sequel we study only regular matrix polynomials). In this case $\sigma(L)$ is either finite (contains at most ml points) or empty. Observe that L(z) is regular provided the leading coefficient A_l is nonsingular, however, this condition is not necessary for the regularity of L(z).

The basic notion is that of a Jordan chain. A sequence of m-dimensional column vectors $\varphi_0, \varphi_1, \dots, \varphi_{k-1}$ is called a <u>right Jordan chain</u> of length k of a regular matrix polynomial L(z) corresponding to $z_0 \in \sigma(L)$ if $\varphi_0 \neq 0$ and the equalities

$$\sum_{p=0}^{i} \frac{1}{p!} L^{(p)}(z_0)\varphi_{i-p} = 0$$

(4.1)

hold true for $i = 0, 1, \dots, k - 1$. Here $L^{(p)}(z)$ denotes the pth derivative of L(z) with respect to z. The initial vector φ_0 in a right Jordan chain of L(z) corresponding to z_0 is called right eigenvector of L(z) corresponding to the same z_0. A prototype case is when $L(z) = A - zI$; in the case equalities (4.1) amount to

$$A\varphi_0 = z_0\varphi_0; \quad (A - z_0 I)\varphi_i = \varphi_{i-1}, \quad i = 1, \dots, k - 1,$$

in other words, $\varphi_0, \varphi_1, \dots, \varphi_{k-1}$ is the Jordan chain of the matrix A with eigenvalue z_0.

A matrix polynomial can generally have many right Jordan chains of various lengths starting with different eigenvectors, all corresponding to the same z_0. We now introduce canonical sets of Jordan chains which serve in a sense as bases in the set of all Jordan chains. An ordered set

$$\Gamma = \{\varphi_0^{(i)}, \varphi_1^{(i)}, \dots, \varphi_{k_i-1}^{(i)}; i = 1, \dots, r\}$$

(4.2)

which consists of r right Jordan chains of L(z) corresponding to the same eigenvalue z_0 is called <u>canonical set</u> if the following properties are valid:

(i)   the right eigenvectors $\varphi_0^{(1)}, \dots, \varphi_0^{(r)}$ are linearly independent;

(ii)   the sum of lengths $k_1 + \dots + k_r$ of the chains in Γ is equal to the multiplicity $m(z_0)$ of z_0 as a zero of the scalar polynomial det L(z).

In connection with the property (ii) observe that for any ordered set of right Jordan chains $\{\psi_0^{(i)}, .., \psi_{p_i-1}^{(i)}; i = 1, .., s\}$ of $L(z)$ corresponding to z_0 with linearly independent $\psi_0^{(1)}, .., \psi_0^{(s)}$ the inequality

$$p_1 + ... + p_s \leq m(z_0)$$

is valid. Thus, a canonical set has largest total length of Jordan chains subject to the linear independence of eigenvectors.

PROPOSITION 4.1. <u>There is a canonical set of right Jordan chains for every eigenvalue z_0 of a regular matrix polynomial $L(z)$. Moreover, the number of chains and their lengths is a canonical set depend only $L(z)$ and on z_0, and do not depend on the choice of a canonical set (corresponding to the same eigenvalue).</u>

In fact, the number of chains in a canonical set is just $\dim \operatorname{Ker} L(z_0)$.

Proposition 4.1 can be proved using the Smith form of matrix polynomials (see, e.g., [BGR, GLR1, GLR2]): every regular matrix polynomial $L(z)$ can be represented in the form

$$L(z) = E(z)\, D(z)\, F(z), \tag{4.3}$$

where $E(z)$ and $F(z)$ are matrix polynomials with constant non-zero determinants (and therefore the inverses $E(z)^{-1}$ and $F(z)^{-1}$ are matrix polynomials as well), and $D(z)$ is a diagonal matrix polynomial. One can further specialize (4.3) to a local Smith form for the eigenvalue z_0:

$$L(z) = E_0(z) \begin{bmatrix} (z-z_0)^{\alpha_1} & 0 & \cdots & 0 \\ 0 & (z-z_0)^{\alpha_2} & \cdots & 0 \\ \vdots & \vdots & & \vdots \\ 0 & 0 & \cdots & (z-z_0)^{\alpha_m} \end{bmatrix} F_0(z), \tag{4.4}$$

where $E_0(z)$ and $F_0(z)$ are rational matrix functions analytic and invertible at z_0, and $\alpha_1 \geq ... \geq \alpha_m \geq 0$ are integers. It turns out that the lengths of right Jordan chains in a canonical set corresponding to a z_0 coincide with the positive numbers among $\alpha_1, ... , \alpha_m$.

We organize a canonical set into a pair of matrices. Let (4.2) be a canonical set of right Jordan chains of $L(z)$ corresponding to z_0. Form two matrices:

$$X(z_0) = [\varphi_0^{(1)}\, \varphi_1^{(1)} ... \varphi_{k_1-1}^{(1)}\, \varphi_0^{(2)}\, \varphi_1^{(2)} ... \varphi_{k_2-1}^{(2)} ... \varphi_0^{(r)}\, \varphi_1^{(r)} ... \varphi_{k_r-1}^{(r)}];$$

$$K(z_0) = K_1 \oplus ... \oplus K_r,$$

where K_i is the (upper triangular) Jordan block of size $k_i \times k_i$ with eigenvalue z_0 (on the main diagonal). Thus, the size of $X(z_0)$ is $m \times m(z_0)$, the size of $K(z_0$ is $m(z_0) \times m(z_0)$. Follow this procedure for every eigenvalue z_0 of $L(z)$, choosing a canonical set of right Jordan chains and constructing the corresponding matrices $X(z_0)$, $K(z_0)$. Finally, put everything together by introducing matrices

$$X = [X(z_1) \; X(z_2) \; ... \; X(z_s)]; \tag{4.5}$$

$$K = K(z_1) \oplus K(z_2) \oplus ... \oplus K(z_s). \tag{4.6}$$

Here $z_1, ... , z_s$ are all the distinct eigenvalues of $L(z)$. The pair (X, K) will be called a right spectral pair of $L(z)$; the size of X is $m \times p$, the size of K is $p \times p$, where p is the degree of the scalar polynomial det $L(z)$.

It is clear from the construction that a right spectral pair is highly non-unique. It depends not only on the choice of a canonical set of right Jordan chains for each eigenvalue, but also on the order of the eigenvalues $z_1, ... , z_s$ that appear in (4.5) and (4.6). Remarkably, it turns out that every two right spectral pairs of $L(z)$ are similar, i.e., if (X_1, K_1) and (X_2, K_2) are right spectral pairs of $L(z)$, then there exists an invertible matrix S such that

$$X_1 = X_2 S, \; K_1 = S^{-1}K_2 S.$$

Without going into details about the proof of this statement, we observe that firstly, the proof is reduced to the case of one eigenvalue (by using the property that the equality $XS = SY$ for square matrices X and Y with no common eigenvalues is possible only when $S = 0$), and secondly, using the local Smith form (4.4) the proof is reduced to the case when $L(z)$ is a diagonal matrix diag$((z - z_0)^{\alpha_1}, (z - z_0)^{\alpha_2}, ... (z - z_0)^{\alpha_n})$. See, e.g., [BGR] for full details.

Because of this property it is natural to extend the definition of a right spectral pair. We say that a pair of matrices (X_0, T_0) is a right spectral pair of $L(z)$ if (X_0, T_0) is similar to a pair of matrices (X, K) obtained as in (4.5), (4.6), i.e.

$$X_0 = XS, \; T_0 = S^{-1}KS$$

for some invertible matrix S. From this definition we see immediately that necessary properties of a right spectral pair (X, T) are that X is $m \times p$, T is $p \times p$ (where p is the degree of det $L(z)$) and that the eigenvalues of T as a matrix coincide with the eigenvalues of $L(z)$ as a matrix polynomial. Deeper properties of right spectral pairs are summarized below.

THEOREM 4.2. Let (X, T) be a right spectral pair of a (regular) $m \times m$ matrix polynomial

$$L(z) = \sum_{j=0}^{l} z^j A_j.$$

The following properties are valid:

(i) $\quad \sum\limits_{j=0}^{l} A_j XT^j = 0;$

(ii) $\quad \bigcap\limits_{j=0}^{\infty} \text{Ker}\,(XT^j) = \{0\};$

(iii) if (X_0, T_0) is any pair of matrices of sizes $m \times q$, $q \times q$ for some q with the properties

$$\sum\limits_{j=0}^{l} A_j X_0 T_0^j = 0, \quad \bigcap\limits_{j=0}^{\infty} \text{Ker}\,(X_0 T_0)^j = \{0\},$$

then (X_0, T_0) is a restriction of (X, T), i.e.,

$$X_0 = XS, \quad ST_0 = TS \tag{4.7}$$

for some left invertible (generally rectangular) matrix S.

A completely analogous construction holds for left Jordan chains. Thus, a sequence of m-dimensional row vectors $\psi_0, \dots, \psi_{k-1}$ is called a left Jordan chain of length k of a regular matrix polynomial $L(z)$ corresponding to $z_0 \in \sigma(L)$ if $\psi_0 \neq 0$ and the equalities

$$\sum\limits_{p=0}^{i} \frac{1}{p!} \psi_{i-p} L^{(p)}(z_0) = 0$$

hold for $i = 0, \dots, k - 1$. Choose a canonical set of left Jordan chains corresponding to z_0

$$\{\psi_0^{(i)}, \psi_1^{(i)}, \dots, \psi_{k_i-1}^{(i)}; i = 1, \dots, r\}$$

(this notion is defined analogously to the canonical set of right Jordan chains), and form two matrices

$$J(z_0) = \begin{bmatrix} J_1^T & 0 & \dots & 0 \\ 0 & J_2^T & \dots & 0 \\ \vdots & \vdots & \ddots & \vdots \\ 0 & 0 & \dots & J_r^T \end{bmatrix}; \quad Y(z_0) = \begin{bmatrix} \psi_0^{(1)} \\ \psi_1^{(1)} \\ \vdots \\ \psi_{k_1-1}^{(1)} \\ \vdots \\ \psi_0^{(r)} \\ \vdots \\ \psi_{k_r-1}^{(r)} \end{bmatrix},$$

where J_i^T is the <u>transpose</u> of the Jordan block of size $k_i \times k_i$ with the eigenvalue z_0. The sizes of $Y(z_0)$ and $J(z_0)$ are $m(z_0) \times m$ and $m(z_0) \times m(z_0)$, respectively, where $m(z_0)$ is the multiplicity of z_0 as a zero of $\det L(z)$. Putting together the matrices $J(z_0)$ and $Y(z_0)$ for all distinct eigenvalues $z_1, \dots, z_s$ of $L(z)$ we obtain the <u>left spectral pair</u> (J, Y) of $L(z)$:

$$
J = \begin{bmatrix} J(z_1) & 0 & \cdots & 0 \\ 0 & J(z_2) & \cdots & 0 \\ \vdots & \vdots & \ddots & \vdots \\ 0 & 0 & \cdots & J(z_s) \end{bmatrix}; \quad Y = \begin{bmatrix} Y(z_1) \\ Y(z_2) \\ \vdots \\ Y(z_s) \end{bmatrix} \tag{4.8}
$$

Finally, we declare that a pair of matrices (T_0, Y_0) is a <u>left spectral pair</u> of $L(z)$ if it is similar to a pair obtained as in (4.8): $J = S^{-1} T_0 S$, $Y = S^{-1} Y_0$ for some invertible S. Many properties of left spectral pairs follow from the corresponding properties of right spectral pairs by using the observation that (T_0, Y_0) is a left spectral pair of $L(z)$ if and only if the pair of transposed matrices (Y_0^T, T_0^T) is a right spectral pair of the matrix polynomial with transposed coefficients $(L(z))^T$.

The importance of spectral pairs stems from the following description of divisibility of matrix polynomials which we state for right spectral pairs only.

THEOREM 4.3. <u>Let</u> $L(z)$ <u>and</u> $M(z)$ <u>be regular</u> $m \times m$ <u>matrix polynomials with right spectral pairs</u> (X_L, T_L) <u>and</u> (X_M, T_M), <u>respectively. Then</u> $M(z)$ <u>is a right divisor of</u> $L(z)$, <u>i.e.</u>

$$L(z) = Q(z) M(z)$$

<u>for some matrix polynomial</u> $Q(z)$, <u>if and only if</u> (X_M, T_M) <u>is a restriction of</u> (X_L, T_L) <u>i.e.,</u>

$$X_M = X_L S, \quad ST_M = T_L S. \tag{4.9}$$

<u>for some left invertible matrix</u> S.

A thorough treatment of this and related results, and also of common multiples and common divisors of matrix polynomials in terms of their spectral pairs is found in [GLR]. In particular, the coprimeness of matrix polynomials (introduced in Section 3) can be expressed and studied via the spectral pairs.

We note another useful interpretation of Theorem 4.3. If (4.9) holds then the range of S is a subspace invariant for T_L: $T_L x \in \text{Range}(S)$ for every $x \in \text{Range}(S)$; and moreover, the restriction of T_L to Range(S) is similar to T_M. Thus, keeping $L(z)$ fixed, one can describe all right divisors $M(z)$ of $L(z)$ in terms of invariant subspaces of T_L.

We now come back to the main result (Theorem 3.1) of the preceding section, and restate it in terms of the Jordan chains:

THEOREM 4.4. (Theorem 5.1 in [GL]). <u>In the notation of Theorem 3.1, the function</u> $F_\Omega(z)$ <u>admits factorization (3.7) if and only if for every symmetric pair of</u>

eigenvalues $z_0 \bar{z}_0^{-1}$ of $A_\Omega(z)$ (if any) and any left Jordan chains $\varphi_0, \ldots, \varphi_\alpha$ and $\psi_0, \psi_1,$

$\ldots, \psi_\beta$ of $A_\Omega(z)$ corresponding to z_0 and $\bar{z}_0^{-1}$ respectively the equations

$$\sum_{i=0}^{k} \varphi_i \psi_{k-i}^* = 0$$

are valid for $k = 0, \ldots, \min(\alpha, \beta)$.

REFERENCES

[A]   N. I. Akhiezer, *The Classical Moment Problem*, Hafner Publishing Company, New York, 1965.

[AG]   D. Alpay, I. Gohberg, On orthogonal matrix polynomials, Operator Theory: Advances and Applications 34(1988), 25-46.

[BGR]   J.A. Ball, I. Gohberg, and L. Rodman, *Interpolation Problems for Rational Matrix Functions*, monograph to be published by Birkhäuser.

[BGK]   H. Bart, I. Gohberg and M.A. Kaashoek, *Minimal Factorizations of Matrix and Operator Functions*, Birkhäuser, Basel, 1979.

[BG]   A. Ben-Artzi and I. Gohberg, Extension of a theorem of M. G. Krein on orthogonal polynomials for the nonstationary case, Operator Theory: Advances and Applications 34(1988), 65-78.

[DG]   P. Delsarte and Y. Genin, Spectral Properties of Finite Toeplitz Matrices, Mathematical Theory of Networks and Systems (Paul A. Fuhrmann, ed.), Lecture Notes in Control and Information Sciences, No. 58, Springer-Verlag, New York (1984), 194-213.

[DGK1]   Ph. Delsarte, Y. Genin and Y. Kamp, Orthogonal polynomial matrices on the unit circle, IEEE Trans. Circuits & Systems 25(1978), 145-160.

[DGK2]   Ph. Delsarte, Y. Genin and Y. Kamp, Schur parametrization of positive definite block-Toeplitz systems, SIAM J. Applied Math. 36(1979), 34-46.

[D]   H. Dym, Hermitian Block Toeplitz matrices, orthogonal polynomials, reproducing kernel Pontryagin spaces, interpolation and extension, Operator Theory: Advances and Applications 34(1988), 79-135.

[EGL]   R. L. Ellis, I. Gohberg and D. C. Lay,. On two theorems of M. G. Krein concerning polynomials orthogonal on the unit circle, Integral Equations and Operator Theory 11(1988), 87-104.

[FK]   B. Fritzsche and B. Kirstein, An extension problem for non-negative Hermitian block Toeplitz matrices III, Math Nachr. 135(1988), 319-341.

[F]   P. A. Fuhrmann, Orthogonal matrix polynomials and system theory, Preprint, 1986.

[Go]   J. S. Geronimo, Matrix orthogonal polynomials on the unit circle, J. Math. Phys. 22(7) (1981), 1359-1365.

[Gs]   L. Ya. Geronimus, *Orthogonal Polynomials*, New York, Consultants Bureau, 1961.

[GLR1] I. Gohberg, P. Lancaster and L. Rodman, *Matrix Polynomials*, Academic Press, 1982.

[GLR2] I. Gohberg, P. Lancaster and L. Rodman, *Invariant Subspaces of Matrices with Applications*, J. Wiley, New York, 1986.

[GL]   I. Gohberg and L. Lerer, Matrix generalizations of M. G. Krein theorems on orthogonal polynomials, Operator Theory: Advances and Applications 34(1988).

[Ka1]   T. Kailath, A view of three decades of linear filtering theory, IEEE trans. Information Theory 20(1974), 145-181.

[Ka2]   T. Kailath, *Linear Systems*, Prentice-Hall, Englewood Cliffs, NJ, 1980.

[KVM]   T. Kailath, A. Vieira and M. Morf, Inverses of Toeplitz operators, innovations, and orthogonal polynomials. SIAM Review 20(1978), 106-119.

[Kr]   M. G. Krein, On the Distribution of the Roots of Polynomials which are Orthogonal on the Unit Circle with Respect to an Alternating Weight, Teor.

        Funkciĭ Funkcional Anal. i Priložen. Resp. Sb. Nr. 2(1966), 131-137 [Russian].

[L]   H. J. Landau, Maximum entropy and the moment problem, Bull. Amer. Math. Soc. 16(1987), 47-77.

[R]   M. Rosenberg, The square-integrability of matrix-valued functions with respect to a non-negative Hermitian measure, Duke Math. J. 31(1964), 291-298.

[S]   G. Szegö, *Orthogonal Polynomial,* Colloquium Publications, No. 23, Amer. Math. Soc., Providence, RI 2nd ed. 1958, 3rd ed. 1967.

[U]   J. V. Uspensky, *Theory of Equations*, McGraw-Hill, New York, 1978.

[YK]   D. C. Youla and N. Kazanjian,. Bauer-type factorization of positive matrices and the theory of matrix polynomials orthogonal on the unit circle, IEEE Trans. on Circuits and Systems, CAS-25(1978), 57-69.

ORTHOGONAL POLYNOMIALS FROM A COMPLEX PERSPECTIVE

E.B. Saff[1]
Institute for Constructive Mathematics
Department of Mathematics
University of South Florida
Tampa, Florida 33620

ABSTRACT. Complex function theory and its close companion - potential theory - provide a wealth of tools for analyzing orthogonal polynomials and orthogonal expansions. This paper is designed to show how the complex perspective leads to insights on the behavior of orthogonal polynomials. In particular, we discuss the location of zeros and the growth of orthogonal polynomials in the complex plane. For some of the basic results we provide proofs that are not typically found in the standard literature on orthogonal polynomials.

1 Introduction

That the theory of complex variables can provide deeper understanding and useful techniques for analyzing real-variable problems should not be surprising to the reader. The computation of (real) integrals via Cauchy's Residue theorem and the analysis of power series are but two instances where the broader view from the complex plane **C** is an invaluable aid. Potential theory in the plane, which is a blend of real and complex analysis, provides an even greater resource for attacking "real problems"; particularly the behavior of orthogonal polynomials and orthogonal expansions. For example, the analysis of polynomials orthogonal on the whole real line **R** with respect to an exponential weight took a quantum step forward when x was replaced by z and potential theoretic arguments were introduced (cf. [MS1],[LS],[R],[GR1]).

Our goal is to illustrate how complex and potential theoretic results can be used to analyze orthogonal polynomials and orthogonal expansions. We assume that the reader has little background in potential theory; consequently we introduce some basic facts from this subject as well as provide references for further study. In Sections 2 and 5 we discuss the location and asymptotic behavior of the zeros of orthogonal polynomials. Bounds for the modulus of these polynomials are considered in Sections 6 and 7.

---

[1]Research supported, in part, by the National Science Foundation under grant DMS-890-6815

363

P. Nevai (ed.), Orthogonal Polynomials, 363–393.
© *1990 by Kluwer Academic Publishers.*

Throughout, μ denotes a finite, positive measure on the Borel subsets of the complex plane **C**. We assume that

$$S = S(\mu) := \text{supp}(\mu)$$

is compact and contains infinitely many points.

The measure μ gives rise to the inner product

$$(f,g) := \int f(z)\overline{g(z)}d\mu \tag{1.1}$$

for functions $f, g \in L_2(\mu)$. Since S is infinite, the monomials $1, z, \ldots, z^n$ are linearly independent in $L_2(\mu)$ for every $n \geq 0$. Hence, by the Gram-Schmidt orthogonalization process, there exist unique polynomials

$$p_n(z) = p_n(z; \mu) = \gamma_n z^n + \cdots \in \mathcal{P}_n, \quad \gamma_n > 0, \tag{1.2}$$

satisfying

$$(p_m, p_n) = \delta_{m,n}, \quad m, n = 0, 1, \ldots,$$

where $\delta_{m,n} = 0$ if $m \neq n, \delta_{m,m} = 1$, and $\mathcal{P}_n$ denotes the collection of all polynomials (with complex coefficients) having degree at most n.

As we shall see, several basic properties of the polynomials $p_n(z)$ are simple consequences of the following extremal property which characterizes orthogonal polynomials.

Theorem 1.1. *The polynomial*

$$P_n(z) := \frac{1}{\gamma_n} p_n(z; \mu) = z^n + \cdots$$

is the unique monic polynomial of degree n *of minimal* $L_2(\mu)-$*norm; that is,* P_n *solves the extremal problem*

$$\min_{z^n + \cdots \in \mathcal{P}_n} \int |z^n + \cdots|^2 d\mu.$$

The proof of this result is straightforward and can be found in [Sz,§2.2].

2 Basic Properties of Zeros

The zeros of orthogonal polynomials play an important role in quadrature formulae, interpolation theory, spectral theory for certain linear operators, and the design of digital filters. Thus it is fundamental to ask: What can be said about the location (in the complex plane) of the n zeros of $p_n(z; \mu)$? As a simple consequence of Theorem 1.1 we shall prove the following result due to Fejér:

Theorem 2.1. *All the zeros of $p_n(z; \mu)$ lie in the convex hull of the support $S = \text{supp}(\mu)$.*

By the convex hull $\text{Co}(S)$ of S we mean the intersection of all closed half-planes containing S.

Proof of Theorem 2.1. It is more convenient to work with the monic orthogonal polynomial $P_n = p_n / \gamma_n$ which has the same zeros as p_n.

Suppose, to the contrary, that $P_n(z_0) = 0$ with $z_0 \notin \text{Co}(S)$. Write $P_n(z) = (z - z_0)q(z)$, where $q \in \mathcal{P}_{n-1}$ is monic. Since $z_0 \notin \text{Co}(S)$, there exists a line $\mathcal{L}$ separating z_0 and S. Let $\hat{z}_0$ be the orthogonal projection of z_0 on $\mathcal{L}$ (see Figure 1). Then

$$|z - \hat{z}_0| < |z - z_0| \qquad \forall \quad z \in S.$$

Hence

$$|(z - \hat{z}_0)q(z)| < |(z - z_0)q(z)| = |P_n(z)|$$

for all $z \in S \setminus \{\text{zeros of } q\}$. Since $S = \text{supp}(\mu)$ is infinite, it follows that

$$\int |(z - \hat{z}_0)q(z)|^2 d\mu < \int |P_n(z)|^2 d\mu,$$

which contradicts Theorem 1.1. ∎

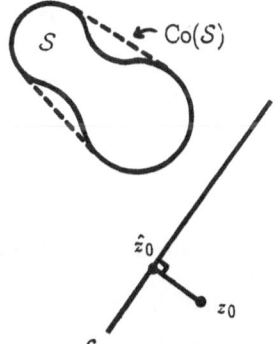

Figure 1

A nice treatment of a generalized version of Theorem 2.1 can be found in the text by P. Davis [D,§10.2].

As the next example illustrates, the zeros of p_n do not, in general, all lie on S (even when $S \subset \mathbf{R}$).

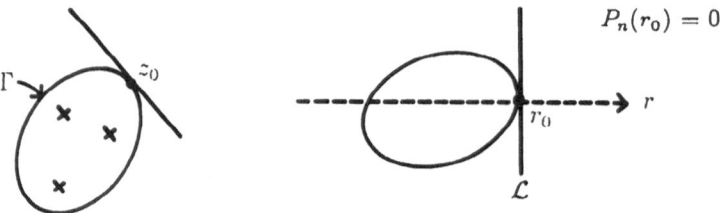

$$P_n(r_0) = 0$$

Figure 2

Example 2.A. Let $d\mu = w(x)dx$, where $w(x)$ is a positive, even, continuous function on $[-2, -1] \cup [1, 2]$. Then Theorem 2.1 asserts that all the zeros of $p_n(z; \mu)$ lie in $\text{Co}(\mathcal{S}) = [-2, 2]$. Now since w is even, it follows from the uniqueness property of orthonormal polynomials that $p_{2n+1}(z; \mu)$ is an odd function for each $n = 0, 1, \ldots$. Thus p_{2n+1} vanishes at $z = 0 \notin \mathcal{S}$.

Except when $\text{Co}(\mathcal{S})$ is an interval, we can strengthen Theorem 2.1 by asserting that the zeros of p_n lie strictly inside the convex hull of $\mathcal{S}$.

Theorem 2.2. *If* $\text{Co}(\mathcal{S}(\mu))$ *is not a line segment, then all the zeros of* $p_n(z; \mu)$ *lie in the interior of* $\text{Co}(\mathcal{S}(\mu))$.

Proof. By Theorem 2.1 we need only show that no zeros of $P_n = p_n/\gamma_n$ lie on the boundary Γ of $\text{Co}(\mathcal{S}(\mu))$.

Suppose that $P_n(z_0) = 0$ for some $z_0 \in \Gamma$. By performing a rotation and translation (see Figure 2) we assume, without loss of generality, that $z_0 = r_0$ is real and that all the points of $\text{Co}(\mathcal{S}(\mu))$ lie on or to the left of the vertical line $\mathcal{L}$ through r_0; i.e. $\mathcal{L}$ is a support line for $\text{Co}(\mathcal{S}(\mu))$. Write $P_n(z) = (z - r_0)q(z)$, $q \in P_{n-1}$, and for $r \in \mathbf{R}$ set

$$I(r) := \int |z - r|^2 |q(z)|^2 d\mu$$
$$= \int (|z|^2 + r^2 - 2r \, \text{Re} \, z)|q(z)|^2 d\mu.$$

By the extremal property of Theorem 1.1, we have

$$I'(r_0) = \int 2(r_0 - \text{Re} \, z)|q(z)|^2 d\mu = 0.$$

But $r_0 - \text{Re} \, z \geq 0$ for all $z \in \mathcal{S}(\mu)$ and so

$$(r_0 - \text{Re} \, z)|q(z)|^2 = 0 \quad d\mu - \text{a.e.}$$

This implies that only finitely many points of $\mathcal{S}(\mu)$ lie to the left of $\mathcal{L}$ and, moreover, q must vanish in these points. Since $\text{Co}(\mathcal{S}(\mu))$ is not an interval, we can therefore find a point $\xi_0 \in \mathcal{S}(\mu) \cap \Gamma$, $\xi_0 \notin \mathcal{L}$, such that $q(\xi_0) = 0$. But then $\xi_0 \in \Gamma$ is a zero of P_n and the preceeding argument (with z_0 replaced by ξ_0) shows that only finitely many points of $\mathcal{S}(\mu)$ can lie in an open half-plane bounded by a support line of $\text{Co}(\mathcal{S}(\mu))$ through ξ_0. Thus $\mathcal{L}$ contains only finitely many points of $\mathcal{S}(\mu)$ and so $\mathcal{S}(\mu)$ is finite, which gives the desired contradiction.

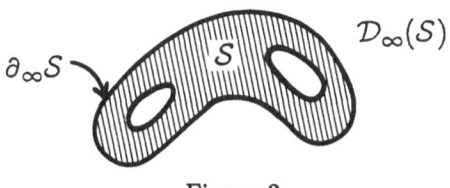

Figure 3

For example, if C is the unit circle $|z| = 1$ and $S(\mu) \subseteq C$, then Theorem 2.2 asserts that all the zeros of $p_n(z; \mu)$ must lie in the open unit disk $|z| < 1$, which is a classical result of Szegő [Sz,§11.4].

How many zeros of $p_n(z; \mu)$ can lie outside S? It is, of course, possible for all the zeros of $p_n(z; \mu)$ to lie off of S. But, as we shall show, only a bounded number (independent of n) can lie on a fixed compact set exterior to S. For this purpose we first introduce some notation.

With $\overline{\mathbf{C}} := \mathbf{C} \cup \{\infty\}$, we let $\mathcal{D}_\infty(S)$ denote the component of $\overline{\mathbf{C}} \backslash S$ containing ∞ (thus $\mathcal{D}_\infty(S)$ is an open, connected, unbounded set). The *outer boundary* of S, denoted by $\partial_\infty S$, is the boundary of $\mathcal{D}_\infty(S)$, i.e.

$$\partial_\infty S := \partial \mathcal{D}_\infty(S)$$

(see Figure 3). Furthermore, we set

$$\mathrm{Pc}(S) := \mathbf{C} \backslash \mathcal{D}_\infty(S),$$

which is called the *polynomial convex hull* of S. Roughly speaking, $\mathrm{Pc}(S)$ is obtained from S by filling in all of its "holes". Since $\mathrm{Pc}(S)$ is a compact set that does not separate the plane, Runge's theorem (cf. [G, p.76]) asserts that any function f analytic on $\mathrm{Pc}(S)$ can be uniformly approximated (as closely as desired) on $\mathrm{Pc}(S)$ by polynomials. Notice also that

$$S \subseteq \mathrm{Pc}(S) \subseteq \mathrm{Co}(S).$$

The following lemma, implicit in the paper by Widom [Wi], will be used to examine the zeros of $p_n(z; \mu)$ that lie in $\mathcal{D}_\infty(S)$.

Lemma 2.3. *If E is a compact set such that $E \cap \mathrm{Pc}(S) = \emptyset$ (i.e. $E \subset \mathcal{D}_\infty(S)$), then there exist an integer m and an α, $0 < \alpha < 1$, with the following property. For any m (not necessarily distinct) points $z_1, \ldots, z_m \in E$, there exist $w_1, \ldots, w_m \in \mathbf{C}$ such that*

$$\prod_{k=1}^{m} \left| \frac{z - w_k}{z - z_k} \right| < \alpha < 1, \qquad \forall \ z \in S. \tag{2.1}$$

Proof. We give a simple argument that is due to V. Totik.

Assume first that E consists of a single point, say $E = \{0\}$ with $0 \notin \mathrm{Pc}(S)$. Then proving (2.1) is equivalent to showing that there exists a monic polynomial $Q(z) = z^m + \cdots \in \mathcal{P}_m$ such that

$$|Q(z)/z^m| < \alpha < 1, \qquad \forall \ z \in \mathcal{S}. \tag{2.2}$$

With the change of variable $\zeta = 1/z$, inequality (2.2) becomes

$$|\zeta^m Q(1/\zeta)| < \alpha < 1, \qquad \forall \ \zeta \in \mathcal{S}^{-1} := \{\zeta | \ 1/\zeta \in \mathcal{S}\}. \tag{2.3}$$

Since $0 \notin \mathrm{Pc}(\mathcal{S})$ it is easy to see that $\mathcal{S}^{-1}$ is compact and that $0 \notin \mathrm{Pc}(\mathcal{S}^{-1})$. Now Q is monic and so (2.3) can be satisfied iff $\exists \ q_{m-1} \in \mathcal{P}_{m-1}$ such that

$$|1 - \zeta q_{m-1}(\zeta)| < \alpha < 1$$

i.e.,

$$|\zeta| |1/\zeta - q_{m-1}(\zeta)| < \alpha < 1, \qquad \forall \ \zeta \in \mathcal{S}^{-1}. \tag{2.4}$$

But the function $1/\zeta$ is analytic on $\mathrm{Pc}(\mathcal{S}^{-1})$, and so q_{m-1} exists by Runge's theorem (the factor $|\zeta|$ causes no difficulty since it is bounded on $\mathcal{S}^{-1}$).

Now we turn to the general case where E is compact and $E \cap \mathrm{Pc}(\mathcal{S}) = \emptyset$. By the first part of the proof, for each $z^* \in E$, there exist $m(z^*), \alpha(z^*)$, and $w_k(z^*), 1 \leq k \leq m(z^*)$, such that

$$\prod_{k=1}^{m(z^*)} \left| \frac{z - w_k(z^*)}{z - z^*} \right| < \alpha(z^*) < 1, \qquad \forall \ z \in \mathcal{S}. \tag{2.5}$$

It then follows, by continuity, that there exists an ϵ–neighborhood $\mathcal{N}(z^*, \epsilon)$ of z^* such that whenever $\{z_1, \ldots, z_{m(z^*)}\} \subset \mathcal{N}(z^*, \epsilon)$, we have

$$\prod_{k=1}^{m(z^*)} \left| \frac{z - w_k(z^*)}{z - z_k} \right| < \alpha(z^*) < 1, \qquad \forall \ z \in \mathcal{S}.$$

It is now possible to complete the proof by using a compactness argument. ∎

With Lemma 2.3 in hand we can easily establish the following theorem of Widom [Wi].

Theorem 2.4. *If E is a closed set such that $E \cap \mathrm{Pc}(\mathcal{S}) = \emptyset$, then the number of zeros of $p_n(z; \mu)$ on E is uniformly bounded in n.*

Proof. By Theorem 2.1, we can assume that E is compact. Let m and α be as in Lemma 2.3 and suppose that $P_n = p_n/\gamma_n$ has $\geq m$ zeros in E, say $z_1, \ldots, z_m$. Then, by Lemma 2.3, $\exists \ w_1, \ldots, w_m$ such that (2.1) holds for all $z \in \mathcal{S}$. Let

$$Q_n(z) := P_n(z) \prod_{k=1}^{m} \left(\frac{z - w_k}{z - z_k} \right),$$

so that Q_n is a monic polynomial of degree n. From (2.1) we get

$$|Q_n(z)| < |P_n(z)|, \qquad \forall \quad z \in \mathcal{S} \setminus \{ \text{ zeros of } P_n \},$$

and so

$$\int |Q_n|^2 d\mu < \int |P_n|^2 d\mu.$$

As the last inequality contradicts Theorem 1.1, it follows that, for each n, the polynomial $p_n(z; \mu)$ has fewer than m zeros on E. ■

Example 2.B. (*Szegő Polynomials*). We use the terminology "Szegő polynomials" to mean orthonormal polynomials $p_n(z; \mu)$ for which $\mathcal{S} = \text{supp}(\mu) \subseteq C := \{ z : |z| = 1 \}$. The zeros of these orthonormal polynomials play an important role in digital filter design. We consider separately the two cases $\mathcal{S} = C$, $\mathcal{S} \neq C$.

Case 1: $\mathcal{S} = C$. Then $\text{Pc}(\mathcal{S}) = \text{Co}(\mathcal{S}) = \{ z : |z| \leq 1 \}$ and, by Theorem 2.2, all the zeros of $p_n(z; \mu)$ lie in the open unit disk.

In Figure 4 we have plotted the zeros of $p_n(z; \mu)$ for $n = 15$ and $n = 25$ for

$$d\mu(e^{i\theta}) = |\sin(\theta/2)|^4 d\theta, \qquad 0 \leq \theta \leq 2\pi. \tag{2.6}$$

Notice that these zeros appear to be approaching the unit circle C and, except near $z = 1$, are close to being equally space in argument.

Figure 5 shows the zeros of $p_{15}(z; \mu)$ and $p_{25}(z; \mu)$ for

$$d\mu(e^{i\theta}) = (5/4 - \cos\theta)d\theta = |1 - e^{i\theta}/2|^2 d\theta, \qquad 0 \leq \theta \leq 2\pi. \tag{2.7}$$

Here the zeros seem to be approaching the circle $|z| = 1/2$, but again, the arguments of the zeros are nearly equally spaced.

We shall see in Section 5 (cf. Theorem 5.3) that the asymptotically uniform spacing of the arguments of the zeros of the $p_n(z; \mu)$ is a phenomenon that can be proved for a large class of measures μ whose support is C.

Case 2: $\mathcal{S} \neq C$. Again, all zeros of $p_n(z; \mu)$ lie in $|z| < 1$. But now $\text{Pc}(\mathcal{S}) = \mathcal{S}$ and so Theorem 2.4 implies that "most zeros" of $p_n(z; \mu)$ tend to $\mathcal{S}$ as $n \to \infty$.

In Figure 6 we have plotted the zeros of $p_6(z; \mu)$ and $p_{16}(z; \mu)$ for

$$d\mu(e^{i\theta}) = |\sin(\theta/2)|^4 d\theta, \quad 0 \leq \theta \leq \pi, \tag{2.8}$$

which is the restriction to the upper half-circle C^+ of the measure in (2.6). Notice that most (in fact, all) of the zeros are approaching $C^+ = \mathcal{S}$, as predicted by Theorem 2.4.

Figure 7 gives analogous plots for the measure that is the restriction of (2.7) to C^+. Again, the zeros are seen to be approaching C^+.

Notice further that the distribution of the zeros in Figures 6 and 7 look very much alike (there is some bunching near ± 1 and the zeros thin out near $z = i$). In Section 5 we shall show that both zero distributions are, in the limit, the equilibrium distribution for C^+. For this purpose we will utilize potential theory and norm comparisons with $L_\infty(\mathcal{S})$.

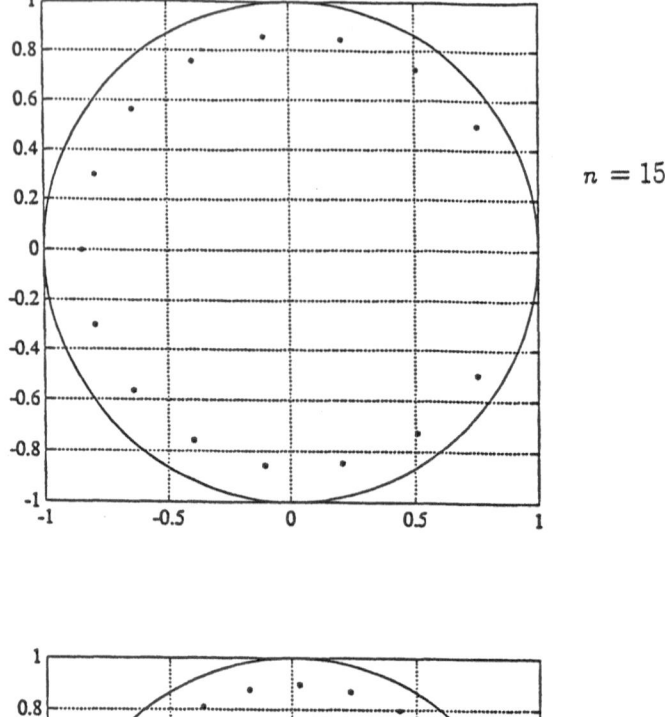

$n = 15$

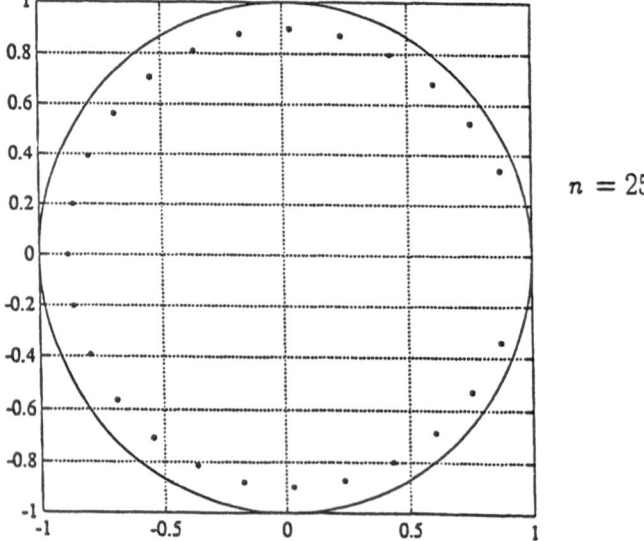

$n = 25$

Figure 4: Zeros of $p_n(z; \mu)$ where $d\mu(e^{i\theta}) = |\sin(\theta/2)|^4 d\theta$, $\quad 0 \le \theta \le 2\pi$.

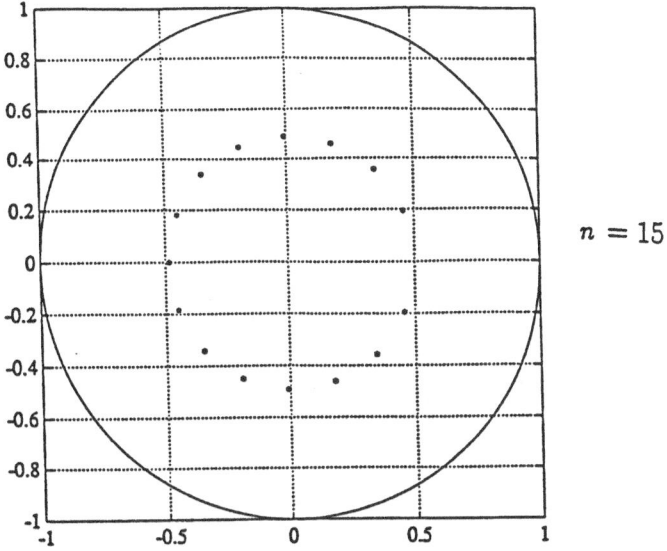

$n = 15$

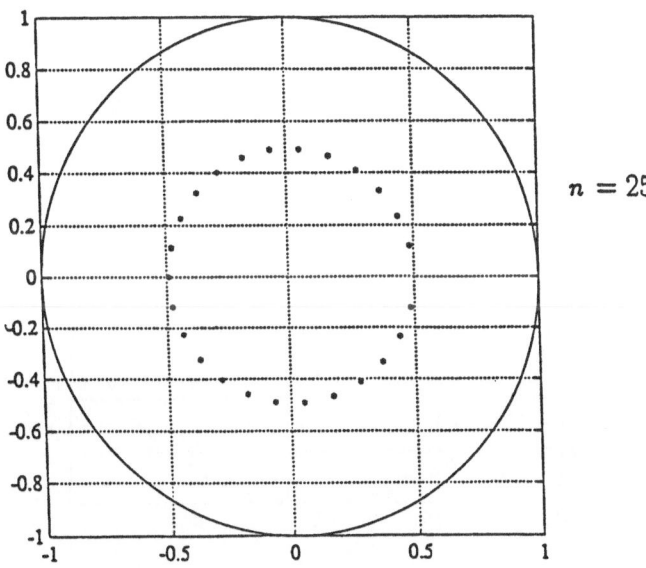

$n = 25$

Figure 5: Zeros of $p_n(z;\mu)$ where $d\mu(e^{i\theta}) = (5/4 - \cos\theta)d\theta,\quad 0 \le \theta \le 2\pi$.

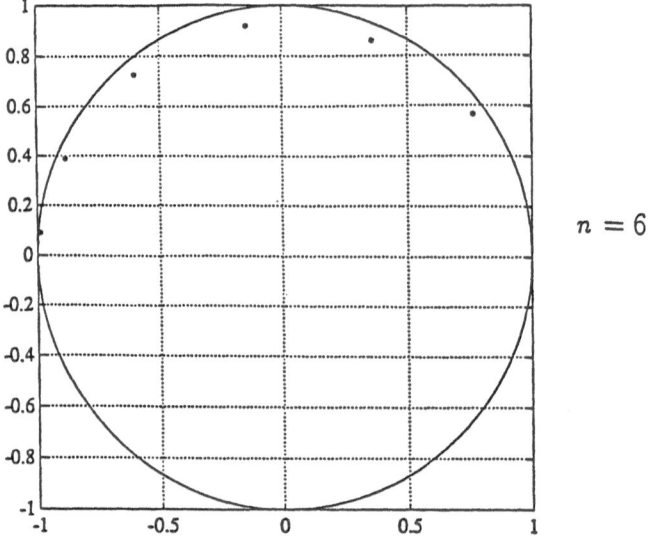

$n = 6$

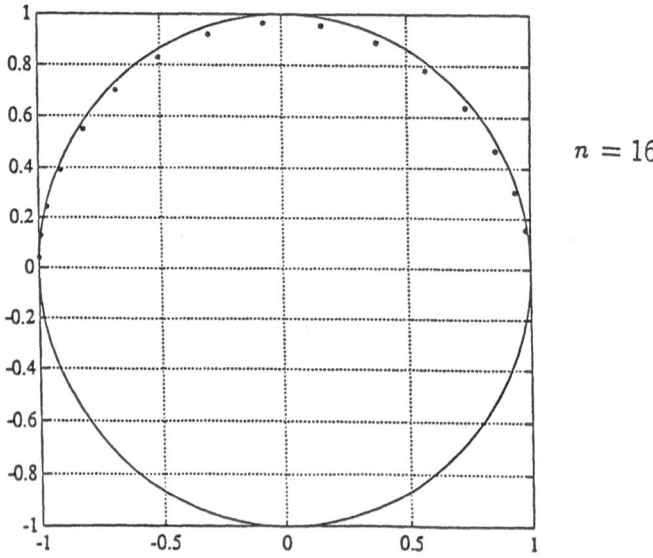

$n = 16$

Figure 6: Zeros of $p_n(z;\mu)$ where $d\mu(e^{i\theta}) = |\sin(\theta/2)|^4 d\theta, \quad 0 \le \theta \le \pi$.

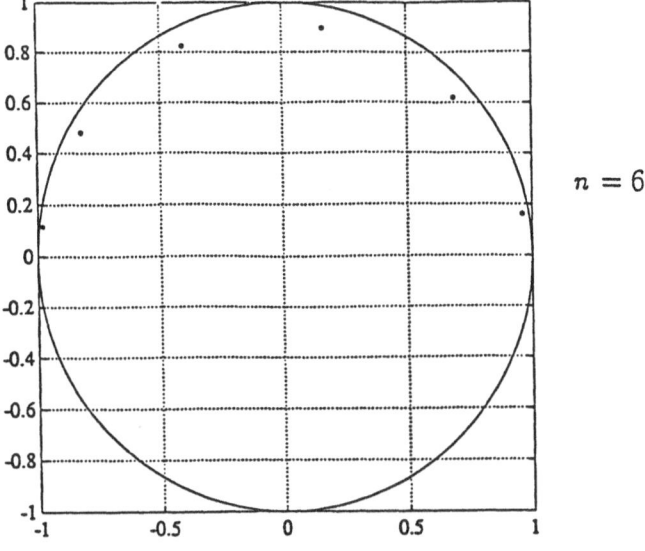

$n = 6$

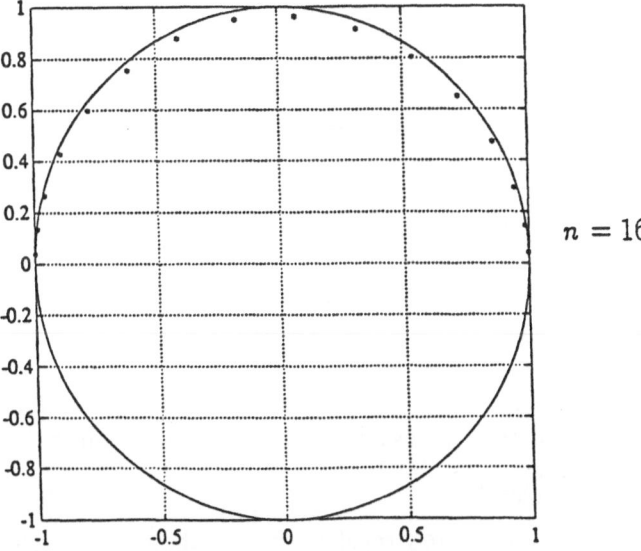

$n = 16$

Figure 7: Zeros of $p_n(z; \mu)$ where $d\mu(e^{i\theta}) = (5/4 - \cos\theta)d\theta$, $0 \le \theta \le \pi$.

3 Completely Regular Measures

To obtain asymptotic results on the zeros of $p_n(z; \mu)$ we will first compare the monic orthogonal polynomials P_n with minimal sup norm polynomials. Let $\| \cdot \|_S$ denote the sup norm over $S = S(\mu)$ and set

$$t_n(S) := \min_{z^n + \cdots \in \mathcal{P}_n} \| z^n + \cdots \|_S, \quad n = 0, 1, \ldots . \tag{3.1}$$

For each $n \geq 0$ there exists a unique monic polynomial $T_n(z) = z^n + \cdots \in \mathcal{P}_n$, called the *Chebyshev polynomial* of degree n, that satisfies

$$t_n(S) = \| T_n \|_S . \tag{3.2}$$

Recalling Theorem 1.1, we note that T_n is just an L_∞ analogue of P_n and that

$$\frac{1}{\gamma_n} = \left(\int |P_n(z)|^2 d\mu \right)^{1/2} \leq \left(\int |T_n(z)|^2 d\mu \right)^{1/2}$$
$$\leq \mu(S)^{1/2} t_n(S), \tag{3.3}$$

where γ_n is the leading coefficient of $p_n(z; \mu)$ (cf. (1.2)). We also remark that the zero results of Theorems 2.1 and 2.4 hold for the polynomials T_n (the same proofs apply).

One advantage of working with the sup norm instead of $L_2(\mu)$ is the fact that $\lim_{n \to \infty} [t_n(S)]^{1/n}$ always exists; this follows from the simple inequality

$$t_{m+n}(S) = \| T_{m+n} \|_S \leq \| T_m \cdot T_n \|_S \leq t_m(S) t_n(S),$$

which shows that $\log t_n(S)$ is a subadditive function of n (cf. [T, §III.5]). In contrast, $\lim_{n \to \infty} \gamma_n^{1/n}$ need not exist.

We write

$$\text{cheb}(S) := \lim_{n \to \infty} [t_n(S)]^{1/n}, \tag{3.4}$$

which is called the *Chebyshev constant* for S. From (3.3) we immediately obtain

$$\liminf_{n \to \infty} \gamma_n^{1/n} \geq \frac{1}{\text{cheb}(S)}. \tag{3.5}$$

Definition 3.1. The measure μ is said to be *completely regular* if

$$\lim_{n \to \infty} \| p_n(z; \mu) \|_{S(\mu)}^{1/n} = 1. \tag{3.6}$$

In other words, for a completely regular measure, there is no essential difference between $L_2(\mu)$ and $L_\infty(S)$ as far as n-th root asymptotics are concerned. Thus we might expect that P_n and T_n have some common limiting properties. For example, it is easy to see that $\lim_{n \to \infty} \gamma_n^{1/n}$ exists for such measures.

Proposition 3.2. *If μ is completely regular, then*

$$\lim_{n\to\infty} \gamma_n^{1/n} = \frac{1}{\text{cheb}(S(\mu))}. \tag{3.7}$$

Proof. From the definition of $t_n(S)$ in (3.1) we have

$$t_n(S) = \| T_n \|_S \leq \| P_n \|_S = \| p_n/\gamma_n \|_S,$$

and so, from (3.6),

$$\limsup_{n\to\infty} \gamma_n^{1/n} \leq \limsup_{n\to\infty} \frac{\| p_n \|_S^{1/n}}{[t_n(S)]^{1/n}} = \frac{1}{\text{cheb}(S)}.$$

Together with (3.5), this proves (3.7). ■

We remark that in the paper of H. Stahl and V. Totik that appears in this Proceedings, property (3.7) is used to define a *regular* measure μ. Thus every completely regular measure is regular. Conversely, if μ is regular and $S(\mu)$ is regular with respect to the Dirichlet problem for $\mathcal{D}_\infty(S(\mu))$, then μ is completely regular.

Although the class of completely regular measures is quite restrictive, it does contain many important measures that arise in applications.

Example 3.A The following are examples of completely regular measures.

(i) $d\mu = w\, dx\, dy$ over a bounded Jordan region R, where the weight w (≥ 0) and some negative power of w are integrable with respect to area over R (cf. [Wa,§5.7]).

(ii) $d\mu = w\, ds$, where ds is arclength over a rectifiable curve Γ and the weight $w \geq 0$ and some negative power of w are integrable with respect to ds (cf. [Wa,§5.7]).

(iii) $S(\mu) = [-1, 1]$ and $\mu' > 0$ a.e. on $[-1, 1]$ (cf. [EF],[ET]).

(iv) $S(\mu) = C : |z| = 1$ and $\lim_{n\to\infty} \gamma_n^{1/n} = 1$ (cf. [LSS]); in particular, if μ belongs to the Szegő class, i.e.

$$\int_0^{2\pi} \log \mu'(e^{i\theta})d\theta > -\infty.$$

To deduce asymptotic properties of orthogonal polynomials with respect to a completely regular measure, we shall appeal to an alternate definition of the constant cheb(S) that comes from potential theory.

4   Basics from Potential Theory

Introductions to potential theory can be found in [He], [Hi] and [T]; a more in depth treatment is given in [La]. Here we provide some basic facts.

Potential theory has its origin in the following

Electrostatics Problem. *Let $E \subset \mathbf{C}$ be compact. Place a unit positive charge on E so that equilibrium is reached in the sense that the energy with respect to the logarithmic potential is minimized.*

To create a mathematical framework for this problem, we let $\mathcal{M}(E)$ denote the collection of all positive, unit Borel measures ν supported on E (so that $\mathcal{M}(E)$ contains all possible distributions of charges placed on E). The *logarithmic potential* associated with $\nu \in \mathcal{M}(E)$ is

$$U^{\nu}(z) := \int \log |z - t|^{-1} d\nu(t), \tag{4.1}$$

which is a superharmonic, lower semi-continuous function on $\mathbf{C}$. The *energy* of such a potential is defined by

$$I[\nu] := \int U^{\nu} d\nu = \int\int \log |z - t|^{-1} d\nu(t) d\nu(z). \tag{4.2}$$

Thus, the electrostatics problem involves the determination of

$$V(E) := \inf_{\nu \in \mathcal{M}(E)} I[\nu], \tag{4.3}$$

which is called the *Robin's constant* for E. The *logarithmic capacity* of E, denoted by $\mathrm{cap}(E)$, is defined by

$$\mathrm{cap}(E) := e^{-V(E)}, \tag{4.4}$$

which is finite and nonnegative.

A fundamental theorem from potential theory asserts that if $\mathrm{cap}(E) > 0$, there exists a unique measure $\nu_E \in \mathcal{M}(E)$ such that

$$I[\nu_E] = V(E). \tag{4.5}$$

The extremal measure ν_E is called the *equilibrium distribution* for E and furnishes the solution to the electrostatics problem.

Some basic facts about $\mathrm{cap}(E)$ and ν_E are:

(i) $\mathcal{S}(\nu_E) = \mathrm{supp}(\nu_E) \subseteq \partial_{\infty} E$; moreover, the set $\partial_{\infty} E \backslash \mathcal{S}(\nu_E)$ has capacity zero.

(ii) The *conductor potential* $U^{\nu_E}(z)$ satisfies $U^{\nu_E}(z) \leq V(E)$ for all $z \in \mathbf{C}$, with equality holding on E except possibly for a set of capacity zero.

(iii) For any compact set E,

$$\mathrm{cap}(E) = \mathrm{cheb}(E). \qquad (4.6)$$

Assertion (iii) provides the alternate interpretation of the Chebyshev constant that will be especially useful for our purposes. That $\mathrm{cheb}(E)$ has a plausible connection with potential theory can be seen from the fact that, for any monic polynomial $Q(z) = z^n + \cdots \in \mathcal{P}_n$, we can write

$$\frac{1}{n} \log \frac{1}{|Q(z)|} = U^{\nu(Q)}(z), \qquad (4.7)$$

where $\nu(Q)$ is the discrete measure with mass $1/n$ at each zero of Q.

To gain some insight into the equilibrium distribution ν_E we turn to the (hopefully) more familiar concept of a Green function. If the outer boundary $\partial_\infty E$ consists of analytic curves, then the *Green function with pole at ∞* for $\mathcal{D}_\infty(E)$ is denoted by $g_E(z, \infty)$ and is defined by the following three properties:

(a) $g_E(z, \infty)$ is harmonic in $\mathcal{D}_\infty(E) \backslash \{\infty\}$.

(b) $g_E(z, \infty) \to 0$ as $z \to \partial_\infty E$, $z \in \mathcal{D}_\infty(E)$.

(c) $\exists$ a constant $\hat{V}$ such that

$$(g_E(z, \infty) - \log |z|) \to \hat{V} \quad \text{as} \quad z \to \infty.$$

Using Green's formula, we can derive the identity (cf. [Wa,§4.2])

$$\begin{aligned}
\hat{V} - g_E(z, \infty) &= \frac{1}{2\pi} \int_{\partial_\infty E} \log|z - t|^{-1} \frac{\partial}{\partial n} g_E(t, \infty)|dt|, \\
&= \int_{\partial_\infty E} \log|z - t|^{-1} d\hat{\nu},
\end{aligned}$$

where n denotes the exterior normal for $\partial_\infty E$ and

$$d\hat{\nu} := \frac{1}{2\pi} \frac{\partial}{\partial n} g_E(t, \infty)|dt|. \qquad (4.8)$$

The relationship between the Green function and the conductor potential is given in

Theorem 4.1. *If $\partial_\infty E$ consists of finitely many analytic curves, then* $\hat{V} = V(E)$, $\hat{\nu} = \nu_E$ *and*

$$\begin{aligned}
U^{\nu_E}(z) &= V(E) - g_E(z, \infty) \\
&= \log \frac{1}{\mathrm{cap}(E)} - g_E(z, \infty). \qquad (4.9)
\end{aligned}$$

It is, of course, possible to define the Green function for the outer domain of more general compact sets. This is done by exhausting $\mathcal{D}_\infty(E)$ by a sequence of open sets $G_1 \subset G_2 \subset \cdots$ containing ∞ and having analytic boundaries, and then taking the limit of the associated Green functions. Provided this limit is not identically infinite, it defines the Green function $g_E(z, \infty)$. Moreover, equation (4.9) persists in this general setting.

It is helpful to keep in mind the following two simple examples.

Example 4.A. Let $E : |z| = R$. Then $\mathrm{cap}(E) = \mathrm{cheb}(E) = R$, $g_E(z, \infty) = \log|z/R|$ for $|z| \geq R$, and $d\nu_E = ds/2\pi R$, where ds is arclength on the circle $|z| = R$. Notice that the formula for $d\nu_E$ follows immediately from (4.8).

Example 4.B. Let $E = [-1, 1]$. Then $\mathrm{cap}(E) = \mathrm{cheb}(E) = 1/2$, $g_E(z, \infty) = \log|z + \sqrt{z^2 - 1}|$ and

$$d\nu_E = \frac{1}{\pi} \frac{dx}{\sqrt{1 - x^2}}, \quad x \in [-1, 1],$$

which is the arcsine measure.

5 Asymptotic Behavior of Zeros

If Q is a polynomial of degree n with zeros $z_1, z_2, \ldots, z_n$, the *normalized zero distribution* associated with Q is defined by

$$\nu(Q) := \frac{1}{n} \sum_{k=1}^{n} \delta_{z_k}, \tag{5.1}$$

where δ_{z_k} denotes the unit point mass at z_k.

As we shall see, the following theorem due to Blatt, Saff and Simkani [BSS] not only leads to asymptotic results for the zeros of certain sequences of orthogonal polynomials, it is useful in many other contexts.

Theorem 5.1. *Let S be a compact set with positive capacity and suppose that the monic polynomials $Q_n(z) = z^n + \cdots \in \mathcal{P}_n$ satisfy the following two conditions:*

(a) $\displaystyle \limsup_{n \to \infty} \| Q_n \|_S^{1/n} \leq \mathrm{cap}(S);$

(b) $\displaystyle \lim_{n \to \infty} \nu(Q_n)(A) = 0$ *for all closed sets A contained in the (2-dimensional) interior of $\mathrm{Pc}(S)$.*

Then, in the weak-star sense,

$$\nu(Q_n) \longrightarrow \nu_S \quad as \quad n \to \infty, \tag{5.2}$$

where ν_S is the equilibrium distribution for S.

By (5.2) we mean that, for all continuous functions f on $\mathbf{C}$ having compact support,

$$\lim_{n\to\infty} \int f d\nu(Q_n) = \int f d\nu_S.$$

Assumption (a) states that the L_∞ norms of the Q_n are asymptotically minimal (recall (3.4) and (4.6)), and (b) means that only $o(n)$ zeros of Q_n can lie on a compact subset of int$(Pc(S))$.

The proof of Theorem 5.1 proceeds roughly along the following lines. First one shows that assumption (a) implies

$$\lim_{n\to\infty} \nu(Q_n)(B) = 0 \quad \forall \quad B \subset \mathcal{D}_\infty(S), \quad B \quad \text{closed}.$$

This fact together with assumption (b) implies that any limit measure of the $\nu(Q_n)$'s is supported on $\partial_\infty S$ (which is the case for the equilibrium measure ν_S). Next, if ν^* is any limit measure of $\{\nu(Q_n)\}_1^\infty$, one can use assumption (a), the representation (4.7), and the minimum principle for superharmonic functions to prove that

$$U^{\nu^*}(z) \le \log \frac{1}{\text{cap}(S)} = V(S) \quad \forall \quad z \in \partial_\infty S. \tag{5.3}$$

Finally, on integrating this last inequality with respect to ν^* we obtain

$$I[\nu^*] = \int U^{\nu^*} d\nu^* \le \int V(S) d\nu^* = V(S) = I[\nu_S],$$

that is, $\nu^* \in \mathcal{M}(S)$ has minimal energy. Consequently, by the uniqueness of the solution to the electrostatics problem, we get $\nu^* = \nu_S$.

Notice that if μ is a completely regular measure, then the monic orthogonal polynomials $P_n(z) = p_n(z; \mu)/\gamma_n$ satisfy condition (a) of Theorem 5.1 for $S = S(\mu)$; indeed from (3.6) and (3.7),

$$\limsup_{n\to\infty} \| P_n \|_{S(\mu)}^{1/n} \le \frac{\displaystyle\limsup_{n\to\infty} \| p_n \|_{S(\mu)}^{1/n}}{\displaystyle\lim_{n\to\infty} \gamma_n^{1/n}} = \text{cheb}(S(\mu)) = \text{cap}(S(\mu)).$$

Moreover, if the interior of $Pc(S(\mu))$ is empty, then condition (b) of Theorem 5.1 is vacuously satisfied. Thus we obtain (compare [BSS, Cor. 2.1])

Theorem 5.2. *Let μ be a completely regular measure. If $S(\mu) = \text{supp}(\mu)$ has positive capacity and $Pc(S(\mu))$ has empty interior, then*

$$\nu(p_n(;\mu)) \longrightarrow \nu_{S(\mu)} \quad as \quad n \to \infty.$$

This result explains the behavior of the zeros plotted in Figures 6 and 7; they have limit distribution equal to ν_{C+}. Moreover, it will follow from Theorem 5.1 that the limiting distribution for the zeros plotted in Figure 4 is $d\nu_C = d\theta/2\pi$, provided one can show that condition (b) is satisfied. That this indeed the case can be seen from the next result due to Mhaskar and Saff [MS4].

Theorem 5.3. *Let* $\{p_n(z;\mu)\}_1^\infty$ *be a sequence of Szegő polynomials* (*i.e.* $S(\mu) \subseteq C :$ $|z| = 1$) *so that* $P_n(0) = p_n(0;\mu)/\gamma_n$ *is the n-th 'reflection coefficient'. Set*

$$\limsup_{n\to\infty} |P_n(0)|^{1/n} =: \rho(\leq 1), \tag{5.4}$$

and let $\Lambda \subset \mathbf{N}$ *satisfy*

$$\lim_{\substack{n\to\infty \\ n\in\Lambda}} |P_n(0)|^{1/n} = \rho.$$

(a) *If* $\rho < 1$, *then* $\nu(p_n(\cdot;\mu)) \to \nu_{C_\rho}$ *as* $n \to \infty$, $n \in \Lambda$, *where* $d\nu_{C_\rho} = ds/(2\pi\rho)$, *for* $\rho > 0$, *is the equilibrium distribution on the circle* $C_\rho : |z| = \rho$ (*cf. Example 4.A*), *and* $\nu_{C_0} := \delta_0$.

(b) *If* $\rho = 1$ *and*

$$\lim_{n\to\infty} \frac{1}{n} \sum_{k=0}^{n} |P_k(0)| = 0, \tag{5.5}$$

then $\nu(p_n(\cdot;\mu)) \to \nu_{C_1}$ *as* $n \to \infty$, $n \in \Lambda$.

The proof of this result follows by applying Theorem 5.1 to the polynomials $Q_n = P_n^*/\overline{P_n(0)}, n \in \Lambda$, on the set $S = C_{1/\rho}$, where $P_n^*(z) := z^n \overline{P_n(1/\bar{z})}$ are the reverse polynomials.

For the Jacobi type weight $d\mu(e^{i\theta}) = |\sin(\theta/2)|^4 d\theta$, $0 \le \theta \le 2\pi$ (cf. (2.8)), the reflection coefficients are $P_n(0) = 2/(n+2)$ so that $\rho = 1$, $\Lambda = \mathbf{N}$, and condition (5.5) is satisfied. Thus by part (b) of the above theorem, $\nu(p_n(\cdot;\mu)) \to \nu_{C_1} = d\theta/2\pi$ as $n \to \infty$, which confirms our expectations from the plots in Figure 4.

Concerning the plots in Figure 5 where

$$d\mu(e^{i\theta}) = (5/4 - \cos\theta)d\theta = |1 - e^{i\theta}/2|^2 d\theta, \quad 0 \le \theta \le 2\pi, \tag{5.6}$$

it turns out that $\rho = 1/2$ and $\Lambda = \mathbf{N}$, so that from part (a) of Theorem 5.3 we get $\nu(p_n(\cdot;\mu)) \to \nu_{C_{1/2}}$, the uniform distribution on the circle of radius $1/2$.

We remark that it is not necessary to have an explicit form for the reflection coefficients in order to determine the constant ρ of (5.4). As observed in [NT], for measures μ belonging to the Szegő class, ρ can be deduced from the analytic properties of the *Szegő function* $D(\mu;z)$ (cf. [Sz, §10.2]). Recall that $D(\mu;z)$ is analytic and nonzero in $|z| < 1$, and satisfies

$$|D(\mu; z)|^2 = \mu'(e^{i\theta}) \quad \text{a.e. on} \quad [0, 2\pi].$$

By considering the orthogonal expansion for the reciprocal $D(\mu; z)^{-1}$, it is easy to see that ρ in (5.4) is the smallest number such that $D(\mu; z)^{-1}$ has an analytic extension to the disk $|z| < 1/\rho$. For example, the Szegő function for the weight (5.6) is $D(\mu; z) = 1 - z/2$, from which we see that $|z| < 2$ is the largest disk for which $D(\mu; z)^{-1} = 1/(1 - z/2)$ is analytic; hence $\rho = 1/2$.

6 Bounds for Polynomials

Another important application of potential theory is in determining bounds for the growth of polynomials in the complex plane; that is, for attacking the following

Problem. *Let $E \subset \mathbf{C}$ be compact with* $\text{cap}(E) > 0$. *Given that $q_n \in \mathcal{P}_n$ and*

$$|q_n(z)| \leq M, \quad \forall z \in E, \tag{6.1}$$

estimate $|q_n(z)|$ for $z \notin E$.

In analyzing this problem we shall make use of the Green function $g_E(z, \infty)$ with pole at ∞ for $\mathcal{D}_\infty(E)$ (cf. Section 4). The level curves of this function shall be denoted by Γ_ρ, that is,

$$\Gamma_\rho := \{z \in \mathbf{C} | \, g_E(z, \infty) = \log \rho\}, \qquad \rho > 1. \tag{6.2}$$

Example 6.A. If $E = [-1, 1]$, then $\Gamma_\rho(\rho > 1)$ is the ellipse with foci at ± 1 and semi-major axis equal to $(\rho + \rho^{-1})/2$ (cf. Example 4.B).

The following result, known as the *Bernstein-Walsh lemma* [Wa, p.77, 87], is a simple application of the maximum principle for subharmonic functions that provides an answer to the above problem.

Lemma 6.1. *Let $E \subset \mathbf{C}$ be compact and have positive capacity. If $q_n \in \mathcal{P}_n$ satisfies (6.1), then*

$$|q_n(z)| \leq M \exp(ng_E(z, \infty)), \quad \forall \; z \in \mathcal{D}_\infty(E).$$

In particular,

$$|q_n(z)| \leq M\rho^n, \qquad \forall \; z \in \Gamma_\rho.$$

Proof. From (6.1) we have

$$\frac{1}{n} \log |q_n(z)| \leq \frac{1}{n} \log M, \quad \forall \quad z \in E,$$

and so

$$\frac{1}{n} \log |q_n(z)| - g_E(z, \infty) \leq \frac{1}{n} \log M \tag{6.3}$$

holds for $z \in \partial_\infty E$ except possibly for a set of capacity zero (recall property (ii) of $U^{\nu_E}(z)$ in Section 4 and the representation (4.9)). But the left-hand side of (6.3) is *subharmonic* in $\mathcal{D}_\infty(E)$ (including ∞), and so the maximum principle (cf. [T, p.77]) implies that (6.3) holds for all $z \in \mathcal{D}_\infty(E)$. ∎

We now show how the above lemma can be used to establish the convergence of certain Fourier expansions. Let $f \in L_2(\mu)$, i.e.

$$\| f \|_{L_2(\mu)} := \left(\int |f|^2 d\mu \right)^{1/2} < \infty.$$

Then the Fourier expansion of f is given by

$$f \sim \sum_{k=0}^\infty a_k p_k(z; \mu), \quad a_k = a_k(f) := \int f \overline{p_k} d\mu, \tag{6.4}$$

and its partial sums

$$s_n(z) = s_n(z; f) := \sum_{k=0}^n a_k p_k(z; \mu) \tag{6.5}$$

are best polynomial approximants to f; more precisely, we have the following well-known result:

Theorem 6.2. *The partial sum s_n is the best $L_2(\mu)$ approximant to f out of $\mathcal{P}_n$ in the sense that*

$$\| f - s_n \|_{L_2(\mu)} \leq \| f - q_n \|_{L_2(\mu)}, \quad \forall \quad q_n \in \mathcal{P}_n.$$

To simplify our discussion (yet still convey the general spirit) we assume throughout the remainder of this section that $\text{supp}(\mu) = [-1, 1]$ and that μ is completely regular; we denote this by writing $\mu \in \text{CR}[-1, 1]$.

The next theorem describes the convergence of the Fourier expansion (6.4) for analytic functions f.

Theorem 6.3. *Let $\mu \in \text{CR}[-1, 1]$ and assume f is analytic inside the ellipse $\Gamma_\rho (\rho > 1)$ of Example 6.A. Then*

$$s_n(z; f) = \sum_0^n a_k p_k(z; \mu) \to f(z) \quad as \quad n \to \infty$$

locally uniformly inside Γ_ρ.

Proof. Let Q_{n-1} denote the unique polynomial in $\mathcal{P}_{n-1}$ that interpolates f in the n points that are the extrema on $[-1,1]$ of the classical Chebyshev polynomial $\cos((n-1)\arccos x)$. Using the analyticity of f and the Hermite error formula (cf. [Wa,§3.1]) for $f - Q_{n-1}$, it is easy to verify the following fact:

$$\limsup_{n \to \infty} \| f - Q_{n-1} \|_{[-1,1]}^{1/n} \leq 1/\rho < 1, \tag{6.6}$$

where $\| \cdot \|_{[-1,1]}$ denotes the sup norm over $[-1,1]$. Since

$$|a_n| = \left| \int_{-1}^1 f p_n d\mu \right| = \left| \int_{-1}^1 (f - Q_{n-1}) p_n d\mu \right|,$$

we deduce from (6.6) that

$$\limsup_{n \to \infty} |a_n|^{1/n} \leq 1/\rho. \tag{6.7}$$

From the assumption that μ is completely regular we also have

$$\lim_{n \to \infty} \| p_n \|_{[-1,1]}^{1/n} = 1,$$

and so, for the sup norm over any ellipse (level curve) Γ_σ, Lemma 6.1 yields

$$\limsup_{n \to \infty} \| p_n \|_{\Gamma_\sigma}^{1/n} \leq \sigma.$$

Hence, for $1 < \sigma < \rho$, we get from (6.7) that

$$\limsup_{n \to \infty} \| a_n p_n \|_{\Gamma_\sigma}^{1/n} \leq \sigma/\rho < 1,$$

which implies that $\sum_0^\infty a_n p_n(z)$ converges uniformly on Γ_σ for each $\sigma < \rho$. Letting $\sigma \to \rho$, the theorem follows. ■

For the case of Jacobi series, the above result is given in [Sz,Chap. 9].

What can be said about the behavior of the partial sums s_n in the complex plane when f is *not* analytic on $[-1,1]$? To gain some insight, let's consider the Chebyshev and Legendre expansions of

$$f(x) = |x| \quad on \quad [-1,1].$$

In Figure 8 we have plotted the zeros of the partial sums $s_{10}(z; f)$ and $s_{20}(z; f)$ for the Chebyshev expansion and, in Figure 9, the analogous plots are given for the

384

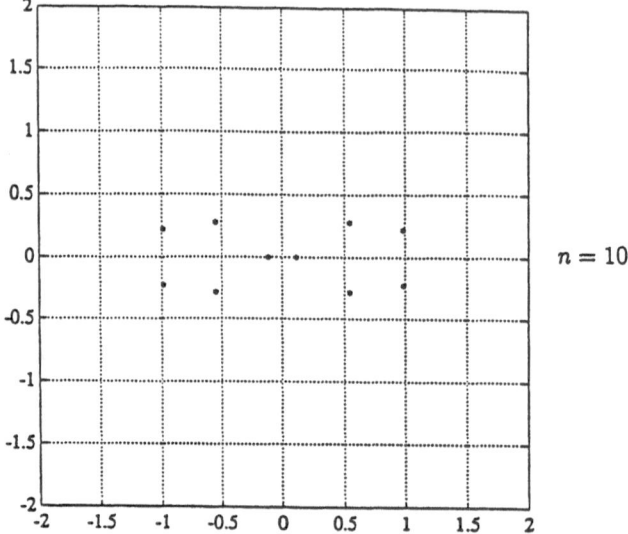

Zeros of n-th partial sum of Chebyshev expansion

$$|x| \sim \frac{2}{\pi} - \frac{4}{\pi} \sum_{k=1}^{\infty} \frac{(-1)^k T_{2k}(x)}{4k^2 - 1}$$

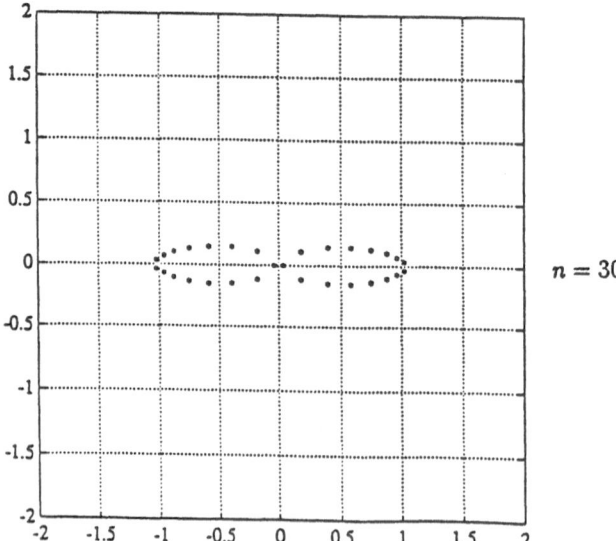

Figure 8

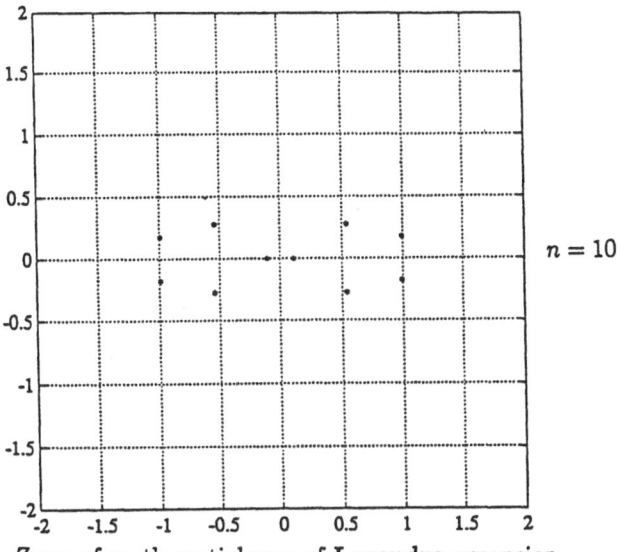

Zeros of n-th partial sum of **Legendre** expansion

$$|x| \sim \frac{1}{2} - \sum_{k=1}^{\infty} (-1)^k \binom{2k}{k-1} \frac{4k+1}{2^{2k+1}k(2k-1)} P_{2k}(x)$$

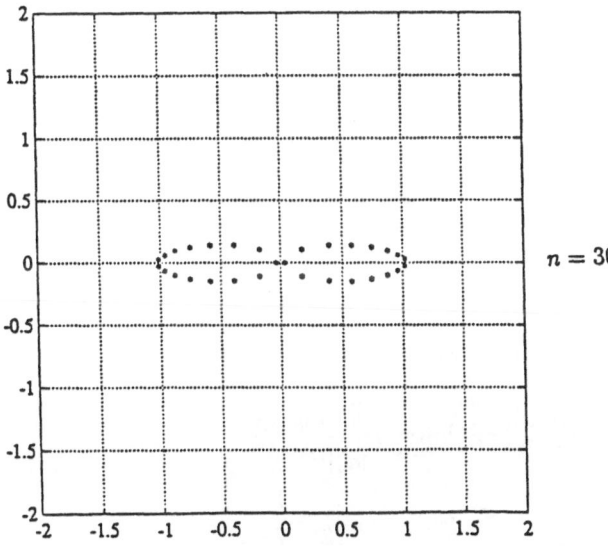

Legendre case

Figure 9

Legendre expansion. We observe that these zeros seem to surround and approach the orthogonality interval $[-1,1]$ as n increases. Moreover, the distributions of the zeros in Figures 8 and 9 are very much alike. This phenomenon is explained in the following theorem due to Li, Saff and Sha [LSS].

Theorem 6.4. *Let $\mu \in CR[-1,1]$. If $f \in L_2(\mu)$ is not equal $d\mu$-a.e. to a function analytic on an open set containing $[-1,1]$, then there exists a subsequence $\Lambda(f) \subseteq$ $\mathbf{N}$ for which the zero measures $\nu(s_n)$ of the partial sums $s_n(z;f)$ satisfy*

$$\nu(s_n) \to \frac{1}{\pi} \frac{dx}{\sqrt{1-x^2}} \quad as \quad n \to \infty, \quad n \in \Lambda(f). \tag{6.8}$$

Proof. We first claim that the Fourier coefficients a_n of f satisfy

$$\limsup_{n\to\infty} |a_n|^{1/n} = 1. \tag{6.9}$$

If not, then $\limsup_{n\to\infty} |a_n|^{1/n} = 1/\rho < 1$ and the argument used in the proof of Theorem 6.3 shows that $\sum_0^\infty a_n p_n$ converges inside the ellipse Γ_ρ to an analytic function. As this contradicts the assumption on f, equation (6.9) follows.

Now choose $\Lambda \subseteq \mathbf{N}$ so that

$$\lim_{\substack{n\to\infty \\ n\in\Lambda}} |a_n|^{1/n} = 1, \tag{6.10}$$

and let

$$Q_n(z) := s_n(z)/(a_n\gamma_n) = z^n + \cdots .$$

Since $\mu \in CR[-1,1]$ and the polynomials s_n have uniformly bounded $L_2(\mu)$-norm, it is easy to see (cf. (3.6)) that

$$\limsup_{n\to\infty} \| s_n \|_{[-1,1]}^{1/n} \le 1.$$

Hence, from (3.7) and (6.10) we get

$$\limsup_{\substack{n\to\infty \\ n\in\Lambda}} \| Q_n \|_{[-1,1]}^{1/n} = \limsup_{\substack{n\to\infty \\ n\in\Lambda}} \frac{\| s_n \|_{[-1,1]}^{1/n}}{|a_n|^{1/n} \gamma_n^{1/n}}$$

$$\le cap([-1,1]) = \frac{1}{2}.$$

But now we are in position to apply Theorem 5.1: we have just verified condition

(a) and condition (b) holds vacuously. Thus

$$\nu(s_n) = \nu(Q_n) \to \nu_{[-1,1]} = \frac{1}{\pi}\frac{dx}{\sqrt{1-x^2}} \quad \text{as} \quad n \to \infty, \quad n \in \Lambda. \qquad \blacksquare$$

From Theorem 6.4 we see that the zeros plotted in Figures 8 and 9 have limiting distribution equal to the arcsine measure. We also obtain the following

Corollary 6.5. *If $\mu \in CR[-1,1]$ and $f \in L_2(\mu)$ is not equal $d\mu$-a.e. to a function analytic on an open set containing $[-1,1]$, then every point of $[-1,1]$ is a limit point of the zeros of the partial sums $\{s_n(z;f)\}_1^\infty$.*

Consequently, $\{s_n(z;f)\}_1^\infty$ does not converge uniformly in any (2-dimensional) neighborhood of a point of $[-1,1]$.

This result illustrates a shortcoming of the partial sums s_n. Although, by Theorem 6.2, they are *globally* best approximants to f on $[-1,1]$, *locally* (in the 2-dimensional sense) the sequence $\{s_n\}_1^\infty$ cannot imitate f; it is useless for the purpose of analytic continuation. We therefore have an example of the following

MORAL: WHAT'S BEST GLOBALLY IS LOCALLY NOT SO GOOD.

To lend further support to this moral we mention a recent result of Li, Saff and Sha dealing with the rate of convergence of the Fourier series on subintervals of $[-1,1]$. Let $s_n(z;f)$ denote, as above, the partial sums of the Fourier expansion for f over $[-1,1]$ and set

$$\| f - s_n \|_{L_2(\mu)} = \left(\int_{-1}^1 |f - s_n|^2 d\mu \right)^{1/2},$$

$$\| f - s_n \|_{L_2(\mu,[a,b])} := \left(\int_a^b |f - s_n|^2 d\mu \right)^{1/2}.$$

Then we have (cf. [LSS])

Theorem 6.6. *If $\mu' > 0$ a.e. on $[-1,1]$ and f is not $(d\mu$- a.e.) a polynomial, then*

$$\sum_{n=0}^\infty \left(\frac{\| f - s_n \|_{L_2(\mu,[a,b])}}{\| f - s_n \|_{L_2(\mu)}} \right)^2 = \infty. \qquad (6.11)$$

for every subinterval $[a,b] \subseteq [-1,1]$ $(a \neq b)$.

Moreover, (6.11) is sharp in the sense that the exponent 2 cannot, in general, be replaced by any larger constant.

The proof of this theorem is based upon a lemma of Máté, Nevai and Totik [MNT].

Notice that the divergence of the series in (6.11) implies that infinitely many of its terms must exceed $1/n^{1+\epsilon}$, $\epsilon > 0$. Indeed we have (cf. [LSS])

Corollary 6.7. *With the assumptions of Theorem 6.6, for each $\epsilon > 0$, there exists a subsequence $\Lambda \subseteq \mathbf{N}$ such that for any $[a,b] \subseteq [-1,1]$ $(a \neq b)$,*

$$\| f - s_n \|_{L_2(\mu,[a,b])} \geq \frac{C}{n^{\frac{1}{2}+\epsilon}} \| f - s_n \|_{L_2(\mu)}, \quad n \in \Lambda, \tag{6.12}$$

where the constant $C > 0$ depends only on $b - a$.

Returning to the moral mentioned above, we see from (6.12) that, with reference to the partial sums s_n, what is globally best cannot locally be much better (only improvements of order $1/\sqrt{n}$ are possible for the rate of convergence on subintervals).

7 Weighted Polynomials over Unbounded Sets

Thus far we have restricted our discussion to orthogonal polynomials over a compact subset of $\mathbf{C}$. In this section we describe an approach for analyzing orthogonal polynomials with respect to unbounded sets; for example, when $\mathrm{supp}(\mu) = \mathbf{R}$.

Let $E \subseteq \mathbf{C}$ be a closed (but not necessarily bounded) set having positive capacity and let $W(z)$ be a nonnegative weight function on E. We now wish to attack the following generalized version of the problem stated at the beginning of Section 6.

Problem. *Given that $q_n \in \mathcal{P}_n$ and*

$$|W(z)q_n(z)| \leq M, \quad \forall \quad z \in E,$$

estimate $|q_n(z)|$ for $z \in \mathbf{C}$.

For example, if

$$|e^{-x^2} q_n(x)| \leq M, \quad \forall \quad x \in \mathbf{R} = (-\infty, \infty), \tag{7.1}$$

then what can be said about $|q_n(z)|$ for $z \in \mathbf{C}$? The first point to observe is that (7.1) contains superfluous information. That is, since $|e^{-x^2} q_n(x)| \to 0$ as $|x| \to \infty$, inequality (7.1) is only needed over some *finite* interval. Indeed, if $q_n \not\equiv 0$ and $\xi \in \mathbf{R}$ is a point for which

$$|e^{-\xi^2} q_n(\xi)| = \max_{\mathbf{R}} |e^{-x^2} q_n(x)|$$

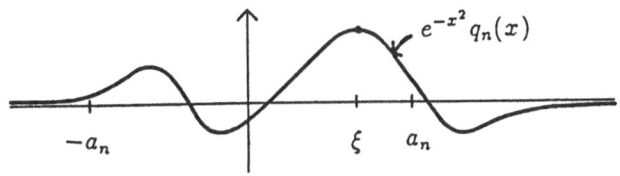

Figure 10

(see Figure 10), then ξ must belong to a finite interval $[-a_n, a_n]$. Moreover, such an a_n can be found that is *independent* of $q_n \in \mathcal{P}_n$. As shown in [MS1], $a_n = \sqrt{n}$ is an asymptotically sharp (smallest) choice. Consequently (7.1) should be replaced by

$$|e^{-x^2} q_n(x)| \leq M, \quad \forall \quad x \in [-\sqrt{n}, \sqrt{n}].$$

What we wish to emphasize is that the above problem is intimately related to the following question: *Where does the sup norm of a weighted polynomial live?*

In answering this question it is convenient to fix n and write $W = w^n$. Our goal is to imitate the Bernstein-Walsh lemma (Lemma 6.1) and in order to carry over the proof we shall assume that the weight w is of a special form that allows us to apply the maximum principle. Namely, we assume that

$$w(z) = \exp(U^\sigma(z) - F_w), \quad z \in E, \tag{7.2}$$

where σ is some probability measure with compact support $S(\sigma) \subseteq E$, $U^\sigma(z)$ is continuous on $\mathbf{C}$, and F_w is a constant. Then the generalization of Lemma 6.1 becomes

Lemma 7.1. *Let* $w : E \to [0, \infty)$ *be of the form (7.2). If* $q_n \in \mathcal{P}_n$ *satisfies*

$$|w(z)^n q_n(z)| \leq M, \quad \forall \quad z \in S(\sigma), \tag{7.3}$$

then $|w(z)^n q_n(z)| \leq M$ *for all* $z \in E$. *Consequently,*

$$\| w^n q_n \|_E = \| w^n q_n \|_{S(\sigma)}. \tag{7.4}$$

Furthermore,

$$|q_n(z)| \leq M \exp\{-n(U^\sigma(z) - F_w)\}, \quad \forall \quad z \in \mathbf{C}. \tag{7.5}$$

Proof. Inequality (7.3) and the representation (7.2) yield

$$U^\sigma(z) - F_w + \frac{1}{n} \log |q_n(z)| \leq \frac{1}{n} \log M, \quad \forall \quad z \in S(\sigma). \tag{7.6}$$

But the left-hand side of (7.6) is subharmonic in $\overline{C}\backslash\mathcal{S}(\sigma)$, even at infinity. Thus, since U^σ is continuous, we can apply the maximum principle to deduce that (7.6) holds for all $z \in \overline{C}\backslash\mathcal{S}(\sigma)$, which completes the proof. ∎

Analyzing the above proof we see that it is not necessary to assume that w is of the form (7.2) for all $z \in E$. Indeed, Lemma 7.1 remains valid provided

$$w(z) = \exp(U^\sigma(z) - F_w), \quad \forall \quad z \in \mathcal{S}(\sigma) \tag{7.7}$$

and

$$w(z) \leq \exp(U^\sigma(z) - F_w), \quad \forall \quad z \in E\backslash\mathcal{S}(\sigma). \tag{7.8}$$

In fact, it's enough to assume that, on the indicated sets, (7.7) and (7.8) hold *quasi-everywhere* (q.e); that is, with the possible exception of a set having capacity zero. Thus to handle general weight functions w we are lead to the following

Question. *Given* $w : E \rightarrow [0, \infty)$, *how do we find a probability measure* σ *and a constant* F_w *so that* (7.7) *and* (7.8) *hold ?*

Readers familiar with the Szegő theory for orthogonal polynomials on the unit circle will recognize that what we seek is essentially a generalized version of the *Szegő function* $D(\mu; z)$.

It turns out that the above question is related to the following

Generalized Electrostatics Problem. *Let* $E \subseteq C$ *be closed. Place a unit charge on* E *so that equilibrium is reached in the presence of an external field due to* w (*see Figure 11*).

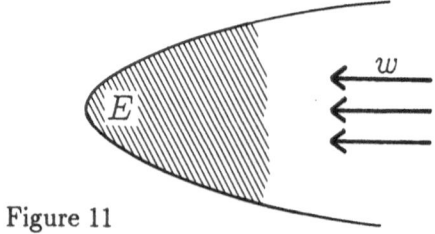

Figure 11

As before, we let $\mathcal{M}(E)$ denote the collection of all positive, unit Borel measures supported on E. Then the energy integral that takes into account the field due to w is

$$I_w[\sigma] := \int\int \log[|z - t|w(z)w(t)]^{-1} d\sigma(z)d\sigma(t)$$

and the generalized Robin's constant is

$$V_w(E) := \inf_{\sigma \in \mathcal{M}(E)} I_w[\sigma]$$

(cf. (4.2) and (4.3)). What we therefore seek is a measure $\sigma_w \in \mathcal{M}(E)$ such that

$$I_w[\sigma_w] = V_w(E). \tag{7.9}$$

For such a measure to exist we need to make some mild assumptions on w.

Definition 7.2. A weight $w : E \to [0, \infty)$ is said to be *admissible* if the following three conditions hold:
(i) w is upper semi-continuous;
(ii) The set $\{z \in E : w(z) > 0\}$ has positive (inner) capacity;
(iii) If E is unbounded, then $|z|w(z) \to 0$ as $|z| \to \infty$, $z \in E$.

Since the external field due to w has a strong repelling effect near points where $w = 0$, the assumption (iii) physically means that this repelling effect is sufficient to prevent charges placed on E from rushing to ∞.

The theory of weighted polynomials and potentials is developed in [GR1],[MS2], [MS3],[GR2] and [STM]. The basis for this theory is the following result (cf. [MS2] for the case when $E \subseteq \mathbf{R}$).

Theorem 7.3. Let $E \subseteq \mathbf{C}$ be closed and $w : E \to [0, \infty)$ be admissible. Then
(a) $\exists$ a unique $\sigma_w \in \mathcal{M}(E)$ such that (7.9) holds.
(b) $S(\sigma_w) = \mathrm{supp}(\sigma_w)$ is compact.
(c) $w(z) = \exp(U^{\sigma_w}(z) - F_w)$ q.e. on $S(\sigma_w)$
and

$$w(z) \leq \exp(U^{\sigma_w}(z) - F_w) \qquad \text{q.e. on} \quad E,$$

where the constant F_w is given by

$$F_w := V_w(E) + \int \log w \, d\sigma_w.$$

Notice that if E is compact, $\mathrm{cap}(E) > 0$ and $w \equiv 1$ on E, then $\sigma_w = \nu_E$, the equilibrium distribution for E.

From Lemma 7.1 and Theorem 7.3 we get

Corollary 7.4. If $w : E \to [0, \infty)$ is admissible and $q_n \in \mathcal{P}_n$, then

$$|q_n(z)| \leq \| w^n q_n \|_E \exp\{-n(U^{\sigma_w}(z) - F_w)\}, \quad \forall \quad z \in \mathbf{C}.$$

We remark that in the above inequality, $\| w^n q_n \|_E$ can be replaced by $\| w^n q_n \|_{S(\sigma_w)}$.

Corollary 7.4 (with $W = w^n$) gives an answer to the problem stated at the beginning of this section. Thus it provides a starting point for the analysis of orthogonal polynomials on unbounded sets. For example, together with some "hard analysis", it leads to the solution of the Freud conjecture dealing with exponential weights on $\mathbf{R}$ (cf. [LMS]).

References

[BSS]   H.–P. Blatt, E.B. Saff and M. Simkani (1988), *Jentzsch–Szegö type theorems for the zeros of best approximants*, J. London Math. Soc., **38**: 307–316.

[D]   P.J. Davis (1963), *Interpolation and Approximation*, Blaisdell Pub. Co, New York.

[EF]   P. Erdös and G. Freud (1974), *On orthogonal polynomials with regularly distributed zeros*, Proc. London Math. Soc., **29**: 521–537.

[ET]   P. Erdös and P. Turán (1940), *On interpolation III*, Ann. of Math., **41**: 510–555.

[F]   G. Freud (1971), *Orthogonal Polynomials*, Pergamon Press, London.

[G]   D. Gaier (1987), *Lectures on Complex Approximation*, Birkhauser, Boston.

[GR1]   A.A. Gonchar and E.A. Rakhmanov (1984), *The equilibrium measure and the distribution of zeros of extremal polynomials*, Math. Sb. **125(167)**: 117–127. Math. USSR. Sbornik **53**(1986), 119–130.

[GR2]   A.A. Gonchar and E.A. Rakhmanov (1987), *Equilibrium distributions and the rate of rational approximation to analytic functions*, Math. Sb. **134(176)**: 305–352. (Russian).

[He]   L.L. Helms (1969), *Introduction to Potential Theory*, Wiley–Interscience (Pure and Applied Mathematics, vol. **XXII**), New York.

[Hi]   E. Hille (1962), *Analytic Function Theory* (Introduction to Higher Mathematics, vol. **II**), Ginn and Co., Boston.

[La]   N.S. Landkof (1972), *Foundations of Modern Potential Theory*, Springer–Verlag, Berlin.

[LMS]   D.S. Lubinsky, H.N. Mhaskar and E.B. Saff (1986), *Freud's conjecture for exponential weights*, Bull. Amer. Math. Soc., **15**: 217–221.

[LS]   D.S. Lubinsky and E.B. Saff (1988), *Strong Asymptotics for Extremal Polynomials Associated with Weights on* $\mathbb{R}$, Lecture Notes in Math, Vol. 1305, Springer–Verlag, Berlin.

[LSS]   X. Li, E.B. Saff and Z. Sha (to appear), *Behavior of best L_p polynomial approximants on the unit interval and on the unit circle*, J. Approx. Theory.

[MNT]   A. Máté, P. Nevai and V. Totik (1987), *Strong and weak convergence of orthogonal polynomials*, Amer. J. Math., **109**: 239–282.

[MS1]   H.N. Mhaskar and E.B. Saff (1984), *Extremal problems for polynomials with exponential weights*, Trans. Amer. Math. Soc. 285: 203–234.

[MS2]   H.N. Mhaskar and E.B. Saff (1985), *Where does the sup norm of a weighted polynomial live?*, Constr. Approx. 1: 71–91.

[MS3]   H.N. Mhaskar and E.B. Saff (1987), *Where does the L_p–norm of a weighted polynomial live?*, Trans. Amer. Math. Soc., 303: 109–124. *Errata* (1988), 303:431.

[MS4]   H.N. Mhaskar and E.B. Saff (to appear), *On the distribution of zeros of polynomials orthogonal on the unit circle*, J. Approx. Theory.

[NT]   P. Nevai and V. Totik (to appear), *Orthogonal polynomials and their zeros*.

[R]   E.A.· Rakhmanov (1984), *On asymptotic properties of polynomials orthogonal on the real axis*, Math. USSR. Sb. 47: 155–193.

[STM]   E.B. Saff, V. Totik and H. Mhaskar (to appear), *Weighted Polynomials and Potentials in the Complex Plane*.

[Sz]   G. Szegő (1967), *Orthogonal Polynomials*, 3rd. ed., Amer. Math. Soc. Colloq. Pub, Vol. 23, Amer. Math. Soc., Providence, R.I.

[T]   M. Tsuji (1959), *Potential Theory in Modern Function Theory*, Dover, New York.

[Wa]   J.L. Walsh (1960), *Interpolation and Approximation by Rational Functions in the Complex Domain*. 3rd ed., Amer. Math. Soc. Colloq. Publ., Vol. 20, Amer. Math. Soc., Providence, R.I.

[Wi]   H. Widom (1967), *Polynomials associated with measures in the complex plane*, J. Math. Mech., 16: 997–1013.

NTH ROOT ASYMPTOTIC BEHAVIOR OF ORTHONORMAL POLYNOMIALS

HERBERT STAHL
Technische Fachhochschule Berlin/FB2
Luxemburger Straße 10
D-1000 Berlin 65, West Germany

VILMOS TOTIK
Bolyai Institute
Aradi Vértanúk tere 1, Szeged
6720 Hungary

ABSTRACT: We are concerned with asymptotic behavior of orthonormal polynomials $p_n(\mu; z)$ as $n \to \infty$. We start from a positive weight measure μ with compact support $S(\mu) \subseteq \mathbf{C}$. The results discussed here include:
 (i) lower and upper asymptotic bounds for the nth root of the orthonormal polynomials,
 (ii) different definitions of regular nth root asymptotic behavior and their equivalence,
(iii) characterizations of regular asymptotic behavior by means of arbitrary sequences of polynomials in $L^2(\mu)$,
(iv) criteria for regular asymptotic behavior and relations between them,
 (v) distribution of zeros of orthonormal polynomials,
(vi) localization,
and some application of these results.

1. Introduction

Throughout this paper μ is a positive Borel measure with compact support $S(\mu) \subseteq \mathbf{C}$. It will be called weight measure. In order to avoid degeneracies we assume that $S(\mu)$ contains infinitely many points. By

$$p_n(\mu; z) = \gamma_n(\mu)z^n + \dots \quad (\gamma_n(\mu) > 0)$$

we denote the uniquely existing orthonormal polynomial of degree $n \in \mathbf{N}$, i.e.,

$$\int p_n(\mu; x)\overline{p_m(\mu; x)}d\mu(x) = \delta_{mn} \quad \text{for all} \quad m, n \in \mathbf{N},$$

where δ_{mn} is Kronecker's symbol.

We investigate the nth root asymptotic behavior of the polynomials $p_n(\mu; z)$ as $n \to \infty$, i.e., we study the sequence

$$\{|p_n(\mu; z)|^{1/n} \,|\, n \in \mathbf{N}\},$$

Based on research supported in part by the Deutsche Fortschungsgemeinschaft, AZ: Sta 299/2-1 (first author) and by the Hungarian Science Foundation for Research, Grant No. 1157 (second author).

P. Nevai (ed.), Orthogonal Polynomials, 395–417.
© *1990 by Kluwer Academic Publishers.*

and ask what can be said about the asymptotic behavior of this sequence in case of a general weight measure μ with compact support $S(\mu) \subseteq \mathbf{C}$. Let us mention at once that considering the nth root of the absolute value of the polynomials $p_n(\mu; z)$ does not restrict generality, for an asymptotic formula on $|p_n(\mu; z)|^{1/n}$ usually automatically implies an asymptotic formula on $p_n(\mu; z)^{1/n}$ itself.

For a classification one can distinguish three types of asymptotic behavior, namely power (or Szegő), ratio, and nth root asymptotics. These mean that the sequences

$$(1.1) \qquad \left\{ \left. \frac{p_n(\mu; z)}{\varphi(z)^n} \right| n \in \mathbf{N} \right\},$$

$$(1.2) \qquad \left\{ \left. \frac{p_{n+1}(\mu; z)}{p_n(\mu; z)} \right| n \in \mathbf{N} \right\},$$

or

$$(1.3) \qquad \left\{ \left. \sqrt[n]{p_n(\mu; z)} \right| n \in \mathbf{N} \right\},$$

respectively, tend to a limit on a certain set of values $z \in \mathbf{C}$ as $n \to \infty$. In (1.1) the existence of a limit also depends on an appropriate choice of the function φ. (For more details on power and ratio asymptotics see [Lu], or §III.7, §V.4, and §V.5 in [Fr]).

It is easy to see that the existence of limits in (1.1), (1.2), and (1.3) forms an hierachy: if there exists a limit for one of the three sequences, then necessarily any later sequence also tends to a limit on the same set of points $z \in \mathbf{C}$. Consequently, from all three types the nth root asymptotic behavior requires the weakest assumptions, and can be investigated under the most general conditions. At the same time it is sufficient for many applications, as for instance the convergence of polynomial (Chebychev-Fourier) expansions based on the system $\{p_n(\mu; z) \mid n \in \mathbf{N}\}$ (see Chapter IV of [Fr]), the convergence of continued fractions or Padé approximants to Markov functions, where orthogonal polynomials appear in the denominator of the approximants (see §68 of [Pe], or Chapter 5 of [BaGM]), or multipoint rational interpolants, to give just some examples.

Apparently the first study of nth root asymptotic behavior was done by Faber [Fa] in 1922 in connection with polynomial expansions. Of course, in the spirit of his time he used quite strong assumptions. In his paper the measure μ is defined on $[-1, 1]$ by a density function which is assumed to be bounded away from zero. Faber's paper appeared about the same time as Szegő's first papers ([Sz1], [Sz2]) on power asymptotics.

Major contributions to the further development of the subject on nth root asymptotic behavior have been given by Erdős and Turán [ErTu], Widom [Wi], Ullman [Ul1], and Erdős and Freud [ErFr]. We shall come back to their contributions at later places. A good survey over recent developments of the whole theory (not only nth root theory) of asymptotic behavior of orthogonal polynomials can be found in [Lu], [Ne1] and [Ne2].

This paper has a survey character; we therefore give no proofs. Some of the results can be found or have been announced in special case in earlier literature, but a great part of our theorems are new. Their proofs are contained in a longer memoir which will be published elsewhere.

2. Upper and lower bounds

In this section we discuss sharp upper and lower bounds for orthonormal polynomials with respect to general weights. The results hold true for any measure μ with compact support in $\mathbf{C}$. We start with some introductory definitions most of which are concerned with potential theory. For reference to potential theory we suggest [dVP],[La] and [Ts].

For a measure ν with compact support $S(\nu)$ in the complex plane $\mathbf{C}$ the logarithmic energy of ν is defined as

$$I(\nu) = \int \int \log \frac{1}{|z-t|} \, d\nu(z) d\nu(t).$$

If K is a compact subset of $\mathbf{C}$, then with

$$V_K = \inf_{S(\nu) \subseteq K} I(\nu)$$

the (logarithmic) capacity of K is defined as

$$\text{cap}(K) = \exp(-V_K).$$

In what follows we need to extend the notion of capacity to Borel sets S: $\text{cap}(S)$ denotes the supremum of the capacities of compact subsets of S (see Chapter 11, §2 of [La]), and we say that a property holds qu.e. (quasi everywhere) on a set $S \subseteq \mathbf{C}$ if it holds on S with possible exceptions on a subset of capacity zero. By $g_B(z; \infty)$ we denote the (generalized) Green function with logarithmic pole at infinity of a Borel set $B \subseteq \overline{\mathbf{C}}$ with bounded complement $\overline{\mathbf{C}} \backslash B$. For our investigation and especially for the definition of the minimal carrier Green function below, it is essential that the Green function is defined for Borel sets $B \subseteq \overline{\mathbf{C}}$, and not only for domains in $\overline{\mathbf{C}}$, as is the case in most text books. The Green function $g_B(z; \infty)$ of a Borel set $B \subseteq \overline{\mathbf{C}}$ with bounded complement has the following defining properties:

(i) $g_B(z; \infty)$ is non-negative and subharmonic in $\mathbf{C}$, and harmonic in $\text{Int}(B) \backslash \{\infty\}$,
(ii) $g_B(z; \infty) = \log|z| - \log \text{cap}(\mathbf{C} \backslash B) + o(1)$ as $|z| \to \infty$, where $o(1)$ tends to 0 as $|z| \to \infty$,
(iii) $g_B(z; \infty) = 0$ for z qu.e. on $\mathbf{C} \backslash B$.

If $\text{cap}(\mathbf{C} \backslash B) = 0$, then $g_B(z; \infty) \equiv \infty$. The function $g_B(z; \infty)$ exists and is uniquely determined by (i) to (iii) for every Borel set $B \subseteq \overline{\mathbf{C}}$ with bounded complement $\overline{\mathbf{C}} \backslash B$,. If the bounded complement $C := \mathbf{C} \backslash B$ is of positive capacity, i.e. if

$$\text{cap}(C) > 0,$$

then there uniquely exists a probability measure ω_C, called the equilibrium distribution of C, with $S(\omega_C) \subseteq \overline{C}$ such that the Green function $g_B(z; \infty)$ has the representation

$$g_B(z; \infty) \equiv -p(\omega_C; z) - \log \text{cap}(C),$$

where $p(\nu; z)$ denotes the logarithmic potential

$$p(\nu; z) = \int \log \frac{1}{|z-t|} \, d\nu(t)$$

of the measure ν.

By μ we will always denote a finite Borel measure on $\mathbf{C}$ with compact support $S(\mu)$. Let $\Omega = \Omega(\mu)$ denote the outer domain of $S(\mu)$, i.e. the unbounded component of $\overline{\mathbf{C}}\backslash S(\mu)$, $Co(S(\mu))$ the convex hull and $Pc(S(\mu))$ the polynomial convex hull of $S(\mu)$, i.e. $Pc(S(\mu)) = \overline{\mathbf{C}}\backslash\Omega$. (The name polynomial convex hull is derived from the fact that for any compact set $S \subseteq \mathbf{C}$ the set $Pc(S)$ is the intersection of all sets $S_p := \{z \in \mathbf{C} \mid |p(z)| \leq \sup_{x \in S} |p(x)|\}$ with p a polynomial not identically zero.) The set $\partial\Omega \subseteq \partial S(\mu)$ is called the outer boundary of $S(\mu)$. If C is a compact set, then $S(\omega_C)$ is contained in the outer boundary $\partial Pc(C)$ of C; in general however, we only know that $S(\omega_C) \subseteq \overline{C}$.

While $\mathrm{cap}(S(\mu))$ and $g_\Omega(z;\infty)$ depend only on the set $S(\mu)$, or more precisely, on the outer boundary $\partial\Omega$ of $S(\mu)$, we now introduce a type of capacity and Green function that depends on the carriers of the measure μ.

Let $\Gamma(\mu)$ be the set of all carriers of the measure μ, i.e.

$$\Gamma(\mu) := \{C \subseteq \mathbf{C} \mid C \text{ a Borel set and } \mu(\mathbf{C}\backslash C) = 0\},$$

then the minimal carrier capacity (of the measure μ) is defined as

$$c_\mu := \inf\{\mathrm{cap}(C) \mid C \in \Gamma(\mu),\ C \text{ bounded}\}$$

and the minimal carrier Green function (of μ) is defined as

$$g_\mu(z;\infty) := \sup\{g_{\overline{\mathbf{C}}\backslash C}(z;\infty) \mid C \in \Gamma(\mu),\ C \text{ bounded}\}, \qquad z \in \mathbf{C}.$$

It is obvious that

$$(2.1) \qquad\qquad c_\mu \leq \mathrm{cap}(S(\mu)),$$

and

$$(2.2) \qquad\qquad g_\mu(z;\infty) \geq g_\Omega(z;\infty)$$

for all $z \in \mathbf{C}$. There are examples showing that in (2.1) and (2.2) proper inequality as well as equality may hold true.

Theorem 2.1. *We have*

$$(2.3) \qquad\qquad \limsup_{n\to\infty} |p_n(\mu;z)|^{1/n} \leq e^{g_\mu(z;\infty)}$$

locally uniformly for all $z \in \mathbf{C}$, and

$$(2.4) \qquad\qquad \liminf_{n\to\infty} |p_n(\mu;z)|^{1/n} \geq e^{g_\Omega(z;\infty)}$$

locally uniformly for all $z \in \mathbf{C}\backslash Co(S(\mu))$. In $Co(S(\mu))\cap\Omega$ the asymptotic lower bound (2.4) holds true only in capacity, i.e. for every compact set $V \subseteq Co(S(\mu)) \cap \Omega$ and every $\varepsilon > 0$ we have

$$(2.5) \qquad \lim_{n\to\infty} \mathrm{cap}\left(\{z \in V \mid |p_n(\mu;z)|^{1/n} < e^{g_\Omega(z;\infty)} - \varepsilon\}\right) = 0.$$

In $Co(S(\mu)) \cap \Omega$ the lower bound can also be given in the following form: For every infinite subsequence $N \subseteq \mathbf{N}$ we have

$$\limsup_{n \to \infty,\; n \in N} |p_n(\mu; z)|^{1/n} \geq e^{g_\Omega(z;\infty)} \qquad \text{for} \quad z \quad \text{qu.e. in} \quad \Omega \cap Co(S(\mu)),$$

and on the outer boundary $\partial \Omega$ of $S(\mu)$ we have

$$\limsup_{n \to \infty,\; n \in N} |p_n(\mu; z)|^{1/n} \geq 1 \qquad \text{for} \quad z \quad \text{qu.e. on} \quad \partial \Omega.$$

Remarks. (1) If the two Green functions $g_\Omega(z; \infty)$ and $g_\mu(z; \infty)$ are identical, then we have proper and identical limits in (2.3) and (2.4). (The existence of a proper limit and equality in (2.4) will be called regular asymptotic behavior in the next section.)

(2) While the upper asymptotic bound (2.3) holds true uniformly on any compact subset of $\mathbf{C}$, the lower bound (2.4) holds true in this strong sense only on compact subsets of $\mathbf{C} \setminus Co(S(\mu))$. This weakness of the lower asymptotic bound in $Co(S(\mu))$ is caused by the zeros that the polynomials $p_n(\mu; z)$ may have everywhere in $Co(S(\mu))$. In Corollary 2.2 the zeros in $Co(S(\mu)) \cap \Omega$ will be factored out.

(3) The upper asymptotic bound (2.3) is not specific for orthonormal polynomials. It holds true for any sequence of polynomials normalized in $L^2(\mu)$, as will be shown in Theorem 2.5 below.

(4) On the outer boundary $\partial \Omega$ of $S(\mu)$ an asymptotic estimate in capacity, like that in (2.5), can, in general, not be true since there the asymptotic density of the zeros can be positive almost everywhere. The special case $\mathrm{Int}(Pc(S(\mu))) = \emptyset$, which includes all weight measures μ on the real axis, is formulated separately in Corollary 2.3.

(5) The simple example of normalized Lebesgue measure on the unit circumference (in which case the n-th orthonormal polynomial is just z^n) shows that in the interior of the polynomial convex hull $Pc(S(\mu))$ we can in general not expect 1 as a lower asymptotic bound.

Corollary 2.2. Let $U, V \subseteq Co(S(\mu)) \cap \Omega$ be two compact sets with U containing V in its interior, and let $x_{n,1}, \ldots, x_{n,m(n)}$ be the zeros of $p_n(\mu; z)$, $n \in \mathbf{N}$, on U. Then we have

$$\liminf_{n \to \infty} \left| \frac{p_n(\mu; z)}{\prod_{j=1}^{m(n)} (z - x_{nj})} \right|^{1/n} \geq e^{g_\Omega(z, \infty)}$$

uniformly on V.

Let us add here (see Theorem 5.1 below) that $m(n)$, the number of zeros in U is bounded as $n \to \infty$.

Corollary 2.3. If the interior of $Pc(S(\mu))$ is empty, then for any infinite subsequence $N \subseteq \mathbf{N}$ we have

$$\limsup_{n \to \infty,\; n \in N} |p_n(\mu; z)|^{1/n} \geq 1 \qquad \text{for} \quad z \quad \text{qu.e. on} \quad S(\mu).$$

Note that $Pc(S(\mu))$ is empty exactly when $S(\mu)$ has connected complement and empty interior.

If we consider the orthonormal polynomials $p_n(\mu; z)$, $n \in \mathbf{N}$, near infinity, then from the upper and lower asymptotic bounds (2.3) and (2.4) in Theorem 2.1 we immediately deduce upper and lower asymptotic bounds for the nth root of the leading coefficient $\gamma_n(\mu)$ of the orthonormal polynomials:

Corollary 2.4. *We have*

$$\frac{1}{\text{cap}(S(\mu))} \leq \liminf_{n\to\infty} \gamma_n(\mu)^{1/n} \leq \limsup_{n\to\infty} \gamma_n(\mu)^{1/n} \leq \frac{1}{c_\mu}.$$

where $\text{cap}(S(\mu)) = 0$ *or* $c_\mu = 0$ *is allowed.*

The next theorem states the upper estimate of Theorem 2.1 for arbitrary sequences of polynomials. The upper bound (2.3) then follows as a corollary.

Theorem 2.5. *For any sequence of polynomials* p_n *not identically zero and of degree at most* $n \in \mathbf{N}$, *we have*

$$\limsup_{n\to\infty} \left[\frac{|p_n(z)|}{\|p_n\|_{L^2(\mu)}} \right]^{1/n} \leq e^{g_\mu(z;\infty)}$$

locally uniformly in $\mathbf{C}$.

Finally, we mention a result that shows the sharpness of the estimates in Theorem 2.1. To do this let us call two measures ν and μ carrier related (in sign $\nu \sim \mu$) if ν and μ have the same carriers, i.e. $\Gamma(\nu) = \Gamma(\mu)$. Obviously, this is the same as their mutual absolute continuity. Since for carrier related measures ν and μ the Green functions g_μ and g_ν are identical, furthermore $S(\nu) = S(\mu)$, it follows that the upper and lower bounds (2.3) and (2.4) are the same for carrier related measures. From this point of view they are sharp as is shown by

Theorem 2.6. *(a) There exists a* $\nu_1 \sim \mu$ *such that*

$$\lim_{n\to\infty} |p_n(\nu_1; z)|^{1/n} = e^{g_\mu(z;\infty)}$$

locally uniformly for $z \notin Co(S(\mu))$.
(b) There exists a $\nu_2 \sim \mu$ *such that*

$$\lim_{n\to\infty} |p_n(\nu_2; z)|^{1/n} = e^{g_\cap(z;\infty)}$$

locally uniformly for $z \notin Co(S(\mu))$.
(c) There exists a $\nu_3 \sim \mu$ *such that*

$$\liminf_{n\to\infty} |p_n(\nu_3; z)|^{1/n} = e^{g_\cap(z;\infty)}$$

and

$$\limsup_{n\to\infty} |p_n(\nu_3; z)|^{1/n} = e^{g_\mu(z;\infty)}$$

for every $z \notin Co(S(\mu))$.

The upper asymptotic bound (2.3) of Theorem 2.1 was proved by Ullman in Theorem 1.5 of [Ul1] for weight measures with $S(\mu) = [-1,1]$, and was announced in Theorem 3 of [Ul3] for general weight measures μ in $\mathbf{C}$.

The lower asymptotic bound (2.4) in Theorem 2.1 has been proved by Widom [Wi] for special weight measures μ in $\mathbf{C}$, which he calls admissible, i.e. measures satisfying Widom's criterion (in its original formulation, as given in Corollary 4.6 below). For these measures the lower bound can be established by a comparison with monic Chebychev polynomials of the set $S(\mu)$. For weight measures μ on $\mathbf{R}$ the lower bound (2.4) has been stated without proof in [Ul2].

The sharpness of the estimates, again for $S(\mu) = [-1, 1]$, was shown in [UWZ] and [UlWi]. In its generality Theorem 2.6 is new.

3. Regular nth root asymptotic behaviour of orthonormal polynomials

When the asumptotic lower bound in Corollary 2.4 is assumed then the measure and the corresponding orthonormal polynomials are called of regular (nth root) asymptotic behaviour. Orthogonal polynomials of regular asymptotic behaviour are the natural generalization of classical orthogonal polynomials to the general case. Many applications use one of the several equivalent forms of regular asymptotic behaviour to be discussed in this section. Thus, for instance, the sequence $\{p_n(\mu; z) \mid n \in \mathbf{N}\}$ is said to have regular exterior asymptotic behavior if

$$(3.1) \qquad \lim_{n \to \infty} |p_n(\mu; z)|^{1/n} = e^{g_\Omega(z;\infty)}$$

locally uniformly for $z \in \overline{\mathbf{C}} \backslash Co(S(\mu))$. This terminology have been used (c.f. [Ul2]) for the special case of weight measures μ with $S(\mu) = [-1, 1]$ in which case (3.1) takes the form

$$(3.2) \qquad \lim_{n \to \infty} |p_n(\mu; z)|^{1/n} = |z + \sqrt{z^2 - 1}|$$

where that branch of the square root is taken which is positive for positive z. The meaning of (3.2) is that $p_n(\mu; z)$ asymptotically behaves like the classical orthogonal polynomials, say Chebyshev or Jacobi polynomials.

If $S(\mu) = [-1, 1]$, then the outer boundary $\partial\Omega$ of $S(\mu)$ is identical with $[-1, 1]$. For this case in [Ul2] it was said that the sequence $\{p_n(\mu; z) \mid n \in \mathbf{N}\}$ has regular interior asymptotic behavior if

$$(3.3) \qquad \limsup_{n \to \infty} |p_n(\mu; z)|^{1/n} = 1 \qquad \text{for} \quad z \quad \text{qu.e. on} \quad \partial\Omega.$$

The notations 'exterior' and 'interior' asymptotic behavior keep their intuitive meaning so long as $\text{Int}(Pc(S(\mu))) = \emptyset$, since then we have $\partial\Omega = S(\mu)$.

A comparison of (3.1) and (3.3) with the asymptotic bounds given in Theorem 2.1 shows that regular asymptotic behavior means that the lower bound (2.4) is assumed asymptotically. This, of course, will always be the case if the weight measure μ is sufficiently dense everywhere on its support $S(\mu)$, and since this is often the case, the name 'regular' for this type of asymptotic behavior is justified. For example, the normalized Jacobi polynomials, which can be seen as the prototypes of orthonormal polynomials associated with a weight measure with compact support, possess the regular asymptotic behavior (3.1) and (3.3). Of course, they also possess the stronger types of regular asymptotic behavior sketched in (1.1) and (1.2) (see for instance [Sz3] or [Fr]).

We remark that $\operatorname{cap}(S(\mu)) = 0$ has not been excluded in (3.1) and (3.3), and neither will it be excluded in the next theorem, although this case is somewhat special: we have $g_\Omega(z; \infty) \equiv \infty$, and therefore the limit (3.1) follows from the lower and upper bounds (2.3) and (2.4) of Theorem 2.1 for any weight measure μ with $\operatorname{cap}(S(\mu)) = 0$.

Our definition of regular (nth root) asymptotic behavior is based on the following theorem:

Theorem 3.1. *The following four assertions are equivalent:*
(i) The limit

$$\lim_{n \to \infty} \gamma_n(\mu)^{1/n} = \frac{1}{\operatorname{cap}(S(\mu))}$$

holds true, where $\gamma_n(\mu)$ is the leading coefficient of the orthonormal polynomial $p_n(\mu; z)$.
(ii) The limit

$$\lim_{n \to \infty} |p_n(\mu; z)|^{1/n} = e^{g_\Omega(z; \infty)}$$

holds true locally uniformly in $\overline{\mathbb{C}} \backslash Co(S(\mu))$.
(iii) The limit

$$\limsup_{n \to \infty} |p_n(\mu; z)|^{1/n} = 1$$

holds true qu.e on $\partial \Omega$.
(iv) The limit in (ii) holds true locally uniformly in $\overline{\mathbb{C}} \backslash Co(S(\mu))$, and in capacity in $\Omega \cap Co(S(\mu))$, i.e. for every compact set $V \subseteq \Omega \cap Co(S(\mu))$ and every $\varepsilon > 0$ we have

$$\lim_{n \to \infty} \operatorname{cap}\left(\left\{ z \in V \;\Big|\; \left| |p_n(\mu; z)|^{1/n} - e^{g_\Omega(z; \infty)} \right| > \varepsilon \right\}\right) = 0,$$

and further the limit in (iii) holds true qu.e. on $\partial \Omega$, and we have

$$\limsup_{n \to \infty} |p_n(\mu; z)|^{1/n} \le e^{g_\Omega(z; \infty)}$$

locally uniformly in $\mathbb{C}$.

If one of the four assertions of Theorem 3.1 holds true, then the orthonormal polynomials $p_n(\mu; z)$, $n \in \mathbb{N}$, associated with the measure μ are said to have regular (n-th root) asymptotic behavior, and we write $\mu \in \mathbf{Reg}$. We shall also often refer to $\mu \in \mathbf{Reg}$ simply as μ is *regular*.

We remark that a measure μ may belong to $\mathbf{Reg}$, although the lower and upper bounds given in Theorem 2.1 are different.

Regular asymptotic behavior discussed in the preceding section is equivalent to several statements relating the size of general polynomials to their $L^2(\mu)$ (or $L^p(\mu)$) norm. We discuss such relations in this section. They have applications in several problems of approximation theory.

Theorem 3.2. *The following five statements are equivalent for a measure μ.*
(i) The sequence $\{p_n(\mu; \cdot)\}_{n=1}^\infty$ has regular (nth root) asymptotic behavior, i.e. $\mu \in \mathbf{Reg}$.
(ii) We have

(3.4)
$$\limsup_{n \to \infty} |p_n(\mu; z)|^{1/n} \le e^{g_\Omega(z; \infty)}$$

locally uniformly on **C**.

(iii) For any sequence $\{P_n\}_{n=1}^{\infty}$ of nonzero polynomials of degree $\deg(P_n) \leq n$ we have

$$\limsup_{n\to\infty} \left(\frac{|P_n(z)|}{\|P_n\|_{L^2(\mu)}} \right)^{1/n} \leq e^{g_\Omega(z;\infty)}$$

*locally uniformly for $z \in$ **C**.*

(iv) We have

$$(3.5) \qquad\qquad \limsup_{n\to\infty} |p_n(\mu; z)|^{1/n} \leq 1$$

for quasi-every $z \in \partial\Omega$.

(v) For any sequence $\{P_n\}$ as in (iii)

$$\limsup_{n\to\infty} \left(\frac{|P_n(z)|}{\|P_n\|_{L^2(\mu)}} \right)^{1/n} \leq 1$$

for quasi-every $z \in \partial\Omega$.

    If we also have $S(\mu) \subseteq \partial\Omega$, in particular, if $\mathrm{Int}(Pc(S(\mu))) = \emptyset$, then each of (i)—(v) is equivalent to

(vi) Let $g_n \geq 0$ be weight functions defined on $S(\mu)$ in such a way that $\lim_{n\to\infty} g_n^{1/n} = g$ exists uniformly on $S(\mu)$, where $g > 0$ and g is continuous on $S(\mu)$. We set $d\mu_n := g_n d\mu$. Then for any sequence of polynomials $\{P_n\}$ as in assertion (iii) we have

$$\limsup_{n\to\infty} \left(\frac{g_n(z)^{1/2}|P_n(z)|}{\|P_n\|_{L^2(\mu_n)}} \right)^{1/n} \leq 1$$

for quasy-every $z \in \partial\Omega = S(\mu)$.

Remarks. (1) We mention that the $L^2(\mu)$ norms in (iii), (v) and (vi) can be replaced by any $L^p(\mu)$, $1 \leq p < \infty$ norms.
(2) Note also that we must automatically have equality in (3.4) for $z \notin Co(S(\mu))$ and in (3.5) for quasy-every $z \in \partial\Omega$ (see Theorem 2.1).

    When the support $S(\mu)$ of μ, more precisely its outer boundary, is regular with respect to the Dirichlet problems in the domain Ω then the conditions in Theorem 3.1 hold uniformly for $\mu \in$ **Reg**. Because of its importance we state this case as a separate theorem.

Theorem 3.3. *Suppose that $S(\mu)$ is a regular set with respect to Dirichlet problems in Ω. Then the following five statements are equivalent.*
*(i) $\mu \in$ **Reg**.*

(ii)
$$\limsup_{n\to\infty} |p_n(\mu; z)|^{1/n} e^{-g_\Omega(z;\infty)} \leq 1$$

uniformly on **C**.

(iii)
$$\lim_{n\to\infty} \left(\sup_{\substack{\deg P_n \leq n \\ P_n \neq 0}} \frac{|P_n(z)|}{\|P_n\|_{L^2(\mu)}} \right)^{1/n} e^{-g_\Omega(z;\infty)} = 1$$

uniformly on **C**.

(iv)
$$\lim_{n\to\infty} \|p_n(\mu;\cdot)\|_{\sup,S(\mu)}^{1/n} = 1,$$

where $\|\cdot\|_{\sup,S(\mu)}$ *denotes the supremum norm on* $S(\mu)$.

(v)
$$\lim_{n\to\infty} \left(\sup_{\substack{\deg P_n \le n \\ P_n \not\equiv 0}} \frac{\|P_n\|_{\sup,S(\mu)}}{\|P_n\|_{L^2(\mu)}} \right)^{1/n} = 1.$$

If in addition $S(\mu) \subseteq \partial\Omega$, then any of (i)—(v) is equivalent to
(vi) if $\{g_n\}$ is a sequence as in assertion (vi) of Theorem 3.1, then

$$\lim_{n\to\infty} \left(\sup_{\substack{\deg P_n \le n \\ P_n \not\equiv 0}} \frac{\|g_n^{1/2} P_n\|_{\sup,S(\mu)}}{\|P_n\|_{L^2(\mu_n)}} \right)^{1/n} = 1.$$

4. Regularity Criteria

The definition of regularity, as for instance

$$\lim_{n\to\infty} \gamma_n(\mu)^{1/n} = \frac{1}{\operatorname{cap}(S(\mu))},$$

(see Theorem 3.1) and also many of the equivalent formulations of it presented in the preceding section refer directly to the orthogonal polynomials in question and therefore, except for a few concrete cases, it is very difficult to verify its validity. Besides this, usually we want to derive different properties of the orthogonal polynomials (say zero asymptotics, asymptotics away from the support etc.) or of general polynomials (c.f. Theorem 3.2,(iii)) from the knowledge of the regularity of μ. Therefore we are looking for criteria that directly connect μ to its regularity.

It follows from Corollary 2.4 and the definition of regularity in the sense of Section 3 that every measure with support of zero capacity is regular. Therefore, in what follows until the end of this section we shall always assume $\operatorname{cap}(S(\mu)) > 0$. This also implies the existence of $\omega_{S(\mu)}$, the equilibrium distribution of $S(\mu)$.

The first regularity criterion was given in an influential paper by P. Erdős and P. Turán [ErTu]. They showed that if μ is a measure on $[-1,1]$ and $\mu'(x) > 0$ almost everywhere on $[-1,1]$, then μ is regular (of course, $\mu'(x)$ denotes the Radon–Nikodym derivative of μ with respect to Lebesgue measure, which, in the present case, coincides a.e. with the usual derivative of the function $\mu([-1,x]))$. Seeing that $\omega_{[-1,1]}$ and the Lebesgue measure on $[-1,1]$ are mutually absolutely continuous with respect to the other one, the following criterion is an obvious generalization:

Erdős–Turán Criterion:

(4.1) $$\frac{d\mu}{d\omega_{S(\mu)}} > 0 \qquad \omega_{S(\mu)} - almost \ everywhere.$$

Theorem 4.1. *The Erdős–Turán Criterion implies* $\mu \in$ **Reg**.

Inequality (4.1) is equivalent to

$$\Gamma(\mu) \subseteq \Gamma(\omega_{S(\mu)}),$$

where $\Gamma(\cdot)$ denotes the set of carriers defined in Section 2.

Corollary 4.2. *If* $S(\mu) = A_1 \cup \ldots \cup A_n \cup E$ *with* $A_1, \ldots, A_n$ *smooth and compact Jordan arcs in* **C**, $A_1 \cup \ldots \cup A_n$ *does not separate* $\overline{\mathbf{C}}$, *and* $E \subseteq \mathbf{C}$ *is a Borel set with* $\mathrm{cap}(E) = 0$, *then*

(4.2) $$\frac{d\mu_1}{ds} > 0 \qquad a.e. \ on \qquad A_j \qquad for \qquad j = 1, \ldots, n$$

implies $\mu \in$ **Reg**. *In (4.2)* ds *is the line element of the arc length on* A_j.

Corollary 4.3. *If* $S(\mu) = [-1, 1]$, *then*

$$\frac{d\mu(x)}{dx} > 0 \qquad a.e. \ on \qquad [-1, 1]$$

implies $\mu \in$ **Reg**.

As has already been mentioned, Corollary 4.3 is the original result of Erdős and Turán [ErTu].

A waker condition than the Erdős–Turán one was introduced by J.L. Ullman in [Ul1]. It claimed for a measure μ with $S(\mu) = [-1, 1]$ that $c_\mu = 1/2$ be satisfied. Thus, we call the following

Ullman's Criterion: $c_\mu = \mathrm{cap}(S(\mu))$.

Recall that c_μ is the minimal carrier capacity associated with μ (see Section 2 above).

Theorem 4.4. *Ullman's Criterion implies the regularity of* μ.

This is obvious from Corollary 2.4. It can be shown that the Erdős–Turán Criterion implies Ullman's Criterion but not vice versa.

The following strengthening of the Erdős–Turán Criterion is, up to a small modification, identical with a regularity criterion introduced by Widom in [Wi]. The original formulation of the criterion is given below as Corollary 4.6

Widom's Criterion: *For every carrier C of the measure μ there exists a sequence of compact sets $V_n \subseteq S(\mu)$, $\text{cap}(V_n) > 0$, $n \in \mathbb{N}$, with*

(i)
$$\text{cap}(V_n) \to \text{cap}(S(\mu)) \quad as \quad n \to \infty$$

and

(ii)
$$\omega_{V_n}(C) \to 1 \quad as \quad n \to \infty.$$

Theorem 4.5. *Widom's Criterion implies $\mu \in$ Reg.*

If all sets V_n are taken to be identical $S(\mu)$, then condition (i) is trivially satisfied and condition (ii) becomes equivalent to condition (4.1) of the Erdős–Turán Criterion. Hence, Widom's Criterion is in general stronger than the Erdős–Turán Criterion. That the converse is not true can be seen e.g. by taking μ as the two dimensional Lebesgue measure on the unit disk.

As a corollary to Theorem 4.5 we state the original formulation of Widom's Criterion from [Wi].

Corollary 4.6. *If there exists a family of compact sets $V_t \subseteq S(\mu)$, $0 \le t < \infty$ with*

(i)
$$\lim_{t \to \infty} \text{cap}(V_t) = \text{cap}(S(\mu)),$$

and for any carrier $C \in \Gamma(\mu)$ we have

(ii)
$$\limsup_{t \to \infty} \omega_{V_t}(C) = 1,$$

then $\mu \in$ Reg.

It turns out that Widom's criterion in the above formulation is equivalent to Ullman's criterion. All relations between the different criteria introduced so far are put together in the next theorem.

Theorem 4.7. *The following dependencies hold:*

$$Erdős\text{-}Turán \ Criterion \Rightarrow$$
$$\Rightarrow Widom's \ Criterion \Leftrightarrow Ullman's \ Criterion \Rightarrow$$
$$\Rightarrow \mu \in \textbf{Reg}.$$

All the three criteria above are rather weak in the sense that they do not take into account the size of the weight measure μ. Even the two stronger criteria, namely Ullman's and Widom's Criterion, refer only to the carriers of μ. Furthermore, they work only when in Theorem 2.1 and Corollary 2.4 the upper and lower bounds coincide, i.e. only in the "trivial" case. For example, they can never be used to check the regularity of a discrete measure. Now we present some new criteria which eliminate this problem.

Let $\Delta_r(z) = \{z' |\ |z' - z| \le r\}$. We start with

CRITERION Λ:

(4.3)
$$\mathrm{cap}\left(\left\{z \,\Big|\, \limsup_{r\to 0+} \frac{\log 1/\mu(\Delta_r(z))}{\log 1/r} < \infty\right\}\right) = \mathrm{cap}(S(\mu)).$$

Theorem 4.8. *Criterion Λ implies $\mu \in$ **Reg**.*

We remark that Λ is weaker than Ullman's Criterion (which is equivalent to Widom's Criterion) because the set

$$\left\{z \,\Big|\, \limsup_{r\to 0+} \frac{\log 1/\mu(\Delta_r(z))}{\log 1/r} \le 2\right\}$$

is always a carrier for μ.

Theorem 4.8 is about as far as anyone can go even if $S(\mu) = [0,1]$, namely one can show that in (4.3) the function $\log 1/r$ cannot be replaced by a much bigger one.

Theorem 4.9. *There exists a measure μ such that $S(\mu) = [0,1]$, the set*

$$\left\{x \,\Big|\, \lim_{r\to 0} \frac{\log 1/\mu([x-r,x+r])}{(\log 1/r)^{1+\eta}} = 0 \quad \text{for every} \quad \eta > 0\right\}$$

*is of capacity $1/4 = \mathrm{cap}([0,1])$ but $\mu \notin$ **Reg**.*

Condition Λ says that the measure μ is sufficiently dense in the points of a set of full capacity. Note that from measure theoretic point of view in general this requires denseness only on a relatively sparse set (consider the fact that a set may have full capacity and zero Lebesgue measure). But actually, we do not need this, all we need is a kind of "denseness in the average", therefore we introduce the following "moving variant" of Λ :

CRITERION Λ^* : *There exists a constant $L > 0$ such that*

(4.4)
$$\lim_{r\to 0} \mathrm{cap}(\{z \,|\, \mu(\Delta_r(z)) \ge r^L\}) = \mathrm{cap}(S(\mu)).$$

Theorem 4.10. *Assume that $S(\mu) = \mathrm{supp}(\mu)$ is regular with respect to the solutions of Dirichlet problems in $\Omega = \mathbf{C}\backslash Pc(S(\mu))$. Then criterion Λ^* implies the regularity of μ.*

It can be shown that Λ^* can be significantly weaker than Λ even for $S(\mu) = [0,1]$, namely one can constructwe a μ with $S(\mu) = [0,1]$ for which (4.4) is true for every $L > 1$ but the set

$$\left\{x \,\Big|\, \limsup_{r\to 0+0} \frac{\log 1/\mu([x-r,x+r])}{\log 1/r} < \infty\right\}$$

is of zero capacity.

Theorem 4.10 is best possible in the following sense:

Theorem 4.11. *If* $\gamma : (0,1) \to (0,1)$ *is any function tending faster to zero than any* r^L, $L > 0$ *as* $r \to 0 + 0$, *then there exists a measure* μ *such that* $S(\mu) = [0,1]$,

$$\lim_{r \to 0} \operatorname{cap}(\{x \mid \mu([x - r, x + r]) \geq \gamma(r)\}) = \frac{1}{4}$$

but $\mu \notin$ **Reg.**

Conditions Λ and $\Lambda^\star$ require that at a relative large set of the points x the μ measure of the disk of radius r around x be at most polynomially small in r. We have already remarked, that this relatively large set must be large only in capacity and not, say, in Lebesgue measure (when e.g. $S(\mu) = [0,1]$). Now we show that the μ measure of the above disks can be almost exponentially small in r provided we assume to hold it in more points. We only consider the classical and most important case $S(\mu) = [0,1]$, and let meas denote Lebesgue measure on $[0,1]$.

CRITERION λ. $S(\mu) = [0,1]$ *and for almost every* $x \in [0,1]$

(4.5) $$\liminf_{r \to 0} r \log \mu([x - r, x + r]) \geq 0.$$

(4.5) is a much weaker – "almost exponentially weaker" – density assumption than the one defining the set in (4.3). The price we pay for it is of course that it must hold in more points (note that $\operatorname{meas}(E) = 1$ for a set $E \subseteq [0,1]$ implies $\operatorname{cap}(E) = 1/4$).

Theorem 4.12. *Criterion* λ *implies* $\mu \in$ **Reg.**

This is best possible in the sense of

Theorem 4.13. *If* $\gamma : (0,1) \to (0,1)$ *is any function with* $\gamma(r)/r \to 0$ *as* $r \to 0 + 0$, *then there exists a measure* μ *such that* $S(\mu) = [0,1]$, *for all* $x \in [0,1]$

$$\lim_{r \to 0+0} \gamma(r) \log \mu([x - r, x + r]) = 0$$

but $\mu \notin$ **Reg.**

The "moving variant" of criterion λ – in the sense of $\Lambda^\star$ – is

CRITERION $\lambda^\star$. $S(\mu) = [0,1]$ *and for every* $\eta > 0$

$$\lim_{n \to \infty} \operatorname{meas}\left(\left\{x \;\middle|\; \mu\left(\left[x - \frac{1}{n}, x + \frac{1}{n}\right]\right) \geq e^{-\eta n}\right\}\right) = 1.$$

Theorem 4.14. *Criterion* $\lambda^\star$ *implies the regularity of* μ.

Since condition $\lambda^\star$ is obviously weaker than λ, we do not have to demonstrate the unimprovability of Theorem 4.14 in the sense of Theorem 4.13 because it follows from the latter one.

It can also be shown that condition λ^* implies condition Λ^* (and so Theorem 4.14 is a consequence of Theorem 4.10).

As a final summary we state

Summary. *The order of the criteria discussed above with regard to decreasing amount of assumptions on μ (and hence increasing strongness in applications) is roughly: (i) Erdős–Turán Criterion, (ii) Ullman's and Widom's Criterion, (iii) Criterion λ, (iv) Criterion λ^*, (v) Criterion Λ, and (vi) Criterion Λ^*.*

5. Zero distribution of orthogonal polynomials

For a polynomial P let ν_P denote the measure that places mass 1 to every zero of P (counting multiplicity). Then $\|\nu_P\| = \deg(P)$ and $\nu_P/\|\nu_P\| = \frac{1}{\deg(P)}\nu_P$ is often referred to as the normalized counting measure on the zeros of P.

In this section we will be primarily interested in the location of the zeros of orthogonal polynomials and in their asymptotic distribution. Of course, by the existence of the asymptotic distribution of the zeros we mean that the sequence of normalized counting measures

$$(5.1) \qquad\qquad \left\{\frac{1}{n}\nu_{p_n(\mu;\cdot)}\right\}$$

has weak* limit. Similarly, by a weak* limit of the zeros we mean a weak* limit of the measures (5.1)

The first result we mention is due to Widom ([Wi]).

Theorem 5.1. *All zeros of the orthonormal polynomials $p_n(\mu; z)$, $n \in \mathbf{N}$ are contained in the convex hull $Co(S(\mu))$ of $S(\mu)$, and for any compact set $V \subseteq Co(S(\mu)) \cap \Omega$ the number of zeros of $p_n(\mu; z)$, $n \in \mathbf{N}$ on V is bounded as $n \to \infty$. Consequently, every weak* limit of the measures (5.1) is supported on the polynomial convex hull $Pc(S(\mu))$ of $S(\mu)$.*

In general nothing more can be said about the zeros in Ω as one can prove that if S is any compact subset of $\mathbf{C}$, then there is a measure μ such that S is the support of μ and every point of $Co(S)$ is a limit point of zeros of the orthogonal polynomials corresponding to μ. In the special case when S is the unit circle the existence of such a measure was a problem of Turán (for a solution see [AlVi] – c.f. also [NeTo]).

Next we are going to characterize the sets of measures that arise as asymptotic distributions (through some subsequence) of the zeros of orthogonal polynomials corresponding to measures μ in the important special case when $S(\mu) = [0,1]$, although our method works finely in many other cases, as well. In other words, we want to determine the sets

$$\mathcal{M}_\mu = \{\nu \mid \nu \text{ is a weak}^* \text{ limit point of the measures } \frac{1}{n}\nu_{p_n(\mu;\cdot)}, \ n = 1, 2, \dots\}$$

when the support of μ is contained in $[0, 1]$. Then of course $\mathcal{M}_\mu \subseteq \mathcal{M}[0, 1]$, where $\mathcal{M}[0, 1]$ denotes the set of unit Borel measures on $[0, 1]$, and we always think of the latter space as equipped with the weak* topology.

Theorem 5.2. $\mathcal{M}_\mu$ *is a closed and connected subset of* $\mathcal{M}[0,1]$, *and if* $\mathcal{M}^\star$ *is any closed and connected subset of* $\mathcal{M}[0,1]$, *then there is a* μ *such that* $\mathcal{M}_\mu = \mathcal{M}^\star$.

To characterize $\mathcal{M}_\mu$ under the assumption that $c_\mu > 0$ is much more difficult and to do so we need potentials $p(\nu; \cdot)$ that are from bounded above and at some point they take on their supremum. We shall write $\mathrm{MAX}p(\mu; \cdot)$ for the set of maximum points of $p(\nu; \cdot)$ with the agreement that $\mathrm{MAX}p(\nu; \cdot)$ is empty if the potential is not bounded from above (or if its supremum is not attained). With this concept we prove

Theorem 5.3. *Let* $S(\mu) \subseteq [0,1]$. *If* $c_\mu > 0$ *and* C *is a minimal carrier of* μ *(i.e.* $\mathrm{cap}(C) = c_\mu$) *then any weak* limit* ν *of the zero distributions* $\{\frac{1}{n}\nu_{p_n(\mu;\cdot)}\}$ *satisfies* $C''\underline{\subseteq}''\mathrm{MAX}p(\nu; \cdot)$ *and* $\mathrm{supp}(\nu) \subseteq \overline{C}$, *where* $''\subseteq''$ *means inclusion except for a set of zero capacity. Conversely, if* $C \subseteq [0,1]$ *is of positive capacity and* $\mathcal{M}_C$ *is the set of probability measures* ν *satisfying* $C''\underline{\subseteq}''\mathrm{MAX}p(\nu; \cdot)$ *and* $\mathrm{supp}(\nu) \subseteq \overline{C}$, *then there is a measure* μ *such that* C *is a minimal carrier of* μ *and* $\mathcal{M}_\mu = \mathcal{M}_C$.

Now we turn to the intrinsic relationship between zero distribution and the asymptotic behavior of the leading coefficients $\gamma_n(\mu)$ of the orthonormal polynomials $p_n(\mu; z) = \gamma_n(\mu)z^n + \cdots$. Since $1/\gamma_n(\mu)$ equals the $L^2(\mu)$ norm of the monic orthogonal polynomial $p_n(\mu; z)/\gamma_n(\mu)$, asymptotics for $\{\gamma_n(\mu)^{1/n}\}$ are often referred to as norm asymptotics for orthogonal polynomials.

Theorem 5.4. *Assume that* $c_\mu > 0$ *and* $\mathrm{Int}(Pc(S(\mu))) = \emptyset$.
(a) The existence of the weak* limit

$$(5.2) \qquad \frac{1}{n}\nu_{p_n(\mu;\cdot)} \xrightarrow{\star} \nu \quad as \quad n \to \infty, \ n \in N,$$

for an infinite subsequence $N \subseteq \mathbf{N}$, *implies the existence of the limit*

$$(5.3) \qquad \lim_{n\to\infty, \ n\in N} \gamma_n(\mu)^{1/n}.$$

Furthermore, if for each of two infinite subsequences N_1, $N_2 \subseteq \mathbf{N}$ *the limit (5.2) exists, and the two limits are identical, then also the two corresponding limits (5.3) are equal.*
(b) If

$$\lim_{\substack{n\to\infty\\n\in N}} \gamma_n(\mu)^{1/n} = \frac{1}{\mathrm{cap}(S(\mu))}$$

for some subsequence $N \subseteq \mathbf{N}$, *then we also have*

$$\frac{1}{n}\nu_{p_n(\mu;\cdot)} \xrightarrow{\star} \omega_{S(\mu)} \quad as \quad n \to \infty, \ n \in N$$

in weak*-sense.
(c) If

$$\lim_{\substack{n\to\infty\\n\in N}} \gamma_n(\mu)^{1/n} = \frac{1}{c_\mu}$$

then

$$\frac{1}{n}\nu_{p_n(\mu;\cdot)} \xrightarrow{\star} \omega_\mu \quad as \quad n \to \infty, \ n \in N$$

in weak*-sense.

Above $\omega_{S(\mu)}$ is the equilibrium measure of $S(\mu)$, while the so called minimal carrier equilibrium measure ω_μ is the equilibrium measure of a carrier of μ of minimal capacity, i.e. $\omega_\mu = \omega_C$, where C is a carrier of μ with $\text{cap}(C) = c_\mu$. It can be shown that such minimal carriers exist and ω_C is the same for all such carriers. We also mention that with this measure ω_μ the minimal carrier Green function has the representation

$$g_\mu(z; \infty) = p(\omega_\mu; z) + \log(1/c_\mu),$$

i.e. the minimal carrier Green function is an ordinary Green function associated with a minimal carrier.

Note that (see Corollary 2.4) $1/\text{cap}(S(\mu))$ and $1/c_\mu$ are the smallest and largest possible limit points of the sequence $\{\gamma_n(\mu)^{1/n}\}$, thus (b) and (c) provide the converse of (a) in these two extreme cases. It can be shown that nothing similar is true if the limit of $\{\gamma_n(\mu)^{1/n}\}$ is in between $1/\text{cap}(S(\mu))$ and $1/c_\mu$.

We also mention that both assumptions $c_\mu > 0$ and $\text{Int}(Pc(S(\mu))) = \emptyset$ are really necessary, examples show that the result is not valid if we allow their converse.

Next we list some corollaries of Theorem 5.1. In each of them we keep the assumptions

(5.4) $$c_\mu > 0, \quad \text{Int}(Pc(S(\mu))) = \emptyset.$$

Corollary 5.5. *With assumption (5.4) the existence of the limit distribution of zeros implies norm asymptotics (i.e. the existence of the limit of $\{\gamma_n(\mu)^{1/n}\}$) for the orthonormal polynomials.*

Corollary 5.6. *Assuming (5.4), the relations*

$$\lim_{n \to \infty} \gamma_n(\mu)^{1/n} = \frac{1}{\text{cap}(S(\mu))}$$

and

$$\lim_{n \to \infty} \frac{1}{n} \nu_{p_n(\mu; \cdot)} = \omega_{S(\mu)}$$

are equivalent.

Corollary 5.7. *Assuming (5.4), the relations*

$$\lim_{n \to \infty} \gamma_n(\mu)^{1/n} = \frac{1}{c_\mu}$$

and

$$\lim_{n \to \infty} \frac{1}{n} \nu_{p_n(\mu; \cdot)} = \omega_\mu$$

are equivalent.

Besides regular (nth root) asymptotic behavior in the sense of Section 3, we consider a closely related type of asymptotic behavior: regular asymptotic zero distribution of the orthonormal polynomials $p_n(\mu; z)$ as $n \to \infty$. (In [ErFr] it is called asymptotically regularly distributed zeros.) It can only be defined if $\text{cap}(S(\mu)) > 0$, since the equilibrium distribution $\omega_{S(\mu)}$ is basic in the definition.

Let $\text{cap}(S(\mu)) > 0$. Then the orthonormal polynomials $p_n(\mu; z)$, $n \in \mathbf{N}$, associated with the measure μ are said to have regular asymptotic zero distribution if

$$\frac{1}{n}\nu_{p_n(\mu;\cdot)} \xrightarrow{\;*\;} \omega_{S(\mu)} \quad \text{as} \quad n \to \infty$$

in the weak* topology. When $S(\mu) = [-1, 1]$, this is the classical arc-sine distribution:

$$\lim_{n\to\infty} \frac{1}{n}\#\{\text{zeros of } p_n(\mu; \cdot) \text{ lying in } (\alpha, \beta)\} = \frac{1}{\pi}(\arccos \alpha - \arccos \beta)$$

for all $(\alpha, \beta) \subseteq [-1, 1]$.

Regular asymptotic zero distribution is very closely related to regular (nth root) asymptotic behavior, however, as the next theorem shows, the two notions are not fully identical. A fairly complete treatise of regular asymptotic zero distribution in case of weight measures μ with $S(\mu) = [-1, 1]$ is contained in [ErFr].

Theorem 5.8. *We assume* $\text{cap}(S(\mu)) > 0$ *and* $\text{Int}(Pc(S(\mu))) = \emptyset$. *Then*
(a) $\mu \in \mathbf{Reg}$ *implies regular asymptotic zero distribution of the orthonormal polynomials* $p_n(\mu; z)$, $n \in \mathbf{N}$.
(b) If in addition $c_\mu > 0$, *then* $\mu \in \mathbf{Reg}$ *and regular asymptotic zero distribution of* $\{p_n(\mu; z), n \in \mathbf{N}\}$ *are equivalent.*

Examples show that Theorem 5.8 is not true if $\text{Int}(Pc(S(\mu))) \neq \emptyset$ is allowed.

In case of a weight measure μ with $S(\mu) \subseteq \mathbf{R}$ we have $\text{Int}(Pc(S(\mu))) = \emptyset$. Hence, for such weight measures regular asymptotic zero distribution follows from regular asymptotic behavior of the orthonormal polynomials $p_n(\mu; z)$, $n \in \mathbf{N}$. That the reverse is in general not true can be shown by examples. Note however, that if $c_\mu > 0$, in particular, if $\mu' > 0$ on a set of positive Lebesgue measure, then regular asymptotic zero distribution is equivalent to $\mu \in \mathbf{Reg}$.

We mention that our results solve the problems of [UlWy].

6. Localization

In this chapter we shall give some results that show that regularity of a measure is a local property. This is more remarkable if one considers that the definition itself of the regularity of μ involves the whole support of μ. For example, if the support of μ is the unit circle, then it turns out that μ is regular if and only if its restriction to every arc is regular, and of course the Green functions associated to the latter ones (that appear in the definition of regularity) are hardly related to the Green function of the exterior of the unit disk. On the other hand, if we consider that regularity means that μ must be sufficiently dense around $\partial\Omega$, then we can see that localization is actually a natural procedure.

First we consider the problem of relating the regularity of the restriction $\mu_K = \mu_{|K}$ of μ to some compact set K to properties of $\{p_n(\mu; \cdot)\}$. That is, we ask if the regularity of the restricted weight measure μ_K is reflected in the behavior of the original orthonormal polynomials corresponding to μ. It turns out that K cannot be arbitrary, what we need of it is that the part of μ carried by K should be essentially carried by the interior K°, as well. More precisely, we will assume

$$(6.1) \qquad \operatorname{cap}(\partial\Omega(\mu_K)\backslash(K^\circ \cap \partial\Omega)) = 0.$$

Recall that $\Omega(\mu_K)$ is the unbounded component of $\mathbf{C}\backslash S(\mu_K) = \mathbf{C}\backslash\operatorname{supp}(\mu_K)$. Typical examples of such K's are closed disks when μ is supported on the real line or on the unit circumference (in these cases the set appearing in (6.1) has at most two points). Note also that for the case $S(\mu) = \partial\Omega$ condition (6.1) is equivalent to

$$\operatorname{cap}(S(\mu_K)\backslash K^\circ) = 0.$$

Our first theorem is

Theorem 6.1. *Let K be a compact set such that the support of $\mu_K := \mu|_K$ is an infinite set and (6.1) holds. Then the following statements are equivalent.*
(i) $\mu_K \in \mathbf{Reg}$, i.e. the sequence $\{p_n(\mu_K;\cdot)\}_{n=0}^\infty$ has regular (nth root) asymptotic behavior.
(ii) We have

$$\limsup_{n\to\infty} |p_n(\mu;z)|^{1/n} \le e^{g_\Omega(\mu_K)(z;\infty)}$$

locally uniformly for $z \in \mathbf{C}$.
(iii) For any sequence $\{P_n\}$ of nonzero polynomials of degree $\deg(P_n) \le n$ we have

$$\limsup_{n\to\infty} \left(\frac{|P_n(z)|}{\|P_n\|_{L^2(\mu)}}\right)^{1/n} \le e^{g_\Omega(\mu_K)(z;\infty)}$$

locally uniformly for $z \in \mathbf{C}$.
(iv) The relation

$$\limsup_{n\to\infty} |p_n(\mu;z)|^{1/n} \le 1$$

holds quasi-everywhere on $\partial\Omega(\mu_K)$.
(v) For any sequence $\{P_n\}$ as in (iii)

$$\limsup_{n\to\infty} \left(\frac{|P_n(z)|}{\|P_n\|_{L^2(\mu)}}\right)^{1/n} \le 1$$

for quasi-every $z \in \partial\Omega(\mu_K)$.
    If in addition we have $S(\mu_K) \subseteq \partial\Omega(\mu_K)$, in particular, if $\operatorname{Int}(Pc(S(\mu_K))) = \emptyset$, then each of (i)—(v) is equivalent to
(vi) for any infinite sequence $\{P_n\}$ of nonzero polynomials of degree $\deg P_n \le n$ and for any sequence of weight functions $g_n \ge 0$, $n \in \mathbf{N}$ defined on $S(\mu)$ such that $\lim g_n^{1/n} = g$ uniformly on $S(\mu_K)$ where $g > 0$ is continuous on $S(\mu_K)$, we have with $d\mu_n = g_n d\mu$

$$\limsup_{n\to\infty} \left(\frac{g_n(z)^{1/2}|P_n(z)|}{\|P_n\|_{L^2(\mu_n)}}\right)^{1/n} \le 1$$

for quasi-every $z \in \partial\Omega(\mu_K) = S(\mu_K)$.

If $S(\mu_K)$ is a regular set with respect to the solution of the Dirichlet problem in the domain $\Omega(\mu_K) = \mathbf{C}\backslash Pc(S(\mu_K))$ then our estimates hold uniformly in the range described.

Corollary 6.2. *Let K be as in Theorem 6.1 and assume that $S(\mu_K)$ is regular with respect to the Dirichlet problem in $\Omega(\mu_K)$. Then μ_K is regular if and only if*

$$\limsup_{n \to \infty} \|p_n(\mu; \cdot)\|_{\sup, S(\mu_K)} = 1.$$

Another equivalent condition is

$$\lim_{n \to \infty} \left(\sup_{\substack{\deg(P_n) \leq n \\ P_n \not\equiv 0}} \frac{\|P_n\|_{\sup, S(\mu_K)}}{\|P_n\|_{L^p(\mu)}} \right)^{1/n} = 1, \qquad 1 \leq p < \infty.$$

We emphasize that everywhere on the left hand sides the original orthonormal polynomials $p_n(\mu; \cdot)$ are used and not the ones with respect to μ_K (which would then be Theorem 3.2 for μ_K).

Now our main localization theorem is

Theorem 6.3. *Let us suppose that quasi every point of $\partial\Omega$ is contained in the union of the interiors of the compact sets $\{K_j\}_{j=1}^{\infty}$ and that each K_j satisfies the condition*

$$(6.2) \qquad \operatorname{cap}(\partial\Omega(\mu_{K_j})\backslash(K_j^{\circ} \cap \partial\Omega)) = 0.$$

Then μ is regular if and only if every $\mu_{K_j} = \mu|_{K_j}$ is regular.
When $\operatorname{Int}(Pc(S(\mu))) = 0$, in particular if μ is supported on $\mathbf{R}$, or even when only $S(\mu) = \partial\Omega$, then condition (6.2) reads as

$$\operatorname{cap}(S(\mu_{K_j})\backslash K_j^{\circ}) = 0.$$

It can be shown that assumption (6.1) is necessary in Theorems 6.1 and 6.3.

7. Regularity and best $L^2(\mu)$ polynomial approximation

In this section we mention an application of the above concepts and results. Other applications include exact rate of rational interpolation and best rational approximation of Markoff functions, determination of the domain of convergence and divergence of Padé approximation to Markoff functions, "where does the L^p–norm of a weighted polynomial live",etc.

Here we shall relate the regularity of μ to best $L^2(\mu)$ approximation of analytic functions. To avoid unnecessary complications we shall always assume that $\partial\Omega$ is regular with respect to the solutions of the Dirichlet problems in Ω.

Let

$$E_n(f)_{L^p(\mu)} = \inf_{\deg P_n \leq n} \|f - P_n\|_{L^p(\mu)}, \qquad 1 \leq p \leq \infty$$

be the best $L^p(\mu)$-approximation of f by polynomials of degree at most n. A well known result of S.N. Bernstein says that $\{E_n(f)_{L^\infty(\mu)}\}_{n=1}^{\infty}$ tends geometrically to zero if and only if f is analytic on $S(\mu)$. More precisely, if

$$\alpha(f) = \inf\{\delta \mid f \text{ is analytic in the region}\{x \mid g_\Omega(z; \infty) < \log 1/\delta\}\}$$

and

$$\varepsilon(f)_{L^p(\mu)} = \limsup_{n\to\infty}(E_n(f)_{L^p(\mu)})^{1/n},$$

then Bernstein's formula

(7.1) $$\alpha(f) = \varepsilon(f)_{L^\infty(\mu)}$$

holds for every f. Note that in this result the measure μ actually does not play any role, the statement only refers to the compact set $S(\mu) = \text{supp}(\mu)$. Of course, the situation will radically change if we consider the case $p < \infty$. We shall show below that the $L^p(\mu)$–analogue of the result above is true if and only if $\mu \in \mathbf{Reg}$.

Besides $\alpha(f)$ and $\varepsilon(f)_{L^p(\mu)}$ we also introduce

$$\varphi(f)_\mu = \limsup_{n\to\infty}|\hat{f}(n)|^{1/n},$$

where

$$\hat{f}(n) = \int f p_n(\mu;\cdot)d\mu$$

are the Fourier coefficients of f with respect to the orthonormal polynomials $\{p_n(\mu;\cdot)\}$.
Let us also agree to write $\|\cdot\|_{S(\mu)}$, $E_n(f)_{S(\mu)}$ etc. instead of $\|\cdot\|_{L^\infty(\mu)}$, $E_n(f)_{L^\infty(\mu)}$.
Consider the following statements.

$A_\mu :$ $$\alpha(f) < 1 \Leftrightarrow \varepsilon(f)_{L^2(\mu)} < 1 \quad for \quad f \in L^2(\mu),$$

that is geometric order of best $L^2(\mu)$ approximation of a function f is equivalent to the analyticity of f (which is to be understood in the sense that f coincides μ-a.e. with a function that is analytic on $S(\mu)$);

$A_\mu^* :$ $$\alpha(f) = \varepsilon(f)_{L^2(\mu)} \quad for\ every \quad f \in L^2(\mu);$$

$F_\mu :$ $$\alpha(f) < 1 \Leftrightarrow \varphi(f)_\mu < 1 \quad for \quad f \in L^2(\mu),$$

that is analyticity is equivalent to geometric decrease of Fourier coefficients;

$F_\mu^* :$ $$\alpha(f) = \varphi(f)_\mu \quad for\ every \quad f \in L^2(\mu);$$

C_μ: if f is analytic on $S(\mu)$, then its Forurier series in $\{p_n(\mu;\cdot)\}$ uniformly converges on $S(\mu)$.

Now each of these is equivalent to the regularity of μ :

Theorem 7.1. *Assume that $\partial\Omega$ is regular with respect to the Dirichlet problem in Ω. Then the six statements $\mu \in \mathbf{Reg}$, A_μ, $A_\mu^\star$, F_μ, $F_\mu^\star$ and C_μ are pairwise equivalent.*

Remarks. (1) Instead of the $L^2(\mu)$ norm we could use the $L^p(\mu)$ norm in the definition of $A_\mu, A_\mu^\star$ so long as $1 \le p < \infty$ and Theorem 7.1 would still hold.
(2) In C_μ we could have requested pointwise convergence of the Fourier series on $S(\mu)$ instead of uniform convergence and Theorem 7.1 would still be valid.

References

[AlVi] M. P. Alfaro and L. Vigil: 'Solution of a problem of P. Turán on zeros of orthonormal polynomials on the unit circle', *J. Approx. Theory*, **53**(1988), 195—197.

[BaGM] G. A. Baker, Jr. and P. R. Graves–Morris: *Padé Approximants, Part I. Basic Theory*, Addison-Wesley Publ. Comp., Reading 1981.

[dVP] Ch. J. de la Vallée–Poussin: *Le Potentiel Logarithmique*, Gauthier–Villars, Paris 1949.

[ErFr] P. Erdős and G. Freud: 'On Orthogonal Polynomials with Regularly Distributed Zeros', *Proc. London Math. Soc.*, **29**(1974), 521—537.

[ErTu] P. Erdős and P. Turán: 'On Interpolation III', *Ann. Math.*, **41**(1940), 510—553.

[Fa] G. Faber: 'Über nach Polynomen fortschreitende Reihen', *Sitzungsberichte der Bayerischen Akademie der Wissenschaften* 1922, 157—178.

[Fr] G. Freud: *Orthogonal Polynomials*, Pergamon Press, Oxford 1971.

[La] N. S. Landkof: *Foundations of Modern Potential Theory*, Springer–Verlag, New York, 1972.

[Lu] D. S. Lubinsky: 'A Survey of General Orthogonal Polynomials for Weights on Finite and Infinite Intervals', *Acta Appl. Math.*, **10**(1987), 237—296.

[Ne1] P. Nevai: 'Géza Freud, Ortogonal Polynomials and Christoffel Functions', *J. Approx. Theory* , **48**(1986), 3—167.

[Ne2] P. Nevai: *Orthogonal Polynomials*, Memoirs Amer. Math. Soc., Vol **213**, Amer. Math. Soc., Providence, R. I., 1979.

[NeTo] P. Nevai and V. Totik: 'Ortogonal polynomials and their zeros', *Acta Sci. Math. (Szeged)*, **53**(1989)

[Pe] O. Perron: *Die Lehre von den Kettenbrüchen*, Chelsea Publ. Comp., New York 1956 (reprint).

[Sz1] G. Szegő: 'Über orthogonale Polynome, die zu einer gegebenen Kurve der komplexen Ebene gehören', *Math. Zeitschr.*, **9**(1921), 218—270 (see also Gábor Szegő, Collected Papers I, Ed. R. Askey, Birkhäuser Verlag, Basel 1982, 315—370).

[Sz2] G. Szegő: 'Über die Entwicklung einer analytischen Funktion nach Polynomen eines Orthogonalsystems', *Math. Ann.*, **82**(1921), 188—212 (see also: Gábor Szegő, Collected Papers I, Ed. R. Askey, Birkhäuser Verlag, Basel 1982, 372—397).

[Sz3] G. Szegö: *Orthogonal Polynomials*, Amer. Math. Soc. Colloq. Publ., Vol. **23**, Amer. Math. Soc., Providence, R. I., 1975.

[Ts] M. Tsuji: *Potential Theory in Modern Function Theory.* Maruzen, Tokyo 1959.

[Ul1] J. L. Ullman: 'On the Regular Behaviour of Orthogonal Polynomials', *Proc. London Math. Soc.*, 24(1972), 119—148.

[Ul2] J. L. Ullman: 'A Survey of Exterior Asymptotics for Orthogonal Polynomials Associated with a Finite Interval and a Study of the Case of general Weight Measures'; in: *Approximation Theory and Spline Functions*, Eds.: S. P. Singh et al., D. Reidel, Dordrecht 1984, 467—478.

[Ul3] J. L. Ullman: 'Orthogonal Polynomials for General Measures I', in: *Rational Approximation and Interpolation*, Eds.: P. R. Graves–Morris et al., Lect. Notes Math. **1105**, Springer–Verlag, New York, 1984, 524-528.

[Ul4] J. L. Ullman: 'Orthogonal Polynomials for General Measures II', in: *Polynomes Orthogonaux et Applications*; Proceedings, Bar-le-Duc 1984, Eds.: C. Brezinski et al., Lect. Notes Math. **1171**, Springer–Verlag, New York, 1986, 247—254.

[UlWy] J. L. Ullman and M. F. Wyneken: 'Weak Limits of Zeros of Orthogonal Polynomials', *Constructive Approx.*, **2**(1986), 339—347.

[UWZ] J. L. Ullman, M. F. Wyneken and L. Ziegler: 'Norm Oscillatory Weight Measures', *J. Approx. Theory*, **46**(1986), 204—212.

[Wi] H. Widom: 'Polynomials Associated with Measures in the Complex Plane', *J. Math. Mech.*, **16**(1967), 997—1013.

AN INTRODUCTION TO GROUP REPRESENTATIONS AND ORTHOGONAL POLYNOMIALS

Dennis Stanton
School of Mathematics
University of Minnesota
Minneapolis, MN 55455, U.S.A.

ABSTRACT. An elementary non-technical introduction to group representations and orthogonal polynomials is given. Orthogonality relations for the spherical functions for the rotation groups in Euclidean space (ultraspherical polynomials), and the matrix elements of $SU(2)$ (Jacobi polynomials) are discussed. A general theory for finite groups acting on graphs, giving a finite set of discrete orthogonal polynomials is given. Explicit examples include graphs giving the Krawtchouk and Hahn polynomials.

Introduction.

The purpose of this paper is to present a friendly, non-technical introduction to group representations and orthogonal polynomials. No previous knowledge of group representations is assumed, but a familiarity with orthogonal polynomials is assumed. In particular, this paper emphasizes the *classical orthogonal polynomials* and their relationship to groups. Other classical special functions can also be studied in this way, e.g. Vilenkin [19] or Miller [14] (which is more elementary). §I of this paper could be considered as a short introduction to the sections of those books relevant to orthogonal polynomials. More modern work on continuous groups and the related analysis has been done by Koornwinder [11] and Dunkl [9]. Some very recent work concerns orthogonal polynomials in several variables [13].

It was not realized until the early 1970's that finite groups could be related to classical orthogonal polynomials. The pioneering work was done by Dunkl [8],[9] and Delsarte [5],[6],[7]. §II is an introduction to the general theory of finite groups. This theory can be generalized to association schemes, which consider relations on a finite set with certain properties. An extensive theory of association schemes can be found in [3] and [5]. A very elementary introduction is given in [16]. A survey of recent work and important problems is given in [4], and in Bannai's paper [2] in this volume.

I. Notation.

The classical orthogonal polynomials can be expressed as hypergeometric series. We will use the usual notation for these series (see [15]). We will be most concerned about three sets of polynomials: Jacobi, Krawtchouk and Hahn.

This material is based upon research supported by the National Science Foundation under grant number DMS-8700995.

P. Nevai (ed.), Orthogonal Polynomials, 419–433.
© *1990 by Kluwer Academic Publishers.*

The Jacobi polynomials

(Ja)
$$P_n^{(\alpha,\beta)}(x) = \frac{(\alpha+1)_n}{n!} \, {}_2F_1 \left(\begin{array}{cc} -n, & n+\alpha+\beta+1; & \frac{1-x}{2} \\ & \alpha+1 & \end{array} \right),$$

are orthogonal on $[-1,1]$ with respect to the weight $w(x) = (1-x)^\alpha(1-x)^\beta$, $\alpha, \beta > -1$. The ultraspherical (or Gegenbauer) polynomials are the special case $\alpha = \beta$ of the Jacobi polynomials.

The Krawtchouk polynomials

(Kr)
$$K_n(x,p,N) = {}_2F_1 \left(\begin{array}{cc} -n, & -x; & \frac{1}{p} \\ & -N & \end{array} \right)$$

are orthogonal on the finite set $x = 0, 1, \cdots, N$ with respect to the binomial distribution $w(x) = \binom{N}{x} p^x (1-p)^{N-x}$, $0 < p < 1$. We will consider $p = 1/2$, and more generally $p = k/(k+1)$ for a positive integer k.

The Hahn polynomials

(Ha)
$$Q_n(x,\alpha,\beta,N) = {}_3F_2 \left(\begin{array}{ccc} -n, & n+\alpha+\beta+1, & -x; \quad 1 \\ & -N, & \alpha+1 \end{array} \right).$$

are orthogonal on the finite set $x = 0, 1, \cdots, N$ with respect to the hypergeometric distribution $w(x) = \binom{\alpha+x}{x}\binom{\beta+N-x}{N-x}$. For a positive weight function, $\alpha, \beta > -1$ or $\alpha, \beta < -N$. We shall be considering the second case.

I.1 Continuous groups.

We begin by considering an elementary problem for polynomials: which polynomials $p(x,y)$ in x and y are harmonic. This means that the Laplacian $\Delta = \partial^2/\partial x^2 + \partial^2/\partial y^2$ annihilates p, $\Delta p = 0$. We can assume that that p is homogeneous of degree n, for some n:

(1)
$$p(x,y) = \sum_{i=0}^n a_i x^i y^{n-i}.$$

Clearly $\Delta p = 0$ is equivalent to

$$(i+2)(i+1)a_{i+2} + (n-i)(n-i-1)a_i = 0 \text{ for } 0 \le i \le n-2.$$

Thus we see that a_0 and a_1 are arbitrary and

$$a_{2i} = (-1)^i \binom{n}{2i} a_0$$

and

$$a_{2i+1} = (-1)^i \binom{n}{2i} a_1.$$

Thus an arbitrary harmonic polynomial, homogeneous of degree n in x and y, has the form

(2)
$$p(x,y) = A[(x+iy)^n + (x-iy)^n] + Bi[(x+iy)^n - (x-iy)^n],$$

for arbitrary real constants A and B.

In polar coordinates, the two independent solutions in (2) are $r^n cos(n\theta)$ and $r^n sin(n\theta)$. Thus we see the Chebyshev polynomials

$$T_n(cos\theta) = cos(n\theta)$$

and

$$U_{n-1}(cos\ \theta) = \frac{sin(n\theta)}{sin(\theta)}$$

appearing in the solutions to our harmonic polynomial problem. This is not surprising, since z^n is a complex analytic function, but what does it have to do with group representations?

Let $O(2)$ be the group of all orthogonal transformations in the x-y plane, so that $O(2)$ consists of all rotations about the origin, and reflections in lines through the origin. The group $O(2)$ acts on the 2-dimensional vector space V of all real harmonic polynomials of degree n. This is clear from (2): if a rotation of α is applied, θ will be increased by α, and the addition formula for $cos(\theta + \alpha)$ will show that $cos(\theta)$ will be mapped to a linear combination of $cos(\theta)$ and $sin(\theta)$. If a reflection through the line $\theta = \beta$ is applied, then θ will be mapped to $2\beta - \theta$, and again we use the addition formula to find the appropriate linear combination.

This vector space V, together with the action of $G = O(2)$, is an example of an irreducible representation of G.

Basically what we have is to "represent" the elements of G as matrices acting on V, in a way that preserves the group multiplication, i.e. a product of group elements is represented by a product of matrices. The representation is irreducible because there is no smaller subspace of V which is fixed by G. Strictly speaking we have the following definition.

DEFINITION 1. *Given a group G and a vector space V, an irreducible representation of G is a continuous homomorphism $\phi : G \rightarrow Aut(V)$ such that the only subspaces of V stable under $\phi(G)$ are V and $\{\vec{0}\}$.*

If V is an inner product space over the complex numbers, and $\phi(g)$ preserves the inner product in V, then we call ϕ a *unitary* representation. Two representations ϕ and ϕ' on V and V' are called *inequivalent* if there is no isomorphism $T : V \rightarrow V'$ such that $T \circ \phi(g) = \phi'(g) \circ T$ for all $g \in G$.

We can give another example of an irreducible representation of a group, this time a finite group. Let G be the dihedral group of order $2m$ which acts on a regular m-gon in the plane. We assume that the center of the m-gon is the origin, and that a vertex lies on the x-axis. Consider functions on the vertices of the m-gon, whose values at vertex k given by $f_1 = cos(n\theta_k)$, where $\theta_k = 2\pi k/m$, $0 \le k \le m - 1$. Define f_2 similarly with sine replacing cosine. These functions are a discrete version of those in (2) and give a 2-dimensional irreducible representation of G.

We now turn to orthogonality relations from representations. Since orthogonal polynomials have orthogonality relations, we should see what kinds of relations are available. The next theorem states three possible orthogonality relations, which may or may not be orthogonality relations for orthogonal polynomials.

ORTHOGONALITY THEOREM. *Let G be a compact or finite group, and let $\phi : G \rightarrow Aut(V)$ and $\phi' : G \rightarrow Aut(V')$, be inequivalent unitary irreducible representations on complex* *vector spaces V and V'. Let $\{v_1, \ldots, v_p\}$ be an orthonormal basis for V. If*

$$T_{ij}^{\phi}(g) = < v_i, \phi(g)v_j >$$

is the ij matrix element of $\phi(g)$ with respect to $\{v_1, \ldots, v_p\}$, then

(1) $< v_i, v_j > = \delta_{ij}$,

(2) $\int_G T_{ij}^{\phi}(g)\overline{T_{i'j'}^{\phi'}(g)}dg = \delta_{ii'}\delta_{jj'}\delta_{\phi\phi'}/dimV$,

(3) $\sum_{j=1}^{p} T_{ij}^{\phi}(g)\overline{T_{kj}^{\phi}(g)} = \delta_{ik}$.

Clearly (1) is just a restatement that is an orthonormal basis for V. For a continuous group G, if we can parametrize G in some way by real numbers, (2) gives us a chance at absolutely continuous measures for orthogonal polynomials. A finite discrete orthogonality is clearly implied by (3). For (1) for $G = O(2)$, we could define the inner product of our two functions to be integration over the circle. Then the orthogonality is the orthogonality of the two Chebyshev polynomials T_n and U_{n-1}. The matrix elements for a rotation by α are again trigonometric functions, and integrating over α again gives this orthogonality. The 2 by 2 matrices are easily seen to have (3). We will explicitly find matrix elements for $G = SU(2)$ in §I.3, and interpret the orthogonalities (2) and (3).

I.2 Spherical functions.

In this section we give the basic properties of *spherical functions*, also called *zonal spherical functions*. There is a great literature on the harmonic analysis of these functions. One example of these functions will be the ultraspherical polynomials.

Suppose G acts transitively on a set X, as $O(2)$ acts on the circle S^1 in the plane. Then we have a representation of G on $L^2(X)$

(1) $$L^2(X) = \bigoplus_{\mu} m_{\mu}V_{\mu}.$$

This representation may not be irreducible, thus it decomposes into many irreducible components, some of which may be *equivalent*. This means that up to a non-singular transformation between these subspaces, the action of G is identical. The integer m_{μ} is the multiplicity of the representation V_{μ} in $L^2(X)$. Thus it is assumed in (1), that different μ correspond to inequivalent representations V_{μ}.

The multiplicities m_{μ} can be found in the following way. Fix $x_0 \in X$, and let $K = \{g \in G : g(x_0) = x_0\}$, so that $X = G/K$. The Frobenius reciprocity theorem gives the following property of m_{μ} and V_{μ}.

PROPOSITION. *The subspace of V_{μ} invariant under K has dimension m_{μ}.*

If each $m_{\mu} = 1$, we call the representation *multiplicity free*. In this case each V_{μ} contains a unique (up to a multiple) vector invariant under K. This suitably normalized vector $\phi_{\mu} \in V_{\mu}$, is the spherical function for V_{μ}. ϕ_{μ} is constant on the K-orbits of G/K, thus constant on the double cosets $K\backslash G/K$. Since these functions are in distinct representations, they are orthogonal by Orthogonality Theorem (2). Thus, if the double cosets can be parametrized by a single real parameter, we will have an absolutely continuous measure for which the spherical functions are orthogonal. Many of the properties of spherical functions can be found in [9]. We mention here that there is (1) an integral representation using the character of the representation, and (2) an integral product formula.

Let $G = O(2)$, $X = S^1$, $x_0 = (1,0)$, and $K =<$ the reflection in the x-axis$>$. Then the K-orbits on X are pairs of points on the unit circle which are complex conjugates. Thus, any function invariant under K on the circle can be considered a function of $cos(\theta)$. For the irreducible representation of harmonic polynomials of degree m in §2, the spherical function clearly is the Chebyshev polynomial $T_m(cos(\theta)) = cos(m\theta)$. The orthogonality of these polynomials is precisely the orthogonality of spherical functions in distinct representations,

$$\int_X \phi_\mu(x)\overline{\phi_{\mu'}(x)}dx = 0, \mu \neq \mu',$$

(2)
$$\int_0^\pi T_m(cos(\theta))\overline{T_{m'}(cos(\theta))}d\theta = 0, m \neq m'.$$

For the dihedral group acting on an n-gon (n odd) the orthogonality is

(3)
$$\frac{1}{n} + \frac{2}{n}\sum_{k=1}^{n/2} T_m(cos(\theta_k))T_{m'}(cos(\theta_k)) = 0, m \neq m'.$$

Clearly, as $n \to \infty$, (3) $\to$ (2).

For the rotation group $SO(3)$ on the sphere S^2, one can choose x_0 to be the north pole, and thus K is the subgroup of rotations about the diameter which passes through x_0. Then any function $f(x)$ on S^2, invariant under K, only depends upon the angle $\theta = x \cdot x_0$. Precisely the same statement can be said about the higher dimensional spheres $S^{n-1} = SO(n)/SO(n-1)$.

THEOREM 1. *Let $Harm(m)$ be the vector space of harmonic polynomials which are homogeneous of degree m. Let $Harm(m)|_{S^{n-1}}$ be the restriction of these functions to the sphere S^{n-1}. Then*

(1) $L^2(S^{n-1}) = \oplus_{m=0}^\infty Harm(m)|_{S^{n-1}}$,
(2) $Harm(m)|_{S^{n-1}}$ *is inequivalent to* $Harm(m')|_{S^{n-1}}$, *for* $m \neq m'$,
(3) *the spherical function for* $Harm(m)|_{S^{n-1}}$ *is the ultraspherical polynomial* $P_m^{((n-3)/2,(n-3)/2)}(cos\ \theta)$, *where* $x \cdot (0,\ldots,0,1) = cos\ \theta$.

Why should the spherical function be an ultraspherical polynomial? It is easy to see that it must be a polynomial of degree m in $cos\theta$. The spherical Laplacian Δ_S commutes with the action of $SO(n)$, and so must be a constant on $Harm(m)|_{S^{n-1}}$. The resulting eigenvalue equation is just the differential equation for the ultraspherical polynomials.

Why should (1) hold in Theorem 1? The vector space of homogeneous polynomials of degree m, $Hom(m)$, decomposes by (see [19, p. 444])

(4)
$$Hom(m) = Harm(m) \oplus r^2 Hom(m-2).$$

Iterating (4) gives

$$Hom(m) = Harm(m) \oplus r^2 Harm(m-2) \oplus \cdots.$$

Thus the vector space of all polynomial functions on S^{n-1} is the vector space of all harmonic polynomial functions.

We can also find $dim(Harm(m))$ from (4). There are $\binom{n+m-1}{n-1}$ distinct monomials in n variables of degree m. This yields

$$dim(Harm(m)) = \binom{n+m-1}{n-1} - \binom{n+m-3}{n-1}.$$

Note that for $n = 2$, $dim(Harm(m)) = 2$, as it is in §2. Also note that every restriction of a non-zero homogeneous polynomial is a non-zero function on S^{n-1}.

Finally we come to the invariant measure for the sphere. On S^2 it is $sin(\theta)d\theta$, which is $sin(\theta)$ times the invariant measure for the circle S^1. It is clear from the Pythagorean theorem, that each dimension gives another factor of $sin(\theta)$, so that the measure for S^{n-1} is $sin^{n-2}(\theta)$. This completes the group theoretic proof that $P_m^{((n-3)/2,(n-3)/2)}(cos(\theta))$ are orthogonal on $[0, \pi]$ with respect to $sin^{n-2}(\theta)$.

I.3 Matrix elements.

In this section we compute the matrix elements for $SU(2)$ and show that they are effectively Jacobi polynomials. From the Orthogonality Theorem and the explicit invariant measure for $SU(2)$, we then have the orthogonality for these polynomials.

We let elements of $g \in SU(2)$ be written as

$$g = \begin{pmatrix} \alpha & \beta \\ -\overline{\beta} & \overline{\alpha} \end{pmatrix},$$

where α and β are complex numbers satisfying $|\alpha|^2 + |\beta|^2 = 1$.

Let V_m be the vector space of homogeneous polynomials $p(x, y)$ in x and y of degree m. The element $g \in SU(2)$ acts by a linear change of variable,

$$p(x,y)g = p(\alpha x - \overline{\beta}y, \beta x + \overline{\alpha}y).$$

We need an appropriate basis for V, and a way of parametrizing the group elements of $SU(2)$. It turns out that appropriate multiples of the monomials give an orthonormal basis, on which the group acts unitarily:

$$v_i = c_i x^{m/2-i} y^{m/2+i}, -m/2 \leq i \leq m/2.$$

(For odd m, i will be half-integral. This slightly unusual way of writing the exponents will simplify the following calculation.) Thus

$$v_i g = c_i(\alpha x - \overline{\beta}y)^{m/2-i}(\beta x + \overline{\alpha}y)^{m/2+i}.$$

Expanding by the binomial theorem we find

(1) $$v_i g = \sum_{s=0}^{m/2-i}\sum_{t=0}^{m/2+i} c_i \binom{m/2-i}{s}\binom{m/2+i}{t}\alpha^{m/2-i-s}(-\overline{\beta})^s \overline{\alpha}^t \beta^{m/2+i-t}.$$

The coefficient of $c_j v_j$ in (1) is a single sum which is a $_2F_1$. This is the basic reason that Jacobi polynomials are the matrix elements.

The group elements g can be parametrized by three parameters (ϕ, ψ, θ), called the Euler angles of $SU(2)$. ϕ and ψ are related to the phase of α and β, and θ gives the modulus of α (and thus β)

$$g(\phi, \psi, \theta) = \begin{pmatrix} e^{i(\phi+\psi)/2}\cos(\theta/2) & ie^{i(\phi-\psi)/2}\sin(\theta/2) \\ ie^{i(-\phi+\psi)/2}\sin(\theta/2) & e^{i(-\phi-\psi)/2}\sin(\theta/2) \end{pmatrix}.$$

A short calculation gives

$$T^m_{ij}(\phi, \psi, \theta) = \text{constant} \cdot (1 - \cos(\theta))^{(i-j)/2} \cdot (1 + \cos(\theta))^{(i+j)/2}$$
$$e^{-i(i\phi+j\psi)} \cdot P^{(i-j,i+j)}_{m/2-i}(\cos(\theta)).$$

The invariant measure on $SU(2)$ in terms of the Euler angles is

$$dg = \frac{d\phi}{2\pi} \frac{d\psi}{4\pi} \frac{\sin(\theta)d\theta}{2}.$$

Clearly the orthogonality in Orthogonality Theorem (2) is trivial if i or j change. If i and j are fixed, then we have the Jacobi polynomial orthogonality.

What is the discrete orthogonality (3) of the Orthogonality Theorem? It becomes

$$\sum_{j=0}^{m} \binom{m}{j}(1 - \cos(\theta))^{j-m/2}(1 + \cos(\theta))^{m/2-j}$$

$$\cdot {}_2F_1\left(\begin{matrix} -(m/2 - i), & -j; \\ & -m \end{matrix} \ \frac{2}{1 - \cos(\theta)}\right)$$

$$\cdot {}_2F_1\left(\begin{matrix} -(m/2 - k), & -j; \\ & -m \end{matrix} \ \frac{2}{1 - \cos(\theta)}\right) = 0 \text{ for } i \neq k.$$

This is precisely the Krawtchouk polynomial orthogonality with $p = (1 - \cos(\theta))/2$.

I.4 Clebsch-Gordan coefficients.

Here we state the basic problem of the Clebsch-Gordan coefficients. These coefficients have orthogonality relations, which turn out to be, for $SU(2)$, the orthogonality relations for Hahn polynomials [1], [12].

The main problem is the following: let V and W be irreducible representations of G. Find the decomposition of the tensor product $V \otimes W$.

For $SU(2)$, let $V = V_{j_1}$ and $W = V_{j_2}$, it can be shown that

$$V_{j_1} \otimes V_{j_2} = \oplus_{j=|j_1-j_2|}^{j_1+j_2} V_j.$$

(This in fact can be proved from the Chebyshev polynomials of the second kind, which are the characters of V_m.)

We have two different orthonormal bases for $V \otimes W$: one made up of tensor products of orthonormal bases from $V = V_{j_1}$ and $W = V_{j_2}$, and another which consists of the basis vectors for V_j inside $V_{j_1} \otimes V_{j_2}$.

(1) $v^{j_1}_{m_1} \otimes v^{j_2}_{m_2}$, $-j_1 \leq m \leq j_1$, $-j_2 \leq m \leq j_2$,
(2) w^j_m, $|j_1 - j_2| \leq j \leq j_1 + j_2$, $-j \leq m \leq j$.

426

The Clebsch-Gordan coefficients are the entries of the matrix which accomplishes this change of basis:

$$v_{m_1}^{j_1} \otimes v_{m_2}^{j_2} = \sum_{m,j} C(j,m|j_1,m_1,j_2,m_2)w_m^j.$$

Since each basis is orthonormal, the Clebsch-Gordan matrix must be orthogonal. The discrete orthogonality is equivalent to Hahn polynomial orthogonality.

II. Finite groups.

In §I at no time did we refer to the three term recurrence relation that all orthogonal polynomials satisfy

(II.1) $$xp_i(x) = A_ip_{i+1}(x) + B_ip_i(x) + C_ip_{i-1}(x).$$

Given (II.1), it is possible to develop a combinatorial theory of general orthogonal polynomials, [18]. In this theory $p_n(x)$ is the generating function for certain weighted lattice paths in the plane, as are the moments μ_n for the measure for $p_n(x)$. We cannot, however, interpret a general set of orthogonal polynomials given by (II.1) as the matrix elements of a group, since matrix elements have special properties that general orthogonal polynomials do not have. Nevertheless, we shall see that (II.1) is the key ingredient for orthogonal polynomials on finite groups.

We give the general theory for finite groups in §II.1. These are applied to the Krawtchouk, Hahn and q-Hahn polynomials in §II.2, §II.3, and §II.4. References to combinatorial applications are given in §II.5.

II.1 Finite groups and graphs.

There is a complete analogy to the action of the continuous group $SO(n)$ on S^{n-1}, and the action of certain finite groups G on certain graphs X. In this section we give the general theory of such an analogy, and prove that the spherical functions are always given by orthogonal polynomials. The definitions in §I.3 and the Orthogonality Theorem hold for finite groups.

Recall that a graph $X = (V,E)$ consists of a set of vertices V, together with some subcollection E of all 2-element subsets of V, called the edges of X. Rather than stick to this cumbersome notation, we let X denote the vertices and the graph simultaneously. The metric $d(x,y)$ on X is defined by the length of the shortest path from x to y in X. If the graph X is connected, i.e. there is some path between any two vertices, then $d(x,y) < \infty$ for all $x,y \in X$.

Let G be the group of automorphisms of X, this means that each $g \in G$ permutes the vertices of X and preserves the edges of X. Thus, each $g \in G$ preserves the metric d. We assume that (*)

(*1) X is connected,
(*2) if $d(v,w) = d(v',w')$, then there exists $g \in G$ such that $(gv,gw) = (v',w')$.

Under these assumptions G acts transitively on X, so that we can think of the action of G on all complex functions on X, $L^2(X)$, as comparable to the action of $SO(3)$ on $L^2(S^2)$. This action of G on $L^2(X)$ is called a permutation representation since G permutes the points of X. One example of such a graph is the n-gon, whose automorphism group is the dihedral group of order $2n$. The following theorem shows the spherical functions of §I.3 are applicable.

THEOREM 1. *The permutation representation of G on X is multiplicity free.*

We will prove Theorem 1 while computing the spherical function for each irreducible representation. To find these subspaces of $L^2(X)$, we need $|X| \times |X|$ matrices which are indexed by X. We let d be the maximum distance in the graph X, and define a matrix A_i for each $0 \le i \le d$ by

$$(A_i)_{xy} = \begin{cases} 1 \text{ if } d(x,y) = i, \\ 0 \text{ otherwise.} \end{cases}$$

The three-term recurrence relation (II.(1)) is given by the multiplication of $A_1 A_i$.

PROPOSITION 1. *For any $0 \le i \le d$,*

$$A_1 A_i = \alpha_i A_{i+1} + \beta_i A_i + \gamma_i A_{i-1},$$

where, for a fixed $(v, w) \in X \times X$

$$\alpha_i = |\{z \in X : d(z,v) = 1, d(z,w) = i\}|, d(v,w) = i + 1,$$
$$\beta_i = |\{z \in X : d(z,v) = 1, d(z,w) = i\}|, d(v,w) = i,$$
$$\gamma_i = |\{z \in X : d(z,v) = 1, d(z,w) = i\}|, d(v,w) = i - 1.$$

PROOF: Find the vw entry of each side by counting the appropriate number of triangles in the graph. By the triangle inequality, the three terms listed are the only ones that contribute. ∎

From Proposition 1, we see that $A_i = p_i(A_1)$, for a finite orthogonal polynomial sequence p_i. Since each A_i is polynomial in A_1, the algebra of symmetric matrices generated by $\{A_0, A_1, \ldots, A_d\}$ is commutative. Thus they are simultaneously diagonalizable, and if $\{\lambda_0 > \lambda_1 > \cdots > \lambda_d\}$ are the distinct eigenvalues of A_1, then $\{p_i(\lambda_0), p_i(\lambda_1), \ldots, p_i(\lambda_d)\}$ are the eigenvalues of A_i.

It was not an accident that there are $d + 1$ distinct eigenvalues of A_1. The dimension of the algebra of matrices generated by $\{A_i : 0 \le i \le d\}$ is clearly $d + 1$. This algebra is a polynomial algebra in A_1, thus A_1 has $d + 1$ distinct eigenvalues.

Let $V_0, V_1, \ldots, V_d$ be the eigenspaces for A_1 corresponding to $\{\lambda_0 > \lambda_1 > \cdots > \lambda_d\}$. These will be the irreducible representations we seek. Thus we need to find how the spherical functions are related to the polynomials $p_i(x)$, and what kind of orthogonality relation they have. To do this, we will use the projection matrices $Pr_0, Pr_1, \ldots, Pr_d$, where Pr_i is the projection of $L^2(X)$ onto V_i. By definition we have

$$(1) \qquad A_i = \sum_{j=0}^{d} p_i(\lambda_j) Pr_j,$$

which we can invert to

$$(2) \qquad Pr_j = \sum_{i=0}^{d} q_j(i) A_i.$$

We use the orthogonality $Pr_j Pr_k = Pr_j \delta_{jk}$, by expanding each Pr_j and Pr_k in terms of the A_i's, then multiplying the A_i by counting triangles in the graph of given side lengths (call this number c_{is}^t), and equating coefficients of A_l. The result is

$$(3) \qquad q_j(l)\delta_{jk} = \sum_{i,s} q_j(i)q_k(s)c_{is}^l.$$

Even though (3) is a double sum, it can easily reduce to a single sum. If $l = 0$, then clearly $c_{is}^0 = 0$ unless $i = s$. So we have

$$(4) \qquad q_j(0)\delta_{jk} = \sum_i q_j(i)q_k(i)c_{ii}^0.$$

This is an orthogonality relation for $q_j(i)$, with weights c_{ii}^0, the sizes of a sphere of radius i centered at any point of the graph. If $q_j(i)$ is a polynomial of degree j evaluated at some μ_i, then we call the graph Q-polynomial (see [2]). It is already P-polynomial.

We now have two orthogonality relations for $q_j(i)$: (4) and the definition (2) of $q_j(i)$ as the inverse matrix to $p_i(\lambda_j)$. Thus, up to constants (which can be explicitly found), we can say that these two quantities are equal. The resulting orthogonality for $p_i(\lambda_j)$ is

$$(5) \qquad \frac{1}{|X|} \sum_{j=0}^{d} p_k(\lambda_j)p_i(\lambda_j)dimV_j = \delta_{ik}c_{kk}^0.$$

We also see that the multiplication formula for $A_i A_j$ gives the following linearization formula for the polynomials $p_i(x)$:

$$p_i(\lambda_m)p_j(\lambda_m) = \sum_k c_{ij}^k p_k(\lambda_m).$$

If $x_0 \in X$ is fixed, and K is the stabilizer of x_0, then any function on X invariant under K must be constant on the K orbits on X. They are

$$\Omega_j = \{x \in X : d(x, x_0) = j\}.$$

Thus the orthogonality relation for spherical functions will be

$$\sum_{j=0}^{d} \phi_m(\Omega_j)\phi_{m'}(\Omega_j)|\Omega_j| = 0 \text{ if } m \neq m'.$$

This appears to agree with (5). Thus the spherical functions $\phi_m(\Omega_j)$ should be $p_j(\lambda_m)$. This is true, and can be verified from the eigenvalue equation

$$(6) \qquad (A_1\phi)(\Omega_j) = \lambda_m\phi(\Omega_j).$$

(6) becomes precisely the three-term recurrence relation (II.(1)), and shows that $\phi_m(\Omega_j)$ is a polynomial of degree j in λ_m. This proves that each V_m has a unique spherical function, so is irreducible by the Proposition in §I.2. We summarize these conclusions in the next theorem.

THEOREM 2. *Let X be a graph satisfying (*) of maximum distance d. Then the permutation representation of G on X is multiplicity free, $L^2(X) = \bigoplus_{m=0}^{d} V_m$, where V_m is the mth eigenspace of the adjacency matrix of X. Moreover, the spherical function $\phi_m(x) \in V_m$ is given by $\phi_m(x) = p_j(\lambda_m)$, $d(x, x_0) = j$, for a finite orthogonal polynomial sequence $p_j(x)$.*

The analogies between the finite case X and the classical case S^{n-1} are summarized in the following list as cases (a) and (b).

(1a) Any differential operator commuting with $SO(n)$ is a polynomial in the spherical Laplacian Δ_S.

(1b) Any operator commuting with G is a polynomial in $A_1 - |\Omega_1|I$.

(2a) The eigenvalues of Δ_S are non-positive.

(2b) The eigenvalues of $A_1 - |\Omega_1|I$ are non-positive.

(3a) $\Delta_S f = \lambda f$ is a second order differential equation for f.

(3b) $A_1 f = \lambda f$ is a second order difference equation for f.

(4a) For $f \in Harm(k)$, the mean value operator T_θ satisfies $T_\theta f = p_k(cos(\theta))f$.

(4b) For $f \in V_k$, the mean value operator A_j satisfies $A_j f = p_j(\lambda_k)f$.

(5a) A limit of the mean value value operator, $\lim_{\theta \to 0} T_\theta - I = c\Delta_S$.

(5b) A difference of the mean value operator is $A_1 - |\Omega_1|I$ is the "closest" difference operator.

II.2 Krawtchouk polynomials.

We next give an appropriate graph X, find the eigenspaces V_j, realize the Krawt-chouk polynomials $K_j(x, 1/2, d)$ as spherical functions.

Let X be the d-dimensional cube: the vertices of X are all 2^d d-tuples of 0's and 1's, and two vertices are connected by an edge if they agree in all but one coordinate. The distance between two vertices is the number of coordinates in which they differ. The group G which acts on X is the hyperoctahedral group: all $d!$ permutations of the coordinates, and all 2^d interchanges of 0's and 1's. This group is a semidirect product of the symmetric group S_d and a product of d cyclic groups of order 2. The conditions (*) are easily verified.

First we find the three-term recurrence relation of Proposition 1. Let $x_0 = 00\cdots0$ be fixed, and let $x = 1\cdots10\cdots0$ be fixed vertex which is distance i from x_0, so that x has i 1's. There are d vertices which are distance 1 from x: i of them are distance $i-1$ from x_0, and $d - i$ of them are distance $i + 1$ from x_0. We find

$$(1) \qquad \lambda p_i(\lambda) = i p_{i-1}(\lambda) + (d - i)p_{i+1}(\lambda).$$

Equation (1) for $i = d$ determines the eigenvalues $\lambda_j = d - 2j$. These can be derived independently by explicitly finding the eigenspaces V_j. For $1 \le i \le d$, let x_i be the ith coordinate function on X defined by $x_i(0) = -1$, and $x_i(1) = 1$. Let V_j be the span of all square free monomials of degree j in the x_i's. It is easy to see that these monomials are orthogonal, so that $dim V_j = \binom{d}{j}$. We now verify that

$$(2) \qquad A_1 x_1 x_2 \cdots x_j = (d - 2j)x_1 x_2 \cdots x_j,$$

so that $d - 2j$ is the eigenvalue of A_1 on V_j. To see this, evaluate both sides of (2) at an arbitrary vertex y of X. If we change one of the first j entries of y, then the value of the monomial changes sign. If we change one of the last $d - j$ entries of y, the value of the monomial remains the same. So the eigenvalue is $d - 2j$.

Next we find an explicit formula for the spherical functions (the Krawtchouk polynomials) as a $_2F_1$. What monomial in V_j is fixed by the subgroup K of G fixing x_0? Clearly $K = S_d$, the symmetric group of the coordinates, so the only polynomial in V_j invariant under S_d is the elementary symmetric function $e_j(x_1, x_2, \ldots, x_d)$ of degree d. We must evaluate this polynomial at a vertex which is distance i from x_0, $y = 1 \cdots 10 \cdots 0$. The function $e_j(x_1, x_2, \ldots, x_d)$ can be thought of as the sum over all subsets S of $\{1, 2, \ldots, d\}$ of size j. Let $|S \cap \{1, 2, \ldots, i\} = k|$, and $|S \cap \{i+1, i+2, \ldots, d\} = j - k|$. The value of the monomial corresponding to S at y is $(-1)^{j-k}$, so we have

$$\phi_j(\Omega_i) = \text{constant} \cdot \sum_{k=0}^{j} \binom{i}{k} \binom{d-i}{j-k} (-1)^{j-k}.$$

This is equivalent (by a $_2F_1$ transformation) to (Kr).

The weight function for $\phi_j(\Omega_i)$ is $|\Omega_i| = \binom{d}{i} = \dim V_i$. Note that if the spherical function is normalized to be 1 at $i = 0$, it is symmetric under the interchange of i and j. This always occurs if the group G is a semidirect product with an abelian normal subgroup.

The Krawtchouk polynomials $K_j(i, k/(k+1), d)$ are the spherical functions for the graph of all d-tuples of 0's, 1's, $\ldots$, k's [8].

II.3 Hahn polynomials.

Next we give an appropriate graph X for the Hahn polynomials.

Let X be the set of all n-subsets of $\{1, 2, \ldots, v\}$, so that $|X| = \binom{v}{n}$. two subsets A and B are connected by an edge if $|A \cap B| = n - 1$. If we assume that $2n \leq v$, then the maximum distance is n, which occurs if two subsets are disjoint. In general, $d(A, B) = n - |A \cap B|$. The symmetric group S_v acts on X and the conditions (*) are satisfied.

First we find the size of the spheres of radius j, which are the weights of the spherical functions. Fix $x_0 = \{1, 2, \ldots, n\}$. The number of n-subsets B of $\{1, 2, \ldots, v\}$ such that $|x_0 \cap B| = n - j$ is

$$\binom{n}{n-j} \binom{v-n}{j}.$$

These are the weights for the Hahn polynomials (see (Ha)) with $\alpha = n - v - 1$, $\beta = -n - 1$, since then

$$\binom{\alpha + j}{j} \binom{\beta + n - j}{n - j} = \binom{n}{n - j} \binom{v - n}{j}.$$

The three term recurrence relation is easily found as

$$\lambda p_i(\lambda) = i^2 p_{i-1}(\lambda) + (n - i)(v - n - i) p_{i+1}(\lambda)$$
$$+ [n(v - n) - i^2 - (n - i)(v - n - i)] p_i(\lambda).$$

This is the second order difference equation for Hahn polynomials, or the three term recurrence relation for the dual Hahn polynomials.

To define the eigenspaces V_j, $0 \leq j \leq n$, we define monomial functions on X by

$$x_{i_1} x_{i_2} \cdots x_{i_j}(A) = \begin{cases} 1 \text{ if } \{i_1, i_2, \cdots, i_j\} \subset A \\ 0 \text{ otherwise.} \end{cases}$$

(These functions are called sections of the zeta function in the theory of posets.) Let $Hom(j)$ be the span of all monomials of degree j, and let $D = \partial/\partial x_1 + \cdots + \partial/\partial x_v$ be the "down" operator. G acts on $Hom(j)$, but the action is not irreducible. Since D commutes with the action of G, G also acts on $Harm(j) = Hom(j) \cap ker D$.

Again to show $Harm(j) = V_j$, i.e. $Harm(j)$ is irreducible, we show that $Harm(j)$ has a unique spherical function ϕ_j, and compute what ϕ_j is. The subgroup fixing x_0 is $K = S_n \times S_{v-n}$. This time the K-invariant polynomials in $Hom(j)$ are spanned by a product of elementary symmetric functions

$$e_s(x_1, \cdots, x_n) e_{j-s}(x_{n+1}, \cdots, x_v), 0 \le s \le j.$$

We find which linear combination is annihilated by D using

$$De_s(x_1, \cdots, x_n) = (n - s + 1)e_{s-1}(x_1, \cdots, x_n)$$
$$De_{j-s}(x_{n+1}, \cdots, x_v) = (v - n - j + s + 1)e_{j-s-1}(x_{n+1}, \cdots, x_v)$$

Suppose

(1)
$$D(\sum_{s=0}^{j} c_s e_s(x_1, \cdots, x_n) e_{j-s}(x_{n+1}, \cdots, x_v)) = 0.$$

From (1) we see that $c_{s+1}(-s + n) = (v - n - j + s + 1)c_s$, so

$$c_s = \frac{(v - n - j + 1)_s}{(-n)_s} c_0.$$

This proves that the K-invariant functions in $Harm(j)$ are 1-dimensional, so $Harm(j)$ is irreducible.

To evaluate $\phi_j(\Omega_i)$, fix $A = \{1, \cdots, n - i, n + 1, \cdots, n + i\} \in \Omega_i$. To evaluate

$$e_s(x_1, \cdots, x_n) e_{j-s}(x_{n+1}, \cdots, x_v)(A)$$

we count the number of pairs of subsets (B, C), where $|B| = s$, $B \subset \{1, \cdots, n - i\}$, $|C| = j - s$, $C \subset \{n + 1, \cdots, n + i\}$. This number is $\binom{n-i}{s}\binom{i}{j-s}$, so

$$\phi_j(\Omega_i) = \text{constant} \sum_{s=0}^{j} \frac{(v - n - j + 1)_s}{(-n)_s} \binom{n - i}{s}\binom{i}{j - s}.$$

This $_3F_2$ is equivalent to (Ha) with $\alpha = n - v - 1$ and $\beta = -n - 1$ by a $_3F_2$ transformation. From $i = 1$ we find $\lambda_j = n(v - n) - j(v + 1 - j)$.

Finally the L^2 norm of the polynomials can be found from $dim V_j = \binom{v}{j} - \binom{v}{j-1}$.

II.4 q-Hahn polynomials.

Next we give a very short introduction to a q-analog of Hahn polynomials, whose explicit formula is in terms of basic hypergeometric series (see Rahman's paper [15] in this volume).

The main idea is that q-binomial coefficients

$$\begin{bmatrix} n \\ k \end{bmatrix}_q = \frac{(q;q)_n}{(q;q)_k(q;q)_{n-k}}$$

will replace the binomial coefficients that were used in §II.3. Thus we need to know what $\begin{bmatrix} n \\ k \end{bmatrix}_q$ counts instead of $\binom{n}{k}$, and how q is involved. The answer is: $\begin{bmatrix} n \\ k \end{bmatrix}_q$ is the number of k-dimensional vector spaces of an n-dimensional vector space over a finite field of order q.

We let X be the set of all n-dimensional vector spaces of an v-dimensional vector space over a finite field of order q. We let the edges of X be pairs of subspaces A-B such that $dim(A \cap B) = n - 1$. The group $G = GL_v(q)$ satisfies the (*) condition. The distance between two subspaces A and B is $d(A,B) = n - dim(A \cap B)$. How many subspaces are distance i from a fixed subspace? The answer is

$$\begin{bmatrix} n \\ n-i \end{bmatrix}_q \begin{bmatrix} v-n \\ i \end{bmatrix}_q q^{i^2},$$

the weights for q-Hahn polynomials.

We state, and do not give the details (see [7],[17]) for the calculation of the spherical function.

THEOREM. *The spherical function $\phi_j(\Omega_i)$ is a q-Hahn polynomial*

$$Q_j(q^{-i}, q^{n-v-1}, q^{-n-1}; n|q) = {}_3\phi_2 \left(\begin{array}{ccc} q^{-j}, & q^{j-v-1}, & q^{-i}; \\ & q^{-n}, & q^{n-v} \end{array} q; q \right).$$

There are also natural q-analogs of the Krawtchouk polynomials. These again are related to classical groups over fields of order q, whose Lie algebra is of type B_n, C_n, and D_n. Details occur in [17]. All known infinite families, whose maximum distance $d \to \infty$ are related to these groups (or their Weyl groups). It is believed that this is it [2].

III. Remarks.

There are many applications of the discrete orthogonal polynomials in §II to coding theory and the theory of designs. The location of the zeros of the polynomials is critical. Bannai's paper [2] in this volume contains many of these results. An example of an analytic theorem motivated by combinatorial constraints is Leonard's theorem. This theorem states that if $p_i(\lambda_j) = q_j(\mu_i)$ for another finite set of orthogonal polynomials $q_j(x)$, then $p_i(x)$ must be a special or limiting case of the Askey-Wilson ${}_4\phi_3$ polynomials.

The continuous version of the groups and graphs in §II. are called rank one symmetric spaces of compact type. They have been classified [20], and the spherical functions are all Jacobi polynomials. There is also work on rank k symmetric spaces, and orthogonal polynomials in several variables. Bannai [2] contains references to this work.

An exciting recent area of research is the relationship of quantum groups to q-orthogonal polynomials. Koornwinder's paper in this volume [10] surveys the work to this day. The quantum groups are not groups, but algebras with a multiplication that in a sense is dual to group multiplication. For $SU(2)$, the irreducible representations of this algebra are analogous to the polynomial representations in §I.3. Explicit calculations can be made, using a form of the binomial theorem for letters A and B which satisfy $BA = qAB$,

$$(A + B)^n = \sum_{k=0}^{n} \begin{bmatrix} n \\ k \end{bmatrix}_q A^k B^{n-k}.$$

When the analogous expansion to (I.3(1)) is made, the matrix elements are $_2\phi_1$'s instead of $_2F_1$'s.

REFERENCES

1. R. Askey and J. Wilson, *A set of orthogonal polynomials that generalize the Racah coefficients or the 6 − j symbols*, SIAM J. Math. Anal. **10** (1979), 1008–1016.
2. E. Bannai, *Orthogonal polynomials in coding theory and algebraic combinatorics*, this volume.
3. E. Bannai and T. Ito, "Algebraic Combinatorics I," Benjamin/Cummins, Menlo Park California, 1984.
4. E. Bannai and T. Ito, *Current research on algebraic combinatorics*, Graphs and Combinatorics **2** (1986), 287–308.
5. P. Delsarte, *An algebraic approach to the association schemes of coding theory*, Philips Res. Repts. Supp. **10**.
6. P. Delsarte, *Properties and applications of the recurrence* $F(i + 1, k + 1, n + 1) = q^{k+1}F(i, k + 1, n) - q^k F(i, k, n)$, SIAM J. Appl. Math. **31** (1976), 262–270.
7. P. Delsarte, *Hahn polynomials, discrete harmonics, and t-designs*, SIAM J. Appl. Math. **34** (1978), 157–166.
8. C. Dunkl, *A Krawtchouk polynomial addition theorem and wreath products of symmetric groups*, Indiana Math. J. **25** (1975), 335–358.
9. C. Dunkl, *Spherical functions on compact groups and applications to special functions*, Symposia Mathematica **22** (1979), 145–161.
10. T. Koornwinder, *Orthogonal polynomials and special functions in connection with quantum groups*, this volume.
11. T. Koornwinder, *Krawtchouk polynomials, a unification of two different group theoretic interpretations*, SIAM J. Math. Anal. **13** (1982), 1011–1023.
12. T. Koornwinder, *Clebsch-Gordan coefficients for SU(2) and Hahn polynomials*, Nieuw. Archief Wisk. (3) **29** (1981), 140–155.
13. I. G. Macdonald, *Orthogonal polynomials associated with root systems*, this volume.
14. W. Miller, "Symmetry groups and Their Applications," Academic Press, New York, 1972.
15. M. Rahman, *Some extensions of the beta integral and the hypergeometric function*, this volume.
16. N. J. A. Sloane, *An introduction to association schemes and coding theory*, in "Theory and Application of Special Functions," R. Askey ed., 1975, pp. 225–260.
17. D. Stanton, *Orthogonal polynomials and Chevalley groups*, in "Special Functions: Group Theoretic Aspects and Applications," R. Askey et al. (eds.), Reidel, Dordrecht, 1984, pp. 87–128.
18. X. Viennot, "Une theorie combinatoire des polynomes orthogonaux generaux," lecture Notes, UQAM, 1983.
19. N. Vilenkin, "Special Functions and the Theory of Group Representations," Translations of the AMS, 1968.
20. H. Wang, *Two-point homogeneous spaces*, Ann. of Math. **55** (1952), 177–191.

ASYMPTOTICS FOR ORTHOGONAL POLYNOMIALS AND THREE-TERM RECURRENCES

WALTER VAN ASSCHE[1]
Katholieke Universiteit Leuven
Department of Mathematics
Celestijnenlaan 200B
B-3030 Leuven (BELGIUM)

ABSTRACT. It is often desirable to obtain (asymptotic) properties of orthogonal polynomials and the measure with respect to which these polynomials are orthogonal. All orthogonal polynomials on the real line (with a positive Borel measure) satisfy a three term recurrence relation. We give a survey showing how properties of the recurrence coefficients reveal properties of the corresponding orthogonal polynomials.

1. Introduction

Given two sequences $\{a_n > 0: n=1,2,...\}$ and $\{b_n \in \mathbb{R}: n=0,1,2,...\}$ one can construct a sequence of polynomials using the three term recurrence formula

$$(1.1) \qquad xp_n(x) = a_{n+1}p_{n+1}(x) + b_n p_n(x) + a_n p_{n-1}(x) \qquad\qquad n=0,1,2,...$$

with initial values $p_0(x) = 1$ and $p_{-1}(x) = 0$ (notice that this choice of p_{-1} makes the value of a_0 irrelevant). It is well known (*Favard's theorem*) [5] that there exists a positive Borel measure μ on the real line such that

$$(1.2) \qquad \int p_n(x)p_m(x)\, \mathrm{d}\mu(x) = \delta_{m,n} \qquad\qquad m,n\geq 0,$$

which means that $\{p_n(x) : n=0,1,2,...\}$ are orthonormal polynomials with orthogonality measure μ. This measure need not be unique (this depends whether or not the Hamburger moment problem is determined) but a sufficient condition for a unique measure is *Carleman's condition*

$$\sum_{n=1}^{\infty} \frac{1}{a_n} = \infty,$$

---

[1] Research Associate of the Belgian National Fund for Scientific Research
E-mail: fgaee03@blekul11.bitnet

P. Nevai (ed.), Orthogonal Polynomials, 435–462.
© *1990 by Kluwer Academic Publishers.*

and this certainly holds whenever $\{a_n\}$ is bounded. The zeros of $p_n(x)$ are all real and simple and we denote them by

$$x_{1,n} < x_{2,n} < \cdots < x_{n,n}.$$

These zeros are all inside the smallest closed interval that contains the support of μ. Let $\hat{p}_n(x)$ be the monic orthogonal polynomial, i.e. $\hat{p}_n(x) = (a_1 a_2 \cdots a_n) p_n(x)$, then the three term recurrence relation becomes

$$\hat{p}_{n+1}(x) = (x-b_n)\hat{p}_n(x) - a_n^2\, \hat{p}_{n-1}(x) \qquad\qquad n=0,1,2,\ldots$$

with $p_{-1} = 0$ and $p_0 = 1$. One can write

$$\hat{p}_n(x) = \det(xI_n - J_n)\,,$$

where I_n is the identity matrix in $\mathbb{R}^{n \times n}$ and J_n is the tridiagonal matrix

$$J_n = \begin{bmatrix} b_0 & a_1 & & & & \\ a_1 & b_1 & a_2 & & & \\ & a_2 & \cdot & \cdot & & \\ & & \cdot & \cdot & \cdot & \\ & & & \cdot & \cdot & a_{n-1} \\ & & & & a_{n-1} & b_{n-1} \end{bmatrix},$$

and therefore the zeros of $p_n(x)$ are the same as the eigenvalues of J_n. If we consider the infinite dimensional Jacobi matrix J, given by

$$J = \begin{bmatrix} b_0 & a_1 & & & \\ a_1 & b_1 & a_2 & & \\ & a_2 & \cdot & \cdot & \cdot \\ & & \cdot & \cdot & \cdot \\ & & & \cdot & \end{bmatrix},$$

then $J: \ell_2 \longrightarrow \ell_2$ is a selfadjoint linear operator whenever the Hamburger moment problem is determined. One can easily verify that the entry in row m and column n is given by

$$J_{m,n} = \int x\, p_m(x) p_n(x)\, d\mu(x)\,,$$

so that $J = \int x\, dE(x)$, with $dE_{m,n}(x) = p_m(x) p_n(x)\, d\mu(x)$. This means that the orthogonality measure μ is also the spectral measure for the operator J. Perturbation techniques for linear operators are therefore very useful to study orthogonal polynomials (see e.g. [1,2,4,30]).

One can use bounds for the eigenvalues of symmetric (tridiagonal) matrices to obtain bounds for the zeros of orthogonal polynomials. By *Gershgorin's theorem* one has

$$x_{j,n} \in \bigcup_{i=0}^{n-1} [b_i - a_i - a_{i+1}, b_i + a_i + a_{i+1}]$$

(where one can put $a_n = 0$), which immediately yields

(1.3) $$\min_{0 \le i \le n-1} (b_i - a_i - a_{i+1}) \le x_{j,n} \le \max_{0 \le i \le n-1} (b_i + a_i + a_{i+1}).$$

These bounds are usually good enough for applications but sharper bounds are available (Freud [13,14], Gilewicz and Leopold [26,29], van Doorn [52]).

2. Perturbations of Chebyshev polynomials

The Chebyshev polynomials of the second kind $U_n(x) = \frac{\sin(n+1)\theta}{\sin\theta}$, where $x = \cos\theta$, have a very simple recurrence formula with $a_n = 1/2$ $(n=1,2,...)$ and $b_n = 0$ $(n=0,1,2,...)$. A system of orthogonal polynomials for which the recurrence coefficients satisfy

(2.1) $$\lim_{n \to \infty} a_n = 1/2, \qquad\qquad \lim_{n \to \infty} b_n = 0$$

will be said to be a perturbation of the Chebyshev polynomials. In a similar way one may consider orthogonal polynomials with recurrence coefficients that satisfy

$$\lim_{n \to \infty} a_n = a/2 > 0, \qquad\qquad \lim_{n \to \infty} b_n = b \in \mathbb{R},$$

and these orthogonal polynomials are perturbations of $U_n((x-b)/a)$ and are said to belong to the class $M(a,b)$ introduced by Nevai [39]. As long as $a \ne 0$ one may, without loss of generality, limit the investigation to $M(1,0)$, which corresponds exactly to (2.1). The following results hold whenever (2.1) is valid (without any assumption on the rate of convergence):

THEOREM 1: *Suppose that (2.1) holds, then $\mu = \mu_c + \mu_d$ with μ_c a continuous measure on $[-1,1]$ and μ_d a discrete measure with a bounded support for which the accumulation points are in $[-1,1]$. If $x + \sqrt{x^2-1}$ is such that $|x + \sqrt{x^2-1}| > 1$ whenever $x \in \mathbb{C} \setminus [-1,1]$, then*

$$\lim_{n \to \infty} \frac{p_{n-1}(x)}{p_n(x)} = x - \sqrt{x^2-1}, \qquad\qquad \lim_{n \to \infty} |p_n(x)|^{1/n} = |x + \sqrt{x^2-1}|$$

hold uniformly for x on compact sets of $\mathbb{C} \setminus supp(\mu)$. If $x \in supp(\mu) \setminus [-1,1]$, then

$$\lim_{n \to \infty} \frac{p_{n-1}(x)}{p_n(x)} = x + \sqrt{x^2 - 1}, \qquad \lim_{n \to \infty} |p_n(x)|^{1/n} = |x - \sqrt{x^2 - 1}|.$$

PROOF: The first part is known as *Blumenthal's theorem* (even though Blumenthal [3] only showed that the zeros of orthogonal polynomials in M(a,b) are dense in in the interval [b-a,b+a]) and is a special case of H. Weyl's theorem on compact perturbations of self-adjoint Hilbert space operators (see e.g. Máté, Nevai and Van Assche [38]). A proof of the ratio asymptotics is given in Section 4, and the root asymptotics are just the geometric mean of this. These limits are solutions of the quadratic equation $z^2 - 2xz + 1 = 0$ and since $|p_{n-1}(x)/p_n(x)|$ tends to zero as $x \to \infty$ we

need to choose the solution $x - \sqrt{x^2 - 1}$ for the ratio asymptotics when $x \in \mathbb{C} \setminus \text{supp}(\mu)$. If $x \in \text{supp}(\mu) \setminus [-1,1]$ then x is a mass point of μ and thus $\Sigma\, p_n^2(x) < \infty$,

and this is only possible if we choose the solution $x + \sqrt{x^2 - 1}$ in the ratio asymptotics for such x. □

For the behaviour of the orthogonal polynomials on the *essential spectrum* [-1,1] we have the following weak asymptotics:

THEOREM 2 (Nevai [39,40], ...): *Suppose that (2.1) holds. Let [A,B] be the smallest interval that contains supp(μ). Then for every function f∈ $C^1[A,B]$ one has*

$$(2.2) \qquad \lim_{n \to \infty} \sum_{j=1}^{n} f(x_{j,n}) - \sum_{j=1}^{n-1} f(x_{j,n-1}) = \frac{1}{\pi} \int_{-1}^{1} \frac{f(t)}{\sqrt{1-t^2}}\, dt.$$

If f∈ C[A,B], then

$$(2.3) \qquad \lim_{n \to \infty} \frac{1}{n} \sum_{j=1}^{n} f(x_{j,n}) = \frac{1}{\pi} \int_{-1}^{1} \frac{f(t)}{\sqrt{1-t^2}}\, dt ,$$

and for every integer k

$$(2.4) \qquad \lim_{n \to \infty} \int f(x)\, p_n(x) p_{n+k}(x)\, d\mu(x) = \frac{1}{\pi} \int_{-1}^{1} \frac{f(t) T_k(t)}{\sqrt{1-t^2}}\, dt ,$$

where $T_k(\cos \theta) = \cos n\theta$ is the Chebyshev polynomial of the first kind of degree k.

PROOF: If one takes the derivative of the ratio asymptotics then one finds the first result for the function $f(t) = (x-t)^{-1}$. Such functions are dense in C[A,B] and thus also in $C^1[A,B]$ and the first result then follows because the left hand side defines a sequence of linear operators on $C^1[A,B]$ which are uniformly bounded. The second result, the famous *arcsin law* for zeros of orthogonal polynomials, follows in a similar way by taking the derivative of the root asymptotics. To prove (2.4) we observe that

$$\int x^m p_n(x) p_{n+k}(x) \, d\mu(x) = (J^m)_{n,n+k} \, ,$$

where J is the infinite Jacobi matrix. Therefore this quantity only depends on a_{n+j} and b_{n+j} for $|j| \leq (k+m)/2$ and the limit (as $n \to \infty$) is the same for all orthogonal polynomials that satisfy (2.1). This quantity can easily be computed when $p_n(x) = U_n(x)$ and gives the result for $f(x) = x^m$ and then also for every $f \in C[A,B]$. □

Conversely if (2.2) or (2.4) holds for $f(x)=x$ and $f(x)=x^2$, then the recurrence coefficients converge as in (2.1). However, the arcsin behaviour (2.3) does not imply that the recurrence coefficients converge; one can show that (2.3) already holds when

$$\lim_{n \to \infty} \frac{1}{n} \sum_{k=0}^{n-1} \left(|1 - 2a_{k+1}| + |b_k| \right) = 0 \, ,$$

(Geronimo, Harrell, Van Assche [20]) and this allows the recurrence coefficients to be unbounded.

The previous results hold without any assumptions on the rate of convergence. If we assume some rate, then much more can be said about the orthogonal polynomials and the spectral measure. Recall that the Chebyshev polynomials of the second kind are the easiest system of orthogonal polynomials for which (2.1) holds. Some important properties of these Chebyshev polynomials of the second kind are

(2.5) $\qquad |U_n(x)| \leq n+1 \qquad\qquad -1 \leq x \leq 1 \, ,$

(2.6) $\qquad |\sqrt{1-x^2} \, U_n(x)| \leq 1 \qquad\qquad -1 \leq x \leq 1 \, ,$

(2.7) $\qquad U_n(x) = \dfrac{z^{n+1} - z^{-n-1}}{z - z^{-1}} \qquad\qquad z = x + \sqrt{x^2 - 1} \, .$

The square root in (2.7) is such that $|x + \sqrt{x^2-1}| > 1$ whenever $x \in \mathbb{C} \setminus [-1,1]$. If the polynomials $p_n(x)$ are a perturbation of these Chebyshev polynomials, then one may wonder whether the polynomials $p_n(x)$ have similar properties. To answer such questions one starts by considering a comparison equation, which is constructed as follows: define $\tilde{p}_n(x) = 2^n(a_1 a_2 ... a_n) p_n(x)$, then $\tilde{p}_n(x)$ has the same leading coefficient as $U_n(x)$ and satisfies the recurrence relation

(2.8) $\qquad 2x \, \tilde{p}_k(x) = \tilde{p}_{k+1}(x) + 2b_k \, \tilde{p}_k(x) + 4a_k^2 \, \tilde{p}_{k-1}(x)$

whereas the Chebyshev polynomials of the second kind satisfy

(2.9) $2x\, U_{n-k-1}(x) = U_{n-k}(x) + U_{n-k-2}(x)$.

Multiply (2.8) by $U_{n-k-1}(x)$ and (2.9) by $\tilde{p}_k(x)$ and substract the obtained equations to obtain

$$0 = U_{n-k-1}(x)\tilde{p}_{k+1}(x) - U_{n-k}(x)\tilde{p}_k(x) + 2b_k\, U_{n-k-1}(x)\tilde{p}_k(x)$$
$$+ 4a_k^2\, U_{n-k-1}(x)\tilde{p}_{k-1}(x) - U_{n-k-2}(x)\tilde{p}_k(x).$$

Summing from k=0 to n-1 gives (a lot of terms cancel out)

(2.10) $\tilde{p}_n(x) = U_n(x) + \displaystyle\sum_{k=0}^{n-1}\left\{(1-4a_{k+1}^2)U_{n-k-2}(x) - 2b_k\, U_{n-k-1}(x)\right\}\tilde{p}_k(x)$,

which shows that $\tilde{p}_n(x)$ is $U_n(x)$ + a remainder which is small (because of (2.1)). An important tool in analyzing (2.10) is:

LEMMA (discrete version of Gronwall's inequality): *Suppose c_n and d_n (n=0,1,2,...) are non-negative real numbers such that*

(2.11) $c_n \leq A + \displaystyle\sum_{k=0}^{n-1} d_k c_k$,

where A is a positive constant, then

(2.12) $c_n \leq A \, exp\left\{\displaystyle\sum_{k=0}^{n-1} d_k\right\}.$

PROOF: Even though this inequality is well known (see e.g. Atkinson [2, p. 455] or Máté and Nevai [32]) we present a proof because of the crucial role this inequality plays in what follows. We use induction on n: if n=0 then the result is true (an empty sum is defined to be zero). Now suppose the result is true for c_0 up to c_{n-1}, then we have from (2.11)

$$c_n \leq A + A \displaystyle\sum_{k=0}^{n-1} d_k \, exp\left\{\sum_{j=0}^{k-1} d_j\right\} .$$

Use the inequality $d_k \leq exp(d_k)-1$, then the right hand side becomes a telescoping sum, which immediately gives (2.12). □

We now have all the tools we need to obtain the analogues of (2.5) and (2.6):

THEOREM 3 (Geronimo and Case [19], Nevai [38]): *For* $-1 \leq x \leq 1$ *one has*

$$(2.13) \qquad |\tilde{p}_n(x)| \leq (n+1) \, exp \left\{ \sum_{k=0}^{n-1} (k+1)(|\,1-4a_{k+1}^2| + 2|b_k|) \right\},$$

$$(2.14) \qquad |\sqrt{1-x^2} \, \tilde{p}_n(x)| \leq 1 + \left\{ \sum_{k=0}^{n-1} (k+1)(|\,1-4a_{k+1}^2| + 2|b_k|) \right\}$$

$$\times \, exp \left\{ \sum_{k=0}^{n-1} (k+1)(|\,1-4a_{k+1}^2| + 2|b_k|) \right\}.$$

PROOF: Use the inequality (2.5) in the equation (2.10), then one finds

$$\frac{|\tilde{p}_n(x)|}{n+1} \leq 1 + \sum_{k=0}^{n-1} (k+1) \left\{ |1-4a_{k+1}^2| + 2|b_k| \right\} \frac{|\tilde{p}_k(x)|}{k+1}.$$

Gronwall's inequality then yields (2.13). In order to obtain (2.14) we use the inequalities (2.6) and (2.13) in the equation (2.10). □

If, in particular

$$(2.15) \qquad \sum_{k=0}^{\infty} (k+1) \left\{ |1-4a_{k+1}^2| + 2|b_k| \right\} = A < \infty,$$

then

$$|\tilde{p}_n(x)| \leq (n+1) \, e^A, \qquad\qquad |\sqrt{1-x^2} \, \tilde{p}_n(x)| \leq 1 + A \, e^A.$$

For *Jacobi polynomials* we have

$$a_n^2 = \frac{4\,n\,(n+\alpha)(n+\beta)\,(n+\alpha+\beta)}{(2n+\alpha+\beta-1)(2n+\alpha+\beta)^2\,(2n+\alpha+\beta+1)}$$

$$= \frac{1}{4} + \frac{1-2(\alpha^2+\beta^2)}{16n^2} + O(n^{-3}),$$

$$b_n = \frac{\beta^2-\alpha^2}{(2n+\alpha+\beta)(2n+\alpha+\beta+2)} = \frac{\beta^2-\alpha^2}{4n^2} + O(n^{-3}),$$

where $\alpha > -1$, $\beta > -1$ and if $\alpha = -\beta$ one has $b_0 = \beta$. Clearly (2.15) is only valid for the four cases $(\alpha,\beta) = (-1/2,-1/2)$, $(1/2,1/2)$, $(-1/2,1/2)$ and $(1/2,-1/2)$ (the first two cases correspond respectively to Chebyshev polynomials of the first and of the second kind). One may however improve the bound (2.14) for the remaining Jacobi polynomials:

THEOREM 4 (Nevai [41,42]): *Suppose that there is a positive constant A such that*

$$\sum_{k=0}^{n-1} (k+1)(|\,1-4a_{k+1}^2| + 2|b_k|) \leq A \, log(n+1),$$

then there exists a constant C and an integer m such that for $-1 \leq x \leq 1$

$$|\tilde{p}_n(x)| \leq C (1-x^2)^{-(m+1)/2}.$$

PROOF: From (2.14) we find

$$|\sqrt{1-x^2}\, \tilde{p}_n(x)| \leq 1 + A\,(n+1)^A \log(n+1) \leq A_1(n+1)^A \log(n+1),$$

where A_1 is a positive constant. Use this new inequality in (2.10), then we find

$$|(1-x^2)\tilde{p}_n(x)| \leq 1 + A_1 \sum_{k=0}^{n-1} (k+1)^A \log(k+1)(|1-4a_{k+1}^2| + 2|b_k|).$$

If $A < 1$ then the sum on the right hand side is bounded and the theorem follows. If $A \geq 1$ then

$$|(1-x^2)\tilde{p}_n(x)| \leq 1 + A_1\, n^{A-1} A \log^2(n+1) \leq A_2(n+1)^{A-1}\log^2(n+1),$$

with A_2 a positive constant. We can repeat this procedure and find

$$|(\sqrt{1-x^2}\,)^m\, \tilde{p}_n(x)| \leq A_m(n+1)^{A+1-m}\log^m(n+1)$$

as long as $A+1 \geq m$. If $A < m$ then one can insert this in (2.10) once more and find that $|(\sqrt{1-x^2}\,)^{m+1}\, \tilde{p}_n(x)|$ is bounded. □

Theorem 3 is only useful when (2.15) is satisfied. Quite often this is too strong. The problems are at $x = \pm 1$ where $U_n(\pm 1) = n+1$ (which tends to infinity) whereas for any other value of x in $(-1,1)$ $U_n(x)$ remains bounded. One can resolve this problem as follows:

THEOREM 5 (Geronimo and Case [19], Nevai [38]): *Let $z = z(x) = x+\sqrt{x^2-1}$, with $|z| > 1$ when $x \in \mathbb{C} \setminus [-1,1]$ and*

$$\lim_{\epsilon \to 0+} z(x+i\epsilon) = e^{i\theta}, \qquad\qquad \lim_{\epsilon \to 0+} z(x-i\epsilon) = e^{-i\theta} \quad (x = \cos \theta).$$

then for every x

$$(2.16) \quad \left|\frac{\tilde{p}_n(x)}{z^n}\right| \leq \frac{3(n+1)}{1+(n+1)|1-z^{-2}|}$$

$$\times \exp\left\{ \sum_{k=0}^{n-1} \frac{3(k+1)}{1+(k+1)|1-z^{-2}|} \left(|1-4a_{k+1}^2| +2|b_k|\right)\right\}.$$

PROOF: The proof works just as in Theorem 3 by using (2.10) and Gronwall's inequality, but now one uses the bound

$$\left|\frac{U_n(x)}{z^n}\right| = \frac{|1-z^{-2n-2}|}{|1-z^{-2}|} \leq \frac{3(n+1)}{1+(n+1)|1-z^{-2}|}.$$

To prove this bound, one observes that

$$|1-z^{-2m}| \leq 2, \qquad\qquad |1-z^{-2m}| \leq m|1-z^{-2}|,$$

the first is good if z is away from ±1, the second is good near ±1. One can combine both bounds to obtain

$$3m|1-z^{-2}| = 2m|1-z^{-2}| + m|1-z^{-2}| \geq m|1-z^{-2}| \, |1-z^{-2m}| + |1-z^{-2m}|,$$

which gives the desired bound. □

Suppose that

$$(2.17) \quad \sum_{k=0}^{\infty} \left(|1-4a_{k+1}^2| + 2|b_k|\right) < \infty,$$

(which holds for instance for all Jacobi polynomials) then one can introduce the function

$$(2.18) \quad \phi_0(x) = 1 + \sum_{k=0}^{\infty} \left(\frac{1-4a_{k+1}^2}{z^2} - \frac{2b_k}{z}\right) \frac{\tilde{p}_k(x)}{z^k}.$$

This function is well defined for every $x \in \mathbb{C} \setminus [-1,1]$ and the series converges uniformly on every compact subset of $\mathbb{C} \setminus [-1,1]$. For $x = \cos\theta$ we have

$$(2.19) \quad \lim_{\epsilon \to 0+} \phi_0(x+i\epsilon) = \phi_0(x) = 1 + \sum_{k=0}^{\infty} \left\{(1-4a_{k+1}^2)e^{-2i\theta} - 2b_ke^{-i\theta}\right\} \tilde{p}_k(x)e^{-ik\theta},$$

$$\lim_{\epsilon \to 0+} \phi_0(x-i\epsilon) = \overline{\phi_0(x)},$$

and these limits exist for every $x \in (-1,1)$. Clearly $\phi_0(x)$ is uniformly continuous on every closed interval of $(-1,1)$. If (2.15) holds then $\phi_0(x)$ is uniformly continuous on $[-1,1]$. This function plays an important role in the asymptotic theory of the orthogonal polynomials $p_n(x)$:

THEOREM 5 (Geronimo and Case [19], Nevai [38]): *Suppose that (2.17) holds, then*

$$(2.20) \qquad \lim_{n \to \infty} \frac{\tilde{p}_n(x)}{z^n} = \frac{z}{z - z^{-1}} \phi_0(x),$$

uniformly for $|x + \sqrt{x^2 - 1}| \geq r > 1$. *If* $x = \cos \theta$ *then*

$$(2.21) \qquad \lim_{n \to \infty} \left\{ [\tilde{p}_n(x) - 4a_n^2 \, e^{-i\theta} \tilde{p}_{n-1}(x)] - e^{in\theta} \phi_0(x) \right\} = 0,$$

uniformly on every compact interval in $(-1,1)$. *In particular we have*

$$(2.22) \qquad \lim_{n \to \infty} \left\{ \sqrt{1 - x^2} \, \tilde{p}_n(x) - |\phi_0(x)| \, \sin((n+1)\theta + \Gamma(x)) \right\} = 0,$$

uniformly on every compact interval in $(-1,1)$, *where* $\Gamma(x) = \arg[\phi_0(x)]$. *If (2.15) holds then (2.21) and (2.22) hold uniformly on* $[-1,1]$.

PROOF: The first result follows easily by using (2.10), (2.7) and dominated convergence. On $(-1,1)$ we may use $U_n(x) - e^{-i\theta} U_{n-1}(x) = e^{in\theta}$ in (2.10) to obtain

$$\tilde{p}_n(x) - e^{-i\theta} \tilde{p}_{n-1}(x) = e^{in\theta} \left\{ 1 + \sum_{k=0}^{n-2} [(1 - 4a_{k+1}^2) e^{-2i\theta} - 2b_k e^{-i\theta}] \, \tilde{p}_k(x) e^{-ik\theta} \right\}$$

$$- 2b_{n-1} \tilde{p}_{n-1}(x).$$

Add $(1 - 4a_n^2) e^{-i\theta} \tilde{p}_{n-1}(x)$ to both sides, then

$$(2.23) \qquad \tilde{p}_n(x) - 4a_n^2 \, e^{-i\theta} \tilde{p}_{n-1}(x)$$

$$= e^{in\theta} \left\{ 1 + \sum_{k=0}^{n-1} [(1 - 4a_{k+1}^2) e^{-2i\theta} - 2b_k e^{-i\theta}] \, \tilde{p}_k(x) e^{-ik\theta} \right\},$$

from which one easily derives (2.21). Multiply both sides of (2.21) by $e^{i\theta}$ and take the imaginary part of both sides, then one has (2.22). □

Notice that from (2.23) and (2.19) one can easily find an upper bound for the rate of convergence by using the bound (2.16), this gives

$$|[\tilde{p}_n(x) - e^{-i\theta}\tilde{p}_{n-1}(x)] - e^{in\theta}\phi_0(x)|$$

$$\leq \sum_{k=n}^{\infty} \frac{3(k+1)}{1+(k+1)\sin\theta}(|1-4a_{k+1}^2| + 2|b_k|)$$

$$\times \exp\left\{\sum_{k=0}^{\infty} \frac{3(k+1)}{1+(k+1)\sin\theta}(|1-4a_{k+1}^2| + 2|b_k|)\right\}.$$

The function $\phi_0(x)$ also contains a lot of information about the spectral measure μ:

THEOREM 6 (Geronimo and Case [19], Nevai [38]): *Suppose (2.17) holds, then μ is absolutely continuous in $(-1,1)$ and $\mu' > 0$ is continuous on $(-1,1)$. Moreover*

$$(2.24) \qquad |\phi_0(x)|^2 = \frac{2\gamma^2}{\pi}\frac{\sqrt{1-x^2}}{\mu'(x)} \qquad\qquad -1 < x < 1,$$

where $\gamma = \prod_{k=1}^{\infty} 2a_k$.

PROOF: If $x \in (-1,1)$ is a mass point of μ then $\Sigma\, p_n^2(x) < \infty$, which implies that $\tilde{p}_n(x)$ tends to zero as $n \to \infty$, and by (2.21) this means that $\phi_0(x) = 0$. We prove that $\phi_0(x) \neq 0$ when $x \in (-1,1)$. Define

$$\psi_n(x) = \tilde{p}_n(x) - 4a_n^2\, e^{-i\theta}\tilde{p}_{n-1}(x) ,$$

then by (2.23)

$$(2.25) \qquad \psi_{n+1}(x) - e^{i\theta}\psi_n(x) = e^{i\theta}\left\{(1-4a_{n+1}^2)e^{-2i\theta} - 2b_n e^{-i\theta}\right\}\tilde{p}_n(x).$$

Observe that

$$|Im\,\psi_{n+1}(x)| = \sqrt{1-x^2}\, 4a_{n+1}^2\, |\tilde{p}_n(x)| \leq |\psi_{n+1}(x)|,$$

therefore (2.25) leads to

$$|\psi_n(x)| \leq |\psi_{n+1}(x)|\left\{1 + \frac{1}{4a_{n+1}^2\sqrt{1-x^2}}(|1-4a_{n+1}^2| + 2|b_n|)\right\}$$

$$\leq |\psi_{n+1}(x)|\, \exp\left\{\frac{1}{4a_{n+1}^2\sqrt{1-x^2}}(|1-4a_{n+1}^2| + 2|b_n|)\right\}.$$

From this we easily find

$$|\psi_n(x)| \le |\psi_m(x)| \exp \left\{ \frac{C}{\sqrt{1-x^2}} \sum_{k=n}^{m-1} (|1-4a_{k+1}^2| + 2|b_k|) \right\}.$$

By letting $m \to \infty$ one finds that $\phi_0(x)$ can only be zero if $\psi_n(x)$ is zero for infinitely many n, and this is impossible because $p_n(x)$ and $p_{n-1}(x)$ cannot be simultaneously zero. Next we prove (2.24). By (2.21) we have

$$\lim_{n \to \infty} |\tilde{p}_n(x) - 4a_n^2 e^{-i\theta} \tilde{p}_{n-1}(x)|^2 = |\phi_0(x)|^2,$$

uniformly on closed intervals of (-1,1). Now on one hand we have

$$|\tilde{p}_n(x) - 4a_n^2 e^{-i\theta} \tilde{p}_{n-1}(x)|^2 = \tilde{p}_n^2(x) + (4a_n^2)^2 \tilde{p}_{n-1}^2(x) - 16a_n^2 x\tilde{p}_n(x)\tilde{p}_{n-1}(x),$$

so that by (2.4) in Theorem 2 we have for every bounded and continuous function f

$$\lim_{n \to \infty} \int |\tilde{p}_n(t) - 4a_n^2 e^{-i\theta} \tilde{p}_{n-1}(t)|^2 f(t) \, d\mu(t) = \frac{2\gamma^2}{\pi} \int_{-1}^{1} f(t)\sqrt{1-t^2} \, dt.$$

On the other hand we have

$$\lim_{n \to \infty} \int |\tilde{p}_n(t) - 4a_n^2 e^{-i\theta} \tilde{p}_{n-1}(t)|^2 f(t) \, d\mu(t) = \int |\phi_0(t)|^2 f(t) \, d\mu(t).$$

By comparing the right hand side of the last two equations (and since μ has no mass points in (-1,1)) we find (2.24). □

As for the mass points of μ, which are all outside (-1,1), we have the following result:

THEOREM 7 (Chihara and Nevai [7], Case and Geronimo [19], Guseinov [27], Marchenko [30]): *Suppose (2.15) holds, then μ has a finite number of mass points outside (-1,1) and there are no mass points at -1 and 1.*

PROOF: We introduce the associated orthogonal polynomials (of order m=0,1,2,...) by the recurrence formula

(2.26) $\quad xp_n^{(m)}(x) = a_{n+m+1}p_{n+1}^{(m)}(x) + b_{n+m}p_n^{(m)}(x) + a_{n+m}p_{n-1}^{(m)}(x)$,

with $p_0^{(m)}(x) = 1$ and $p_{-1}^{(m)}(x) = 0$. Clearly $p_{n-m}^{(m)}(x)$ is, for fixed m, another solution of the recurrence formula (1.1). Let

$$\tilde{p}_n^{(m)}(x) = 2^n(a_{m+1}a_{m+2}\cdots a_{m+n}) \, p_n^{(m)}(x),$$

and define

$$(2.27) \qquad \phi_m(x) = 1 + \sum_{k=m}^{\infty} \left\{ \frac{1-4a_{k+1}^2}{z^2} - \frac{2b_k}{z} \right\} \frac{\tilde{p}\{_{-m}^{m}\}(x)}{z^{k-m}} .$$

If $p_n(x)$ are orthogonal polynomials satisfying (2.1), then clearly also $p_n^{(m)}(x)$ satisfy (2.1). The function $\phi_m(x)$ is the limit function in (2.20) and (2.21) for the associated polynomials of order m. We will show that for sufficiently large m $\phi_m(x)$ has no zeros. From (2.13) we find

$$|\tilde{p}\{_{-m}^{m}\}(x)| \leq (n+1) \exp \left\{ \sum_{k=m}^{\infty} (k+1)(|1-4a_{k+1}^2| + 2|b_k|) \right\} ,$$

which, when inserted in (2.27) gives

$$|\phi_m(x) - 1| \leq \sum_{k=m}^{\infty} (k+1)(|1-4a_{k+1}^2| + 2|b_k|)$$

$$\times \exp \left\{ \sum_{k=m}^{\infty} (k+1)(|1-4a_{k+1}^2| + 2|b_k|) \right\} .$$

Clearly the right hand side can be made as small as possible by choosing m large enough, which implies that for such m ϕ_m cannot become zero for any x. We already know that ϕ_0 has no zeros in $(-1,1)$, so $\phi_0(x)$ can only become zero for $x \in \mathbb{R} \setminus (-1,1)$. A mass point of μ is always a zero of ϕ_0 because $p_n(x)$ is square summable at every mass point of x, which implies that $|\tilde{p}_n(x)| \to 0$ as $n \to \infty$. If ϕ_m has no zeros, then this means that the orthogonality measure $\mu^{(m)}$ for the associated orthogonal polynomials has no mass points and in particular that all the zeros of these associated polynomials are in $(-1,1)$. The zeros of $p_n^{(m-1)}$ and $p_{n+1}^{(m)}$ interlace, consequently $\mu^{(m-1)}$ can have at most 2 mass points outside $(-1,1)$. If we continue like this, we find that $\mu^{(0)} = \mu$ can have at most 2m mass points outside $(-1,1)$. $\square$

It is possible to give an upperbound on the number of mass points: let $N(a,b)$ be the number of mass points in (a,b) and suppose that (2.15) holds, then

$$N(1,\infty) \leq \sum_{k=0}^{\infty} (k+1) \left\{ (4a_{k+1}^2-1)^+ + 2b_k^+ \right\} ,$$

$$N(-\infty,-1) \leq \sum_{k=0}^{\infty} (k+1) \left\{ (4a_{k+1}^2-1)^+ - 2b_k^- \right\}$$

(Geronimo [16,17]). Máté and Nevai [33] and Nikishin [44] have shown that the number of mass points is finite whenever

$$\sum_{k=n}^{\infty} |4a_{k+1}^2 - 1| = o(1/n), \qquad \sum_{k=n}^{\infty} (|b_k+b_{k+1}| + |b_k b_{k+1}|) = o(1/n),$$

which is a stronger result than Theorem 7, but which is not the strongest possible. Indeed, Chihara [6] proved that for the special case $a_n = 1/2 + d/n^2$ and $b_n = c/n^2$ there will be a finite number of mass points when $2d+c < 1/8$ and $2d-c < 1/8$.

We end this section by stating (without proof) two more results which are directly related to perturbations of Chebyshev polynomials:

THEOREM 8 (Geronimo and Nevai [15,18,21]): *Suppose* v_n *($n \in \mathbb{Z}$) is such that*

$$v_0 = 1, \ v_{-n} = v_n$$

$$v_n \le v_{n+1}, \qquad v_n \le v_m v_{n-m} \qquad n,m \ge 0,$$

$$\lim_{n\to\infty} v_n^{1/n} = 1,$$

then

$$(2.28) \qquad \sum_{k=1}^{\infty} k\, v_{2k} \left(|1-4a_k^2| + 2|b_{k-1}| \right) < \infty$$

and $\phi_0(\pm 1) \ne 0$ *if and only if the orthogonality measure* μ *has a finite number of mass points outside* $[-1,1]$ *and is absolutely continuous in* $(-1,1)$ *with* μ' *such that*

$$\frac{\mu'(\cos\theta)}{\sin\theta} = \sum_{n=-\infty}^{\infty} q_n e^{in\theta},$$

where $\sum_{n=-\infty}^{\infty} |n| v_n |q_n - q_{n-2}| < \infty$. *If* $\lim_{n\to\infty} v_n^{1/n} = R > 1$ *then* $\phi_0(x) \ne 0$ *for* $|x| \le (R+R^{-1})/2$ *and* (2.28) *hold if and only if the orthogonality measure* μ *has a finite number of mass points outside* $[-(R+R^{-1})/2,(R+R^{-1})/2]$, *no mass points in* $[-(R+R^{-1})/2,(R+R^{-1})/2]$, *and* μ *is absolutely continuous in* $(-1,1)$ *with* μ' *analytic in the ellips* $|x + \sqrt{x^2-1}| < R$ *and*

$$\frac{\mu'\left(\frac{z+z^{-1}}{2}\right)}{z - z^{-1}} = \sum_{n=-\infty}^{\infty} q_n z^n, \qquad\qquad 1/R \le |z| \le R,$$

where $\sum_{n=-\infty}^{\infty} |n| v_n |q_n - q_{n-2}| < \infty$.

This last result (with $R > 1$) applies when one is considering q-orthogonal polynomials.

Some of the techniques used in this Section have also been applied succesfully to orthogonal polynomials with asymptotically periodic recurrence coefficients (orthogonal polynomials on several intervals) and recurrence coefficients that converge to their asymptotic values in an oscillating way (as is the case for sieved orthogonal polynomials) (Geronimus [24,25], Geronimo and Van Assche [22,49], Máté, Nevai and Totik [37,41]).

3. Turán Determinants

In a 1950 paper [48] Paul Turán made a study of the zeros of Legendre polynomials, and in order to show some properties of these zeros he proved the inequality

$$P_n^2(x) - P_{n-1}(x)P_{n+1}(x) \geq 0 \qquad -1 \leq x \leq 1.$$

(Szegő [47] and Karlin and Szegő [28] have some alternative proofs). The polynomial on the left hand side can easily be written as a determinant and similar formulas for other orthogonal polynomials are now known as Turán determinants. The positivity of Turán determinants for orthogonal polynomials satisfying (2.1) is not too surprising (at least as $n \to \infty$) because, as we shall show, these Turán determinants are very closely related to the weight function of the corresponding orthogonal polynomials.

In what follows we define the Turán determinant as

$$(3.1) \qquad D_n(x) = p_n^2(x) - \frac{a_{n+1}}{a_n} p_{n+1}(x)p_{n-1}(x) .$$

This normalization is such that the coefficient of z^{2n} vanishes. Other normalizations are also useful (Turán's normalization for Legendre polynomials was chosen in such a way that the determinant vanishes at ± 1).

THEOREM 9: *Suppose that the recurrence coefficients converge as in (2.1) and let [A,B] be the smallest interval that contains supp(μ), then for every $f \in C[A,B]$*

$$\lim_{n \to \infty} \int f(t) \, D_n(t) \, d\mu(t) = \frac{2}{\pi} \int_{-1}^{1} f(t) \sqrt{1-t^2} \, dt .$$

PROOF: This follows immediately from (2.4). □

The previous result gives weak asymptotics for the Turán determinant. We are going to show that one can also obtain uniform convergence for Turán determinants, if one imposes some very weak conditions on the recurrence coefficients. First we give an alternative formula for $D_n(x)$:

LEMMA (Van Assche): *Turán determinants satisfy the recurrence relation*

$$(3.2) \qquad D_k(x) = D_{k-1}(x) + \frac{b_k - b_{k-1}}{a_k} p_k(x)p_{k-1}(x) + \frac{a_k^2 - a_{k-1}^2}{a_k\, a_{k-1}} p_k(x)p_{k-2}(x) \, ,$$

from which one has

$$(3.3) \qquad D_n(x) = 1 + \sum_{k=1}^{n} \left\{ \frac{b_k - b_{k-1}}{a_k} p_k(x)p_{k-1}(x) + \frac{a_k^2 - a_{k-1}^2}{a_k\, a_{k-1}} p_k(x)p_{k-2}(x) \right\}.$$

PROOF: This follows by applying the recurrence relation a couple of times (see e.g. Van Assche [49], p. 117). □

We want to allow $n \to \infty$ in (3.3). In order to do so we need some bounds on the orthogonal polynomials in terms of the differences $|b_k - b_{k-1}|$ and $|a_k - a_{k-1}|$. The idea in Section 2 was to compare the orthogonal polynomials with the Chebyshev polynomials of the second kind, which in the limit have the same recurrence coefficients. Now we will use a different comparison system, namely

$$(3.4) \qquad G(k,m) = \sum_{i=k+1}^{m} \left(\prod_{j=i+1}^{m} 1/\rho_j \right) \left(\prod_{j=k+1}^{i-1} \rho_j \right),$$

where $\rho_j = \rho_j(x) = \frac{x-b_j}{2a_{j+1}} + \sqrt{\left(\frac{x-b_j}{2a_{j+1}}\right)^2 - 1}$. Notice that when $a_j = 1/2$ and $b_j = 0$ for every j, then $G(k,m) = U_{m-k-1}(x)$. An empty sum is always defined to be zero, so that $G(k,m) = 0$ for $k \geq m$. One easily verifies that $G(k,m)$ satisfies

$$(3.5) \qquad G(k+1,m) + G(k-1,m) = (1/\rho_{k+1} + \rho_k) G(k,m) + \delta_{k,m} \, ,$$

whereas the recurrence formula for the orthogonal polynomials can be rewritten as

$$(3.6) \qquad p_{k+1}(x) + a_k/a_{k+1} p_{k-1}(x) = (\rho_k + 1/\rho_k) p_k(x) \, .$$

Multiply (3.5) by $p_k(x)$ and (3.6) by $G(k,m)$, substract the obtained equations and take the sum from $k=0$ to $k=m$ to find

$$(3.7) \qquad p_m(x) = G(-1,m)$$
$$+ \sum_{k=0}^{m-1} \left\{ (1/\rho_k - 1/\rho_{k+1})G(k,m) + (1 - \frac{a_{k+1}}{a_{k+2}})G(k+1,m) \right\} p_k(x),$$

which is the appropriate comparison equation. We are now ready to give the relevant bounds for the orthogonal polynomials:

THEOREM 10 (Máté, Nevai and Totik [31,34], Van Assche and Geronimo [50]): *Suppose K is a compact set in $\mathbb{C} \setminus \{-1,1\}$ (the complex plane with the two points ± 1 excluded) and assume that (2.1) holds together with*

$$(3.8) \qquad \sum_{k=0}^{\infty} \left(|b_k - b_{k+1}| + |a_{k+1} - a_{k+2}| \right) < \infty ,$$

then there exist positive constants A and B (depending on K) such that

$$(3.9) \qquad | p_n(x)/ \prod_{j=0}^{n+1} \rho_j | \leq A \, exp(B \sum_{k=0}^{n-1} e_k) ,$$

where $e_k = |b_k - b_{k+1}| + |a_{k+1} - a_{k+2}|$.

PROOF: first one needs a good bound for G(k,m). Observe that

$$(3.10) \qquad G(k,m) - \frac{1}{\rho_m} G(k,m-1) = \prod_{j=k+1}^{m-1} \rho_j .$$

If we define

$$\tilde{G}(k,m) = (\prod_{j=k+1}^{m+1} \rho_k^{-1}) \, G(k,m),$$

then one finds for k+1<m

$$(3.11) \qquad \tilde{G}(k,m-1) - \rho_m \rho_{m+1} \tilde{G}(k,m) = -1,$$

which means that the left hand side is independent of m. The equality of these expressions for m and m−1 gives

$$\tilde{G}(k,m-1) - \tilde{G}(k,m) = \frac{\tilde{G}(k,m-2) - \tilde{G}(k,m-1)}{\rho_m \rho_{m+1}} + \frac{\rho_{m+1} - \rho_{m-1}}{\rho_{m+1}} \tilde{G}(k,m-1)$$

and since $|\rho_k| \geq 1$ one has

$$|\tilde{G}(k,m-1) - \tilde{G}(k,m)| \leq |\tilde{G}(k,m-2) - \tilde{G}(k,m-1)|$$
$$+ |\rho_{m+1} - \rho_{m-1}| \, |\tilde{G}(k,m-1)|,$$

which iterates to

$$|\tilde{G}(k,m-1) - \tilde{G}(k,m)| \leq 1 + \sum_{i=k+1}^{m-1} |\rho_{i+2} - \rho_i| \, |\tilde{G}(k,i)|.$$

From (3.11) one has

$$|\tilde{G}(k,m-1)-\tilde{G}(k,m)| \geq |\tilde{G}(k,m)|\ |1-\rho_m\rho_{m+1}| - 1\ ,$$

which gives

$$|\tilde{G}(k,m)|\ |1-\rho_m\rho_{m+1}| \leq 2 + \sum_{i=k+1}^{m-1} |\rho_{i+2}-\rho_i|\ |\tilde{G}(k,i)|\ .$$

If $x \in K$ then there exists an integer n_0 and a constant C_1 such that

$$|\rho_{i+1}-\rho_i| \leq C_1\, e_i \qquad\qquad i \geq n_0,$$

and a constant C_2 such that

$$|1-\rho_m\rho_{m+1}|^{-1} \leq C_2 \qquad\qquad m \geq n_0.$$

This means that for $k \geq n_0$

$$|\tilde{G}(k,m)| \leq 2C_2 + C_1 C_2 \sum_{i=k+1}^{m-1} (e_i+e_{i+1})\ |\tilde{G}(k,i)|\ ,$$

and by Gronwall's inequality this implies that

$$(3.12) \qquad |\tilde{G}(k,m)| \leq C \exp\left\{ D \sum_{i=k+1}^{m} e_i \right\}$$

holds for $k \geq n_0$ and some constants C and D. The inequality can easily be extended to hold for $k \geq -1$ by using the recurrence relation (3.5). The inequality (3.9) finally follows by using the inequality (3.12) in the comparison equation (3.7) and by Gronwall's inequality. □

If (2.1) and (3.8) hold, then one can define the function

$$(3.13) \qquad \varphi(x) = 1 + \sum_{k=0}^{\infty} \left\{ (1/\rho_k - 1/\rho_{k+1}) + (1 - \frac{a_{k+1}}{a_{k+2}})/\rho_{k+1} \right\} p_k(x)/ \prod_{j=0}^{k} \rho_j\ ,$$

and this series converges uniformly on every compact set K of $\mathbb{C} \setminus \{-1,1\}$. This function is relevant because of the following asymptotic results:

THEOREM 11 (Máté, Nevai, Totik [31,34], Van Assche and Geronimo [50]): *Suppose that (2.1) and (3.8) hold, then uniformly for x on a compact set $K \subset \mathbb{C} \setminus \{-1,1\}$*

$$(3.14) \qquad \lim_{n \to \infty} \frac{p_n(x) - \frac{a_n}{a_{n+1}} \rho_n^{-1} p_{n-1}(x)}{\prod\limits_{j=0}^{n-1} \rho_j} = \varphi(x).$$

In particular this implies that uniformly for x on compact sets of $\mathbb{C} \setminus [-1,1]$

$$\lim_{n \to \infty} p_n(x) / \prod_{j=0}^{n-1} \rho_j = \frac{z}{z-z^{-1}} \varphi(x),$$

and for x on closed intervals in $(-1,1)$

$$\lim_{n \to \infty} \left\{ \sqrt{1-x^2}\, p_n(x) - |\varphi(x)| \prod_{j=0}^{n} |\rho_j|\, \sin\!\left[\sum_{j=0}^{n} \arg(\rho_j) + \Gamma(x)\right] \right\} = 0,$$

where $\Gamma(x) = \arg[\varphi(x)]$.

PROOF: Use (3.7) for $p_n(x)$ and $p_{n-1}(x)$, then one finds

$$p_n(x) - \rho_n^{-1} p_{n-1}(x) = G(-1,n) - \rho_n^{-1} G(-1,n-1)$$

$$+ \sum_{k=0}^{n-2} \left\{ (1/\rho_k - 1/\rho_{k+1})[G(k,n) - \rho_n^{-1} G(k,n-1)] \right.$$

$$\left. + (1 - \frac{a_{k+1}}{a_{k+2}})[G(k+1,n) - \rho_n^{-1} G(k+1,n-1)] \right\} p_k(x)$$

$$+ (\rho_{n-1}^{-1} - \rho_n^{-1}) p_{n-1}(x).$$

Add $\rho_n^{-1}(1 - \frac{a_n}{a_{n+1}}) p_{n-1}(x)$ to both sides of this equation and use (3.9) to find

$$(3.15) \qquad p_n(x) - \frac{a_n}{a_{n+1}} \rho_n^{-1} p_{n-1}(x)$$

$$= \prod_{j=0}^{n-1} \rho_j \left\{ 1 + \sum_{k=0}^{n-1} \left[(1/\rho_k - 1/\rho_{k+1}) + (1 - \frac{a_{k+1}}{a_{k+2}})/\rho_{k+1} \right] p_k(x) / \prod_{j=0}^{k} \rho_j \right\},$$

from which the asymptotic formulas follow. $\square$

It remains to show how $\varphi(x)$ reflects the properties of the spectral measure μ:

THEOREM 12 (Dombrowski [8–12], Máté, Nevai, Totik [31,34], Van Assche and Geronimo [50]): *Suppose that (2.1) and (3.8) hold, then μ is absolutely continuous in (−1,1), μ′ is strictly positive and continuous on (−1,1) and*

$$(3.16) \qquad |\varphi(x)|^2 \prod_{j=0}^{\infty} |\rho_j|^2 = \frac{2}{\pi} \frac{\sqrt{1-x^2}}{\mu'(x)}$$

$$= 1 + \sum_{k=1}^{\infty} \left\{ \frac{b_k - b_{k-1}}{a_k} p_k(x)p_{k-1}(x) + \frac{a_k^2 - a_{k-1}^2}{a_k \, a_{k-1}} p_k(x)p_{k-2}(x) \right\}.$$

holds for −1<x<1.

PROOF: The proof is similar to the proof of Theorem 6. The last equation follows by using Theorem 9 and by letting n → ∞ in (3.3). □

Notice that from (3.9) and (3.3) one finds that the Turán determinant D_n converges uniformly on every compact set K in (−1,1), and that

$$|D_n(x) - \frac{2}{\pi} \frac{\sqrt{1-x^2}}{\mu'(x)}| \le C \sum_{k=n+1}^{\infty} (|b_k - b_{k-1}| + |a_k - a_{k-1}|) ,$$

where C is a constant that depends on the closed set K. Typical examples of orthogonal polynomials for which the results in this Section hold are the *Pollaczek polynomials* for which

$$a_n^2 = \frac{1}{4} \frac{n(n+2\lambda-1)}{(n+\lambda+a-1)(n+\lambda+a)} = \frac{1}{4} - \frac{a}{2n} + O(n^{-2}) ,$$

$$b_n = \frac{-b}{n+\lambda+a} = -\frac{b}{n} + O(n^{-2}) ,$$

where $\lambda > 0$ and $a+\lambda > 0$ or $-1/2 < \lambda < 0$ and $-1 < \lambda+a < 0$. With the help of the trace formula (3.16)(or similar such formulas as given by Dombrowski [8–12]) one may derive properties of the orthogonal polynomials and their spectral measure when the recurrence coefficients converge to their limits in a monotonic way [9,11,12,45].

4. Unbounded Recurrence Coefficients

Up to now we have been dealing with bounded recurrence coefficients. It turns out that some of the techniques of the previous sections can be used to investigate orthogonal polynomials with unbounded recurrence coefficients, provided one works with $p_n(c_n x)$ rather then $p_n(x)$, where c_n (n=0,1,2,...) is a positive non-decreasing sequence such that $(a_n + |b_n|)/c_n$ remains bounded. There is still a lot of work to be done for obtaining relevant asymptotic properties of the non-scaled orthogonal polynomials $p_n(x)$ when the recurrence coefficients are unbounded.

What we need is a convenient way to describe the unboundedness of the recurrence coefficients. We will asume that there is a positive and non-decreasing sequence c_n such that

$$(4.1) \qquad \lim_{n \to \infty} a_n/c_n = a > 0, \qquad\qquad \lim_{n \to \infty} b_n/c_n = b \in \mathbb{R}.$$

An extra condition on the contraction sequence c_n is needed. Two conditions are very useful (and relevant), the first condition (introduced in [43]) is

$$(4.2) \qquad \lim_{n \to \infty} c_{n+k}/c_n = 1 \qquad\qquad \text{for } k=0,1,2,\dots,$$

and another condition is that c_n is a *regularly varying* sequence with index $\alpha > 0$:

$$(4.3) \qquad c_n = n^{\alpha} L(n),$$

where $L: \mathbb{R}^+ \longrightarrow \mathbb{R}^+$ is *slowly varying*, meaning

$$(4.4) \qquad \lim_{x \to \infty} L(xt)/L(x) = 1, \qquad\qquad \text{for every } t > 0.$$

One can easily show that (4.3) implies (4.2), but the sequence $\exp(\sqrt{n})$ satisfies (4.2) and is not regularly varying. If (4.1) holds with an increasing sequence c_n then it follows from (1.3) that

$$\min(0,b-2a) \leq \liminf_{n \to \infty} x_{1,n}/c_n \leq \limsup_{n \to \infty} x_{n,n}/c_n \leq \max(0,b+2a)$$

which means that the *contracted zeros* $x_{j,n}/c_n$ for $n \to \infty$ are in the interval $[A,B]$, where $A = \min(0,b-2a)$ and $B = \max(0,b+2a)$. This interval $[A,B]$ is the smallest closed interval that contains both $\{0\}$ and $[b-2a,b+2a]$. We are now ready to formulate the analog of Theorem 1:

THEOREM 13 (Van Assche [49]): *Suppose that (4.1) and (4.2) are valid with c_n non-decreasing, then uniformly for x on compact sets of $\mathbb{C} \setminus [A,B]$*

$$(4.5) \qquad \lim_{n \to \infty} \frac{p_{n-1}(c_n x)}{p_n(c_n x)} = \frac{x-b}{2a} - \sqrt{\left(\frac{x-b}{2a}\right)^2 - 1}.$$

If (4.1) and (4.3) are valid, with $\alpha > 0$, then

$$(4.6) \qquad \lim_{n \to \infty} |p_n(c_n x)|^{1/n} = \exp\left\{ -\int_0^1 \log\left| \frac{x-bt^{\alpha}}{2at^{\alpha}} - \sqrt{\left(\frac{x-bt^{\alpha}}{2at^{\alpha}}\right)^2 - 1} \right| dt \right\},$$

uniformly for x on compact sets of $\mathbb{C} \setminus [A,B]$.

PROOF: Let $\delta > 0$ be the distance between the compact set $K \subset \mathbb{C} \setminus [A,B]$ and the closed interval $[A,B]$, then one can show that for some positive constant C

$$\left| \frac{p_{k-1}(c_n x)}{p_k(c_n x)} \right| \le \frac{C}{\delta} \qquad\qquad k \le n,$$

for n large enough. By (3.2) we then have

$$\left| \frac{D_k(c_n x)}{p_k^2(c_n x)} \right| \le \left| \frac{D_{k-1}(c_n x)}{p_{k-1}^2(c_n x)} \right| (C/\delta)^2 + A_k$$

with $A_k \to 0$ as $k \to \infty$. This implies that $D_n(c_n x)/p_n^2(c_n x)$ tends to zero uniformly in x whenever δ is large enough, which in turn implies that $p_{n-1}(c_n x)/p_n(c_n x)$ and $p_n(c_n x)/p_{n+1}(c_n x)$ have the same accumulation points for δ large enough. If we use this in the recurrence formula (1.1), then one finds that the possible limits of $p_{n-1}(c_n x)/p_n(c_n x)$ are solutions of the quadratic equation $az^2-(x-b)z+a=0$, and (4.5) then follows for δ large enough because $p_{n-1}(c_n x)/p_n(c_n x)$ tends to zero as $x \to \infty$. The restriction on δ can be removed since the ratio's are, for large n, analytic functions on K which are uniformly bounded. The nth root behaviour follows from the ratio behaviour because

$$|p_n(c_n x)|^{1/n} = \exp\left\{ -\frac{1}{n} \sum_{k=1}^{n} \log \left| \frac{p_{k-1}(c_n x)}{p_k(c_n x)} \right| \right\} . \quad \square$$

It follows from Theorem 13 that the contracted zeros $x_{j,n}/c_n$ are dense in $[A,B]$ and that for every $f \in C[A,B]$

$$(4.7) \qquad \lim_{n \to \infty} \frac{1}{n} \sum_{j=1}^{n} f(x_{j,n}/c_n) = \int_A^B f(t) v(t; \alpha, a, b) \, dt ,$$

where $v(t; \alpha, a, b)$ is the *Nevai–Ullman weight* with parameter α corresponding to the interval $[b-2a, b+2a]$. Notice that the support of this Nevai-Ullman weight is $[A,B]$ which may be different from $[b-2a, b+2a]$.

In Section 2 we were able to obtain precise results for orthogonal polynomials by comparing them to Chebyshev polynomials of the second kind. Unfortunately such a comparison system is not available for unbounded recurrence coefficients. The Hermite polynomials (corresponding to $\alpha = 1/2$) and the Laguerre polynomials (corresponding to $\alpha = 1$) are the only orthogonal polynomials with unbounded recurrence coefficients that have been investigated thoroughly and for other values of α no such systems are known in sufficient detail. Máté, Nevai and Totik [35,36] however succeeded in giving precise asymptotic results for the greatest zeros of orthogonal polynomials with $b_n = 0$ and $a_n = cn^\alpha(1+o(n^{-2/3}))$ by comparing these zeros with the zeros of Hermite polynomials: they showed that for such orthogonal polynomials

$$x_{n-j+1,n}/n^\alpha = 2c - c3^{-1/3}(2\alpha)^{2/3}i_j/n^{2/3} + o(n^{-2/3}),$$

where i_j is the jth zero of *Airy's function*, which is defined as the unique solution of $y'' - xy/3 = 0$ which remains bounded as $x \to \infty$.

The method used in Section 3 can be used with some minor modifications to obtain asymptotic results for the orthogonal polynomials. The appropriate comparison system now becomes

$$(4.8) \qquad G_n(k,m) = \sum_{i=k+1}^{m} \left(\prod_{j=i+1}^{m} 1/\rho_{j,n} \right) \left(\prod_{j=k+1}^{i-1} \rho_{j,n} \right),$$

where $\rho_{j,n} = \rho_{j,n}(x) = \frac{c_n x - b_j}{2a_{j+1}} + \sqrt{(\frac{c_n x - b_j}{2a_{j+1}})^2 - 1}$. The bound (3.12) now becomes

$$(4.9) \qquad \left| \frac{G_n(k,m)}{\prod\limits_{j=k+1}^{m+1} \rho_{j,n}} \right| \le C \exp \left\{ D \sum_{i=k+1}^{m} (|b_i - b_{i+1}| + |a_{i+1} - a_{i+2}|)/c_n \right\}.$$

In Section 3 we needed to be careful at the branch points of ρ_j (the points $b_j \pm 2a_{j+1}$) because the order estimates do not hold there. We were able to take care of this by keeping away of ± 1, which are the accumulation points of these branch points. In the case under consideration the branch points of $\rho_{j,n}$ are $(b_j \pm 2a_{j+1})/c_n$ which for $j \le n$ and $n \to \infty$ are dense in $[A,B]$. Therefore the bound (4.9) only holds on compact sets $K \subset \mathbb{C} \setminus [A,B]$ and the constant C and D depend on K. The analog of Theorem 11 is:

THEOREM 14 (Van Assche and Geronimo [51]): *Suppose that (4.1) and (4.3) hold with $\alpha > 0$ and assume moreover that*

$$(4.10) \qquad \lim_{n\to\infty} n(a_{n+1} - a_n)/c_n = a\alpha, \qquad\qquad \lim_{n\to\infty} n(b_{n+1} - b_n)/c_n = b\alpha,$$

then uniformly for x on compact sets of $\mathbb{C} \setminus [A,B]$

$$(4.11) \qquad \lim_{n \to \infty} \frac{p_n(c_n x) - \frac{a_n}{a_{n+1}} \rho_{n;n}^{-1} \, p_{n-1}(c_n x)}{\prod_{j=0}^{n-1} \rho_{j,n}}$$

$$= \frac{x}{a\rho} \left\{ \frac{(x-b)^2 - 4a^2}{x^2} \right\}^{1/4} exp \left\{ -\frac{b}{2} \int_0^1 \frac{ds}{\sqrt{(x-bs)^2 - 4a^2 s^2}} \right\},$$

with $\rho = \frac{x-b}{2a} + \sqrt{(\frac{x-b}{2a})^2 - 1}$, and in particular

$$(4.12) \qquad \lim_{n \to \infty} \frac{p_n(c_n x)}{\prod_{j=0}^{n-1} \rho_{j,n}} = \left\{ \frac{(x-b)^2 - 4a^2}{x^2} \right\}^{-1/4} exp \left\{ -\frac{b}{2} \int_0^1 \frac{ds}{\sqrt{(x-bs)^2 - 4a^2 s^2}} \right\}.$$

PROOF: we present a proof which is different from the proof in [51]. If (4.10) is valid then

$$\lim_{n \to \infty} \sum_{i=0}^{n} (|b_i - b_{i+1}| + |a_{i+1} - a_{i+2}|)/c_n = (|b| + a)\alpha \int_0^1 t^{\alpha - 1} dt = |b| + a ,$$

and by using this in (4.9) and (3.7) one can show that there is a constant C (which depends on the compact set in $\mathbb{C} \setminus [A,B]$ such that for every n and for $m \leq n$

$$| p_m(c_n x) / \prod_{j=0}^{m-1} \rho_{j,n} | \leq C.$$

Let $\psi_{k,n}(x) = p_k(c_n x) - (a_k/a_{k+1})\rho_{k;n}^{-1} p_{k-1}(c_n x)$, then by (3.15) we have for $k > m$

$$(4.13) \qquad |\psi_{k,n}(x) - \psi_{m,n}(x)| \leq C \sum_{i=m}^{k-1} (|1/\rho_{i,n} - 1/\rho_{i+1,n}| + |1 - \frac{a_{i+1}}{a_{i+2}}|) ,$$

and

$$(4.14) \qquad |\psi_{k,n}(x) - 1| \leq C \sum_{i=0}^{k-1} (|1/\rho_{i,n} - 1/\rho_{i+1,n}| + |1 - \frac{a_{i+1}}{a_{i+2}}|) .$$

If we set $\psi_n(t) = \psi_{\lceil nt \rceil, n}(x)$, where $\lceil nt \rceil$ is the largest integer $\leq nt$, then (4.10), (4.13) and (4.14) imply that ψ_n is an equicontinuous family of Borel functions on [0,1], and thus there exists an index set Δ such that $\psi_n(t) \longrightarrow \psi(t)$ uniformly for $t \in$ [0,1] when $n \to \infty$ in the index set Δ. Let $n \in \Delta$ tend to infinity in (3.15), then we find an integral equation for ψ, namely

$$\psi(t) = 1 + \int_0^t h(s)\psi(s)\, ds \,,$$

with

$$h(s) = \frac{\alpha}{s}\left\{ 1 - \frac{x}{as^\alpha}\, \frac{1}{\rho(s)-\rho^{-1}(s)} \right\} \frac{\rho^{-1}(s)}{\rho(s)-\rho^{-1}(s)} \,,$$

and $\rho(s) = \frac{x-bs^\alpha}{2as^\alpha} + \sqrt{\left(\frac{x-bs^\alpha}{2as^\alpha}\right)^2-1}$. The unique solution of this equation is

$$\psi(t) = \exp \int_0^t h(s)\, ds \,,$$

and with some calculus one may show that $\psi(1)$ is the right hand side in (4.11). This limit is independent of the index set Δ and therefore the full sequence converges to this limit. □

Now one can use orthogonal polynomials with recurrence coefficients satisfying (4.1), (4.3) and (4.10) as a comparison system and use the techniques of Section 2 to obtain asymptotic properties.

THEOREM 15 (Geronimo and Van Assche [23]): *Suppose that (4.1) and (4.3) hold with* $\alpha > 0$. *Assume moreover that* a_n^0 *(n=1,2,...) and* b_n^0 *(n=0,1,2,...) are recurrence coefficients of orthogonal polynomials* $q_n(x)$ *that satisfy (4.1) and (4.3) and that*

(4.15) $\qquad \lim_{n\to\infty} n(a_n^0-a_n)/c_n = C \,, \qquad\qquad \lim_{n\to\infty} n(b_n^0-b_n)/c_n = D \,,$

where C and D are real numbers, then

(4.16) $\qquad \lim_{n\to\infty} \left(\prod_{k=1}^{n} \frac{a_k}{a_k^0} \right) \frac{p_n(c_n x)}{q_n(c_n x)}$

$$= \exp \int_0^1 \frac{t^{\alpha-1}}{\sqrt{(x-bt^\alpha)^2-4a^2t^{2\alpha}}}\left\{ D + \frac{4aCt^\alpha}{x-bt^\alpha+\sqrt{(x-bt^\alpha)^2-4a^2t^{2\alpha}}} \right\} dt \,,$$

uniformly on compact subsets of $\mathbb{C} \setminus [A,B]$.

460

This result can be applied to many orthogonal polynomials with unbounded recurrence coefficients, such as Hermite polynomials, Laguerre polynomials, Meixner polynomials, Meixner-Pollaczek polynomials and orthogonal polynomials with weight function $\exp(-x^{2m})$ (Freud polynomials) with m a positive integer. A crucial next step would be to analyse the asymptotic behaviour on the interval [A,B] but, as explained earlier, our techniques fail because the available bounds do not hold in the neighborhood of the branch points of $\rho_{j,n}$ which are dense in [A,B].

References

[1] Z.S. Agranovich, V.A. Marchenko, *The Inverse Problem of Scattering Theory*, Gordon and Breach, New York, 1963.

[2] F.V. Atkinson, *Discrete and Continuous Boundary Problems*, Academic Press, New York, 1964.

[3] O. Blumenthal, *Über die Entwicklung einer willkurlichen Funktion nach den Nennern des Kettenbruches für $\int_{-\infty}^{0} [\phi(\xi)/(z-\xi)]\, d\xi$* , Inaugural Dissertation, Gottingen, 1898.

[4] K.M. Case, 'Orthogonal Polynomials revisited', in *"Theory and Application of Special Functions"* (R.A. Askey, ed.), Academic Press, New York, 1975, pp. 289-304.

[5] T.S. Chihara, *An Introduction to Orthogonal Polynomials*, Gordon and Breach, New York, 1978.

[6] T.S. Chihara, 'Orthogonal polynomials whose distribution functions have finite point spectra', *SIAM J. Math. Anal.* 11 (1980), 358-364.

[7] T.S. Chihara, P.G. Nevai, 'Orthogonal polynomials and measures with finitely many point masses', *J. Approx. Theory* 35 (1982), 370-380.

[8] J. Dombrowski, 'Spectral properties of phase operators', *J. Math. Phys.* 15 (1974), 576-577.

[9] J. Dombrowski, 'Spectral properties of real parts of weighted shift operators', *Indiana Univ. Math. J.* 29 (1980), 249-259.

[10] J. Dombrowski, 'Tridiagonal matrix representations of cyclic self-adjoint operators, I, II', *Pacific J. Math.* 114 (1984), 325-334; 120 (1985), 47-53.

[11] J. Dombrowski, G.H. Fricke, 'The absolute continuity of phase operators', *Trans. Amer. Math. Soc.* 213 (1975), 363-372.

[12] J. Dombrowski, P. Nevai, 'Orthogonal polynomials, measures and recurrence relations', *SIAM J. Math. Anal.* 17 (1986), 752-759.

[13] G. Freud, 'On the greatest zeros of an orthogonal polynomial, I, II', *Acta Sci. Math. (Szeged)* 34 (1973), 91-97; 36 (1974), 49-54.

[14] G. Freud, 'On the greatest zero of an orthogonal polynomial', *J. Approx. Theory* 46 (1986), 16-24.

[15] J.S. Geronimo, 'A relation between the coefficients in the recurrence formula and the spectral function for orthogonal polynomials', *Trans. Amer. Math. Soc.* 260 (1980), 65-82.

[16] J.S. Geronimo, 'An upper bound on the number of eigenvalues of an infinite dimensional Jacobi matrix', *J. Math. Phys.* 23 (1982), 917-921.

[17] J.S. Geronimo, 'On the spectra of infinite-dimensional Jacobi matrices', *J. Approx. Theory* 53 (1988), 251-265.

[18] J.S. Geronimo, 'Scattering theory, orthogonal polynomials and q-series', *SIAM J. Math. Anal.*

[19] J.S. Geronimo, K.M. Case, 'Scattering theory and polynomials orthogonal on the real line', *Trans. Amer. Math. Soc.* **258** (1980), 467-494.

[20] J.S. Geronimo, E.M. Harrell II, W. Van Assche, 'On the asymptotic distribution of eigenvalues of banded matrices', *Constr. Approx.* **4** (1988), 403-417.

[21] J.S. Geronimo, P.G. Nevai, 'Necessary and sufficient conditions relating the coefficients in the recurrence formula to the spectral function for orthogonal polynomials', *SIAM J. Math. Anal.* **14** (1983), 622-637.

[22] J.S. Geronimo, W. Van Assche, 'Orthogonal polynomials with asymptotically periodic recurrence coefficients', *J. Approx. Theory* **46** (1986), 251-283.

[23] J.S. Geronimo, W. Van Assche, 'Relative asymptotics for orthogonal polynomials with unbounded recurrence coefficients', *J. Approx. Theory*

[24] Ya.L. Geronimus, 'On the character of the solutions of the moment problem in the case of a limit-periodic associated fraction', *Izv. Akad. Nauk CCCP* **5** (1941), 203-210 (Russian).

[25] Ya.L. Geronimus, 'On some finite difference equations and corresponding systems of orthogonal polynomials', *Zap. Mat. Otd. Fiz.-Mat. Fak. i Kharkov Mat. Obsc.* (4) **25** (1957), 87-100 (Russian).

[26] J. Gilewicz, E. Leopold, 'On the sharpness of results in the theory of location of zeros of polynomials defined by three term recurrence relations' in *"Polynômes Orthogonaux et Applications"*, Lecture Notes in Mathematics 1171, Springer, Berlin, 1985, pp. 259-266.

[27] G.S. Guseinov, 'The determination of an infinite Jacobi matrix from the scattering data', *Dokl. Akad. Nauk. CCCP* **227** (6) (1976), 1289-1292 (Russian); *Soviet Math. Dokl.* **17** (1976), 596-600.

[28] S. Karlin, G. Szegő, 'On certain determinants whose elements are orthogonal polynomials', *J. Analyse Math.* **8** (1960), 1-157.

[29] E. Leopold, 'Location of the zeros of polynomials satisfying three-term recurrence relations III. Positive coefficients case', *J. Approx. Theory* **43** (1985), 15-24.

[30] V.A. Marchenko, *Spectral Theory of Sturm-Liouville Operators*, Naukova Dumka, Kiev, 1972.

[31] A. Máté, P. Nevai, 'Orthogonal polynomials and absolutely continuous measures', in *"Approximation IV"* (C.K. Chui et al., eds.), Academic Press, New York, 1983, pp. 611-617.

[32] A. Máté, P. Nevai, 'Sublinear perturbations of the differential equation $y^{(n)}=0$ and of the analogous difference equation', *J. Differential Equations* **53** (1984), 234-257.

[33] A. Máté, P. Nevai, 'Eigenvalues of finite band-width Hilbert space operators and their applications to orthogonal polynomials', *Canad. J. Math.*

[34] A. Máté, P. Nevai, V. Totik, 'Asymptotics for orthogonal polynomials defined by a recurrence relation', *Constr. Approx.* **1** (1985), 231-248.

[35] A. Máté, P. Nevai, V. Totik, 'Asymptotics for the greatest zeros of orthogonal polynomials', *SIAM J. Math. Anal.* **17** (1986), 745-751.

[36] A. Máté, P. Nevai, V. Totik, 'Asymptotics for the zeros of orthogonal polynomials associated with infinite intervals', *J. London Math. Soc.* (2) **33** (1986), 303-310.

[37] A. Máté, P. Nevai, V. Totik, 'Twisted difference operators and perturbed Chebyshev polynomials', *Duke Math. J.* **57** (1988), 301-331.

[38] A. Máté, P. Nevai, W. Van Assche, 'The supports of measures associated with orthogonal polynomials and the spectra of the related self adjoint operators', *Rocky Mountain J. Math.*

[39] P.G. Nevai, *Orthogonal Polynomials*, Memoirs Amer. Math. Soc. **213**, Providence,R.I., 1979.

[40] P.G. Nevai, 'Distribution of zeros of orthogonal polynomials', *Trans. Amer. Math. Soc.* **249** (1979), 341-361.

[41] P.G. Nevai, 'Orthogonal polynomials defined by a recurrence relation', *Trans. Amer. Math. Soc.* **250** (1979), 369-384.

[42] P.G. Nevai, 'On orthogonal polynomials', *J. Approx. Theory* **25** (1979), 34-37.

[43] P.G. Nevai, J.S. Dehesa, 'On asymptotic average properties of zeros of orthogonal polynomials', *SIAM J. Math. Anal.* **10** (1979), 1184-1192.

[44] E.M. Nikishin, 'Discrete Sturm-Liouville operators and some problems of function theory', *Trudy Sem. Petrovsk.* **10** (1984), 3-77 (Russian); *J. Soviet Math.* **35** (1986), 2679-2744.

[45] J.A. Shohat, *Théorie Générale des Polynomes Orthogonaux de Tchebichef*, Mémorial des Sciences Mathématiques **66**, Gauthier-Villars, Paris, 1934.

[46] G. Szegő, *Orthogonal Polynomials*, Amer. Math. Soc. Colloq. Publ. **23**, Providence, R.I., 4th edition 1975.

[47] G. Szegő, 'On an inequality of P. Turán concerning Legendre polynomials', *Bull. Amer. Math. Soc.* **54** (1948), 401-405.

[48] P. Turán, 'On the zeros of the polynomials of Legendre', *Časopis Pěst. Mat. a Fys.* **75** (1950), 113-122.

[49] W. Van Assche, *Asymptotics for Orthogonal Polynomials*, Lecture Notes in Mathematics **1265**, Springer, Berlin, 1987.

[50] W. Van Assche, J.S. Geronimo, 'Asymptotics for orthogonal polynomials on and off the essential spectrum', *J. Approx. Theory* **55** (1988), 220-231.

[51] W. Van Assche, J.S. Geronimo, 'Asymptotics for orthogonal polynomials with regularly varying recurrence coefficients', *Rocky Mountain J. Math.* **19** (1989)

[52] E.A. van Doorn, 'Representations and bounds for zeros of orthogonal polynomials and eigenvalues of sign-symmetric tri-diagonal matrices', *J. Approx. Theory* **51** (1987), 254-266.

Al-Salam, Nadhla, ASI participant, Department of Mathematics, University of Edmonton, Edmonton, Alberta T6G 2G1, Canada

Al-Salam, Waleed, invited ASI lecture, Department of Mathematics, University of Edmonton, Edmonton, Alberta T6G 2G1, Canada

Alfaro Garcia, Manuel, ASI participant, Departamento de Matemáticas, Facultad de Ciencias, Universidad de Zaragoza, 50009-Zaragoza, Spain

Allaway, William, ASI participant, Department of Mathematics, Lakehead University, Thunder Bay, Ontario P7B 5E1, Canada

Andrews, George, invited lecture, Department of Mathematics, Pennsylvania State University, University Park, PA 16802, U.S.A.

Antolin Coma, Juan, ASI participant, Departamento de Fisica Teorica, Facultad de Ciencias, Universidad de Zaragoza, 50009-Zaragoza, Spain

Ari, Nasit, guest, Department of Engineeing Science, Lafayette College, Easton PA 18042, U.S.A.

Askey, Richard, invited ASI lecture, Department of Mathematics, University of Wisconsin, 480 Lincoln Drive, Madison, WI 53706, U.S.A.

Baker, George, invited lecture, T-11, MS-B262, Los Alamos National Laboratory, Los Alamos NM 87545, U.S.A.

Bannai, Eiichi, invited lecture, Department of Mathematics, Ohio State University, 231 West 18th Avenue, Columbus, OH 43210-1174, U.S.A.

Bauldry, William C., guest, Department of Mathematics, Appalachian State University, Boone, NC 28608, U.S.A.

Beerends, René J., ASI participant, Brueghelstraat 109, 2525 RC Den Haag, Netherlands

Belmehdi, Said, ASI participant, 3, Rue Jouye-Roure, 7520 Paris, France

Berg, Christian, ASI participant, Mathematics Insitute. University of Copenhagen, Universitetsparken 5, 2100 Copenhagen 0, Denmark

Berndt, Bruce, guest, Department of Mathematics, University of Illinois, Urbana, IL 61801, U.S.A.

Bessis, Daniel, principal ASI lecturer, Service de Physique Théorique, Centre d'Études Nucléaires de Saclay, 91191 Gif-sur-Yvette Cédex, France

Blachman, Nelson M., guest, GTE Government Systems Co., Mountain View, CA 94039-7188, U.S.A.

Bonan, Stanford S., guest, 53-16 37th Road, Woodside, NY 11377, U.S.A.

Bressoud, David, invited lecture, Department of Mathematics, Pennsylvania State University, University Park, PA 16802, U.S.A.

Brezinski, Claude, ASI participant, Laboratorie d'Analyse Numérique et d'Optimisation, Université de Lille 1, 59655-Villeneuve d'Ascq Cédex, France

Bultheel, Adhemar, invited ASI lecture, Department of Computer Sci., Katholieke Universiteit Leuven, Clestijnenlaan 200 A, B-3030 Leuven (Heverlee) , Belgium

Bustoz, Joaquin, guest, Department of Mathematics, Arizona State University, Tempe, AZ 85287, U.S.A.

Carroll, Francis, guest, Department of Mathematics, Ohio State University, 231 West 18th Avenue, Columbus, OH 43210-1174, U.S.A.

Charris, Jairo A., guest, Department of Mathematics, National Univesity of Columbia, Bogotá, Columbia

Charron, Richard J., ASI participant, Department of Mathematics & Stat., Memorial University of Newfoundland, St. John's, Newfoundland A1C 5S7, Canada

Chen, Li-Chen, guest, Department of Mathematics, University of South Florida, Tampa, FL 33620-5700, U.S.A.

Chihara, Laura, guest, Department of Mathematics, St. Olaf College, Northfield, MN 55057, U.S.A.

Chihara, Theodore S., invited ASI lecture, Department of Mathematics, Purdue University Calumet, Hammond, IN 46323, U.S.A.

Cooper, Sandra Clement, guest, Department of Pure and Applied Mathematics, Washington State University, Pullman, WA 99164-2930, U.S.A.

de Bruin, Marcel G., ASI participant, Faculty of Tech. Mathematics and Informatics, Delft University of Technology, P.O. Box 356, 2600 AJ Delft, Netherlands

Dehesa, Jesus Sanchez, ASI participant, Departamento de Fisica Moderna, Facultad de Ciencias, Universidad de Granada, E-18071 Granada, Spain

Divis, Zita, guest, Department of Mathematics, Ohio State University, 231 West 18th Avenue, Columbus, OH 43210-1174, U.S.A.

Dombrowski, Joanne M., invited lecture, Department of Mathematics, Wright State University, Dayton, OH 45435, U.S.A.

Dunkl, Charles, invited lecture, Department of Mathematics, University of Virginia, Charlottesville, VA 22904, U.S.A.

Edgar, Gerald, guest, Department of Mathematics, Ohio State University, 231 West 18th Avenue, Columbus, OH 43210-1174, U.S.A.

Erdélyi, Tamas, guest, Department of Mathematics, Ohio State University, 231 West 18th Avenue, Columbus, OH 43210-1174, U.S.A.

Feinsilver, Philip, guest, Department of Mathematics, Southern Illinois University, Carbondale, IL 62901, U.S.A.

Fischer, Ismor, guest, Department of Mathematics, University of Wisconsin, 480 Lincoln Drive, Madison, WI 53706, U.S.A.

Fishel, SU.S.A.nna, guest, School of Mathematics, University of Minnesota, 206 Church Street S.E., Minnesota, MN 55455, U.S.A.

Gabbert, Mark. J., guest, 2920 Chapel Hill Road, 26-B, Durham, NC 27707, U.S.A.

Garcia-Lazaro, Paloma, ASI participant, Departamento de Matemática Aplicada, E.T.S de Ingenieros Industriales, C/José Gutierrez Abascal 2, 28006 Madrid, Spain

Gasper, George, invited lecture, Department of Mathematics, Northwestern University, Evanston, IL 60201, U.S.A.

Gatteschi, Luigi, ASI participant, Department of Mathematics, University of Turin, Via Carlo Alberto 10, I-10123 Torino, Italy

Gautschi, Walter, principal ASI lecturer, Department of Computer Science, Purdue University, West Lafayette, IN 47907, U.S.A.

Gawronski, Wolfgang, ASI participant, Abteilung für Mathematik, Universität Trier, Postfach 3825, D-5500 Trier, Federal Republic of Germany

Genin, Yves, principal ASI lecturer, Philips Laboratoy, Av. Van Becelaese 2, Box 8, B-1170 Brussels, Belgium

Geronimo, Jeffrey, guest, School of Math., Georgia Inst. of Technology, Atlanta, GA 30332, U.S.A.

Gilewicz, Jacek, ASI participant, Centre de Physique Théorique, CNRS, Luminy - case 907, 13288 Marseille Cédex 9, France

Godoy Malvar, Eduardo, ASI participant, Departamento de Matematica Aplicada, E.T.S.I.I de Vigo, Apartado de Correos, 62, 36280 Vigo (Pontevedra), Spain

Gökmen, Güzin, ASI participant, Dokuz Eylül Universitesi, Faculty of Engineering, Applied Mathematics Division, Bornova, Izmir, Turkey

Gori, Laura, ASI participant, Dipartimento Metodi e Modelli Matem. Sci. Appl., Universita di Roma "La Sapienza", Via A. Scarpa 10, 00161 Roma, Italy

Guadalupe, José Hernandez, ASI participant, Departamento de Matemáticas, Facultad de Ciencias, Universidad de Zaragoza, 50009-Zaragoza, Spain

Gustafson, Robert, guest, Department of Mathematics, Texas A&M University, College Station, TX 77843, U.S.A.

Habsieger, Laurent, ASI participant, Département de Mathematiques, Univ. Québec á Montréal, Case Postale 8888, Succursale A, Montréal, P.Q. H3C 3P8, Canada

Haimo, Deborah Tepper, guest, Department of Mathematics, Univ. of Missouri at St. Louis, St. Louis, MO 63121, U.S.A.

Haydock, Roger, principal ASI lecturer, Materials Science Institute, University Oregon, Eugene, OR 97403, U.S.A.

He, Xisheng, guest, Department of Mathematics, University of South Florida, Tampa, FL 33620-5700, U.S.A.

Hoggar, Stuart G., ASI participant, Department of Mathematics, Glasgow University, University Avenue, Glasgow G12 8QW, Scotland, United Kingdom

Iserles, Arieh, invited ASI lecture, University of Cambridge, DAMTP, Silver Street, Cambridge CB3 9EW, England, United Kingdom

Ismail, Mourad, Associate Director and invited ASI lecture, Department of Mathematics, University of South Florida, Tampa, FL 33620-5700, U.S.A.

Jacobsen, Lisa, invited ASI lecture, Department of Mathematics, University of Trondheim, N-7034 Trondheim - NTH, Norway

Johnson, Warren, guest, Department of Mathematics, University of Wisconsin, 480 Lincoln Drive, Madison, WI 53706, U.S.A.

Kadell, Kevin W. J., guest, Department of Mathematics, Arizona State University, Tempe, AZ 85287, U.S.A.

Kalia, R. N., guest, Department of Mathematics, St. Cloud State University, St. Cloud, MN 56301, U.S.A.

Kato, Yoshifumi, guest, Department of Mathematics, Fac. Sci. & Eng., Meijo Univ., Shiogamaguchi 1-501, Tenpaku-Ku, Nagoya 418, Japan

Kim, Dongsu, guest, School of Mathematics, University of Minnesota, 206 Church Street S.E., Minnesota, MN 55455, U.S.A.

Koekoek, Roelof, ASI participant, Faculty of Tech. Mathematics and Informatics, Delft University of Technology, P.O. Box 356, 2600 AJ Delft, Netherlands

Koornwinder, Tom, principal ASI lecturer, Centre for Mathematics and Computer Science, P.O. Box 4079, 1009 AB Amsterdam, Netherlands

Kuo, Yueh-er, guest, Department of Mathematics, University of Tennessee, Knoxville, TN 37996-1300, U.S.A.

Lachance, Michael, guest, 305 Oak Street, Ypsilanti, MI 48197, U.S.A.

Landau, Henry, invited lecture, Mathematical Sciences Research Center, AT&T Bell Labs, 600 Mountain Avenue, Murray Hill, NJ 07974, U.S.A.

Letessier, Jean, ASI participant, Physique Théorique et Hautes Énergies, Tour-14, 5e étage, 2, place Jussieau, 75251 Paris Cédex 05, France

Leviatan, Dany, guest, School of Mathematical Sciences, University of Tel-Aviv, Ramat-Aviv, 69978 Tel-Aviv, Israel

Levin, Eli, guest, Department of Mathematics, The Open Univ. of Israel, P.O. Box 39328, 16 Klausner Street, Tel-Aviv 61392, Israel

Li, Ninghua, ASI participant, Department of Physiscs, The University of Western Ontario, London, Ontario, N6A 3K7, Canada

Li, Xin, guest, Department of Mathematics, University of South Florida, Tampa, FL 33620-5700, U.S.A.

Lilly, Glenn, guest, Department of Mathematics, University of Kentucky, Patterson Office Tower, Lexington, KY 40506, U.S.A.

Littlejohn, Lance, guest, Department of Mathematics, UMC 41, Utah State University, Logan, UT 84322, U.S.A.

Logan, Ben, guest, Mathematical Sciences Research Center, AT&T Bell Labs, 600 Mountain Avenue, Murray Hill, NJ 07974, U.S.A.

Lubinsky, Doron Shaul, principal ASI lecturer, Department of Mathematics, Witwatersrand University, P.O. Wits 2050, South Africa

Luh, Wolfgang, ASI participant, Abteilung für Mathematik, Universität Trier, Postfach 3825, D-5500 Trier, Federal Republic of Germany

Macdonald, Ian, principal ASI lecturer, School of Mathematical Sciences, Queen Mary College, University of London, Mile End Road, London E1, United Kingdom

Marcellán, Francisco, ASI participant, Departamento de Matemática Aplicada, E.T.S de Ingenieros Industriales, C/José Gutierrez Abascal 2, 28006 Madrid, Spain

Markett, Clemens, invited ASI lecture, Lehrstuhl A für Mathematik, Technische Hochschule Aachen, Tempelgraben 55, D-5100 Aachen, Federal Republic of Germany

Maroni, Pascal, ASI participant, Lab. d'Analyse Numérique, Tour-55-56, 5e étage, Univ. P. et M. Curie, 4, place Jussieau, 75230 Paris Cédex 05, France

Marxer, Hermann, ASI participant, Fakultät für Physik, Albert Ludwigs Universität, Hermann Herder Strasse 3, D-7800 Freiburg I. Br., Federal Republic of Germany

Masson, David R., ASI participant, Department of Mathematics, University of Toronto, Toronto, Ontario M5S 1A4, Canada

Mastroianni, Giuseppe, ASI participant, Instituto per Applicazioni della Matematica - CNR, Via P. Castellino, 111, 80131 Napoli, Italy

Máté, Attila, guest, Department of Mathematics, Brooklyn College - CUNY, Bredford Avenue & Avenue H, Brooklyn, NY 11210, U.S.A.

Matos, Ana Cristina, ASI participant, Grupo de Matematica Aplicada, Faculdade de Ciencias, Universidade do Porto, 4000 Porto, Portugal

Meyers, Leroy, guest, Department of Mathematics, Ohio State University, 231 West 18th Avenue, Columbus, OH 43210-1174, U.S.A.

Mhaskar, Hrushiekesh, guest, Department of Mathematics, California State University, Los Angeles, CA 90032, U.S.A.

Milne, Stephen C., guest, Department of Mathematics, University of Kentucky, Patterson Office Tower, Lexington, KY 40506, U.S.A.

Mimachi, Katsuhisa, guest, Department of Mathematics, Nagoya University, Furou-Cho, ChikU.S.A.-Ku, Nagoya 464, Japan

Mityagin, Boris, guest, Department of Mathematics, Ohio State University, 231 West 18th Avenue, Columbus, OH 43210-1174, U.S.A.

Mthembu, Thandwa, guest, Department of Mathematics, Witwatersrand University, P.O. Wits 2050, South Africa

Muldoon, Martin E., invited ASI lecture, Department of Mathematics, York University, North York, Ontario M3J 1P3, Canada

Nevai, Paul, Director of ASI, Department of Mathematics, Ohio State University, 231 West 18th Avenue, Columbus, OH 43210-1174, U.S.A.

Nex, Chris, ASI participant, Department of Physics, Cavendish laboratory, Madingley Road, Cambridge CB3 0HE, England, United Kingdom

Njastad, Olav, ASI participant, Department of Mathematics, University of Trondheim, N-7034 Trondheim - NTH, Norway

Notaris, Sotirios, guest, Department of Mathematical Sciences, Indiana Univ. - Purdue Univ., 1125 East 38th St., Indianapolis, IN 46205, U.S.A.

Noumi, Masatoshi, guest, Department of Mathematics, Sophia University, Kioi-Cho 7, Chiyoda-Ku, Tokyo 102, Japan

Obermaier, Josef, ASI participant, MEDIS-Institut der GSF, Ingolstädter Landstrasse 1, D-8042 Neuherberg, Federal Republic of Germany

Osilenker, Boris Petrovich, guest, 2-oj Krestovskii Pereulok, dom 4, kv. 55, 129041 Moscow U-41, USSR

Peherstorfer, Franz, guest, Institut für Mathematik, J. Kepler University Linz, A-4040 Linz, Austria

Pérez Riera, Mario, ASI participant, Departamento de Matemáticas, Facultad de Ciencias, Universidad de Zaragoza, 50009-Zaragoza, Spain

Perez-Grasa, Isabel, ASI participant, Departamento Analisis Economico, Fac Ciencias Economicas Y Emr., Doctor Cerrada, 3, 50005 Zaragoza, Spain

Prevost, Marc, ASI participant, Laboratorie d'Analyse Numérique et d'Optimisation, Université de Lille 1, 59655-Villeneuve d'Ascq Cédex, France

Qiao, Hongzhu, guest, Department of Mathematics, University of South Florida, Tampa, FL 33620-2643, U.S.A.

Rahman, Mizan, invited ASI short course, Department of Mathematics, Carleton University, Ottawa, Ontario K1S 5B6, Canada

Ramos Alves da Rocha, Maria Zelia, ASI participant, Grupo de Matematica Aplicada, Faculdade de Ciencias, Universidade do Porto, 4000 Porto, Portugal

Reichel, Lothar, ASI participant, Bergen Scientific Centre, Allengaten 36, N-5007 Bergen, Norway

Reti, Zoltan, guest, Department of Mathematics, Ohio State University, 231 West 18th Avenue, Columbus, OH 43210-1174, U.S.A.

Ricci, Paolo Emilio, ASI participant, Dipartimento Metodi e Modelli Matem. Sci. Appl., Universita di Roma "La Sapienza", Via A. Scarpa 10, 00161 Roma, Italy

Rodman, Leiba, invited ASI short course, Department of Mathematics, The College of William & Mary, Williamsburg, VA 23185, U.S.A.

Rodrigues, Maria Joao, ASI participant, Grupo de Matematica Aplicada, Faculdade de Ciencias, Universidade do Porto, 4000 Porto, Portugal

Ronveaux, André, ASI participant, Département de Physique, Facultés Universitares de Namur, 61, rue de Bruxelles, B-5000 Namur, Belgium

Rooney, Paul G., ASI participant, Department of Mathematics, University of Toronto, Toronto, Ontario M5S 1A4, Canada

Roy, Ranjan, guest, Department of Mathematics, Beloit College, Beloit, WI 53511, U.S.A.

Ruedeman, Richard, guest, Department of Mathematics, University of South Florida, Tampa, FL 33620-5700, U.S.A.

Ruiz Blasco, Francisco, ASI participant, Departamento de Matemáticas, Facultad de Ciencias, Universidad de Zaragoza, 50009-Zaragoza, Spain

Saff, Edward B., principal ASI lecturer, Department of Mathematics, University of South Florida, Tampa, FL 33620-5700, U.S.A.

Saidi, Samira , guest, Department of Mathematics, University of Bahrain, P.O. Box 32038, Bahrain

Sansigre Vidal, Gabriela, ASI participant, Departamento de Matemática Aplicada, E.T.S de Ingenieros Industriales, C/José Gutierrez Abascal 2, 28006 Madrid, Spain

Santi, Elisabetta, ASI participant, Dipartimento Energetica, Univ. degli Studi de l'Aquila, Fac. de Ingegneria, I-67040 Monteluco, Roio Poggio, Italy

Schmidt, Asmus, L., ASI participant, Mathematics Insitute. University of Copenhagen, Universitetsparken 5, 2100 Copenhagen 0, Denmark

Siafarikas, Panayiotis D., ASI participant, Department of Mathematics, University of Patras, 265 10 Patras, Greece

Singh, Raghvendra, guest, Department Mechanical Engineering, The Ohio State University, Columbus, OH 43210, U.S.A.

Singh, Sankatha P., ASI participant, Department of Mathematics & Stat., Memorial University of Newfoundland, St. John's, Newfoundland A1C 5S7, Canada

Solé, Patrick, ASI participant, School of Computer Science, Syracuse University, Syracuse, NY 13210, France

Somali, Sennur, ASI participant, Dokuz Eylül Universitesi, Elektronik Bölümü, Bornova, Izmir, Turkey

Sonnevend, György, guest, Institute for Applied Mathemtics, University of Würzburg, Am Hubland, 8700 Würzburg, Federal Republic of Germany

Stahl, Herbert, principal ASI lecturer, Stuttgarter Platz 20, D-1000 Berlin 12, Federal Republic of Germany

Stanton, Dennis, Associate Director and invited ASI short course, School of Mathematics, University of Minnesota, 206 Church Street S.E., Minnesota, MN 55455, U.S.A.

Strasser, Wolfgang, ASI participant, MEDIS-Institut der GSF, Ingolstädter Landstrasse 1, D-8042 Neuherberg, Federal Republic of Germany

Szabados, József, guest, Mathematicsl Institute, Hungarian Acadademy of Sciences,1053 Budapest, Reáltanoda u 13/15, Hungary

Szabó, László, guest, Department of Mathematics, Ohio State University, 231 West 18th Avenue, Columbus, OH 43210-1174, U.S.A.

Temme, Nico, ASI participant, Centre for Mathematics and Computer Science, P.O. Box 4079, 1009 AB Amsterdam, Netherlands

Totik, Vilmos, principal ASI lecturer, Bolyai Institute, József Attila University, Aradi Vértanúk Tere 1, 6720 Szeged , Hungary

Urbina, Wilfredo O., guest, Department of Mathematics #038-16, Temple University, Philadelphia, PA 19122, U.S.A.

Uyanto, Stanislaus Suryadi, guest, Department of Math. & Stat., SUNY at Albany, P.O. Box 22622, 1400 Washington Ave, Albany, NY 12222, U.S.A.

Van Assche, Walter, invited ASI lecture, Department of Mathematics, Katholieke Universiteit Leuven, Clestijnenlaan 200 B, B-3030 Leuven (Heverlee) , Belgium

Van Iseghem, Jeannette, ASI participant, Laboratorie d'Analyse Numérique et d'Optimisation, Université de Lille 1, 59655-Villeneuve d'Ascq Cédex, France

Varma, Arun, guest, Department of Mathematics, University of Florida, Gainesville, FL 32611, U.S.A.

Varona Malumbres, Juan Luis, ASI participant, Colegio Universitario de la Rioja, c/Obispo Bustamante no. 3, 26001 Logroño, Spain

Viennot, Xavier Gérard, principal ASI lecturer, UER Mathématique et Informatique, Université Bordeaux I, 351 Cours de la Libération, 33405 Talence Cédex, France

Voit, Michael, ASI participant, Math. Institut, Technische Universität München, Arcisstrasse 21, Postfach 20 24 20, D-8000 München 2, Federal Republic of Germany

Walter, Gilbert, guest, Department of Mathematics, University of Wisconsin at Milwaukee, Milwaukee, WI 53201, U.S.A.

Weinstein, Lenard, guest, 24 West 85th Street, New York, NY 10024, U.S.A.

Wieneken, Matthew F., guest, Department of Mathematics, University Michigan at Flint, Flint, MI 48502-2186, U.S.A.

Wierld, Máté, guest, Department of Mathematics, Ohio State University, 231 West 18th Avenue, Columbus, OH 43210-1174, U.S.A.

Yuan, Xu, guest, Department of Mathematics, University of Texas, AU.S.A.tin, TX 78712, U.S.A.

Zarzo Altarejos, Alejandro, ASI participant, Departamento de Matemática Aplicada, E.T.S de Ingenieros Industriales, C/José Gutierrez Abascal 2, 28006 Madrid, Spain

Zayed, Ahmed, guest, Department of Mathematics, California Politechnic State University, San Luis Obispo, CA 93407, U.S.A.

Zhang, Hu, guest, Department of Mathematics (Applied Mathematics), University of Arizona, Tucson, AZ 85721, U.S.A.

Zhang, John, guest, Department of Mathematics, Ohio State University, 231 West 18th Avenue, Columbus, OH 43210-1174, U.S.A.

Zhang, Ruiming, guest, Department of Mathematics, University of South Florida, Tampa, FL 33620-5700, U.S.A.

Zhang, Weijian, guest, Department of Electrical Engineering, 6731 Boelter, University of California, Los Angeles, CA 90024, U.S.A.

Zheng, Ligang, ASI participant, Department of Mathematics, University of Ottawa, 585 King Ave, Ottawa, Ontario, K1N 6N5, Canada

Scientific program for the NATO Advanced Study Institute on "Orthogonal Polynomials and Their Applications"

Monday May 22, 1989	Tuesday May 23, 1989	Wednesday May 24, 1989	Thursday May 25, 1989	Friday May 26, 1989	Saturday May 27, 1989
Paul Nevai (Chairman)	*Mourad Ismail (Chairman)*	*Marcel de Bruin (Chairman)*	*Charles Dunkl (Chairman)*	*Lance Littlejohn (Chairman)*	*H. N. Mhaskar (Chairman)*
Edward B. Saff (9:30 AM) Orthogonal Polynomials from a Complex Perspective (Behavior of Zeros)	Edward B. Saff (9 AM) Orthogonal Polynomials from a Complex Perspective (Growth of Polynomials)	Lisa Jacobsen (9 AM) Continued Fractions and Their Relation to Orthogonal Polynomials	Xavier Gérard Viennot (9 AM) Orthogonal Polynomials in Combinatorics	Xavier Gérard Viennot (9 AM) Orthogonal Polynomials in Combinatorics	Xavier Gérard Viennot (9 AM) Orthogonal Polynomials in Combinatorics
Yves Genin (11:00 AM) Orthogonal Polynomials in Signal Processing	Yves Genin (10:30 AM) Orthogonal Polynomials in Signal Processing	Roger Haydock (10:30 AM) The Recursion Method and the Schrödinger Equation	Roger Haydock (10:30 AM) The Recursion Method and the Schrödinger Equation	Walter Gautschi (10:30 AM) Computational Aspects of Orthogonal Polynomials	Walter Gautschi (10:30 AM) Computational Aspects of Orthogonal Polynomials
Henry Landau (12:00 AM) Continuous Analogues of Orthogonal Polynomials	Clemens Markett (11:30 AM) Riemann's Integration Method for PDE's: An Approach to Convolution Structures for Orthogonal Expansions	Martin Muldoon (11:30 AM) Differential Equations and Zeros of Orthogonal Polynomials	George Baker (11:30 AM) Integral Polynomials	David Bressoud (11:30 AM) The Land of OZ	Waleed Al-Salam (11:30 AM) Characterization Theorems for Classical Orthogonal Polynomials
Short Course (55 mins) Mizanur Rahman (2:30 PM) Some Extensions of the Beta Integral and the Hypergeometric Function	*Short Course (55 mins)* Mizanur Rahman (2:30 PM) Some Extensions of the Beta Integral and the Hypergeometric Function	*Short Course (55 mins)* Mizanur Rahman (2:30 PM) Some Extensions of the Beta Integral and the Hypergeometric Function	Theodore S. Chihara (2:30 PM) Spectral Properties of Measures Associated with Orthogonal Polynomials	*Panel Discussion (2:30 PM)* Moderator: Eiichi Bannai Applications of Orthogonal Polynomials in Coding Theory	
Short Course (55 mins) Dennis Stanton (3:30 PM) Orthogonal Polynomials and Representation Theory	*Short Course (55 mins)* Dennis Stanton (3:30 PM) Orthogonal Polynomials and Representation Theory	*Short Course (55 mins)* Dennis Stanton (3:30 PM) Orthogonal Polynomials and Representation Theory	Joanne Dombrowski (3:30 PM) Applications of Functional Analysis in Orthogonal Polynomials	*Participants: E. Bannai, S. Hoggar, A. Musemesa, D. Stanton & A. Zayed*	
Short Course (50 mins) Leiba Rodman (4:30 PM) Orthogonal Matrix Polynomials	*Short Course (50 mins)* Leiba Rodman (4:30 PM) Orthogonal Matrix Polynomials	*Short Course (50 mins)* Leiba Rodman (4:30 PM) Orthogonal Matrix Polynomials	Walter Van Assche (4:30 PM) Asymptotics for Orthogonal Polynomials and Three-Term Recurrences		

Scientific program for the NATO Advanced Study Institute on "Orthogonal Polynomials and Their Applications"

Monday May 29, 1989
Ted Chihara (Chairman)

- Herbert Stahl (9 AM) Orthogonal Polynomials and Potential Theory
- Daniel Bessis (10:30 AM) Orthogonal Polynomials, Padé Approximations and Julia Sets
- Arieh Iserles (11:30 AM) Orthogonality in a Sobolev Space

Tuesday May 30, 1989
Arieh Iserles (Chairman)

- Herbert Stahl (10 AM) Orthogonal Polynomials and Potential Theory
- Daniel Bessis (11:30 AM) Orthogonal Polynomials, Padé Approximations and Julia Sets
- Adhemar Bultheel (2:45 PM) Formal Orthogonal Polynomials and Laurent-Padé Approximation
- Michael Voit (3:15 PM) Limit Theorems for Random Walks on Polynomial Hyper-Groups
- Josef Obermaier (4 PM) The Dirichlet and Fejér Kernels for Orthogonal Polynomials related to Hypergroups
- Nelson Blachman (4:50 PM) Random Processes and Orthogonal Linear Forms
- Yoshifumi Kato (5:10 PM) The Theory of q-Analogue - the Infinite Matrix Method

Wednesday May 31, 1989
Panos Stafarikas (Chairman)

- Tom Koornwinder (9 AM) Orthogonal Polynomials and Special Functions in connection with Quantum Groups
- Daniel Bessis (10:30 AM) Orthogonal Polynomials, Padé Approximations and Julia Sets
- Charles Dunkl (11:30 AM) Reflection Groups and Orthogonal Polynomials
- Panel Discussion (2:30 PM) Moderator: Mourad Ismail Orthogonal Polynomials and Birth and Death Processes
- Participants: M. Ismail J. Letessier & D. Masson

Thursday June 1, 1989
Marcel de Bruin (Chairman)

- Tom Koornwinder (9 AM) Orthogonal Polynomials and Special Functions in connection with Quantum Groups
- Doron S. Lubinsky (10:30 AM) Orthogonal Polynomials and Weighted Approximation on $(-\infty,\infty)$
- George Andrews (11:30 AM) Ramanujan and his Continued Fractions
- Panel Discussion (2:30 PM) Moderator: George Gasper Orthogonal Polynomials and Symbolic Computer Algebraic Systems
- Participants: G. Andrews Nasri Ari, G. Edgar, G. Gasper & A. Ronveaux

Friday June 2, 1989
Lisa Jacobsen (Chairman)

- Ian Macdonald (9 AM) Orthogonal Polynomials Associated with Root Systems
- Doron S. Lubinsky (10:30 AM) Orthogonal Polynomials and Weighted Approximation on $(-\infty,\infty)$
- Vilmos Totik (11:30 AM) Orthogonal Polynomials and Potential Theory
- Katsuhisa Mimachi (2:30 PM) Representation Theory of Quantum Groups
- Panel Discussion (3 PM) Moderator: Richard Askey History of Orthogonal Polynomials
- Participants: R. Askey C. Brezinski, T. S. Chihara, C. Dunkl, W. Gautschi & H. Stahl

Saturday June 3, 1989
Paul Nevai (Chairman)

- Ian Macdonald (9 AM) Orthogonal Polynomials Associated with Root Systems
- Vilmos Totik (10:30 AM) Orthogonal Polynomials and Potential Theory
- Leonard Weinstein (11:30 AM) The Bieberbach Conjecture
- Boris Osilenker (Noon) Orthogonal Fourier Series
- Richard Askey (12:30 PM) Overview